Guide to Energy Management

Seventh Edition

Guide to Energy Management
Seventh Edition

by Barney L. Capehart, Ph.D., CEM
      Wayne C. Turner, Ph.D. PE, CEM
      William J. Kennedy, Ph.D., PE

THE FAIRMONT PRESS, INC.

CRC Press
Taylor & Francis Group

Library of Congress Cataloging-in-Publication Data

Capehart, B. L. (Barney L.)
    Guide to energy management / by Barney L. Capehart, Wayne
C. Turner, William J. Kennedy. -- 7th ed.
      p. cm.
    Includes bibliographical references and index.
    ISBN-10: 0-88173-671-6 (alk. paper)
    ISBN-10: 0-88173-672-4 (electronic)
    ISBN-13: 978-1-4398-8348-8 (alk. paper)
    1. Energy conservation--Handbooks, manuals, etc. 2. Energy con-
sumption--Handbooks, manuals, etc. I. Turner, Wayne C., 1942- II.
Kennedy, William J., 1938- III. Title.

TJ163.3.C37 2011
621.042--dc23   2011021960

Published by The Fairmont Press, Inc.
700 Indian Trail
Lilburn, GA 30047
tel: 770-925-9388; fax: 770-381-9865
http://www.fairmontpress.com

Distributed by Taylor & Francis Ltd.
6000 Broken Sound Parkway NW, Suite 300
Boca Raton, FL 33487, USA
E-mail: orders@crcpress.com

Distributed by Taylor & Francis Ltd.
23-25 Blades Court
Deodar Road
London SW15 2NU, UK
E-mail: uk.tandf@thomsonpublishingservices.co.uk

Printed in the United States of America
10 9 8 7 6 5 4 3 2

ISBN-10: 0-88173-671-6 (The Fairmont Press, Inc.)
ISBN-13: 978-1-4398-8348-8 (Taylor & Francis Ltd.)

Contents

Preface to the Seventh Edition

The wild ride on the roller coaster of energy prices continues with the price of oil having soared to almost $150 a barrel in early 2008, and plunged to $35 a barrel in late 2008. For 2010, oil prices have averaged about $80 a barrel. In the spring of 2011, gasoline prices are now over $4.00 a gallon again, and both people and businesses are suffering some economic consequences. The cost of oil is now over $100 a barrel again. With the cost of most of our energy sources remaining at high levels, and having significantly greater federal expenditures for energy efficiency and renewable energy over the last two years, our work as energy managers, facility managers, and other energy professionals has continued in high gear. Using our new opportunities for implementing more energy cost reduction projects, results have come in the form of huge cost savings for our companies and organizations. However, all of these past successes have not eliminated—or really even slowed—the continuing need to install new equipment, new technology and new processes to produce energy savings as well as help reduce pollution and improve quality and productivity. Energy managers and energy professionals are not going to work themselves out of a job!

One more reason that energy managers and energy professionals are not going to work themselves out of a job is that "the job" keeps changing. First it was just energy and energy cost, then it expanded to include water and sewer use and cost. Now our responsibilities have greatly expanded to include construction and operational aspects involving sustainability, green, LEED, Energy Star, renewable energy, and low carbon footprints. All of these new parts of our jobs are intimately related to our energy use, so we are the "usual suspects" to be asked to accomplish these tasks, too! Now we have a large set of additional drivers for our "old work" of making our facilities and operations more energy efficient and using more renewable energy. While this increases our work load and our need to learn new things, it also greatly expands our opportunities to find ways to make some of our "energy projects" far more cost effective. There will be many more win-win projects for us in the future.

The *Guide to Energy Management* continues as one of the leading educational resources for the person who is active as an energy manager or energy professional, as well as helping new people enter the fascinating and important field of energy management and energy engineering. It is the most widely used college and university textbook in this field, as well as one of the most widely used books for professional development training in the field. At the end of 2010 over 15,000 energy professionals had been trained using the first six editions of the *Guide to Energy Management*. In this seventh edition, we have added one new chapter with the extremely timely topic of green house gas emissions management, and we have updated two chapters that help prepare people to address the new areas of web based energy information and control systems, and creating green buildings.

The authors wish to thank Dr. Eric Woodroof for his contribution of the new Chapter 17, where he discusses the importance of green house gas emissions impact on global climate change, and our need to manage these emissions from our buildings and other facilities. He explains how to measure and report green house gas emissions, how to reduce them, and how to mitigate or offset them. The authors wish to thank Dr. Diane Schaub for her contribution of the Compressed Air section to Chapter 12. Also, thanks go to Mr. Paul Allen for updating Chapter 15 on Web Based Building Automation Controls and Energy Information Systems; and Dr. Steven Roosa for updating Chapter 16 on Creating Green Buildings. And thanks to Mr. Klaus Pawlik for his help in coordinating the *Solutions Manual* with the problems contained in Appendix I of this book.

Thanks to the many energy professionals who have suggested improvements to this book, and have helped point out errors or inconsistencies. Special thanks go to Mr. Joseph Dussault for his careful review and identification of some errors in this book. There is always room for improvement, so please let us know if you find any parts of the book needing improvement. We always appreciate hearing constructive criticism.

Good luck to all of you in your search for new, green, energy cost savings opportunities! And may we all be successful in providing an energy future for our country and our grandchildren that is efficient and sustainable.

Barney L. Capehart
Wayne C. Turner
William J. Kennedy
May 2011

Chapter 1

Introduction to Energy Management

1.0 ENERGY MANAGEMENT

The phrase energy management means different things to different people. To us, energy management is:

> The efficient and effective use of energy to maximize profits (minimize costs) and enhance competitive positions

This rather broad definition covers many operations from services to product and equipment design through product shipment. Waste minimization and disposal also presents many energy management opportunities. Our main focus in this book is energy management in buildings, manufacturing, and industry.

A whole systems viewpoint to energy management is required to ensure that many important activities will be examined and optimized. Presently, many businesses and industries are adopting a Total Quality Management (TQM) strategy for improving their operations. Any TQM approach should include an energy management component to reduce energy costs.

The primary objective of energy management is to maximize profits or minimize costs. Some desirable subobjectives of energy management programs include:

1. Improving energy efficiency and reducing energy use, thereby reducing costs

2. Reduce greenhouse gas emissions and improve air quality.

3. Cultivating good communications on energy matters

1

4. Developing and maintaining effective monitoring, reporting, and management strategies for wise energy usage

5. Finding new and better ways to increase returns from energy investments through research and development

6. Developing interest in and dedication to the energy management program from all employees

7. Reducing the impacts of curtailments, brownouts, or any interruption in energy supplies

Although this list is not exhaustive, these seven are sufficient for our purposes. However, the seventh objective requires a little more explanation.

Curtailments occur when a major supplier of an energy source is forced to reduce shipments or allocations (sometimes drastically) because of severe weather conditions and/or distribution problems. For example, natural gas is often sold to industry relatively inexpensively, but on an interruptible basis. That is, residential customers and others on noninterruptible schedules have priority, and those on interruptible schedules receive what is left. This residual supply is normally sufficient to meet industry needs, but periodically gas deliveries must be curtailed.

Even though curtailments do not occur frequently, the cost associated with them is so high—sometimes a complete shutdown is necessary—that management needs to be alert in order to minimize the negative effects. There are several ways of doing this, but the method most often employed is the storage and use of a secondary or standby fuel. Number 2 fuel oil is often stored on site and used in boilers capable of burning either natural gas (primary fuel) or fuel oil (secondary fuel). Then when curtailments are imposed, fuel oil can be used. Naturally, the cost of equipping boilers with dual fire capability is high, as is the cost of storing the fuel oil. However, these costs are often minuscule compared to the cost of forced shutdowns. Other methods of planning for curtailments include production scheduling to build up inventories, planned plant shutdowns, or vacations during curtailment-likely periods, and contingency plans whereby certain equipment, departments, etc., can be shut down so critical areas can keep operating. All these activities must be included in an energy management program.

Although energy conservation is certainly an important part of energy management, it is not the only consideration. Curtailment-contingency planning is certainly not conservation, and neither are load shedding or power factor improvement, both of which will be discussed later on in this book. To concentrate solely on conservation would preclude some of the most important activities—often those with the largest savings opportunity.

1.1   THE NEED FOR ENERGY MANAGEMENT

1.1.1 Economics

The American free enterprise system operates on the necessity of profits, or budget allocations in the case of nonprofit organizations. Thus, any new activity can be justified only if it is cost effective; that is, the net result must show a profit improvement or cost reduction greater than the cost of the activity. Energy management has proven time and time again that it is cost effective.

An energy cost savings of 5-15 percent is usually obtained quickly with little to no required capital expenditure when an aggressive energy management program is launched. An eventual savings of 30 percent is common, and savings of 50, 60, and even 70 percent have been obtained. These savings all result from retrofit activities. New buildings designed to be energy efficient can operate on 20 percent of the energy (with a corresponding 80 percent savings) normally required by existing buildings. In fact, for most manufacturing, industrial, and other commercial organizations *energy management is one of the most promising profit improvement-cost reduction programs available today.*

1.1.2 National Good

Energy management programs are vitally needed today. One important reason is that energy management helps the nation face some of its biggest problems. The following statistics will help make this point.*

*   Growth in U.S. energy use:
    It took 50 years (1900-1950) for total annual U.S. energy consumption to go from 4 million barrels of oil equivalent (MBOE) per day to 16 MBOE. It took only 20 years (1950-1970) to go from 16 to 32 MBOE. This rapid growth in energy use slowed in the early 1970's, but took a spurt in the late 1970's, reaching 40 MBOE in 1979. Energy use slowed again in the early 1980's and dropped to about 37 MBOE in 1983. Economic growth in the mid 1980's returned the use to 40 MBOE in 1987. Energy use remained fairly steady at just over 42 MBOE in the late 1980's, but started growing in the 1990s. By the end of 1994, energy use was up to

---

*These statistics come from numerous sources, mostly government publications from the Energy Information Administration or from the U.S. Statistical Abstract.

almost 45 MBOE, and in 2004, just under 50 MBOE per day. Energy use remained around 50 MBOE per day for 2005 and 2006. After that, the world-wide economic slowdown dropped energy use to 47 MBOE in 2009. For 2010, this increased to 49 MBOE, per day.

- Comparison with other countries:
  With only 4.5 percent of the world's population, the United States consumes about 25 percent of its energy and produces about 22 percent of the world's gross national product (GNP). However, some nations such as Japan and Germany produce the same or greater GNP per capita with significantly less energy than the United States.

- U.S. energy production:
  Domestic crude oil production peaked in 1970 at just under 10 million barrels per day (MBD), and has fallen slowly since then to about 5.6 MBD in 2006, and 5.3 MBD in 2009. For 2010, this increased to about 5.85 MBD. Domestic gas production peaked in 1973 at just over 21.7 trillion cubic feet (TCF) per year. Gas production slowly declined until 1987 when it fell to 16.1 TCF. Since 1988, gas production increased very slowly, and in 2003 was 19.7 TCF, and in 2006 it was 19.1. Deregulation has improved our domestic production. Since 1988, gas imports have been over 1 TCF per year, and have been increasing rapidly. In 2006, we imported over 4 TCF of natural gas.

  In 2009, net U.S. imports of natural gas were the lowest since 1994, representing just 12 percent of total consumption. The primary underlying cause for the lower level of net imports was continued strong levels of natural gas production in the lower 48 states. Dry natural gas production increased 3.3 percent compared with 2008 and was nearly 9 percent higher than in 2007. With these recent gains in domestic production, the United States is now the largest producer of natural gas in the world. U.S. domestic consumption decreased in 2009, which in turn contributed to a reduced demand for imports. Although liquefied natural gas (LNG) gross imports increased almost 30 percent (from a 5-year low established in 2008), LNG remains a very small source of supplies for the United States, accounting for less than 2 percent of consumption.

  In 2010, the domestic production of natural gas reached 22.1 TCF due to expanded shale gas production. Natural gas imports were down slightly to 3.78 TCF.

- Cost of imported oil:
  Annual average prices per barrel for imported crude oil rapidly escalated from $3.00 in the early 1970's to $12 in 1973-1974 and to $36 in

1981. After 1981, prices dropped to about $12 in 1986. From 1986 to 1999, prices ranged from about $12 to $20 a barrel, with a short spike in prices during the 1989-90 Gulf War. Prices dropped to $10 in 1998, and rose back to about $26 in 2003. The 2004 cost was about $24, but in early 2005, spot prices shot up to $58 a barrel. In 2006 prices averaged around $60 a barrel, and in 2007 spot prices for oil reached almost $100 a barrel. In 2008 spot prices for oil spiked to almost $150 a barrel before crashing back to $35 a barrel later in the year. For 2009 the average cost for oil was almost $60 per barrel. For 2010, the average cost of oil was about $80 per barrel.

- Reliance on imported oil:
  The United States has been a net importer of oil since 1947. In 1970 the bill for this importation was only $3 billion; by 1977 it was $42 billion; by 1979, $57 billion; and by 1980, almost $80 billion. This imported oil bill has severely damaged our trade balance and weakened the dollar in international markets. In 1986 the bill for oil imports fell to a low of $35 billion. It climbed to almost $62 billion in 1990. In 1996 it was just over $72 billion, and with lower prices after 1996, it was just over $50 billion in 1998. But, with higher prices starting in 2000, it was $119 billion in 2000, $132 billion in 2003, and $179 billion in 2004. With the higher prices in 2005 and 2006, the total was $300 billion in 2006! The cost continued to rise to over $350 billion in 2008, but then dropped to less than $200 billion in 2009 with the U.S. and global recession significantly reducing the use and cost of oil. In 2010, our cost of imported oil jumped 33% to $251 billion.

In addition to these discouraging statistics, there are a host of major environmental problems, as well as economic and industrial competitiveness problems, that have come to the forefront of public concern. Reducing energy use can help minimize these problems by:

- Reducing acid rain. Acid rain has been reduced mainly through national and regional environmental policies around the world. The United States adopted the Clean Air Act Amendments of 1990, and the Clear Skies Act of 2003 which established a new trade program for sulfur oxides and nitrous oxides and mercury emissions from power plants. Canada has worked closely with the United States on acid rain reduction. The EU has adopted several policies related to the reduction of acid rain. The Asian development Bank has worked closely with China to help it make significant reductions in its potential for acid rain.

- Limiting global climate change. Carbon dioxide, the main contributor to potential global climate change, is produced by the combustion of fossil fuel, primarily to provide transportation and energy services. In 1992, many countries of the world adopted limitations on carbon dioxide emissions. Reducing fossil energy use through energy efficiency improvements and the use of renewable energy is without doubt the quickest, most effective, and most cost-effective manner for reducing greenhouse gas emissions, as well as improving air quality, in particular in densely populated areas.

  The first international treaty to address climate change was the United Nations Framework Convention on Climate Change (UNFCCC), which entered into force in 1994 and has been ratified by 186 countries, including the United States. Delegates to the UNFCCC then met in Kyoto, Japan, in 1997 to adopt a more significant treaty calling for binding targets and timetables, eventually agreeing on the Kyoto Protocol (KP). Delegates rejected language requiring participation by developing countries, thus damping U.S. enthusiasm. Nevertheless, the Kyoto Protocol entered into force in 2005, having been ratified by EU countries, Canada, Japan, Russia, and most developing countries. The United States and Australia are currently not parties to the KP.

  The U.S. considered legislation to control carbon emissions with a cap and trade approach in 2009 and 2010, but no formal control was approved. Currently, this legislation appears very unlikely to be approved anytime in the near future.

- Reducing ozone depletion. The Montréal Protocol of 1987, and its subsequent updates, is one of the most successful environmental protection agreements in the world. The Protocol sets out a mandatory timetable for the phase out of ozone depleting substances. This timetable has been under constant revision, with phaseout dates accelerated in accordance with scientific understanding and technological advances. Chlorofluorocarbon (CFC) production by developed countries was phased out at the end of 1995. Starting in 1996, hydrochlorofluorocarbons (HCFCs) began to be phased out by developed countries with a 65% reduction by 2010, and complete phase out by 2020.

- Improving national security. Oil imports directly affect the energy security and balance of payments of our country. These oil imports must be reduced for a secure future, both politically and economically. In 1973 our net imports were 34.8% of our use. In 1977 they were 46.5%, and in 1998 hit 50%. The percentage of our oil use that is imported continues to climb—56.1% in 2003, 57.8% in 2004, and 59.5% in 2006. In 2009, import levels reached almost 63%. Preliminary data for 2010 show a 61% level of oil imports.

- Improving U.S. competitiveness. The U.S. spends about 9 percent of its gross national product for energy—a higher percentage than many of its foreign competitors. This higher energy cost amounts to a surtax on U.S. goods and services.

- Helping other countries. The fall of the Berlin Wall in 1989 and the emergence of market economies in most Eastern European countries have led to major changes in world energy supplies and demands. These changes significantly affect our nation, and provide us an economic impetus to help these countries greatly improve their own energy efficiencies and reduce their energy bills.

There are no easy answers. Each of the possibilities discussed below has its own problems.

- Many look to coal as the answer. Yet coal burning produces sulfur dioxide, and carbon dioxide which produce acid rain and potential global climate change. Research and development on "clean coal" technology is currently underway.

- Synfuels require strip mining, incur large costs, and place large demands for water in arid areas. On-site coal gasification plants associated with gas-fired, combined-cycle power plants are presently being demonstrated by several electric utilities. However, it remains to be seen if these units can be built and operated in a cost-effective and environmentally acceptable manner.

- Solar-generated electricity, whether generated through photovoltaics or thermal processes, is still more expensive than conventional sources and has large land requirements. Technological improvements are occurring in both these areas, and costs are decreasing. Sometime in the near future, these approaches may become cost-effective, with the current cost of large-scale solar PV generation under $4000 per kW in 2010.

- Biomass energy is also expensive, and any sort of monoculture would require large amounts of land. Some fear total devastation of forests. At best, biomass can provide only a few percentage points of our total needs without large problems.

- Wind energy is only feasible in limited geographical areas where the wind velocity is consistently high, and there are also some noise and aesthetic problems. However, the cost of wind generation systems has come down to $1000-$2000 per kW, and they are cost-effective in windy areas of the U.S. Operating costs are very low, and with new wind turbine technologies, large wind farms are being constructed in the midwest, southwest, and western regions of the U.S.

- Fuel cells and their ability to cleanly produce electricity from hydrogen and oxygen are what make them and hydrogen attractive. However, hydrogen is not a primary source of energy. It is made from other forms of energy; most hydrogen production today is by steam reforming natural gas. Natural gas is a fossil fuel, so the carbon dioxide released in the reformation process adds to the greenhouse effect. Only when hydrogen is made cost effectively from renewable energy sources does it have any significant value as a fuel source for a fuel cell. Then fuel cells will be of great interest and use. 2010 prices for fuel cells were around $4500 per kW.

- Alcohol production from agricultural products raises perplexing questions about using food products for energy when large parts of the world are starving. Newer processes for producing alcohol from wood waste are still being tested, and may offer some significant improvements in this limitation. In the meantime, quite a few new ethanol plants are being started up to produce this alternate energy fuel. Brazil has a very large and successful ethanol production industry from sugar cane.

- Fission has the well-known problems of waste disposal, safety, and a short time span with existing technology. Without breeder reactors or nuclear fuel reprocessing, we will soon run out of fuel, but breeder reactors dramatically increase the production of plutonium—a raw material for nuclear bombs. Nuclear fuel reprocessing could provide many years of fuel by recycling partially used fuel now being kept in storage. Newer reactor designs appear to be safer and porentially cheaper.

- Fusion seems to be everyone's hope for the future, but many claim that we do not know the area well enough yet to predict its problems. When available commercially, fusion may very well have its own style of environmental-economical problems.

The preceding discussion paints a rather bleak picture. Our nation and our world are facing severe energy problems and there appears to be no simple answers.

Time and again energy management has shown that it can substantially reduce energy costs and energy consumption through improved energy efficiencies. This saved energy can be used elsewhere, so one energy source not mentioned in the preceding list is energy management. In fact, energy available from energy management activities has almost always proven to be the most economical source of "new" energy. Furthermore, energy management activities are more gentle to the environment than large-scale energy production, and they certainly lead to less consumption of scarce and valuable resources. Thus, although energy management cannot solve all the nation's problems, *perhaps it can ease the strain on our environment and give us time to develop new energy sources.*

The value of energy management is clear. There is an increased need for engineers who are adequately trained in the field of energy management, and a large number of energy management jobs are available. This text will help you prepare for a career which will be both exciting and challenging.

Table 1-1
Energy Units and Energy Content of Fuels

1 kWh	3412 Btu
1 ft^3 natural gas	1000 Btu
1 Ccf natural gas	100 ft^3 natural gas
1 Mcf natural gas	1000 ft^3 natural gas
1 therm	100,000 Btu
1 barrel crude oil	5,100,000 Btu
1 ton coal	25,000,000 Btu
1 gallon gasoline	125,000 Btu
1 gallon #2 fuel oil	140,000 Btu
1 gallon LP gas	95,000 Btu
1 cord of wood	30,000,000 Btu
1 MBtu	1000 Btu
1 MMBtu	10^6 Btu
1 Quad	10^{15} Btu
1 MW	10^6 watts

1.2   ENERGY BASICS FOR ENERGY MANAGERS

An energy manager must be familiar with energy terminology and units of measure. Different energy types are measured in different units. Knowing how to convert from one measurement system to another is essential for making valid comparisons. The energy manager must also be informed about the national energy picture. The historical use patterns as well as the current trends are important to an understanding of options available to many facilities.

1.2.1 Energy Terminology, Units and Conversions

Knowing the terminology of energy use and the units of measure is essential to developing a strong energy management background. Energy represents the ability to do work, and the standard engineering measure for energy used in this book is the British thermal unit, or Btu. One Btu is the amount of energy needed to raise the temperature of one pound of water one degree Fahrenheit. In more concrete terms, one Btu is the energy released by burning one kitchen match head, according to the U.S. Energy Information Agency. The energy content of most common fuels is well known, and can be found in many reference handbooks. For example, a gallon of gasoline contains about 125,000 Btu and a barrel of oil contains about 5,100,000 Btu. A short listing of the average energy contained in a number of the most common fuels, as well as some energy unit conversions is shown below in Table 1-1.

Electrical energy is also measured by its ability to do work. The traditional unit of measure of electrical energy is the kilowatt-hour; in terms of Btu's, one kilowatt-hour (kWh) is equivalent to 3412 Btu. However, when electrical energy is generated from steam turbines with boilers fired by fossil fuels such as coal, oil or gas, the large thermal losses in the process mean that it takes about 10,000 Btu of primary fuel to produce one kWh of electrical energy. Further losses occur when this electrical energy is then transmitted to its point of ultimate use. Thus, although the electrical energy at its point of end-use always contains 3412 Btu per kWh, it takes considerably more than 3412 Btus of fuel to produce a kWh of electrical energy.

1.2.2 Energy Supply and Use Statistics

Any energy manager should have a basic knowledge of the sources of energy and the uses of energy in the United States. Both our national energy policy and much of our economic policy are dictated by these supply and use statistics. Figure 1-1 shows the share of total U.S. energy supply provided by each major source. Figure 1-2 represents the percentage of total energy consumption by each major end-use sector.

1.2.3 Energy Use in Commercial Businesses

One question frequently asked by facility energy managers is "How does energy use at my facility compare to other facilities in general, and to other facilities that are engaged in the same type of operation?" Figure 1-3 shows general energy usage in commercial facilities, and Figure 1-4 shows their electricity use. While individual facilities may differ significantly from these averages, it is still helpful to know what activities are likely to consume the most energy. This provides some basis for a comparison to other facilities—both energy wasting and energy efficient. In terms of priority of action for an energy management program, the largest areas of energy consumption should be examined first. The greatest savings will almost always occur from examining and improving the areas of greatest use.

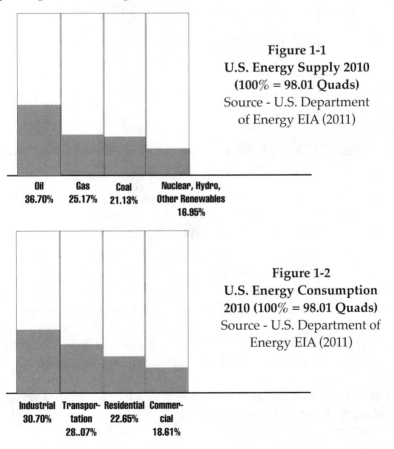

Figure 1-1
U.S. Energy Supply 2010
(100% = 98.01 Quads)
Source - U.S. Department
of Energy EIA (2011)

Oil	Gas	Coal	Nuclear, Hydro, Other Renewables
36.70%	25.17%	21.13%	16.95%

Figure 1-2
U.S. Energy Consumption
2010 (100% = 98.01 Quads)
Source - U.S. Department of
Energy EIA (2011)

Industrial	Transportation	Residential	Commercial
30.70%	28..07%	22.65%	18.61%

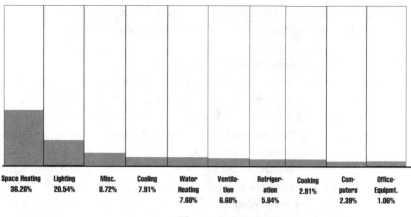

Space Heating	Lighting	Misc.	Cooling	Water Heating	Ventila-tion	Refriger-ation	Cooking	Com-puters	Office-Equipmt.
36.26%	20.54%	8.72%	7.91%	7.68%	6.68%	5.84%	2.91%	2.39%	1.06%

Figure 1-3
Commercial Energy Use 2003 (end-use basis) for all Commercial Buildings
Source - U.S. Department of Energy EIA (2011)

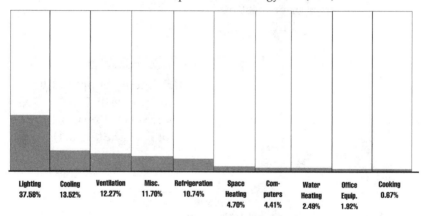

Lighting	Cooling	Ventilation	Misc.	Refrigeration	Space Heating	Com-puters	Water Heating	Office Equip.	Cooking
37.58%	13.52%	12.27%	11.70%	10.74%	4.70%	4.41%	2.49%	1.92%	0.67%

Figure 1-4.Commercial Electric Use 1999 (end-use basis)
Source - U.S. Department of Energy EIA (2011)

The commercial sector uses about 19 percent of all the primary energy consumed in the United States, at a cost of over $192 billion in 2008 [1]. On an end-use basis, natural gas, oil and district heating constitute about 45.4 percent of the commercial energy use, mainly for space heating. Over 64 percent of the energy use is in the form of electricity for lighting, air conditioning, and ventilation. Electricity provides over half of the end-use energy used by a commercial facility, but it represents about two thirds of the cost of the energy needed to operate the facility. Lighting is the predominant use of electricity in commercial buildings, and accounts for over one-third of the cost of electricity.

Commercial activity is very diverse, and this leads to greatly varying energy intensities depending on the nature of the commercial facility. Recording energy use in a building or a facility of any kind and providing a history of this use is necessary for the successful implementation of an energy management program. A time record of energy use allows analysis and comparison so that results of energy productivity programs can be determined and evaluated.

1.2.4 Energy Use in Industry

The industrial sector—consisting of manufacturing, mining, agriculture and construction activities—consumes almost 31% of the nation's primary energy use, at an annual cost of over $206 billion in 2008 [2]. Industrial energy use is shown in Figure 1-5 and industrial electricity use is shown in Figure 1-6.

Manufacturing companies, which use mechanical or chemical processes to transform materials or substances into new products, account for about 85 percent of the total industrial sector use. The "big three" in energy use are chemicals, petroleum, and pulp and paper; these industries together consume over one-half of all industrial energy. The "big five," which add the primary metals industry, as well as the food and kindred products group, together account for 70 percent of all industrial sector energy consumption.

According to the U.S. Energy Information Administration, energy efficiency in the manufacturing sector improved by 25 percent over the period 1980 to 1985 [3]. During that time, manufacturing energy use declined 19 percent, and output increased 8 percent. These changes resulted in an

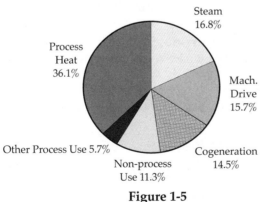

Figure 1-5
Industrial Energy Use 2002 (end-use basis)
Source - U.S. Department of Energy EIA (2005)

Figure 1-6
Industrial Electricity Use (end-use basis)
Source - Federal Energy Management Agency

overall improvement in energy efficiency of 25 percent. However, the "big five" did not match this overall improvement; although their energy use declined 21 percent, their output decreased by 5 percent—resulting in only a 17 percent improvement in energy efficiency during 1980-1985. This overall five-year record of reduction in energy use of the manufacturing sector came to an end in 1986, with total energy use in the sector growing by 10 percent from 1986 to 1988. Manufacturing energy use has continued to grow at a slower rate since 1988, but industrial output has grown at a faster rate [9]. Use of new energy efficient technology, and the changing production mix from the manufacture of energy-intensive products to less intensive products has accounted for much of this difference.

Between 1985 and 2004, industrial energy intensity, based on energy used per unit of gross domestic product, dropped almost 20%. These data came from the Energy Efficiency and Renewable Energy Office of the U.S. Department of Energy. In addition, some data have become available on the energy use intensity for two subsectors of the primary metals category. These data are for the energy used to produce $2000 of shipped product value in 1998, 2002, and 2006.

Intensity of Fuel Use, Btu per 2000 Dollar Value of Shipment

	1998	2002	2006
Alumina and Aluminum	13,065	11,418	8,342
Iron and Steel Mills	32,177	30,176	18,468

Sources: EIA, Manufacturing Energy Consumption Surveys, U.S. Department of Commerce, Bureau of Economic Analysis, Value of Shipments and Price Indexes by Detailed Industry (1998-2006)

The energy used to produce $2000 of alumina and aluminum dropped 36% from 1998 to 2006, and the energy used to produce $2000 of products from iron and steel mills dropped 43% in that same time period.

Continuing this record of energy efficiency improvements in manufacturing will require both re-establishing emphasis on energy management and making capital investments in new plant processes and facilities improvements. Reducing our energy costs per unit of manufactured product is one way that our country can become more competitive in the global industrial market. It is interesting to note that Japan—one of our major industrial competitors—has a law that every industrial plant must have a full-time energy manager [4].

1.3   DESIGNING AN ENERGY MANAGEMENT PROGRAM

1.3.1 Management Commitment

The most important single ingredient for successful implementation and operation of an energy management program is commitment to the program by top management. Without this commitment, the program will likely fail to reach its objectives. Thus, the role of the energy manager is crucial in ensuring that management is committed to the program.

Two situations are likely to occur with equal probability when designing an energy management program. In the first, management has decided that energy management is necessary and wants a program implemented. This puts you—the energy manager—in the *response* mode. In the second, you—an employee—have decided to convince management of the need for the program so you are in the aggressive mode. Obviously, the most desirable situation is the response mode as much of your sales effort is unnecessary; nonetheless, a large number of energy management programs have been started through the *aggressive* mode. Let's consider each of these modes.

In a typical scenario of the response mode, management has seen rapidly rising energy prices and/or curtailments, has heard of the results of other energy management programs, and has then initiated action to start the program. In this case, the management commitment already exists, and all that needs to be done is to cultivate that commitment periodically and to be sure the commitment is evident to all people affected by the program. We will discuss this aspect more when we talk about demonstrating the commitment.

In the aggressive mode, you, the employee, know that energy costs are rising dramatically and that sources are less secure. You may have taken a course in energy management, attended professional conferences, and/or read papers on the subject. At any rate, you are now convinced that the company needs an energy management program. All that remains is to convince management and obtain their commitment.

The best way to convince management is with facts and statistics. Sometimes the most startling way to show the facts is through graphs such as Figure 1-7. Note that different goals of energy cost reduction are shown. This graph can be done in total for all energy sources, or several graphs can be used—one for each source. The latter is probably better as savings goals can be identified by energy source. You must have accurate data. Past figures can use actual utility bills, but future figures call for forecasting. Local utilities and various state energy agencies can help you provide management with more accurate data.

Follow this data with quotes on programs from other companies showing these goals are realistic. Other company experiences are widely published in the literature; results can also be obtained through direct contacts with the energy manager in each company. Typical cost avoidance figures are shown in Table 1-2. However, as time progresses and the technology matures, these figures tend to change. For example, a short time ago only a few people believed that an office building could reduce

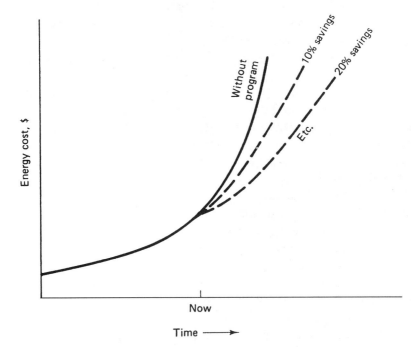

Figure 1-7. Energy Costs—Past and Future.

energy consumption by 70 percent or that manufacturing plants could operate on half the energy previously required, yet both are now occurring on a regular basis.

Table 1-2
Typical Energy Savings

Low cost, no cost changes	5-10%
Dedicated programs (3 years or so)	25-35%
Long-range goal	40-50%

As the proponent of an energy management program, you could then talk about the likelihood of energy curtailments or brownouts and what they would mean to the company. Follow this with a discussion of what the energy management program can do to minimize the impacts of curtailments and brownouts.

Finally, your presentation should discuss the competition and what they are doing. Accurate statistics on this can be obtained from trade and professional organizations as well as the U.S. Department of Energy. The savings obtained by competitors can also be used in developing the goals for your facility.

1.3.2 Energy Management Coordinator/Energy Manager

To develop and maintain vitality for the energy management program, a company must designate a single person who has responsibility for coordinating the program. If no one person has energy management as a specific part of his or her job assignment, management is likely to find that the energy management efforts are given a lower priority than other job responsibilities. Consequently, little or nothing may get done.

The energy management coordinator (EMC) should be strong, dynamic, goal oriented, and a good manager. Most important, management should support that person with resources including a staff. The energy management coordinator should report as high as possible in the organization without losing line orientation. A multiplant or multidivisional corporation may need several such coordinators—one for each plant and one for each level of organization. Typical scenarios are illustrated in Figure 1-8.

1.3.3 Backup Talent

Unfortunately, not all the talent necessary for a successful program resides in one person or discipline. For example, several engineering disci-

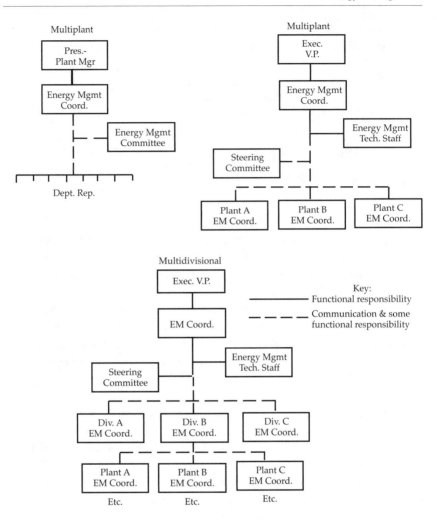

Figure 1-8
Typical Organization Designs for Energy Management Programs

plines may be necessary to accomplish a full-scale study of the plant steam production, distribution, usage, and condensate return system. For this reason, most successful energy management programs have an energy management committee. Two subcommittees that are often desirable are the technical and steering subcommittees.

The technical committee is usually composed of several persons with strong technical background in their discipline. Chemical, industrial, electrical, civil, and mechanical engineers as well as others may all be represented on this committee. Their responsibility is to provide technical assistance for the coordinator and plant-level people. For example, the committee can keep up with developing technology and research into po-tential applications company-wide. The results can then be filtered down.

While the energy management coordinator may be a full-time position, the technical committee is likely to operate part-time, being called upon as necessary. In a multiplant or multidivisional organization, the technical committee may also be full time.

The steering committee has an entirely different purpose from the technical committee. It helps guide the activities of the energy management program and aids in communications through all organizational levels. The steering committee also helps ensure that all plant personnel are aware of the program. The steering committee members are usually chosen so that all major areas of the company are represented. A typical organization is presented in Figure 1-9.

Steering committee members should be selected because of their widespread interests and a sincere desire to aid in solving the energy problems. Departmental and hourly representatives can be chosen on a rotating basis. Such a committee should be able to develop a good composite picture of plant energy consumption which will help the energy management coordinator choose and manage his/her activities.

1.3.4 Cost Allocation

One of the most difficult problems for the energy manager is to try to reduce energy costs for a facility when the energy costs are accounted for as part of the general overhead. In that case, the individual managers and supervisors do not consider themselves responsible for controlling the energy costs. This is because they do not see any direct benefit from reducing costs that are part of the total company overhead. The best solution to this problem is for top management to allocate energy costs down to "cost centers" in the company or facility. Once energy costs are charged to production centers in the same way that materials and labor

Figure 1-9
Energy Management Steering Committee.

are charged, then the managers have a direct incentive to control those energy costs because this will improve the overall cost-effectiveness of the production center.

For a building, this allocation of energy costs means that each of the tenants are given information on their energy consumption, and that they individually pay for that energy consumption. Even if a large building is "master metered" to reduce utility fixed charges, there should be a division of the utility cost down to the individual customers.

1.3.5 Reporting and Monitoring

It is critical for the energy management coordinator and the steering committee to have their fingers on the "pulse of energy consumption" in the plant. This is best achieved through an effective and efficient system of energy reporting.

The objective of an energy reporting system is to measure energy consumption and compare it either to company goals or to some *standard of energy consumption.* Ideally, this should be done for each operation or production cost center in the plant, but most facilities simply do not have the required metering devices. Many plants only meter energy consumption at one place—where the various sources enter the plant. Most plants are attempting to remedy this, however, by installing additional metering devices when the opportunity arises (steam system shutdowns, vacation downtime, etc.). Systems that should be metered include steam, compressed air, and chilled and hot water.

As always, the reporting scheme needs to be reviewed periodically to ensure that only necessary material is being generated, that all needed data is available, and that the system is efficient and effective. For a more complete description of this method and its applications, see section 1.7 on Energy Monitoring, Targeting and Reporting.

1.3.6 Training

Most energy management coordinators find that substantial training is necessary. This training can be broken down as shown in Figure 1-10.

Training cannot be accomplished overnight, nor is it ever "completed." Changes occur in energy management staff and employees at all levels, as well as new technology and production methods. All these precipitate training or retraining. The energy management coordinator must assume responsibility for this training.

Personnel involved	Type of necessary training	Source of required training
1. Technical committee	1. Sensitivity to EM	1. In house (with outside help ?)
	2. Technology developments	2. Professional societies universities, consulting groups, journals
2. Steering committee	1. Sensitivity to EM 2. Other Industries' experience	1. See 1 above 2. Trade journals, energy sharing groups, consultants
3. Plant-wide	1. Sensitivity to EM 2. What's expected, goals to be obtained, etc.	1. In house 2. In house

Figure 1-10
Energy Management Training

1.4   STARTING AN ENERGY MANAGEMENT PROGRAM

Several items contribute to the successful start of an energy management program. They include:

1.   Visibility of the program start-up
2.   Demonstration of management commitment to the program
3.   Selection of a good initial energy management project

1.4.1 Visibility of Start-up

To be successful, an energy management program must have the backing of the people involved. Obtaining this support is often not an easy task, so careful planning is necessary. The people must:

1.   Understand why the program exists and what its goals are;
2.   See how the program will affect their jobs and income;
3.   Know that the program has full management support; and
4.   Know what is expected of them.

Communicating this information to the employees is a joint task of management and the energy management coordinator. The company must take advantage of all existing communications channels while also taking into consideration the preceding four points. Some methods that have proven useful in most companies include:

- Memos. Memos announcing the program can be sent to all employees. A comprehensive memo giving fairly complete details of the program can be sent to all management personnel from first-line supervision up. A more succinct one can be sent to all other employees that briefly states why the program is being formed and what is expected of them. These memos should be signed by local top management.

- News releases. Considerable publicity often accompanies the program start-up. Radio, TV, posters, newspapers, and billboards can all be used. The objective here is to obtain as much visibility for the program as possible and to reap any favorable public relations that might be available. News releases should contain information of interest to the general public as well as employees.

- Meetings. Corporate, plant, and department meetings are sometimes used, in conjunction with or in lieu of memos, to announce the program and provide details. Top management can demonstrate commitment by attending these meetings. The meeting agenda must provide time for discussion and interaction.

- Films, video tapes. Whether produced in-house or purchased, films and video tapes can add another dimension to the presentation. They can also be reused later for new employee training.

1.4.2 Demonstration of Management Commitment

As stressed earlier, management commitment to the program is essential, and this commitment must be obvious to all employees if the program is to reach its full potential. Management participation in the program start-up demonstrates this commitment, but it should also be emphasized in other ways. For example:

- Reward participating individuals. Recognition is highly motivating for most employees. An employee who has been a staunch supporter of the program should be recognized by a "pat on the back," a letter in the files, acknowledgment at performance appraisal time, etc. When the employee has made a suggestion that led to large energy savings, his/her activities should be recognized through monetary rewards, publicity, or both. Public recognition can be given in the company newsletter, on bulletin boards, or in plant-department meetings.

- Reinforce commitment. Management must realize that they are continually watched by employees. Lip service to the program is not enough—personal commitment must be demonstrated. Management should reinforce its commitment periodically, although the visibility scale can be lower than before. Existing newsletters, or a separate one for the energy management program, can include a short column or letter from management on the current results of the program and the plans for the future. This same newsletter can report on outstanding suggestions from employees.

- Fund cost-effective proposals. All companies have capital budgeting problems in varying degrees of severity, and unfortunately energy projects do not receive the same priority as front-line items such as equipment acquisition. However, management must realize that turning down the proposals of the energy management team while accepting others with less economic attractiveness is a sure way to kill enthusiasm. Energy management projects need to compete with others fairly. If an energy management project is cost effective, it should be funded. If money is not available for capital expenditures, then management should make this clear at the outset of the program and ask the team to develop a program which does not require capital expenditures.

1.4.3 Early Project Selection

The energy management program is on treacherous footing in the beginning. Most employees are afraid their heat is going to be set back, their air conditioning turned off, and their lighting reduced. If any of these actions do occur, it's little wonder employee support wanes. These things might occur eventually, but wouldn't it be smarter to have less controversial actions as the early projects.

An early failure can also be harmful, if not disastrous, to the program. Consequently, the astute energy management coordinator will "stack the deck" in his or her first set of projects. These projects should have a rapid payback, a high probability of success, and few negative consequences.

These ideal projects are not as difficult to find as you might expect. Every plant has a few good opportunities, and the energy management coordinator should be looking for them.

One good example involved a rather dimly lit refrigerator warehouse area. Mercury vapor lamps were used in this area. The local energy management coordinator did a relamping project. He switched from mercury vapor lamps to high pressure sodium lamps (a significantly more efficient source) and carefully designed the system to improve the lighting levels. Savings were quite large; less energy was needed for lighting; less "heat of light" had to be refrigerated; and, most important, the employees liked it. Their environment was improved since light levels were higher than before.

Other examples you should consider include:

1.   Repairing steam leaks. Even a small leak can be very expensive over a year and quite uncomfortable for employees working in the area.

2.   Insulating steam, hot water, and other heated fluid lines and tanks. Heat loss through an uninsulated steam line can be quite large, and the surrounding air may be heated unnecessarily.

3.   Install high efficiency motors. This saves dramatically on the electrical utility cost in many cases, and has no negative employee consequences. However, the employees should be told about the savings since motor efficiency improvement has no physically discernible effect, unlike the lighting example above.

This list only begins to touch on the possibilities, and what may be glamorous for one facility might not be for another. All facilities, however, do have such opportunities. Remember, highly successful projects should be accompanied by publicity at all stages of the program—especially at the beginning.

1.5   MANAGEMENT OF THE PROGRAM

1.5.1 Establishing objectives in an Energy Management Program

Creativity is a vital element in the successful execution of an energy management program, and management should do all it can to encourage creativity rather than stifle it. Normally, this implies a *laissez-faire* approach by management with adequate monitoring. Management by objectives (MBO) is often utilized. If TQM is being implemented in a facility, then employee teams should foster this interest and creativity.

Goals need to be set, and these goals should be tough but achievable, measurable, and specific. They must also include a deadline for accomplishment. Once management and the energy management coordinator have agreed on the goals and established a good monitoring or reporting system, the coordinator should be left alone to do his/her job.

The following list provides some examples of such goals:

- Total energy per unit of production will drop by 10 percent the first year and an additional 5 percent the second.

- Within 2 years all energy consumers of 5 million British thermal units per hour or larger will be separately metered for monitoring purposes.

- Each plant in the division will have an active energy management program by the end of the first year.

- All plants will have contingency plans for gas curtailments of varying duration by the end of the first year.

- All boilers of 50,000 lb/hour or larger will be examined for waste heat recovery potential the first year.

The energy management coordinator must quickly establish the reporting systems to measure progress toward the goals and must develop the strategy plans to ensure progress. Gantt or CPM charting is often used to aid in the planning and assignment of responsibilities.

Some concepts or principles that aid the EMC in this execution are the following:

- Energy costs, not just Btus, are to be controlled. This means that any action that reduces energy costs is fair game. Demand shedding or leveling is an example activity that saves dollars but does not directly save Btus.

- Energy needs should be recognized and billed as a direct cost, not as overhead. Until the energy flow can be measured and charged to operating cost centers, the program will not reach its ultimate potential.

- Only the main energy functions need to be metered and monitored. The Pareto or ABC principle states that the majority of the energy costs are incurred by only a few machines. These high-use machines should be watched carefully.

1.5.2 A Model Energy Management Program

Headquartered in St. Paul, MN [5], the 3M Company is an excellent example of a longtime successful energy management program in a large corporation. 3M is a diversified manufacturing company with more than 50 major product lines; it makes some 50,000 products at over 50 factory locations in the U.S. Their energy management objective is to use energy as efficiently as possible in all operations; management believes that all companies have an obligation to conserve energy and all natural resources.

Energy productivity at 3M's US operations improved over 80% from 1973-2010. Their energy management programs saved over $70 million in 1996, and more than $1.5 billion in energy expenses from 1973-2004. From 1998—2005, they reduced overall energy use by 29% in their worldwide operations. From 2006-2011, 3M increased their yearly energy use reduction goal to 5% per year. Also they reported a cumulative savings of $100 million from 2005-2011 [8]. Their program is staffed by six people who educate and motivate all levels of personnel on the benefits of energy management. The categories of programs implemented by 3M include: conservation, maintenance procedures, utility operation optimization, efficient new designs, retrofits through energy surveys, and process changes.

Energy efficiency goals at 3M are set and then the results are measured against a set standard in order to determine the success of the programs. The technologies that have resulted in the most dramatic improvement in energy efficiency include: heat recovery systems, high efficiency motors, variable speed drives, computerized facility management systems, steam trap maintenance, combustion improvements, variable air volume systems, thermal insulation, cogeneration, waste steam utilization, and process improvements. Integrated manufacturing techniques, better equipment utilization and shifting to non-hazardous solvents have also resulted in major process improvements.

The energy management program at 3M has worked well, but management is not yet satisfied. Their goal is to further improve energy efficiency at a rate of 3 percent per year for the next five years, 2006-2010. They expect to substantially reduce their emissions of waste gases and liquids, to increase the energy recovered from wastes, and to constantly increase the profitability of their operations. 3M continues to stress the importance that efficient use of energy can have on industrial productivity.

1.6   ENERGY ACCOUNTING

Energy accounting is a system used to keep track of energy consumption and costs. "Successful corporate-level energy managers usually rank energy accounting systems right behind commitment from top corporate officials when they list the fundamentals of an ongoing energy conservation program. If commitment from the top is motherhood, careful accounting is apple pie."*

A basic energy accounting system has three parts: energy use monitoring, an energy use record, and a performance measure. The performance measure may range from a simple index of Btu/ft^2 or $Btu/unit$ of production to a complex standard cost system complete with variance reports. In all cases, energy accounting requires metering. Monitoring the energy flow through a cost center, no matter how large or small, requires the ability to measure incoming and outgoing energy. The lack of necessary meters is probably the largest single deterrent to the widespread utilization of energy accounting systems.

1.6.1 Levels of Energy Accounting

As in financial accounting, the level of sophistication or detail of energy accounting systems varies considerably from company to company. A very close correlation can be developed between the levels of sophistication of financial accounting systems and those of energy accounting systems. This is outlined in Figure 1-11.

Most companies with successful energy management programs have passed level 1 and are working toward the necessary submetering and reporting systems for level 2. In most cases, the subsequent data are compared to previous years or to a particular base year. However, few companies have developed systems that will calculate variations and find causes for those variations (level 3). Two notable exceptions are General Motors and Carborundum. To our knowledge, few companies have yet completely developed the data and procedures necessary for level 4, a standard Btu accounting system. Some examples of detailed energy accounting can be found in [6].

1.6.2 Performance Measures

1.6.2.1 Energy Utilization Index

A very basic measure of a facility's energy performance is called the Energy Utilization Index (EUI). This is a statement of the number of Btu's of energy used annually per square foot of conditioned space. To compute

---

*"Accounting of Energy Seen Corporate Must," *Energy User News*, Aug. 27, 1979, p. 1.

Financial	*Energy*
1. General accounting	1. Effective metering, development of reports, calculation of energy efficiency indices
2. Cost accounting	2. Calculation of energy flows and efficiency of utilization for various cost centers; requires substantial metering
3. Standard cost accounting historical standards	3. Effective cost center metering of energy and comparison to historical data; complete with variance reports and calculation of reasons for variation
4. Standard cost accounting engineered standards	4. Same as 3 except that standards for energy consumption are determined through accurate engineering models

Figure 1-11. Comparison between Financial and Energy Accounting

the EUI, all of the energy used in the facility must be identified, the total Btu content tabulated, and the total number of square feet of conditioned space determined. The EUI is then found as the ratio of the total Btu consumed to the total number of square feet of conditioned space.

Example 1.1—Consider a building with 100,000 square feet of floor space. It uses 1.76 million kWh and 6.5 million cubic feet of natural gas in one year. Find the Energy Utilization Index (EUI) for this facility.

Solution: Each kWh contains 3412 Btu and each cubic foot of gas contains about 1000 Btu. Therefore the total annual energy use is:

$$
\begin{aligned}
\text{Total energy use} \quad &= (1.76 \times 10^6 \text{ kWh}) \times (3412 \text{ Btu}/\text{kWh}) \\
&\quad + (6.5 \times 10^6 \text{ ft}^3) \times (1000 \text{ Btu}/\text{ft}^3) \\[4pt]
&= 6.0 \times 10^9 + 6.5 \times 10^9 \\
&= 1.25 \times 10^{10} \text{ Btu}/\text{yr}
\end{aligned}
$$

Dividing the total energy use by 10^5 ft^2 gives the EUI:

$$\text{EUI} = (1.25 \times 10^{10}\ \text{Btu/yr})/(10^5\ \text{ft}^2)$$
$$= 125{,}000\ \text{Btu/ft}^2/\text{yr}$$

The average building EUI is 89,800 Btu/ft^2/yr; the average office building EUI is 92,900 Btu/ft^2/yr. Figure 1-12 shows the range of energy intensiveness in 1000 Btu/ft^2/yr for the twelve different types of commercial facilities listed [7].

1.6.2.2 Energy Cost Index

Another useful performance index is the Energy Cost Index or ECI. This is a statement of the dollar cost of energy used annually per square foot of conditioned space. To compute the ECI, all of the energy used in the facility must be identified, the total cost of that energy tabulated, and the total number of square feet of conditioned space determined. The ECI is then found as the ratio of the total annual energy cost for a facility to the total number of square feet of conditioned floor space of the facility.

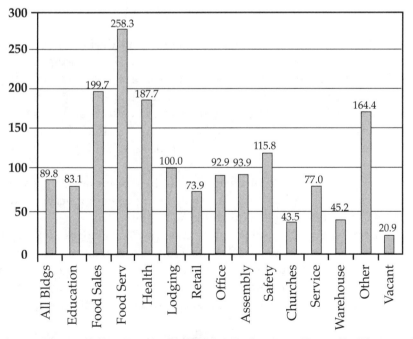

Figure 1-12. Building Energy Utilization Index, 2003 Data. (In Thousand Btu per Sq Ft per Year). Source—U.S. Department of Energy EIA 4/2011)

Example 1.2 Consider the building in Example 1.1. The annual cost for electric energy is $115,000 and the annual cost for natural gas is $32,500. Find the Energy Cost Index (ECI) for this facility.

Solution: The ECI is the total annual energy cost divided by the total number of conditioned square feet of floor space.

$$\text{Total energy cost} = \$115,000 + \$32,500 = \$147,500/yr$$

Dividing this total energy cost by 100,000 square feet of space gives:

$$\text{ECI} = (\$147,500/yr)/(100,000\ ft^2) = \$1.48/ft^2/yr$$

The Energy Information Administration data for 2003 showed a value of the ECI for the average building as $1.51/ft^2/yr. The ECI for an average office building was $1.71/ft^2/yr.

1.6.2.2 One-Shot Productivity Measures

The purpose of a one-shot productivity measure is illustrated in Figure 1-13. Here the energy utilization index is plotted over time, and trends can be noted.

Significant deviations from the same period during the previous year should be noted and explanations sought. This measure is often used to justify energy management activities or at least to show their effect. For example, in Figure 1-13 an energy management (EM) program was started at the beginning of year 2. Its effect can be noted by comparing peak summer consumption in year 2 to that of year 1. The decrease in peaks indicates that this has been a good program (or a mild summer, or both).

Table 1-3 shows some often-used indices. Some advantages and disadvantages of each index are listed, but specific applications will require careful study to determine the best index.

Table 1-4 proposes some newer concepts. Advantages and disadvantages are shown, but since most of these concepts have not been utilized in a large number of companies, there are probably other advantages and disadvantages not yet identified. Also, there are an infinite number of possible indices, and only three are shown here.

1.6.3 An Example Energy Accounting System

General Motors Corporation has a strong energy accounting sys-

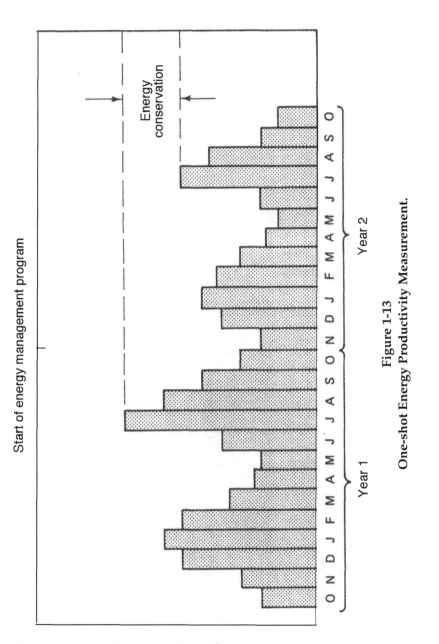

Figure 1-13
One-shot Energy Productivity Measurement.

Table 1-3. Commonly Used Indices

Productivity indicator	Advantages	Disadvantages
1. Btu/unit of production	1. Concise, neat 2. Often accurate when process energy needs are high 3. Good for interplant and company comparison when appropriate	1. Difficult to define and measure "units" 2. Often not accurate (high HVAC* and lighting makes energy nonlinear to production)
2. Btu/degree day	1. Concise, neat, best used when HVAC* is a majority of energy bill 2. Often accurate when process needs are low or constant 3. Very consistent between plants, companies, etc. (all mfg can measure degree days)	1. Often not accurate (disregards process needs) 2. Thermally heavy buildings such as mfg plants usually do not respond to degree days
3. Btu/ft²	1. Concise, neat 2. Accurate when process needs are low or constant and weather is consistent 3. Very consistent (all mfg can measure square feet) 4. Expansions can be incorporated directly	1. No measure of production or weather 2. Energy not usually linearly proportional to floor space (piecewise linear?)
4. Combination, e.g., Btu/ unit-degree day-ft² or Btu/unit-degree day	1. Measures several variables 2. Somewhat consistent, more accurate than above measures 3. More tailor-made for specific needs	1. Harder to comprehend

*Heating, ventilating, and air conditioning.

Table 1-4
Proposed Indices

Productivity indicator	Advantages	Disadvantages
1. Btu/sales dollar	1. Easy to compute	1. Impact of inflation
2. $\dfrac{\$\text{ energy}}{(\$\text{ sales}) \text{ or } (\$\text{ profit}) \text{ or } (\$\text{ value added})}$	1. Really what's desired 2. Inflation cancels or shows changing relative energy costs 3. Shows energy management results, not just conservation (e.g., fuel switching, demand leveling, contingency planning)	1. Very complex, e.g., lots of variables affect profit including accounting procedures 2. Not good for general employee distribution
3. Btu/DL hour (or machine hour or shift) where DL = direct labor	1. Almost a measure of production (same advantage as in Table 1-3) 2. Data easily obtained when already available 3. Comparable between plants or industries 4. Good for high process energy needs	1. More complex, e.g., can't treat a DL hour like a unit of production 2. Energy often not proportional to labor or machine input, e.g., high HVAC and lighting

tem which uses an energy responsibility method. According to General Motors, a good energy accounting system is implemented in three phases: (1) design and installation of accurate metering, (2) development of an energy budget, and (3) publication of regular performance reports including variances. Each phase is an important element of the complete system.

1.6.3.1 The GM system

Phase 1—Metering. For execution of a successful energy accounting program, energy flow must be measured by cost center. The designing of cost center boundaries requires care; the cost centers must not be too large or too small. However, the primary design criterion is how much energy is involved. For example, a bank of large electric induction heat-treating furnaces might need separate metering even if the area involved is relatively small, but a large assembly area with only a few energy-consuming devices may require only one meter. Flexibility is important since a cost center that is too small today may not be too small tomorrow as energy costs change.

The choice of meters is also important. Meters should be accurate, rugged, and cost effective. They should have a good turndown ratio; a turndown ratio is defined as the ability to measure accurately over the entire range of energy flow involved.

Having the meters is not enough. A system must be designed to gather and record the data in a useful form. Meters can be read manually, they can record information on charts for permanent records, and/or they can be interfaced with microcomputers for real-time reporting and control. Many energy accounting systems fail because the data collection system is not adequately designed or utilized.

Phase 2—Energy Budget. The unique and perhaps vital aspect of General Motors' approach is the development of an energy budget. The GM energy responsibility accounting system is somewhere between levels 3 and 4 of Figure 1-11. If a budget is determined through engineering models, then it is a standard cost system and it is at level 4. There are two ways to develop the energy budget: statistical manipulation of historical data or utilization of engineering models.

The Statistical Model. Using historical data, the statistical model shows how much energy was utilized and how it compared to the standard year(s), but it does not show how efficiently the energy was used. For example, consider the data shown in Table 1-5.

Table 1-5. Energy Data for Statistical Model[a]

	1995	1996	1997
Total energy (units)	1,000	1,100	1,050
Production (units)	600	650	650
Square feet	150,000	150,000	170,000
Degree days (heating)	6,750	6,800	6,800

[a]Taken, in part, from R.P. Greene, (see the Bibliography).

The statistical model assumes that the base years are characteristic of all future years. Consequently, if 1996 produced 600 units with the same square footage and degree days as 1995, 1000 units of energy would be required. If 970 units of energy were used, the difference (30 units) would be due to conservation.

We could use multiple linear regression to develop the parameters for our model, given as follows:

$$\text{energy forecast} = a(\text{production level}) + b(\text{ft}^2) + c(\text{degree days}) \quad (1\text{-}1)$$

We can rewrite this in the following form:

$$X_4 = aX_1 + bX_2 + cX_3$$

where X_1 = production (units) X_3 = weather data (degree days)

X_2 = floor space (ft^2) X_4 = energy forecast (Btu)

Degree days are explained in detail in Chapter Two, section 2.1.1.2. Their use provides a simple way to account for the severity of the weather, and thus the amount of energy needed for heating and cooling a facility. Of course, the actual factors included in the model will vary between companies and need to be examined carefully.

Multiple linear regression estimates the parameters in the universal regression model in Equation 1-1 from a set of sample data. Using the base years, the procedure estimates values for parameters a, b, and c in Equation 1-1 in order to minimize the squared error where

$$\text{squared error} = \sum_{\substack{\text{base} \\ \text{years}}} \left(X_4^i - X_4 \right)^2 \quad (1\text{-}2)$$

with X_4 = energy forecast by model

X_4^i = actual energy usage

The development and execution of this statistical model is beyond the scope of this book. However, regardless of the analytical method used, a statistical model does not determine the amount of energy that ought to be used. It only forecasts consumption based on previous years' data.

The engineering model. The engineering model attempts to remedy the deficiency in the statistical model by developing complete energy balance calculations to determine the amount of energy theoretically required. By using the first law of thermodynamics, energy and mass balances can be completed for any process. The result is the energy required for production. Similarly, HVAC and lighting energy needs could be developed using heat loss equations and other simple calculations. Advantages of the engineering model include improved accuracy and flexibility in reacting to changes in building structures, production schedules, etc. Also, computer programs exist that will calculate the needs for HVAC and lighting.

Phase 3—Performance Reports. The next step is the publication of energy performance reports that compare actual energy consumption with that predicted by the models. The manager of each cost center should be evaluated on his or her performance as shown in these reports. The publication of these reports is the final step in the effort to transfer energy costs from an overhead category to a direct cost or at least to a direct overhead item. One example report is shown in Figure 1-14.

Sometimes more detail on variance is needed. For example, if consumption were shown in dollars, the variation could be shown in dollars and broken into price and consumption variation. Price variation is calculated as the difference between the budget and the actual unit price times the present actual consumption. The remaining variation would be due to a change in consumption and would be equal to the change in consumption times the budget price. This is illustrated in Example 1.3. Other categories of variation could include fuel switching, pollution control, and new equipment.

However, had energy consumption not been reduced, the total energy cost would have been:

$$2125(4.50) + 6400(3.12) + 2571(4.46) = \$40,997.$$

The total cost avoidance therefore was:

$$\$40,997 - \$34,000 = \$6997$$

	Actual	Budget	Variance	% variance
Department A				
Electricity	2000	1500	+500	+33.3%
Natural gas	3000	3300	−300	−9.1%
Steam	3500	3750	−250	−6.7%
Total	8500	8550	−50	−0.6%
Department B				
Electricity	1500	1600	−100	−6.2%
Natural gas	2000	2400	−400	−16.7%
Fuel oil	1100	1300	− 200	−15.4%
Coal	3500	3900	− 400	−10.2%
Total	8100	9200	− 1100	− 11.9%
Department C				
•				
•				
•				

Figure 1-14. Energy Performance Report (10^6Btu)

which is the drop in consumption times the actual price or

$$(2125 - 2000)\ 4.5 + (6400 - 4808)\ 3.12 + (2571 - 2242)\ 4.46 = \$6997$$

This problem of increased energy costs despite energy management savings can arise in a number of ways. Increased production, plant expansion, or increased energy costs can all cause this result.

1.7   ENERGY MONITORING, TARGETING AND REPORTING*

1.7.1 Introduction

Energy Monitoring, Targeting and Reporting (MT&R) is a powerful management technique for

- analyzing the historical energy performance of industrial, commercial, and institutional facilities
- setting energy reduction targets

---

*This section was written by Mr. Doug Tripp, P. Eng., Executive Director, Canadian Institute for Energy Training,; and Mr. Stephen Dixon, President, TdS Dixon Inc.

Example 1.3

The table shown in Figure 1-15 portrays a common problem in energy management reporting. The energy management program in this heat treating department was quite successful. When you examine the totals, you see that the total consumption (at old prices) was reduced by $5631. The total energy cost, however, went up by $500, which was due to a substantial price variation of $6131. Consequently, total energy costs increased to $34,000.

Heat Treating Department	[A] Actual $	[B] Budget $	[C] Unit price (budget)	[D] Unit price (actual)	[E] A - B[a] variance	[F] (D - C)A[b] price variance	[G] E – F or (B[b] – A[b])C consumption variance
	10^6 Btu	10^6 Btu	$/10^6$ Btu	$/10^6$ Btu			
Electricity							
Cost	$9,000	$8,500	$4.00	$4.50	+$500	+$1000	-$500
Use	2,000	2,125	—	—	—	—	—
Natural gas							
Cost	15,000	16,000	2.50	3.12	-1000	+2980	-3980
Use	4,808	6,400	—	—	—	—	—
Steam							
Cost	10,000	9,000	3.50	4.46	+1000	+2151	-1151
Use	2,242	2,571	—	—	—	—	—
Total cost	$34,000	$33,500	—	—	+$500	+$6131	-$5631

[a]Measured in $
[b]Measured in 10^6 Btu

Figure 1-15. Energy Cost in Dollars by Department with Variance Analysis

- controlling current energy performance
- and, projecting future energy budgets.

It is a technique that has proven its effectiveness in achieving energy cost savings in the range five to fifteen percent as a direct consequence of effective performance monitoring, and in creating the management information needed to identify and implement energy efficiency measures. Further, it provides a framework for savings verification when measures are implemented.

The working definitions that commonly apply are the following:

- **Energy Monitoring** is the regular collection and analysis of information on energy use. Its purpose is to establish a basis of management control, to determine when and why energy consumption is deviating from an established pattern, and to provide a basis for taking management action where necessary.

- **Targeting** is the identification of levels of energy consumption towards which it is desirable, as a management objective, to work.

- **Reporting** closes the loop, by putting the management information generated in a form that enables ongoing control of energy use, the achievement of reduction targets, and the verification of savings.

MT&R is built around one key statistical technique: CUSUM (Cumulative Sum of Differences) analysis of the variance between energy consumption predicted by an energy performance model (EPM), and the actual measured consumption. Ancillary functions that are derived from the CUSUM analysis are a target-setting methodology, and the application of energy control charts for real-time management of performance.

The key steps in an effective MT&R process are:

- measurement of energy consumption over time

- measurement of the independent variables that influence energy consumption (weather, production, occupancy) over corresponding time intervals

- development of a relationship (the energy performance model) between energy and the independent variables

- historical analysis of energy performance using CUSUM, and application of the CUSUM trend into the future

- definition of reduction targets

- frequent comparison of actual consumption to targets

- reporting of consumption and target variances

- taking action to address variances and ensure targets are met.

The achievement of energy cost savings is the primary objective of MT&R, but there are other benefits as well, including:

- improved budgeting and forecasting

- improved product/service costing

- tracking and verification of energy efficiency retrofits

- opportunities for improved operation and maintenance practices.

1.7.1.1 MT&R and Continuous Improvement

Monitoring and target setting have elements in common and they share much of the same information. As a general rule, however, monitoring comes before target setting because without monitoring you cannot know precisely where you are starting from or decide if a target has been achieved. The reporting phase not only supports management control, but also provides for accountability in the relationship between performance and targets.

MT&R is consistent with other continuous improvement techniques applied in organizations, and should be viewed as an ongoing, cyclical process, as Figure 1-16 suggests.

The cycle begins with any measured energy data presently available, typically energy bills or invoices. Once assembled the data can be analyzed to reveal patterns, trends and consumption statistics. The reporting of the information resulting from this analysis can be used to prompt actions that produce results, typically the reduction of consumption and costs. Subsequent measurements and analysis reveal the actual result of the actions. The process then enters another cycle of measurement, analysis and action.

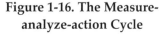

Figure 1-16. The Measure-analyze-action Cycle

1.7.1.2 Energy Cost Center

The organizational basis for MT&R is the energy cost center (ECC). An ECC is a unit for which energy use may be measured along with other factors that influence the energy consumption. For example, the ECC might be a single building in a portfolio of properties, a production unit or department in a plant, or a major energy consuming system such as the heating plant.

Basic criteria for the designation of an ECC are:

- energy consumption can be measured in isolation
- the cost of measurement can be justified by potential savings
- the ECC must correspond to existing business structures
- someone must be accountable for the ECC
- a factor of influence must be measurable.

1.7.1.3 Units of Measure

An energy monitoring and targeting system must measure data for all energy forms utilized in the ECC. While the goal of the system is to reduce costs, dollars are not good universal measurement units for basic data. Unit prices tend to fluctuate over time making comparison of present to previous energy usage difficult if not impossible. Current unit prices can be applied to calculated energy savings to determine total cost avoidance.

Wherever possible, data should be recorded in physically measurable units (typically the units of purchase such as gallons, litres, therms, kWh, ft^3, M^3 and so on). If more than one energy source is being used by the ECC, it is necessary to convert the units of purchase to a common unit using the appropriate energy equivalence factors for each form.

1.7.2 Principles of Energy Monitoring

Energy monitoring involves the development of an energy performance model (EPM) that quantifies a relationship between consumption and the applicable independent variables, and the comparison of performance predicted by the model to actual performance by means of CUSUM analysis. Although the EPM can be developed in a number of ways, it most commonly involves the use of regression analysis.

1.7.2.1 Independent Variables

In addition to energy data for each ECC, the applicable factors of influence, or independent variables, must be measured. Some examples include:

- **Operating Hours/Occupancy**—In office buildings, schools, hospitals, retail stores, warehouses, accommodation facilities, energy consumption depends primarily on the weather through the heating and cooling systems, and secondarily on patterns of use such as operating hours and occupancy. Electricity and fuel each have their own dependencies or combination of dependencies specific to each facility's systems.

- **Production Level**—Energy use in manufacturing and industrial facilities depends strongly on production level or a measure of the facility's output, in units, tons, or some other appropriate unit. Also, energy use in these operations often exhibits a strong dependence upon weather.

- **Degree Days**—Degree days are a measure of the outside temperature for a given period and can often be used to predict the amount of heating and cooling energy a building requires.

Heating degree days (HDD), for the winter season indicate the difference between the outside temperature and a baseline (typically 65°F or 18°C) and how long that difference exists for. Likewise cooling degree days (CDD) indicate the difference above a base temperature (typically 72°C or 22°C) for the cooling season.

Degree days can be derived from temperature measurements at the facility, or can be obtained from weather services and some utilities. For a basic MT&R system, degree days from a nearby site are often adequate although they may only be available on a monthly basis.

1.7.2.2 Functional Relationship between Energy and the Independent Variables

1.7.2.2.1 The degree-day

The energy consumed to maintain building temperature at its desired set point is approximately equivalent to heat lost through the building envelope and ventilation in the heating season or gained in the cooling season. It can be shown from standard heat loss/gain relationships that heat is related to the difference between indoor and outdoor temperature according to Equation 1-3:

$$H = UA + (C_p NV)(T_i - T_o) \qquad (1\text{-}3)$$

where:

H   is the rate of heat flow

U   is the heat transfer coefficient of the building envelope, taking into account its components such as glazing, interior wall finish, insulation, exterior wall, etc.

A   is the external area of the building envelope
C_p is the specific heat of air
N   is the number of air changes of per unit of time
V   is the volume of the building being ventilated
T_i-T_o is the difference between the inside and outside temperatures.

U, A, C_p, N, and V are all characteristic constants of the building, and usually T_i is controlled at a constant set value. Therefore, the only variable in the equation is T_o and it may be higher or lower than the inside temperature depending on the season. When T_o is less than T_i, H is a positive number indicating that heat must be added to the building in the heating season, and when T_o is greater than T_i, H is a negative number indicating that heat must be removed from the building in the cooling season.

When degree-days are used as the determinant of heat load, equation 1-3 becomes:

$$H = (UA + C_pNV) \times \text{degree-days} + c \qquad (1\text{-}4)$$

which is the equation of a straight line, as in Figure 1-17, when H is plotted against degree-days, having a slope = $(UA + C_pNV)$ and an intercept on the y-axis = c. This constant c is the 'no load' energy provided whatever the weather conditions by such things as office equipment, the losses from the boiler, lighting, and people.

1.7.2.2.2 Production

Energy used in production processes typically heats, cools, changes the state of, or moves material. Obviously it is impossible to generalize as industrial processes are both complex and widely varied. However, a similar theoretical assessment of specific processes as that done for degree-days will yield a similar conclusion: that is, there is reason to expect that energy plotted against production will also produce a straight line of the general form:

$$y = mx + c \qquad (1\text{-}5)$$

where c, the intercept (and, no load or zero production energy consumption), and m, the slope are empirical coefficients, characteristic of the system being analyzed. See Figure 1-17.

1.7.2.3 Establishing the Energy Performance Model

Energy use data alone are of very limited usefulness in understand-

ing the nature of the
energy system, identify-
ing opportunities for ef-
ficiency improvement,
and controlling energy
use in the future. Refin-
ing data to informa-
tion that facilitates these
functions involves anal-
ysis, following steps
illustrated here.

Figure 1-17. Relationship between Degree-days and Heat Load

Three basic meth-
ods exist for establish-
ing a model:

- **previous year's data**—simply using last year as a
  predictor of this year's consumption. Typically only useful when
  there are no significant factors of influence.

- **regression analysis**—a statistical approach based upon historical
  consumption and the factors of influence.

- **simulation model**—using complex numerical computer models to
  simulate the energy consumption.

The most common method for a basic system is regression analysis.

In many instances linear regression of energy consumption against a
single independent variable (degree-days or production) generates a valid
energy performance model. In some cases, multivariant linear regression,
for example against degree-days and production in a plant for which there
is significant dependence on weather, is a better representation of the energy relationship.

To illustrate, two data sets are shown below, one for a building with
energy consumption driven by degree days, and one for a plant in which
the driver is production. In each case, the scatter plot produced when consumption is plotted against the independent variable is subjected to linear
regression, with the results shown in Figure 1-18.

The graph produces an energy performance model equation as
shown. That is,

Table 1-6. Sample Energy Data for a School

Month	Heating Degree Days	Total Electricity kWh
Feb 01	577	82,800
Mar 01	613	80,640
Apr 01	490	73,440
May 01	279	43,920
Jun 01	179	33,840
Jul 01	29	18,720
Aug 01	31	10,080
Sep 01	40	11,520
Oct 01	172	26,640
Nov 01	316	50,400
Dec 01	456	62,640
Jan 02	579	84,240

Figure 1-18. Regression Analysis of School Energy Data

Electricity (kWh) = 124.54 × HDD + 9206.7   (1-6)

The two parameters in the equation have a physical meaning:

• the slope of the line, 124.54 represents the incremental energy consumption per heating degree day

• the intercept, 9206.7, represents the non-heating or weather-independent load.

Assessment of building performance may involve an examination of each of these components of load separately.

Similarly, a plant production data-set in Table 1-7 generates a regression model as in Figure 1-19.

Table 1-7. Sample Energy Data for a Food Processing Plant

Month	Production lb.	Total Process Energy, MMBtu
Jan 02	39,600	137,243
Feb 02	21,120	107,620
Mar 02	15,840	94,630
Apr 02	13,200	102,649
May 02	44,880	152,845
Jun 02	47,520	171,792
Jul 02	31,680	126,754
Aug 02	10,560	84,905
Sep 02	29,040	120,510
Oct 02	23,760	108,051
Nov 02	10,560	87,491
Dec 02	13,200	89,379
Jan 03	36,960	131,255
Feb 03	40,920	144,886
Mar 03	43,560	145,882
Apr 03	50,160	158,760
May 03	10,560	81,102
Jun 03	14,520	86,234
Jul 03	39,600	129,613
Aug 03	21,120	98,710
Sep 03	16,632	88,233
Oct 03	29,040	112,643
Nov 03	18,480	92,912
Dec 03	44,880	142,198
Jan	50,160	147,453
Feb 04	42,240	147,231
Mar 04	31,680	123,359

Figure 1-19. Regression Analysis of Food Processing Plant Energy Consumption

Here the energy performance model is

Process Energy (MMBtu) = 1.8873 × Production (lb.) + 63,726   (1-7)

in which, once again, we see the two components of energy consumption, the production-related component (1.9929 MMBtu/lb.) and the production-unrelated base load (65,546 MMBtu).

In each of these examples, the R^2 value indicates the level of confidence we have in the fit of the regression line to the scatter of points.

In the industrial example, it is important to note that many points lie above and below the regression line. This may indicate that energy performance has changed at some point within the 27 months considered. If the points were plotted chronologically, it might become evident that the early points fall above the line (i.e. at relatively higher energy consumption for given production levels) while later points fall below the line (i.e. at relatively lower energy consumption), or vice versa.

If there has been a change in performance, either due to a deliberate action or for an as yet unknown reason, the regression model for the entire data set is not a useful basis for comparison; that is, we need a "baseline" period that is characterized by consistent performance or efficiency.

1.7.2.3.1 Defining the Baseline

    Finding a baseline period may involve trial and error analysis of the data, or it may be defined as a result of knowledge about plant operations. For the purposes of this illustration, let us suppose that it is known that the plant performed consistently for the first 12 months, at which point an improvement was implemented. The regression of the first 12 points in the data set yields Figure 1-20 and a new energy performance model for the baseline period.

The baseline relationship is

Process Energy (MMBtu) = 2.0078 × Production (lb.) + 64,966   (1-8)

Comparison of Equations 1-7 and 1-8 immediately indicates two important findings:

•   the production-related energy is lower for the entire data set than it was in the first 12 months (1.8873 vs. 2.0078 MMBtu/lb.)

•   the production-unrelated energy is lower for the entire data set than it was in the first 12 months (63,726 vs. 64,966 MMBtu)

Figure 1-20. Baseline Model for Food Processing Plant

Both of these findings suggest that performance improvements have taken place to lower the overall energy consumption rates from what they were in the first year.

1.7.2.4 CUSUM Analysis

The baseline EPM, for the shaded months in Table 1-8, is used in CUSUM analysis.

- Predicted values of energy consumption are calculated from Equation 1-8 the actual production values.

- Variance is simply actual consumption-predicted consumption.

- CUSUM values are, as the name indicates, the cumulative algebraic sum of the variances.

So, for example:
the CUSUM value for Oct '02
= 5,389 (the cumulative sum for the previous month) +
(−4,621) (the variance for Oct '02).

The CUSUM values are plotted in a time series shown in Figure 1-22.

1.7.2.4.1 Interpreting the CUSUM Graph

The CUSUM graph reveals changes in energy performance at any point where there is a significant change in the slope of the line. A downward trending line indicates energy saved in comparison to the baseline performance, while an upward trending line indicates a higher rate of consumption.

After the first 12 months of the data set, a downward trend that continues until approximately month 18 is noted. At that point, the downward trend increases in rate, indicating that energy is being saved at a higher rate than in the previous 6 months; this trend continues until month 25. At month 25, another change in slope is observed, but this time to a lower rate of saving; this change indicates that one of the improvements, probably the second one, has stopped functioning, and that action is required to correct the malfunction. Comparison of the slopes for line segments 12-18 and 25-27 indicates that they appear to be approximately the same; that is, the rate of savings is the same in these two periods.

Overall, the graph indicates that a total of approximately 130,000

Table 1-8. CUSUM Calculations for Food Processing Plant

Month	Production lb.	Total Process Energy, MMBtu	Predicted Process Energy MMBtu	Variance Actual— Predicted MMBtu	CUSUM MMBtu
Jan 02	39,600	137,243	144,476	−7,233	−7,233
Feb 02	21,120	107,620	107,371	248	−6,985
Mar 02	15,840	94,630	96,770	−2,140	−9,125
Apr 02	13,200	102,649	91,469	11,179	2,054
May 02	44,880	152,845	155,077	−2,233	−178
Jun 02	47,520	171,792	160,378	11,414	11,235
Jul 02	31,680	126,754	128,574	−1,820	9,416
Aug 02	10,560	84,905	86,169	−1,263	8,152
Sep 02	29,040	120,510	123,273	−2,763	5,389
Oct 02	23,760	108,051	112,672	−4,621	768
Nov 02	10,560	87,491	86,169	1,322	2,090
Dec 02	13,200	89,379	91,469	−2,090	−0
Jan 03	36,960	131,255	139,175	−7,921	−7,921
Feb 03	40,920	144,886	147,126	−2,240	−10,161
Mar 03	43,560	145,882	152,427	−6,545	−16,706
Apr 03	50,160	158,760	165,679	−6,919	−23,625
May 03	10,560	81,102	86,169	−5,067	−28,691
Jun 03	14,520	86,234	94,120	−7,886	−36,577
Jul 03	39,600	129,613	144,476	−14,863	−51,440
Aug 03	21,120	98,710	107,371	−8,661	−60,101
Sep 03	16,632	88,233	98,360	−10,127	−70,228
Oct 03	29,040	112,643	123,273	−10,630	−80,858
Nov 03	18,480	92,912	102,071	−9,159	−90,018
Dec 03	44,880	142,198	155,077	−12,879	−102,897
Jan 04	50,160	147,453	165,679	−18,226	−121,123
Feb 04	42,240	147,231	149,777	−2,546	−123,669
Mar 04	31,680	123,359	128,574	−5,215	−128,884

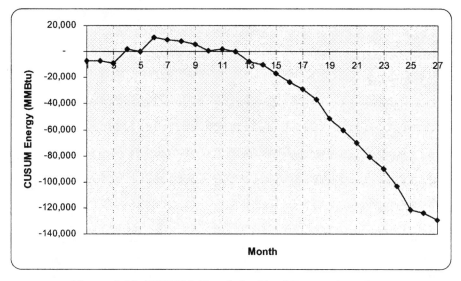

Figure 1-21. CUSUM Graph for Food Processing Plant

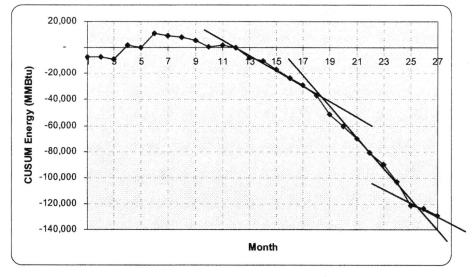

Figure 1-22. CUSUM Trends

MMBtu (actually 128,884 MMBtu from Table 1-8) has been saved in comparison to what would have been consumed had the baseline performance continued for the entire period.

1.7.2.4.2 Source of the Savings

The CUSUM graph indicates when performance changes occurred, and what they achieved in terms of energy saved or wasted. It does not directly indicate how or why those changes occurred. However, further examination of the period of best performance, months 18 through 24 in the example, does give some further information. Figure 1-23 is the regression line for those months.

The performance parameters for this period compared to the baseline period indicate the relative improvements:

Table 1-9. Comparison of Peak Performance Period to Baseline Period

Parameter	Baseline	Months 18-24	% Improvement
Slope—production related consumption	2.0078	1.8231	9.20
Intercept—production unrelated consumption	64,966	59,228	8.83

Figure 1-23. Regression for Months 18-24

The improvement in the production-related consumption is 9.20%, while the production-unrelated baseload has been reduced by 8.83%. That is, there has been an improvement in operating efficiency as well as a reduction in baseload waste.

1.7.3 Setting Performance Targets

MT&R provides a statistical basis for setting performance targets. The parameters of the energy performance model typically quantify incremental and base loads, as seen above, and each of these components of load should be addressed separately.

Since an EPM can be defined for any period of historical performance, initial targets are often based on periods of peak performance. In the industrial example, a reasonable target is to sustain the performance of the 18- to 24-month period, as defined in Table 1-9.

Other strategies for setting targets include:

- **eliminating the highest or least efficient points** from the data set; in Figure 1-18, this would involve deleting the points that fall above the regression line from the data set and developing a new EPM on what remains;

- **defining best historical performance as the target;** in Figure 1-19, this would involve selecting only the points that fall well below the regression line and developing a new EPM on these points alone. This is equivalent to selecting the period of best performance on the CUSUM graph, as indicated above;

- **assessing the base and incremental loads** to identify specific actions that can be taken to reduce them, and defining a new EPM that adjusts the incremental and base load parameters accordingly.

Whichever method is used, target setting should be viewed as a continuous improvement strategy in order to maximize and sustain savings.

1.7.4 Controlling Performance

Once a target has been set, the challenge is to manage performance to achieve and sustain that performance. Control charts similar to those used to manage other functions can be constructed from the energy monitoring analysis to control energy performance.

Using the production case to illustrate, the data of Table 1-8 can be re-examined, now in light of a target EPM. For example, the 18 to 24 month period, the shaded cells in Table 1-10, is a suitable initial target since it represents superior performance that has already been demonstrated. The EPM for this period has been determined to be:

$$\text{Process Energy (MMBtu)} = 1.8231 \times \text{Production} + 59{,}228 \qquad (1\text{-}9)$$

A new set of predicted energy consumption values can be calculated, the "target" consumption as shown in Table 1-10. The variance between actual and target consumption is then calculated; positive values indicate consumption above target, negative below. Control charts are a plot of variance from standard (not cumulative variance as in CUSUM). The variance values calculated in Table 1-10 are plotted in Figure 1-24, the control chart.

Control limits can be applied to the control chart as shown; in Figure 1-24 they are arbitrarily set at +/2,500. The upper and lower control limits may reflect knowledge about what variance is tolerable or reasonable in the system being controlled (in this case, it is the entire process operation). There are also statistical criteria for setting control limits; a good value for the control level is 1.4 times the average of the variances in the target period ignoring the signs, in this example, +/–1,250 (as calculated from Table 1-10 for the period June through December, 2003).

While the CUSUM graph can also be used to monitor performance into the future, the control chart is a tool that lends itself to production floor or mechanical room applications for real-time control. Where in the example the time increment for the analysis is monthly, the same approach can be applied on a daily, weekly, shift, or batch-by-batch basis.

1.7.5 Reporting

Reporting within a MT&R system has a number of functions:
- to create motivation for energy saving actions;
- to report regularly on performance;
- to monitor overall utility costs;
- to monitor cost savings.

Within most organizations, the need for the type of information generated by a monitoring and targeting system varies with level and responsibility. Typically as the need moves from the operational level in the plant or building operation to the senior management level the requirement for detail diminishes, as does the frequency of reporting.

Operations staff need energy control information to stimulate

Table 1-10. Control Chart Calculations for Production Example

Month	Production lb.	Total Process Energy, MMBtu	Predicted Process Energy MMBtu	Variance Actual-Predicted MMBtu	CUSUM MMBtu	Target Process Energy MMBtu	Variance Actual-Target MMBtu
Jan 02	39,600	137,243	144,476	-7,233	-7,233	131,423	5,820
Feb 02	21,120	107,620	107,371	248 1	-6,985	97,732	9,888
Mar 02	15,840	94,63-0	96,770	-2,140	-9,125	88,106	6,524
Apr 02	13,200	102,649	91,469	11,179	2,054	83,293	19,356
May 02	44,880	152,845	155,077	-2,233	-178	141,049	11,796
Jun 02	47,520	171,792	160,378	11,414	11,235	145,862	25,930
Jul 02	31,680	126,754	128,574	-1,820	9,416	116,984	9,771
Aug 02	10,560	84,905	86,169	-1,263	8,152	78,480	6,426
Sep 02	29,040	120,510	123,273	-2,763	5,389	112,171	8,339
Oct 02	23,760	108,051	112,672	-4,621	768	102,545	5,556
Nov 02	10,560	87,491	86,169	1,322	2,090	78,480	9,011
Dec 02	13,200	89,379	91,469	-2,090	-0	83,293	6,086
Jan 03	36,960	131,255	139,175	-7,921	-7,921	126,610	4,645
Feb 03	40,920	144,886	147,126	-2,240	-10,161	133,829	11,057
Mar 03	43,560	145,882	152,427	-6,545	-16,706	138,642	7,240
Apr 03	50,160	158,760	165,679	-6,919	-23,625	150,675	8,086
May 03	10,560	81,102	1 86,169	-5,067	-28,691	78,480	2,622
Jun 03	14,520	86,234	94,120	-7,886	-36,577	85,699	535
Jul 03	39,600	129,613	144,476	-14,863	-51,440	131,423	-1,810
Aug 03	21,120	98,710	107,371	-8,661	-60,101	97,732	979
Sep 03	16,632	88,233	98,360	-10,127	-70,228	89,550	-1,317
Oct 03	29,040	112,643	123,273	-10,630	-80,858	112,171	472
Nov 03	18,480	92,912	102,071	-9,159	-90,018	92,919	-7
Dec 03	44,880	142,198	155,077	-12,879	-102,897	141,049	1,149
Jan 04	50,160	147,453	165,679	-18,226	-121,123	150,675	-3,222
Feb 04	42,240	147,231	149,777	-2,546	-123,669	136,236	10,995
Mar 04	31,680	123,359	128,574	-5,215	-128,884	116,984	6,376

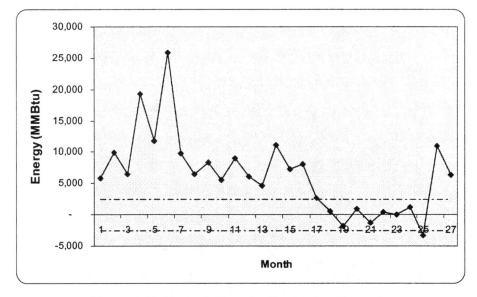

Figure 1-24. Control Chart for Production Example

specific energy savings actions; while senior managers need summary information with which to guide the organization's energy management effort. This is depicted in Figure 1-25. One report for all will not result in the appropriate decisions being made, and actions being taken.

The format of the reports is unique to each organization. In general, reporting of information should be integrated into the organizational management information system.

1.7.6 Savings Verification

There is a well-established protocol for verifying the savings that result from energy management measures. The IPMVP (International Performance Measurement and Verification Protocol) defines the fundamental relationship for comparing post-installation to pre-installation performance as:

Savings = (Baseline Energy Use$_{adjusted}$ – (Post-installation energy use)

The complicating factors are:
- what **adjustments** to the baseline performance are required, and how are they carded out;
- what **measurements** are required to determine post-installation performance, and how are they carried out.

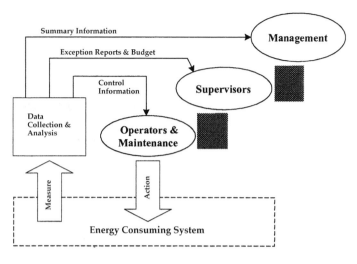

Figure 1-25. Reporting Pathways for a MT&R System

The adjustment of baseline energy use is derived in part from the same energy performance model that is used to predict consumption for the purpose of monitoring and control. That is, adjustments need to account for varying weather factors (HDD and CDD), production, occupancy, and so on, just as was required for MT&R.

Organizations that routinely conduct MT&R are prepared to apply the same management information to the verification of savings resulting from energy management measures.

1.8   SUMMARY

This chapter has discussed the need for energy management, the historical use of energy, and the design, initiation, and management of energy management programs. The chapter emphasizes energy accounting, especially cost center accounting and necessary submetering.

We defined an energy management activity as any decision that involves energy and affects the profit level. Anything that improves profits and/or enhances competitive positions is considered effective energy management, and anything else is poor energy management. The motivation for starting energy management programs is multi-faceted and varies among companies. The following outline lists the major reasons:

• Economic—Energy management will improve profits and enhance competitive positions.

- National good—Energy management is good for the U.S. economy as the balance of payments becomes more favorable and the dollar stronger.

  Energy management makes us less vulnerable to energy cutoffs or curtailments due to political unrest or natural disasters elsewhere.

  Energy management is kind to our environment as it eases some of the strain on our natural resources and may leave a better world for future generations.

  In designing an energy management program, several ingredients are vital:

- Top Management commitment. Commitment from the top must be strong and highly visible.

- One-person responsibility. The responsibility for the energy management program must lie in one person who reports as high in the organization structure as possible.

- Committee backup. The energy management coordinator must have the support of two committees. The first is a steering committee, which provides direction for the program. The second is a technical committee, which provides technical backup in the necessary engineering disciplines.

- Reporting and monitoring. An effective monitoring and reporting system for energy consumption must be provided.

- Training. Energy management is a unique undertaking. Hence, training and retraining at all levels is required.

To successfully start an energy management program, some publicity must accompany the early stages. This can be achieved with news releases, films, plant meetings, or a combination of them. Early project selection is a critical step. Early projects should be visible, and should have good monetary returns, with few negative consequences.

Management and creative personnel are always critical components of an energy management program. Tough, specific, and measurable goals need to be developed. Once the goals are established, management should carefully monitor the results, but the energy management staff should be allowed to perform its functions. Staff and management need to realize that (1) energy costs, not consumption, are to be controlled (2) energy should be a direct cost—not an overhead item, and (3) only the main energy consumers need be metered and monitored closely.

Energy accounting is the art and science of tracing Btu and energy dollar flow through an organization. Cost center orientation is important, as are comparison to some standard or base and calculations of variances. Causes for variances must then be sought. General Motor's energy responsibility accounting system was discussed in some detail. However, no accounting system is a panacea, and any system is only as accurate as the metering and reporting systems allow it to be.

References
1. Energy Information Administration, Commercial Sector Energy Price and Expenditure Estimates, for 2008. U.S. Department of Energy, Washington, D.C., 2010.
2. Energy Information Administration, Industrial Sector Energy Price and Expenditure Estimates for 2008. U.S. Department of Energy, Washington, D.C., 2010.
3. Energy Information Administration, *Manufacturing Energy Consumption Survey: Changes in Energy Efficiency 1985-1991*, U.S. Department of Energy, Washington, D.C., October, 1995.
4. The National Energy Strategy, *Chapter on Industrial Energy Use*, U.S. Department of Energy, Washington, D.C., April 2, 1990.
5. Aspenson, R.L., *Testimony to the U.S. Department of Energy on the National Energy Strategy, Hearing on Energy and Productivity*, Providence, RI, December 1, 1989.
6. Applebaum, Bruce and Mark Oven, "Energy Management Skills for Energy Engineers," *Energy Engineering*, Vol. 89, No. 5, 1992, pp. 52-80.
7. Energy Information Administration, *Commercial Buildings Consumption and Expenditures 2003*, Energy Information Administration (EIA), U.S. Department of Energy, Washington, D.C., April, 2011.
8. 3M Company Website, Improving Energy Efficiency, www.mmm.com, accessed August 28, 2007, and April 8, 2011.
9. Energy Information Administration, "Energy Efficiency Report, Chapter 6, The Industrial Sector," EIA website accessed at www.eia.doe.gov/emeu/efficiency/ee_ch6.htm, on June 15, 2005.
10. EIA Monthly Energy Review, March, 2011. Accessed April 8, 2011.

Bibliography
Allen, Paul J., "How Disney Saves Energy and Operating Costs," *Heating/Piping/ Air Conditioning (HPAC) Engineering*, January 2005.
Barrett, John A., "Energy Accounting Systems for Existing Buildings," ASHRAE Journal, January 1979, p. 56.
"Btu Accounting," client newsletter, Seidman and Seidman, Certified Public Accountants, Grand Rapids, MI, 1980.

Capehart, Barney L., "Energy Management," Chapter 70 in the *Handbook of Industrial Engineering*, Second Edition, John Wiley and Sons, Inc., New York, 1992.

Capehart, Barney, Editor, *Encyclopedia of Energy Engineering and Technology*," Taylor and Francis/CRC Publishing Company, Boca Raton, FL, July, 2007.

Capehart, Barney L. and Mark Spiller, "Energy Auditing," Chapter 3 in the *Energy Management Handbook*, Fifth Edition, Fairmont Press, 2005.

Energy Conservation and Program Guide for Industry and Commerce, GPO, Washington, DC, September 1974.

Energy Conservation Trends—Understanding the Factors That Affect Conservation in the U.S. Economy, U.S. Department of Energy, DOE/PE-0092, September, 1989.

Energy Management for the Year 2000, A Handbook prepared by the National Electrical Contractors Association and the National Electrical Manufacturers Association, Washington, DC, First Edition, 1993.

Grant, Eugene L. and Richard S. Leavenworth, Statistical Quality Control, Sixth Edition, McGraw-Hill Company, New York, NY, 1988.

Greene, Richard P., "Energy Responsibility Accounting, an Energy Conservation Tool for Industrial Facilities," unpublished working paper, General Motors Corporation, date unknown.

Millhone, John P., John A. Beldock, Barney L. Capehart, and Jerome Lamontagne, "United States Energy Efficiency: Strategy for the Future," Strategic Planning for Energy and the Environment, Vol. 11, No. 3, Winter, 1992.

Schultz, Steven, 3M Corporation, St. Paul, MN, personal correspondence, 1997, 2001, 2005.

Shirley, James M., and W.C. Turner, "Industrial Energy Management—First Steps Are Easy," Industrial Engineering, May 1978, pp. 3441.

Stobaugh, Robert, and Daniel Yergin, *Energy Future*, Second Edition, Random House, New York, 1983.

Turner, W.C., Editor, *Energy Management Handbook*, Fifth Edition, Fairmont Press, Atlanta, GA, 2005.

Turner, W.C., and B.W. Tompkins, "Productivity Measures for Energy Management," in IIE National Meeting, Detroit, MI, Proceedings, May 1981.

Witte, Larry C., Phillip S. Schmidt, and David R. Brown, *Industrial Energy Management and Utilization*, Hemisphere Publishing Corporation, Washington, DC, 1988.

Chapter 2

The Energy Audit Process: An Overview

2.0 INTRODUCTION

Once a commercial or industrial facility has designated its energy manager and given that person the support and authority necessary to develop an adequate energy management program, the first step the energy manager should take is to conduct an energy audit. Also called an energy survey, energy analysis, or energy evaluation, the energy audit examines the ways energy is currently used in that facility and identifies some alternatives for reducing energy costs. The goals of the audit are:

- to clearly identify the types and costs of energy use,
- to understand how that energy is being used—and possibly wasted,
- to identify and analyze alternatives such as improved operational techniques and/or new equipment that could substantially reduce energy costs*, and
- to perform an economic analysis on those alternatives and determine which ones are cost-effective for the business or industry involved.

This chapter addresses the three phases of an energy audit: preparing for the audit visit; performing the facility survey and implementing the audit recommendations. In the first phase, data from the energy bills are analyzed in detail to determine what energy is being used and how the use varies with time. Preliminary information on the facility is compiled, the necessary auditing tools are gathered, and an audit team is assembled.

---

*In most cases the energy cost savings will result from reduced consumption, but occasionally a cost savings will be associated with increased energy use. For example, a thermal storage system for heating and/or cooling may save on electric bills, but may actually increase the use of electric energy due to the losses in the storage system. While the primary goal of energy management programs is to reduce energy costs, some proposed alternatives may not always produce greater energy efficiency. However, an overall improvement in a facility's energy efficiency should be the overriding goal for any company's energy management team.

Phase two starts after a safety briefing when the team performs a walk-through inspection, looking carefully at each of the physical systems within the facility and recording the information for later use. After the plant survey, the audit team must develop an energy balance to account for the energy use in the facility. Once all energy uses have been identified and quantified, the team can begin analyzing alternatives. The final step of phase two is the audit report which recommends changes in equipment, processes or operations to produce energy cost savings.

Phase Three—the implementation phase—begins when the energy manager and the facility management agree on specific energy savings goals and initiate some or all of the actions recommended to achieve those goals. Setting up a monitoring system will allow management to assess the degree to which the chosen goals have been accomplished and to show which measures have been successful and which have failed. The results of the monitoring should feed back to the beginning of the audit cycle and thus potentially initiate more analysis, implementation, and monitoring.

2.1   PHASE ONE—PREPARING FOR AN ENERGY AUDIT

The energy audit process starts with an examination of the historical and descriptive energy data for the facility. Specific data that should be gathered in this preliminary phase include the energy bills for the past twelve months, descriptive information about the facility such as a plant layout, and a list of each piece of equipment that significantly affects the energy consumption. Before the audit begins, the auditor must know what special measurement tools will be needed. A briefing on safety procedures is also a wise precaution.

2.1.1 Gathering Preliminary Data on the Facility

Before performing the facility audit, the auditors should gather information on the historical energy use at the facility and on the factors likely to affect the energy use in the facility. Past energy bills, geographic location, weather data, facility layout and construction, operating hours, and equipment lists are all part of the data needed.

2.1.1.1 Analysis of Bills

The audit must begin with a detailed analysis of the energy bills for the previous twelve months. This is important for several reasons: the bills show the proportionate use of each different energy source when compared to the total energy bill; an examination of where energy is used can point out previously unknown energy wastes; and, the total amount spent on energy puts an obvious upper limit on the amount that can be saved. The data from the energy bills can be conveniently entered onto a form such as shown in Figure 2-1. Note that the most significant billing factors are shown, including peak demand for electricity.

| Location/Meter # _____ |
| From _____ to _____ |
| (Mo./Yr.)   (Mo./Yr.) |

	Electrical use			Gas use			Fuel oil	
Month	Peak kW	Usage: kWh	Cost	MMCF[a]	Dth[b]	Cost	Gallons	Cost
___	___	___	___	___	___	___	___	___
___	___	___	___	___	___	___	___	___
___	___	___	___	___	___	___	___	___
___	___	___	___	___	___	___	___	___
___	___	___	___	___	___	___	___	___
___	___	___	___	___	___	___	___	___

[a] 1 MMCF = 10^6 ft^3
[b] 1Dth = 10 therms = 10^6 Btu

Figure 2-1. Summary Form for Energy Use

The energy bill data must be analyzed by energy source and billed location. The data can be tabulated as illustrated in Figure 2-2 of Example 2-1.

Example 2-1. This example demonstrates the importance of analyzing energy bills. As Figure 2-2 shows, most of the gas used at this facility is used by the main heating plant. Therefore most of the energy management effort and money should be concentrated on the main heating plant.

Building	Energy Costs	Percentage of total
Heating plant	$38,742.34	83.2%
East dormitory	4,035.92	8.7%
Married student apartments	1,370.79	2.9%
Undergraduate dormitory	768.42	1.7%
Greenhouse	560.21	1.2%
Child development center	551.05	1.2%
President's home	398.53	0.9%
Art barn	104.77	0.2%

Figure 2-2. Natural Gas Bills for a Small College

The data in this example should raise some questions for the auditor before he ever visits the facility. The greenhouse appears to use a lot of energy; it uses more than the child development center and almost as much as the undergraduate dormitory. Since the greenhouse is not particularly large, these data raised a red flag for the energy auditor. Because one should never make assumptions about what is actually using the energy, the auditor checked the gas consumption meter at the greenhouse to make sure it was not measuring gas consumption from somewhere else as well. Subsequent investigation revealed that the heating and cooling in the greenhouse were controlled by different thermostats. One thermostat turned the cooling on when the temperature got too high—but before the second thermostat had turned the heat off! If it had turned out that the gas use from several other buildings had been metered by the same meter as the greenhouse, it would have been necessary to find a way to allocate gas consumption to each building.

In this example, the amount of energy used in the president's house could also be questioned; it uses nearly as much gas as the child development center. Perhaps the president's home is used for activities that would warrant this much gas use, but some equipment problem might also be causing this difference, so an energy audit of this facility might be worthwhile.

Another way to present the data is in graph form. A sample of the type of graph that should be made for each type of energy is shown in Figure 2-3. Each area of the country and each different industry type has a unique pattern of energy consumption, and presenting the data as shown in Figure 2-3 helps in defining and analyzing these patterns. In the facility from which this example came, natural gas is used in the winter for space heating, so the January peak is not surprising. For electrical consumption, if a peak demand charge is based on the *annual* peak, the energy auditor must know the time and size of this peak in order to address measures to reduce it.

A complete analysis of the energy bills for a facility requires a detailed knowledge of the rate structures in effect for the facility. To accurately determine the costs of operating individual pieces of equipment, the energy bills must be broken down into their components, such as demand charge and energy charges for the electric bill. This breakdown is also necessary to be able to calculate the savings from Energy Management Opportunities (EMOs) such as high-efficiency lights and high-efficiency motors, and off-peak electrical use by rescheduling some opera-

Figure 2-3
Graph of Energy Consumption Over Time. (Courtesy of U.S. Department of Energy Conservation, *Instructions for Energy Auditors*, DOE/CS-0041/12 and 0041/13, Sept. 1978.)

tions. This examination of energy rate structures is explained in detail in Chapter 3.

2.1.1.2 Geographic Location/Degree Days/Weather Data

The geographic location of the facility should be noted, together with the weather data for that location. The local weather station, the local utility or the state energy office can provide the average degree days for heating and cooling for that location for the past twelve months. This degree-day data will be very useful in analyzing the energy needed for heating or cooling the facility.

Heating degree days (HDD) and *cooling degree days* (CDD) are given separately, and are specific to a particular geographic location. The degree day concept assumes that the average building has a desired indoor temperature of 70°F, and that 5°F of this is supplied by internal heat sources such as lights, appliances, equipment, and people. Thus, the base for computing HDD is 65°F.

Example 2-2 If there were a period of three days when the outside temperature averaged 50°F each day, then the number of HDD for this three day period would be

$$HDD = (65° - 50°) \times 3 \text{ days} = 45 \text{ degree days.}$$

The actual calculation of HDD for an entire year is more involved than this example. HDD for a year are found by taking the outside temperature each hour of the heating season, subtracting that temperature from 65°F, and summing up all of these hourly increments to find the total number of degree hours. This total is then divided by 24 to get the number of HDD. Cooling degree days are similar, using 65°F as the base, and finding the number of hours that the outside temperature is above 65°F, and dividing this by 24 to get the total CDD [1].

Bin weather data are also useful if a thermal envelope simulation of the facility is going to be performed as part of the audit. Weather data for a specific geographic location have been statistically analyzed and the results grouped in cells or bins. These bins contain the number of hours that the outside temperature was within a certain range (e.g. 500 hours at 90-95°). These data are called bin weather data and are available for a large number of locations throughout the United States [2].

2.1.1.3 Facility Layout

Next the facility layout or plan should be obtained, and reviewed to determine the facility size, floor plan, and construction features such as wall and roof material and insulation levels, as well as door and window sizes and construction. A set of building plans could supply this information in sufficient detail. It is important to make sure the plans reflect the "as-built" features of the facility, since the original building plans seldom are completed without alterations.

2.1.1.4 Operating Hours

Operating hours for the facility should also be obtained. How many shifts does the facility run? Is there only a single shift? Are there two? Three? Knowing the operating hours in advance allows some determination as to whether any loads could be shifted to off-peak times. Adding a second shift can often reduce energy bills because the energy costs during second and third shifts are usually substantially cheaper. See Chapter Three for an explanation of on-peak and off-peak electric rates.

2.1.1.5 Equipment List

Finally, the auditor should get an equipment list for the facility and review it before conducting the audit. All large pieces of energy-consuming equipment such as heaters, boilers, air conditioners, chillers, water heaters, and specific process-related equipment should be identified. This list, together with data on operational uses of the equipment allows the auditor to gain a good understanding of the major energy-consuming tasks or equipment at the facility.

The equipment found at an audit location will depend greatly on the type of facility involved. Residential audits for single-family dwellings generally involve small lighting, heating, air conditioning and refrigeration systems. Commercial operations such as grocery stores, office buildings and shopping centers usually have equipment similar to residences, but the equipment is much larger in size and in energy use. Large residential structures such as apartment buildings have heating, air conditioning and lighting systems that are closer in size to commercial systems. Some commercial audits will require an examination of specialized business equipment that is substantially different from the equipment found in residences.

Industrial auditors encounter the most complex equipment. Commercial-scale lighting, heating, air conditioning and refrigeration, as well as office business equipment, is generally used at most industrial facilities. The major difference is in the highly specialized equipment used for the industrial production processes. This can include equipment that is used for such processes as chemical mixing and blending, metal plating and treatment, welding, plastic injection molding, paper making and printing, metal refining, electronic assembly, or making glass.

2.1.2 Tools for the Audit

To obtain the best information for a successful energy cost control program, the auditor must make some measurements during the audit visit. The amount of equipment needed depends on the type of energy-consuming equipment used at the facility, and on the range of potential EMOs that might be considered. For example, if waste heat recovery is being considered, then the auditor must take substantial temperature measurement data from potential heat sources. Tools commonly needed for energy audits include the following:

- *Tape measures*—The most basic measuring device is the tape measure. A 25-foot tape measure and a 100-foot tape measure are used to check the dimensions of the walls, ceilings, doors, and windows, and the distances between pieces of equipment for purposes such as determining the length of a pipe for transferring waste heat from one piece of equipment to another.

- *Lightmeter*—A portable lightmeter that can fit into a pocket is extremely useful. This instrument is used to measure illumination levels in facilities. A lightmeter that reads in footcandles allows direct analysis of lighting systems and comparison with recommended light levels specified by the Illuminating Engineering Society [3]. Many areas in buildings and plants are still significantly over-lighted, and measuring this excess illumination allows the auditor to recommend a reduction in lighting levels through lamp removal programs or by replacing inefficient lamps with high efficiency lamps that may supply slightly less illumination than the old inefficient lamps.

- *Thermometers*—Several thermometers are generally needed to measure temperatures in offices and other worker areas, and to measure the temperature of operating equipment. Knowing process temperatures allows the auditor to determine process equipment efficiencies, and also to identify waste heat sources for potential heat recovery programs. Inexpensive electronic thermometers with interchangeable probes are available to measure temperatures in both these areas. Some common types include an immersion probe, a surface temperature probe, and a radiation-shielded probe for measuring true air temperature. Other types of infra-red thermometers and thermographic equipment are also available. An infra-red "gun" can measure temperatures of steam lines that are inaccessible without a ladder.

- *Voltmeter*—A voltmeter is useful for determining operating voltages on electrical equipment, and especially useful when the nameplate has worn off of a piece of equipment or is otherwise unreadable or missing. The most versatile instrument is a combined volt-ohm-ammeter with a clamp-on feature for measuring currents in conductors that are easily accessible. This type of multi-meter is convenient and relatively inexpensive.

- *Wattmeter/Power Factor Meter*—A portable hand-held wattmeter and power factor meter is very handy for determining the power consumption and power factor of individual motors and other inductive devices, and the load factors of motors. This meter typically has a clamp-on feature which allows an easy and safe connection to the current-carrying conductor, and has probes for voltage connections.

- *Combustion Analyzer*—Combustion analyzers are portable devices which estimate the combustion efficiency of furnaces, boilers, or other fossil fuel burning machines. Digital combustion analysis equipment performs the measurements and reads out combustion efficiency in percent. These instruments are very reliable, easy to use, and inexpensive.

- *Ultrasonic Leak Detector*—Ultrasonic compressed air leak detectors are electronic ultrasonic receivers that are tuned very precisely to the frequency of the hissing sound of an air leak. These devices are reasonably priced, and are extremely sensitive to the noise a small air leak makes. The detectors can screen out background noise and pick up the sound of an air leak. All facilities which use compressed air for applications beyond pneumatic controls should have one of these devices, and should use it routinely to identify wasteful air leaks.

  When equipped with an optional probe attachment, some ultrasonic receivers can also be used to test steam traps. The sound pattern will allow a determination or whether the steam trap is functioning properly, or whether it is stuck open or stuck shut. More expensive models of ultrasonic detectors have an interface to a portable computer which can analyze the sound signal and display a status of the steam trap.

- *Airflow Measurement Devices*—Measuring air flow from heating, air conditioning or ventilating ducts, or from other sources of air flow is one of the energy auditor's tasks. Airflow measurement devices can be used to identify problems with air flows, such as whether the combustion air flow into a gas heater is correct. Typical airflow measuring devices include a velometer, an anemometer, or an airflow hood.

- *Blower Door Attachment*—Building or structure tightness can be measured with a blower door attachment. This device is frequently used in residences and in office buildings to determine the air leakage rate or the number of air changes per hour in the facility. This often helps determine whether the facility has substantial structural or duct leaks that need to be found and sealed.

- *Smoke Generator*—A simple smoke generator can be used in residences, offices and other buildings to find air infiltration and leakage around doors, windows, ducts and other structural features. Care must be taken in using this device, since the chemical "smoke" produced may be hazardous, and breathing protection masks may be needed.

- *Safety Equipment*—The use of safety equipment is a vital precaution for any energy auditor. A good pair of safety glasses is an absolute necessity for almost any audit visit. Hearing protectors may also be required on audit visits to noisy plants or in areas where high horsepower motors are used to drive fans and pumps. Electrically insulated gloves should also be used if electrical measurements will be taken, and insulated gloves should be used for working around boilers and heaters. Breathing protection masks may also be needed when hazardous fumes are present from processes or materials used. Steel-toe and steel-shank safety shoes may be needed on audits of plants where heavy, hot, sharp or hazardous materials are being used.

2.1.3 Safety Considerations

Safety is a critical part of any energy audit. The auditor and the audit team should have a basic knowledge of safety equipment and procedures. Before starting the facility tour, the auditor or audit team should be thoroughly briefed on any specialized safety equipment and procedures for the facility. They should never place themselves in a position where they could injure themselves or other people at the facility.

Adequate safety equipment should be worn at all appropriate times. Auditors should be extremely careful making any measurements on electrical systems, or on high temperature devices such as boilers, heaters, cookers, etc. Electrical gloves or insulated gloves should be worn as appropriate. If a trained electrician is available at the facility, they should be asked to make any electrical measurements.

The auditor should be cautious when examining any operating piece of equipment, especially those with open drive shafts, belts, gears, or any form of rotating machinery. The equipment operator or supervisor should be notified that the auditor is going to look at that piece of equipment and might need to get information from some part of the device. If necessary, the auditor may need to return when the machine or device is idle in order to get the data safely. The auditor should never approach a piece of equipment and inspect it without notifying the operator or supervisor first.

Safety Checklist

Electrical:
- Avoid working on live circuits, if possible.
- Securely lock circuits and switches in the off position before working on a piece of equipment.
- Always keep one hand in your pocket while making measurements on live circuits to help prevent accidental electrical shocks.

Respiratory:
- When necessary, wear a full face respirator mask with adequate filtration particle size.
- Use activated carbon cartridges in the mask when working around low concentrations of noxious gases. Change the cartridges on a regular basis.
- Use a self-contained breathing apparatus for work in toxic environments.

Hearing:
- Use foam insert plugs while working around loud machinery to reduce sound levels by nearly 30 decibels.

2.2   PHASE TWO—THE FACILITY INSPECTION

Once all of the basic data have been collected and analyzed, the audit team should tour the entire facility to examine the operational patterns and equipment usage, and should collect detailed data on the facility itself as well as on all energy using equipment. This facility inspection should systematically examine the nine major systems within a facility, using portable instrumentation and common sense guided by an anticipation of what can go wrong. These systems are: the building envelope; the boiler and steam distribution system; the heating, ventilating, and air conditioning system; the electrical supply system; the lighting system, including all lights, windows, and adjacent surfaces; the hot water distribution system; the compressed air distribution system; the motors; and the manufacturing system. Together, these systems account for all the energy used in any facility; examining all of them is a necessary step toward understanding and managing energy utilization within the facility. We briefly describe these systems later in this chapter; we also cover most of them in detail in separate chapters.

The facility inspection can often provide valuable information on ways to reduce energy use at no cost or at a low cost. Actually, several inspections should be made at different times and on different days to discover if lights or other equipment are left on unnecessarily, or to target process waste streams that should be eliminated or minimized.* These inspections can also help identify maintenance tasks that could reduce energy use. Broken windows should be fixed, holes and cracks should be filled, lights should be cleaned, and HVAC filters should be cleaned or replaced.

The facility inspection is an important part of the overall audit process. Data gathered on this tour, together with an extensive analysis of these data will result in an audit report that includes a complete description of the time-varying energy consumption patterns of the facility, a list of each piece of equipment that affects the energy consumption together with an assessment of its condition, a chronology of normal operating and

---

*Two preliminary energy inspections are sometimes performed to make sure that the full-blown audit will be worthwhile. The first is done under usual working conditions and is aimed at uncovering practices that are consistently expensive or wasteful. The second (the 2 a.m. survey) is done at midnight or later when energy consumption should be at a minimum. The objective of this second inspection is to find lights that are left on for no reason, motors that are running but not being used, rooms that are warm but not warming people, air being cooled unnecessarily, and air and steam leaks that might not be detected under the noise conditions of daily operations.

maintenance practices, and a list of recommended energy management ideas for possible implementation.

2.2.1 Introductory Meeting

The audit leader should start the audit by meeting with the facility manager and the maintenance supervisor. He should briefly explain the purpose of the audit and indicate the kind of information the team needs to obtain during the facility tour. If possible, a facility employee who is in a position to authorize expenditures or make operating policy decisions should be at this initial meeting.

2.2.2 Audit Interviews

Getting the correct information on facility equipment and operation is important if the audit is going to be most successful in identifying ways to save money on energy bills. The company philosophy towards investments, the impetus behind requesting the audit, and the expectations from the audit can be determined by interviewing the general manager, chief operating officer, or other executives. The facility manager or plant manager should have access to much of the operational data on the facility, and a file of data on facility equipment. The finance officer can provide any necessary financial records, such as utility bills for electric, gas, oil, other fuels, water and wastewater, expenditures for maintenance and repair, etc.

The auditor must also interview the floor supervisors and equipment operators to understand the building and process problems. Line or area supervisors usually have the best information on the times their equipment is used. The maintenance supervisor is often the primary person to talk to about types of lighting and lamps, sizes of motors, sizes of air conditioners and space heaters, and electrical loads of specialized process equipment. Finally, the maintenance staff must be interviewed to find the equipment and performance problems.

The auditor should write down these people's names, job functions and telephone numbers, since additional information is often needed after the initial audit visit.

2.2.3 Initial Walk-through Tour

An initial facility/plant tour should be conducted by the facility/plant manager, and should allow the auditor or audit team to see the major operational and equipment features of the facility. The main purpose of the initial tour is to obtain general information, and to obtain a general understanding of the facility's operation. More specific information should be obtained from the maintenance and operational people during a second, and more detailed data collection tour.

2.2.4 Gathering Detailed Data

Following the initial facility or plant tour, the auditor or audit team should acquire the detailed data on facility equipment and operation that will lead to identifying the significant Energy Management Opportunities (EMOs) that may be appropriate for this facility. These data are gathered by examining the nine major energy-using systems in the facility.

As each of these systems are examined, the following questions should be asked:

1. What function(s) does this system serve?
2. How does this system serve its function(s)?
3. What is the energy consumption of this system?
4. What are the indications that this system is probably working?
5. If this system is not working, how can it be restored to good working condition?
6. How can the energy cost of this system be reduced?
7. How should this system be maintained?
8. Who has direct responsibility for maintaining and improving the operation and energy efficiency of this system?

As each system is inspected, these data should be recorded on individualized data sheets that have been prepared in advance. Manual entry data forms for handling these energy data are available from several sources, including the energy management handbook from the National Electrical Manufacturers Association [4]. Some energy analysis procedures in current use are computer-based, and data are entered directly into the computer.

2.2.5.1 The building envelope.

The building envelope includes all building components that are directly exposed to the outside environment. Its main function is to protect employees and materials from outside weather conditions and temperature variations; in addition, it provides privacy for the business and can serve other psychological functions. The components of the building envelope are outside doors, windows, and walls; the roof; and, in some cases, the floor. The heating and cooling loads for the building envelope are discussed in Chapter Six on the HVAC system.

As you examine the building envelope, you should record information on the insulation levels in the various parts of the facility, the condition of the roof and walls, the location and size of any leaks or holes, and the location and size of any door or windows that open from conditioned to unconditioned space. Insulation is discussed in Chapter Eleven. Figure 2-4 shows a sample data form for the building envelope.

System: <u>Envelope</u>

Component	Location	Maintenance condition	Est. air gap (total)
Door	North side	Poor	0.2 ft^2
	South	OK	0.05 ft^2
	Gymnasium	Good	None
Windows	North	Some broken	2.2 ft^2
	East	OK	None
Roof	Main building	No insulation	

Figure 2-4. Completed Inspection Form for Building Envelope

<u>2.2.5.2 The steam boiler and steam distribution system.</u>

A *steam boiler* burns fuel to produce heat that converts water into steam, and the *steam distribution system* takes the steam from the boiler to the point of use. Boilers consume much of the fuel used in many production facilities. The boiler is thus the first place to look when attempting to reduce natural gas or oil consumption. The steam distribution system is also a very important place to look for energy savings, since every pound of steam lost is another pound of steam that the boiler must produce. A detailed description of boilers and the steam distribution system, including operating and maintenance recommendations, is covered in Chapters 7 and 8.

<u>2.2.5.3 The heating, ventilating, and air conditioning system.</u>

All *heating, air conditioning and ventilation* (HVAC) equipment should be inventoried. Prepared data sheets can be used to record type, size, model numbers, age, electrical specifications or fuel use specifications, and estimated hours of operation. The equipment should be inspected to determine the condition of the evaporator and condenser coils, the air filters, and the insulation on the refrigerant lines. Air velocity measurements may also be made and recorded to assess operating efficiencies or to discover conditioned air leaks. These data will allow later analysis to examine alternative equipment and operations that would reduce energy costs for heating, ventilating, and air conditioning. The HVAC system is discussed in detail in Chapter Six.

2.2.5.4 The electrical supply system.

This system consists of transformers, wiring, switches, and fuses—all the components needed to enable electricity to move from the utility-owned wires at the facility boundary to its point of use within the company. By our definition, this supply system does not include lights, motors, or electrical controls. Most energy problems associated with the distribution of electricity are also safety problems, and solving the energy problems helps to solve the safety-related problems.

Electricity from a utility enters a facility at a service transformer. The area around the transformer should be dry, the transformer fins should be free from leaves and debris so that they can perform their cooling function, and the transformer should not be leaking oil. If a transformer fails to meet any one of these conditions there is a serious problem which should justify a call to the local electrical utility, or, if the transformer is company-owned, to the person or department in charge of maintaining the electrical system.

A more detailed audit of transformers should also include drawing a small (1-pint) sample of transformer dielectric fluid and examining it both visually and for dielectric strength. If the fluid is brown, the dielectric has been contaminated by acid; if it is cloudy, it is contaminated with water. The dielectric strength should be 20,000 v/cm or equivalent. Both the color and the dielectric strength should be recorded for comparison against future readings.

In examining transformers, also check to see whether any company-owned transformer is serving an area that is not currently used. A transformer that is connected to the utility lines but not supplying power to the facility is wasting one to three percent of its rated capacity in core losses. These losses can be avoided by disconnecting the transformer or by installing switching between the transformer and the electrical lines from the utility.*

A person performing an energy audit should examine the electrical supply panels and switch boxes. Danger signs and symptoms of wasted energy include signs of arcing such as burned spots on contacts, burned insulation, arcing sounds, and frayed wire. Other concerns are warm

---

*Personal communication from Mr. Bryan Drennan, Customer Services, Utah Power and Light Company, Salt Lake City, Utah.

spots around fuse boxes and switches and the smell of warm insulation. Any of these symptoms can indicate a fire hazard and should be checked in more detail immediately. Safety considerations are paramount when inspecting live electrical systems.

2.2.5.5 Lights, windows, and reflective surfaces.

The functions of this system are to provide sufficient light for necessary work, to enable people to see where they are going, to assist in building and area security at night, to illuminate advertising, and to provide decoration. Making a detailed inventory of all lighting systems is important. Data should be recorded on numbers of each type of light fixture and lamp, the wattages of the lamps, and the hours of operation of each group of lights. A lighting inventory data sheet should be used to record these data. See Figure 2-5 for a sample lighting data sheet. Lighting is discussed in detail in Chapter 5. Windows and reflective surfaces are discussed in Chapter Six on the HVAC System and in Chapter Thirteen on Renewable Energy Sources.

2.2.5.6 The hot water distribution system

The hot water system distributes hot water for washing, for use in industrial cleaning, and for use in kitchens. Its main components are hot water heaters, storage tanks, piping, and faucets. Electric boilers and radiators are also found in some facilities. Boilers and the hot water distribution system are discussed in Chapter Eight.

All water heaters should be examined, and data recorded on their type, size, age, model number, electrical characteristics or fuel use. What the hot water is used for, how much is used, and what time it is used should all be noted. The temperature of the hot water should be measured and recorded.

2.2.5.7 Air compressors and the air distribution system.

Air compressors and the *air distribution system* provide motive power for tools and some machinery, and often provide air to operate the heating, ventilating, and air conditioning system. If you use compressed air to run equipment, look for leaks and for places where compressed air is purposely allowed to vent into the air. Such leaks can be expensive.

2.2.5.8 Motors

Electric motors account for between two-thirds and three-fourths of all the electric energy used by industry and about two-fifths of all electric energy use by commercial facilities. Replacement of existing motors with

Area	Type of lighting (e.g., HPS)	Watts per fixture	Number of fixtures	Total kW	Operating hours	Operating days	kWh/month
Interior							
Exterior							

Figure 2-5. Data Collection Form for Lighting System

more efficient models is usually cost effective for applications where the motor is heavily used. Motors and drives are discussed in Chapter Twelve on Process Energy Improvements.

All electric motors over 1 horsepower should be inventoried. Prepared data sheets can be used to record motor size, use, age, model number, estimated hours of operation, other electrical characteristics, and possibly the full load power factor. Measurement of voltages, currents, power factors, and load factors may be appropriate for larger motors. Notes should be taken on the use of motors, particularly recording those that are infrequently used and might be candidates for peak load control or shifting use to off-peak times. All motors over 1 hp and with times of use of 2000 hours per year or greater, are likely candidates for replacement by high-efficiency motors—at least when they fail and must be replaced. It should be noted that few motors run at full load. Typical motor load factors are around 40-60%.

2.2.5.9 Manufacturing processes

Each manufacturing process has opportunities for energy management, and each offers ways for the unwary to create operating problems in the name of energy management. The best way to avoid such operating problems is to include operating personnel in the energy audit process (see Section 2.2.2) and to avoid rigid insistence on energy conservation as the most important goal.

The generic industrial processes that use the most energy are combustion for process steam and self-generated electricity, electrolytic processes, chemical reactors, combustion for direct heat in furnaces and kilns, and direct motor drive. Process energy efficiency improvement is discussed in Chapter Twelve.

Any other equipment that consumes a substantial amount of energy should be inventoried and examined. Commercial facilities may have extensive computer and copying equipment, refrigeration and cooling equipment, cooking devices, printing equipment, water heaters, etc. Industrial facilities will have many highly specialized process and production operations and machines. Data on types, sizes, capacities, fuel use, electrical characteristics, age, and operating hours should be recorded for all of this equipment.

2.2.6 Preliminary Identification of Energy Management Opportunities

As the audit is being conducted, the auditor should take notes on potential EMOs that are evident. As a general rule, the greatest effort should be devoted to analyzing and implementing the EMOs which show the greatest savings, and the least effort to those with the smallest savings potential. Therefore, the largest energy and cost activities should be examined carefully to see where savings could be achieved.

Identifying EMOs requires a good knowledge of the available energy efficiency technologies that can accomplish the same job with less energy and less cost. For example, over-lighting indicates a potential lamp removal or lamp change EMO, and inefficient lamps indicate a potential lamp technology change. Motors with high use times are potential EMOs for high efficiency replacements. Notes on waste heat sources should indicate what other heating sources they might replace, and how far away they are from the end use point. Identifying any potential EMOs during the walk-through will make it easier later on to analyze the data and to determine the final EMO recommendations.

2.2.7 The Energy Audit Report

The next step in the energy audit process is to prepare a report which details the final results of the energy analyses and provides energy cost saving recommendations. The length and detail of this report will vary depending on the type of facility audited. A residential audit may result in a computer printout from the utility. An industrial audit is more likely to have a detailed explanation of the EMOs and benefit-cost analyses. The following discussion covers the more detailed audit reports.

The report should begin with an executive summary that provides the owners/managers of the audited facility with a brief synopsis of the total savings available and the highlights of each EMO. The report should then describe the facility that has been audited, and provide information on the operation of the facility that relates to its energy costs. The energy bills should be presented, with tables and plots showing the costs and consumption. Following the energy cost analysis, the recommended EMOs should be presented, along with the calculations for the costs and benefits, and the cost-effectiveness criterion.

Regardless of the audience for the audit report, it should be written in a clear, concise and easy-to understand format and style. An executive summary should be tailored to non-technical personnel, and technical jargon should be minimized. The reader who understands the report is more likely to implement the recommended EMOs. An outline for a complete energy audit report is shown in Figure 2-6 below. See reference [5] for a suggested approach to writing energy audit reports.

Energy Audit Report Format

Executive Summary
    A brief summary of the recommendations and cost savings

Table of Contents

Introduction
    Purpose of the energy audit
    Need for a continuing energy cost control program

Facility Description
    Product or service, and materials flow
    Size, construction, facility layout, and hours of operation
    Equipment list, with specifications

Energy Bill Analysis
    Utility rate structures
    Tables and graphs of energy consumptions and costs
    Discussion of energy costs and energy bills

Energy Management Opportunities
    Listing of potential EMOs
    Cost and savings analysis
    Economic evaluation

Energy Action Plan
    Recommended EMOs and an implementation schedule
    Designation of an energy monitor and ongoing program

Conclusion
    Additional comments not otherwise covered

Figure 2-6. Outline of Energy Audit Report

2.2.8 The Energy Action Plan

An important part of the energy audit report is the recommended action plan for the facility. Some companies will have an energy audit conducted by their electric utility or by an independent consulting firm, and will then make changes to reduce their energy bills. They may not spend any further effort in the energy cost control area until several years in the future when another energy audit is conducted. In contrast to this is the company which establishes a permanent energy cost control program, and assigns one person—or a team of people—to continually monitor and improve the energy efficiency and energy productivity of the company. Similar to a Total Quality Management program where a company seeks to continually improve the quality of its products, services and operation, an energy cost control program seeks continual improvement in the amount of product produced for a given expenditure for energy.

The energy action plan lists the EMOs which should be implemented first, and suggests an overall implementation schedule. Often, one or more of the recommended EMOs provides an immediate or very short payback period, so savings from that EMO—or those EMOs—can be used to generate capital to pay for implementing the other EMOs. In addition, the action plan also suggests that a company designate one person as the energy monitor or energy manager for the facility if it has not already done so. This person can look at the monthly energy bills and see whether any unusual costs are occurring, and can verify that the energy savings from EMOs is really being seen. Finally, this person can continue to look for other ways the company can save on energy costs, and can be seen as evidence that the company is interested in a future program of energy cost control.

2.3   IMPLEMENTING THE AUDIT RECOMMENDATIONS

After the energy consumption data hve been collected and analyzed, the energy-related systems have been carefully examined, the ideas for improvement have been collected, and management commitment has been obtained, the next steps are to obtain company support for the program, to choose goals, and to initiate action.

2.3.1 The Energy Action Team

Now that the preliminary audits have uncovered some energy management measures that can save significant amounts of money or can substantially improve production, funding for the changes and employee support are two additional critical ingredients for success. These can best be obtained with the help of a committee, preferably called something like the energy action team. The functions of this committee are given in Table 2-1.

No program will work within a company without employee support, particularly such a program as energy management which seems

Table 2-1. Functions of the Energy Action Committee

1.   Create support within the company for energy management.
2.   Generate new ideas.
3.   Evaluate suggestions
4.   Set goals.
5.   Implement the most promising ideas.

to promise employee discomfort at no visible increase in production. Therefore, one function of the energy action committee is to give representation to every important political group within the company. For this purpose, the committee must include people from unions, management, and every major group that could hinder the implementation of an energy management plan. The committee must also include at least one person with financial knowledge of the company, a person in charge of the daily operation of the facility, and line personnel in each area of the facility that will be affected by energy management. In a hospital, for example, the committee would have to include a registered nurse, a physician, someone from hospital administration, and at least one person directly involved in the operation of the building. In a university, the committee should include a budget officer, at least one department chairperson, a faculty member, a senior secretary, someone from buildings and grounds, and one or more students.

In addition to providing representation, a broadly based committee provides a forum for the evaluation of suggestions. The committee should decide on evaluation criteria as soon as possible after it is organized. These criteria should include first cost, estimated payback period or (for projects with a payback period longer than 2 years) the constant-dollar return on investment (see Chapter 4), the effects on production, the effects on acceptance of the entire program, and any mitigating effect on problems of energy curtailment.

The committee has the additional duty to be a source of ideas. These ideas can be stimulated by the detailed energy audit which clearly shows problems and areas for improvement. The energy manager should be aware, however, that most maintenance personnel become quickly defensive and that their cooperation, and hopefully their support, may be important. The specific tasks of this committee are to set goals, implement changes, and monitor results.

2.3.1.1 Goals

At least three different kinds of goals can be identified. First, *performance goals*, such as a reduction of 10 percent in Btu/unit product, can be chosen. Such goals should be modest at first so that they can be accomplished—in general, 10-30 percent reduction in energy usage for companies with little energy management experience and 8-15 percent for companies with more. These goals can be accompanied by goals for the reduction of projected energy costs by a similar amount. The more experienced the company is in energy management, the fewer easy saving possibilities exist; thus lower goals are more realistic in that case.

A second type of goal that can be established is an accounting goal. The ultimate objective in an energy accounting system is to be able to allocate the cost of energy to a product in the same way that other direct costs are allocated, and this objective guides the establishment of preliminary energy accounting goals. A preliminary goal would therefore be to determine the amount of electricity and the contribution to the electrical peak from each of the major departments within the company. This will probably require some additional metering, but the authors have found that such metering pays for itself in energy saving (induced by a better knowledge of the energy consumption patterns) in six months or less.

The third type of goal is that of employee participation. Even if an energy management program has the backing of the management, it will still fail without the support and participation of the employees. Ways to measure this include the number of suggestions per month; the dollar value of improvements adopted as a result of employee suggestions, per month; and the number of lights left on or machines left running unnecessarily, on a spot inspection. Work sampling has been used to estimate the percentage of time that people are working at various tasks—it can be used equally well on machines.

2.3.1.2 Implementing Recommendations

In addition to providing and evaluating ideas, setting goals, and establishing employee support, the energy action committee has the duty of implementing the most promising ideas that have emerged from the energy evaluation process. Members of the committee have the responsibility to see that people are assigned to each project, that timetables are established, that money is provided, and that progress reporting procedures are set up and followed. It is then the committee responsibility to follow up on the progress of each project; this monitoring process is described in detail in the next section.

2.3.1.3 Monitoring

Energy management is not complete without monitoring and its associated feedback, and neither is the energy audit process. In an energy audit, monitoring discloses what measures contributed toward the company goals, what measures were counterproductive, and whether the goals themselves were too low or too high.

Monitoring consists of collecting and interpreting data. The data to collect are defined by the objectives chosen by the energy action committee. At the very least, the electrical and gas bills and those of other relevant energy sources must be examined and their data graphed each month. Monthly graphs should include: the total energy used of each type (kWh of electricity, therms [10^5 Btu] of gas, etc.); the peaks, if they determine part of the cost of electricity or gas; and any other factors that contribute to the bills. At the same time, other output-related measures, such as Btu/ton, should also be calculated, recorded, and graphed.

The monitoring data should provide direct feedback to those most able to implement the changes. Often this requires that recording instruments be installed in a number of departments in addition to the meters required by the utility company. The additional expense is justified by increased employee awareness of the timing and amounts of energy consumed, and usually this awareness leads to a reduction in energy costs. Metering at each department also enables management to determine where the energy is consumed and, possibly, what is causing the energy consumption. Such metering also helps each department manager to understand and control the consumption of his or her own department.

Monitoring should result in more action. Find what is good, and copy it elsewhere. Find what is bad, and avoid it elsewhere. If the goals are too high, lower them. If the goals are too low, raise them. Wherever the difference between the planned objectives and the achievements is great, initiate an analysis to determine the reasons and then develop new objectives, initiate new action, and monitor the new results. In this way, the analysis, action, and monitoring process repeats itself.

2.4 SUMMARY

In this chapter, we have explained the mechanics of performing and implementing the energy audit as well as the reasoning behind the energy audit process. The energy audit process should be a dynamic feedback loop. The process starts with the analysis of past data. Then each energy system is examined for savings potential. Recommendations for energy cost saving actions are made and an energy action committee is formed. Next, attractive energy cost savings projects are implemented. The last step is the monitoring program which necessarily leads back to the first step of analysis, thus renewing the cycle.

References
1. *2005 ASHRAE Handbook on Fundamentals*, American Society of Heating, Refrigerating and Air Conditioning Engineers, Inc., Atlanta, GA, May, *2005*.
2. Simplified Energy Analysis Using the Modified Bin Method, American Society of Heating, Refrigerating and Air Conditioning Engineers, Inc., Atlanta, GA, *1995*.
3. *IES Lighting Handbook*, Illuminating Engineering Society, New York, NY, 2000.
4. *Energy Management for the Year 2000*, A handbook prepared by the National Electrical Contractors Association and the National Electrical Manufacturers Association, Washington, DC, First Edition, 1993.
5. Capehart, Lynne C. and Barney L. Capehart, "Writing User-Friendly Energy Audit Reports," *Strategic Planning for Energy and the Environment*, Vo. 14, No. 4, 1994.

Bibliography
Baumeister, Theodore, Eugene A. Avallone, and Theodore Baumeister, III, Marks' *Standard Handbook for Mechanical Engineers*, 10th edition, McGraw-Hill, New York, 1998.

Capehart, Barney, Editor, *Encyclopedia of Energy Engineering and Technology*," Taylor and Francis/CRC Publishing Company, Boca Raton, FL, July, 2007.

Energy Conservation Guide for Industry and Commerce, National Bureau of Standards Handbook 115 and Supplement, 1978. Available through U.S. Government Printing Office, Washington, DC.

Facility Design and Planning Engineering Weather Data, Departments of the Air Force, the Army, and the Navy, 1978.

Instructions For Energy Auditors, Volumes I and II, U.S. Department of Energy, DOE/CS-0041/12&13, September, 1978. Available through National Technical Information Service, Springfield, VA.

Kennedy, W.J., Jr., "How to Conduct an Energy Audit," in Proceedings of the 1978 Fall AIIE Conference, Institute of Industrial Engineers, Atlanta, GA, 1978.

Kennedy, W.J., Jr., "An Introduction to HVAC Systems," in Proceedings of the 1980 Fall AIIE Conference, Institute of Industrial Engineers, Atlanta, GA, 1980.

McQuiston, Faye C., Jerald D. Parker, and Jeffrey D. Spitler, *Heating, Ventilating, and Air-Conditioning, Analysis and Design*, Wiley, New York, 2004.

Peterson, Carl, *Energy Survey and Conservation Manual*, Granite School District, Salt Lake City, 1980.

Turner, Wayne C., "The Impact of Energy on Material Handling," Proceedings of the 1980 Conference on Material Handling, Materials Handling Institute, Pittsburgh, PA, 1980.

Turner, Wayne C., and Steve Doty, Editor, *Energy Management Handbook*, Sixth Edition, Fairmont Press, Atlanta, GA, 2007.

Threshold Limit Values for Chemical Substances and Physical Agents and Biological Exposure Indices, 1990-91 American Conference of Governmental Industrial Hygienists.

Thumann, Albert, and William J. Younger, *Handbook of Energy Audits*, Sixth Edition, Fairmont Press, Atlanta, GA, 2003.

Variable Air Volume Systems Manual, 1980 edition, Manual No. 74-5461, Honeywell, Inc., Minneapolis, MN, 1980.

Ventilation for Acceptable Indoor Air Quality, ASHRAE 62.1-2007, American Society of Heating, Refrigerating and Air-Conditioning Engineers, Inc., 2007.

Witte, Larry C., Philip S. Schmidt, and David R. Brown, *Industrial Energy Management and Utilization*, Hemisphere Publishing Corporation, Washington, DC, 1988.

2004 Handbook—HVAC Systems and Equipment, American Society of Heating, Refrigerating, and Air-Conditioning Engineers, Inc., Atlanta, GA, 2004.

Chapter 3

Understanding Energy Bills

3.0 INTRODUCTION

Although temporary increases in energy supplies may cause short-term rate decreases or rate stability, the long-term prospect is for energy costs to continue to increase. This is particularly true in the case of electric power where the costs are associated as much with the cost of building the new generation facilities as with the cost of producing the electricity. The impact of an increase in energy costs can be easily seen by examining the rate schedules for the various fuel sources, yet few managers take the time to peruse and understand their utility's billing procedures.

Why? The reasons are many, but the main ones seem to be the following:

- Rate schedules are sometimes very complicated. They are difficult to understand, and the explanations developed by state utility regulatory boards and by the utilities themselves often confuse the customer rather than clarify the bill.

- Energy is too often treated as an overhead item. Even though energy is frequently a substantial component of a product's cost, the cost of the energy is almost always included as an overhead item rather than a direct cost. This makes the energy cost more difficult to account for and control. Consequently, management does not give energy costs the attention they deserve.

This must change and, in fact, is changing. More and more managers are trying to understand their rate schedules, and sometimes they even participate in utility rate increase hearings.

Managers should know what electric rate schedule they are under and how much they are charged for the various components of their electric bill: demand, consumption, power factor, sales tax, etc. They should also know the details of their other energy rate structures. This chapter covers rate schedules for the major energy sources utilized in this country. While the majority of the discussion focuses on electricity, attention is also given to gas, fuel oil, coal, and steam/chilled water.

3.1   ELECTRIC RATE STRUCTURES

3.1.1 Utility Costs

Perhaps the best way to understand electric utility billing is to examine the costs faced by the utility. The major utility cost categories are the following:

- *Physical Plant.* This is often the single biggest cost category. Because electric power plants have become larger and more technologically sophisticated with more pollution control requirements, the cost of building and operating an electric power generation facility continues to increase. Furthermore, the utility is required to have sufficient capacity to supply the peak needs of its customers while maintaining some equipment in reserve in case of equipment failures. Otherwise brownouts or even blackouts may occur. This added capacity can be provided with expensive new generating facilities. Alternatively, instead of building new facilities, many utilities are urging their customers to reduce their peak demand so that the existing facilities will provide sufficient capacity.

- *Transmission lines.* Another major cost category is the cost of transmission lines to carry the electricity from where it is produced to the general area where it is needed. Electricity is transmitted at relatively high voltages—often 500 to 1,000 kilovolts—to minimize resistance (I^2R) losses. This loss can be large or small depending on the transmission distances involved.

- *Substations.* Once the electricity reaches the general area where it is needed, the voltages must be reduced to the lower levels which can be safely distributed to customers. This is done with step-down transformers at substations. A few customers may receive voltage at transmission levels, but the vast majority do not.

- *Distribution systems.* After the voltage is reduced at a substation, the electricity is delivered to the individual customers through a local distribution system typically at a voltage level around 12 kilovolts. Most residential customers are supplied electricity at 120 and 240 volts, single-phase. In addition to these two voltages, commercial customers often take 230-volt, three-phase service. Some larger customers must also have 480-volt, three-phase service in order to power their large motors, ovens, and process equipment. The desired voltages are provided through the use of appropriate step-down transformers at the customer's specific location. Components of the distribution system which contribute to the utility's costs include utility poles, lines, transformers, and capacitors.

- *Meters.* Meters form the interface between the utility company and customer. Although the meter costs are relatively small, they are considered a separate item by the utility and are usually included in the part of the bill called the customer charge. The cost of a meter can range from under $50 for a residential customer to $1500 or more for an industrial customer requiring information on consumption, demand and power factor.

- *Administrative.* Administrative costs include salaries for executives, middle management, technical and office staff, as well as for maintenance staff. Office space and office equipment, taxes, insurance, and maintenance equipment and vehicles are also part of the administrative costs.

- *Energy.* Once the generation, transmission and distribution systems are in place, some form of primary energy must be purchased to fuel the boilers and generate the electricity. In the case of hydroelectric plants, the turbine generators are run by water power and the primary energy costs are small. Similarly, fuel costs for wind, solar and biomass power are small. Fossil fuel electric plants have experienced dramatic fluctuations in fuel costs depending on how national and world events alter the availability of oil, gas, and coal. The cost of fuel for nuclear power plants is reasonable, but the costs of disposing of the radioactive spent fuel rods, while still unknown, are expected to be relatively large.

- *Interest on debt.* This cost category can be quite large. For example, the interest on debt for a large power plant costing $500 million to $1 billion is substantial. Utilities commonly sell bonds to generate capital, and these bonds represent debt that the company must pay interest on.

- *Profit.* Finally the utility must generate enough additional revenue above costs to provide a reasonable profit to stockholders. The profit level for private utility companies is determined by the state utility regulatory commission and is called the *rate of return*. Public-owned utilities such as municipal utilities or rural electric cooperatives usually set their own rates and their profit goes back to their customers in the form of reduced municipal taxes or customer rebates.

Once you understand what costs contribute to an electric bill, the next step is to learn how these costs are allocated to the various customers. The *billing procedure*, also called the *rate schedule*, should be designed to reflect the true costs of generating the electric power. If the customers understand the problems faced by the utilities, they can help the utilities minimize these costs. Recent rate schedules and proposed new ones capture the true costs of generation much better than has been done in the past, but more changes are still needed.

3.1.2 Regulatory Agencies

Private electric and gas utilities are chartered and regulated by individual states, and are also subject to some federal regulation. The state utility regulatory agencies are most often called Public Utility Commissions or Public Service Commissions. Private utilities are called *Investor Owned Utilities* (IOUs), and their retail rates for residential, commercial and industrial customers are subject to review and approval by the state utility regulatory agencies.

Utility rates are set in two steps: first, the revenue requirements to cover costs plus profit is determined; second, rates are designed and set to recover these costs or revenue requirements [1]. The state regulatory agencies set a *rate of return* for utilities. The rate of return is the level of profit a utility is allowed to make on its investment in producing and selling energy. In developing rates, the costs of serving different classes of customers must be determined and allocated to the customer classes. Rates are then structured to recover these costs from the appropriate customer class. Such rate designs are called *cost-based rates*. Often, these costs are *average*, or *embedded costs*, and do not consider the *marginal costs* associated with providing electricity at different times of the day and different seasons of the year. Rate design is subject to many competing viewpoints, and there are many different objectives possible in rate setting.

When an IOU requests a rate increase, the state regulatory agency holds a public hearing to review the proposed rate increase, and to take testimony from the utility staff, consulting engineers, customers and the public at large. The utility presents its case for why it needs a rate increase, and explains what its additional costs are. If these costs are judged "prudent" by the state regulatory agency and approved, the utility is allowed to recover the costs, plus adding some of that cost to its rate base—which is the accumulated capital cost of facilities purchased or installed to serve the customers and on which the utility can earn its rate of return.

Many large utility customers participate actively in the rate hearings for their utility. Some state regulatory agencies are very interested in comments from utility customers regarding quality of service, reliability, lengths of outages, and other utility service factors. State regulatory agencies vary greatly in their attitude toward utility rate increases. Some states favor the utilities and consider their interests to be first priority, while other states consider the interests of the customers and the public as paramount.

Two other major categories of utilities exist: *public* or *municipal utilities* owned and operated by cities and local government entities; and *Rural Electric Cooperatives* (RECs) established under the Rural Electrification Administration and operated by customer Boards of Directors. State regulatory agencies generally do not exercise the same degree of control over public utilities and RECs, since these utilities have citizens and customers controlling the rates and making the operating decisions, whereas the IOUs have stockholders making those decisions. Municipals and RECs also hold public hearings or public meetings whenever rate increases are contemplated. Customers who have an interest in participating in these meetings are usually encouraged to do so.

Interstate transactions involving the wholesale sale or purchase of electricity between utilities in different states are subject to regulation by the Federal Energy Regulatory Commission (FERC) in Washington, DC. FERC also regulates the designation of some cogenerators and renewable electric energy suppliers as Qualifying Facilities (QFs) or Small Power Producers (SPPs). Few retail customers outside of those engaged in self-generation would have any reason to participate in the regulatory process at this level. FERC also licenses non-federal hydroelectric facilities.

3.1.3 Customer Classes and Rate Schedules

An electric utility must serve several classes of customers. These classes vary in complexity of energy use, amount of consumption, and priority of need. The typical customer class categories are residential, commercial and industrial. Some utilities combine commercial and industrial customers into one class while other utilities divide the industrial class into heavy industrial and light industrial customers.

The state regulatory agencies and utilities develop different rate schedules for each customer class. Electric rate structures vary greatly from utility to utility, but they all have a series of common features. The most common components of rate schedules are described below, but not all of these components are included in the rate schedule for every customer class.

- *Administrative/Customer Charge:* This fee covers the utility's fixed cost of serving the customer including such costs as providing a meter, reading the meter, sending a bill, etc. This charge is a flat monthly fee per customer regardless of the number of kWh of electricity consumed.

- *Energy charge:* This charge covers the actual amount of electricity consumed measured in kilowatt-hours. The energy charge is based on an average cost, or base rate, for the fuel (natural gas, fuel oil, coal, etc.) consumed to produce each kWh of electricity. The energy charge also includes a charge for the utility's operating and maintenance expenses.

  Many utilities charge a constant rate for all energy used, and this is called a *flat rate* structure. A *declining block* approach may also be used. A declining block schedule charges one price for the first block of energy (kWh) used and less for the next increment(s) of energy as more energy is used. Another approach is the *increasing block* rate where more is charged per increment as the consumption level increases. Although this approach would tend to discourage electric energy waste, it does not meet the *cost-based rate* criterion and is therefore not widely used.

- *Fuel cost adjustment:* If the utility has to pay more than its expected cost for primary fuel, the increased cost is "passed on" to the customer through use of a prescribed formula for a fuel adjustment cost. In times of rapidly increasing fuel prices, the fuel adjustment cost can be a substantial proportion of the bill. This concept was adopted when fuel costs were escalating faster than utility commissions could grant rate increases. However, utilities can also use the fuel adjustment cost to reduce rates when fuel costs are lower than the cost included in the base rate.

- *Demand charge:* The demand charge is used to allocate the cost of the capital facilities which provide the electric service. The demand charge may be "hidden" in the energy charge or it may be a separate charge; for example it may be expressed as $6.25 per kW per month for all kW above 10 kW. For large customers, the demand charge is generally based on their kilowatt demand load. For smaller users such as residential and small commercial customers this charge is usually averaged into the energy charge. The demand charge is explained in more detail in Section 3.1.6.2.

Understanding the difference between *electric demand*, or power in kilowatts (kW), and *electric energy*, or consumption in kilowatt-hours (kWh), will help you understand how electric bills are computed. A helpful analogy is to think of an automobile where the speedometer measures the rate of travel in miles per hour, and the odometer measures the total miles traveled. In this instance, speed is analogous to electric power, and miles traveled is analogous to total energy consumed. In analytical terms, *power is the rate of use of electric energy*, and conversely, *energy is the time integral of the power*. Finally, the value of the power or demand a utility uses to compute an electric bill is obtained from a peak power measurement that is averaged over a short period of time. Typical averaging times used by various electric utilities are 15 minutes, 30 minutes and one hour. The averaging time prevents unreasonable charges from occurring because of very short, transient peaks in power consumption. Demand is measured by a demand meter.

- *Demand ratchet:* An industrial or commercial rate structure may also have a demand ratchet component. This component allows the utility to adequately charge a customer for creating a large kilowatt demand in only a few months of the year. Under the demand ratchet, a customer will not necessarily be charged for the actual demand for a given month. Instead the customer will be charged a percentage of the largest kW value during the last 11 months, or the current month's demand, whichever is higher.

- *Power factor:* If a large customer has a poor power factor, the utility may impose another charge, assessed as a function of that power factor. Power factor is discussed in detail in section 3.1.6.4.

All of these factors are considered when a utility sets its *base rates*— the rates the utility must charge to recover its general cost of doing business. The term "base rates" should not be confused with the term "rate base" which was previously defined. The base rates contain an energy charge that is estimated to cover the average cost of fuel in the future. The fuel adjustment charge keeps the utility from losing money when the price of their purchased primary fuel is higher than was estimated in their base rates.

Figure 3-1 presents a generalized breakdown of these rate components by customer class.

In addition, there are also a number of other features of electric rates incorporated in the *rate structure* which includes the relationship and form of prices within particular customer classes. The rate structure is set to maintain equity between and within customer classes, ensuring that there is no discrimination against or preferential treatment of any particular customer group. Some of the factors considered in the rate structure are: season of use; time of use; quantity of energy used—and whether increased consumption is encouraged, discouraged, or considered neutral; and social aspects such as the desire for a "lifeline" rate for low-income or elderly customers. A number of these factors are illustrated in the description of the specific rate structures shown in examples for particular customer classes.

3.1.4 Residential Rate Schedules

As shown in Figure 3-1, there are many residential users, but each is a relatively small consumer. A typical residential bill includes an administrative/customer charge, an energy charge which is large enough to cover both the actual energy charge and an implicit demand charge, and a fuel adjustment charge. Residential rates do not usually include an explicit demand charge because the individual demand is relatively inconsequential and expensive to meter.

3.1.4.1 Standard residential rate schedule

Figure 3-2 presents a typical monthly rate schedule for a residential customer.

3.1.4.2 Low-use residential rate schedule

A typical low-use residential service rate is shown in Figure 3-3. This schedule, which is an attempt to meet the needs of those on fixed incomes, is used for customers whose monthly consumption never exceeds 500 kWh. In addition, it cannot exceed 400 kWh more than twice a year. This rate is sometimes referred to as a *lifeline rate*.

3.1.4.3 Residential rate schedules to control peak uses

Although individual residential demand is small, collectively residential users place a peak demand burden on the utility system because the majority of them use their electricity at the same times of the day during the same months of the year. Some utilities charge more for energy during peaking months in an attempt to solve this problem. Many utilities

Typical schedule bills for:

Customer Class	Comments	Consumption (kWh)	Demand (kW)	Power factor (kVAR)
1. Residential	Small user but large numbers of them	√		
2. Commercial	Small to moderate user; relatively large numbers	√		
3. Small industrial	Small to moderate user; fewer customers	√	√	
4. Large industrial	Large user with low priority; typically, only a few customers in this class, but they consume a large percentage of the electricity produced.	√	√	√

Figure 3-1.
Generalized Breakdown of Electric Rate Schedule Components

Customer charge:	$8.00/month
Energy Charge:	All kWh @ 6.972 ¢/kWh
Fuel adjustment:	(A formula is provided by the utility to calculate the fuel adjustment charge each month. It is rather complex and will not be covered here.)

Figure 3-2. Typical Residential Rate Schedule

Customer charge:	$5.45/month
Energy Charge:	5.865 ¢/kWh
Fuel adjustment:	(A formula is provided for calculating this charge.)

Figure 3-3. Low-use Residential Schedule
(Courtesy of Oklahoma Gas and Electric Company)

have an optional *time-of-day* or *time-of-use* rate which is supposed to help alleviate the daily peaking problem by charging customers more for electric use during these peak periods. A number of utilities also have a load management program to control customers' appliances.

Figure 3-4 provides examples of a Florida utility's residential demand profile over a given 24-hour period during the weekdays. Figure 3-4 (a) shows the residential winter peak demand profile. This utility experiences one large peak around 9:00 a.m. and another somewhat smaller

Typical Winter Residential Peak

Figure 3-4 (a)

Typical Summer Residential Peak

Figure 3-4 (b)

peak near 9:00 p.m. The first peak occurs when people get up in the morning and start using electricity. They all turn up their electric heat, cook breakfast, take a shower, and dry their hair at about the same time on weekday mornings. Then in the evening, they all come home from work, start cooking dinner, turn the heat back up (or use it more because nights are colder) and turn on the TV set at about the same time. Figure 3-4 (b) shows the residential summer peak demand profile for the same utility.

- **Seasonal use rate schedule.** Figure 3-5 presents a residential rate schedule where the season of use is a factor in the rate structure. This utility has chosen to attack its residential peaking problem by charging more for electricity consumed in the summer months when the highest peaks occur.

Customer charge:	$6.50/month
Energy Charge:	On-peak season (June through October)
	All kWh @ 7.728¢/kWh
	Off-peak season (November through May)
	First 600 kWh @ 7.728¢/kWh
	All additional kWh @ 3.898¢/kWh
Fuel adjustment:	(Calculated by a formula provided by the utility.)

Figure 3-5. Seasonal Use Residential Rate Schedule
(Courtesy of Oklahoma Gas and Electric Company)

During the summer peak season this utility uses a constant charge or flat rate for all energy (7.728 cents/kWh) regardless of the amount consumed. In the off-peak season, however, the utility uses a declining block approach and charges a higher rate for the first 600 kWh of energy than it does for the remaining kilowatt-hour use.

- **Time-of-day or time-of-use pricing.** To handle the daily peaking problem, some utilities charge more for energy consumed during peak times. This requires the utility to install relatively sophisticated meters. It also requires some customer habit changes. Time-of-use pricing for residential customers is not very popular today; however, most utilities are required by their state regulatory agencies to provide a time-of-use rate for customers who desire one, so most utilities have some form of time-of-use pricing.

A sample time-of-day rate for residential customers served by Florida Power Corporation is shown in Figure 3-6.

- **Peak shaving.** Some utilities offer a discount to residential customers if the utility can hook up a remote control unit to cycle large electricity using appliances in the home (usually electric heaters, air conditioners and water heaters). This utility load control program is also called *load management*. This way the utility can cycle large appliance loads on and off periodically to help reduce demand. Since the cycling is performed over short periods of time, most customers experience little to no discomfort. This approach is rapidly gaining in popularity.

A sample load management rate for residential customers served by the Clay Electric Cooperative in Keystone Heights, FL, is shown in Figure 3-7. This rate provides a rebate to customers who agree to allow the utility to turn off their electric water heaters or air conditioners for short periods of time during peak hours. Note that the Clay Electric rate also includes an inclining block feature.

3.1.5 General Service Rate Schedules

A *general service rate schedule* is used for commercial and small industrial users. This is a simple schedule usually involving only consumption (kWh) charges and customer charges. Sometimes, demand (kW) charges

---

Customer charge:	$16.00/month
Energy charge:	
On-peak energy	10.857¢/kWh
Off-peak energy	5.80¢/kWh
On-peak hours:	
November through March:	
Monday through Friday	6:00 a.m. to 10:00 a.m.
	6:00 p.m. to 10:00 p.m.
April through October:	
Monday through Friday	12:00 noon to 9:00 p.m.
Off-peak hours:	All other hours

Figure 3-6. Sample Time-of-day Electric Rate.
(Courtesy Florida Power Corporation, St. Petersburg, FL)

Customer charge: $9.00/month

Energy charge: First 1000 kWh @ $0.0825/kWh
                 Over 1000 kWh @ $0.0930/kWh

Load management credit per month: Credit will be applied to the bill of all customers with load management switches who use 500 kWh or more per month as follows:

Electric water heater controlled January-December	$4.00
Electric central heating controlled October-March for 5 to 7.5 minutes of each 25-minute period	$3.00
Electric central air conditioner controlled April-September for 5 to 7.5 minutes of each 25-minute period	$3.00
Electric central heating controlled October-March for 12.5 minutes of each 25-minute period	$8.00
Electric central air conditioner controlled April-September for 5 to 7.5 minutes of each 25-minute period	$8.00

Figure 3-7. Sample Load Management Rate for Residential Service
(Courtesy of Clay Electric Cooperative, Keystone Heights, FL)

are used; this requires a demand meter. (See Section 3.1.3 for a more detailed discussion of demand charges.)

The energy charge for this customer class is often substantially higher than for residential users for various noneconomic reasons. Some of these reasons include the fact that many businesses have widely varying loads depending on the health of the economy, and many businesses close after only a few months of operation—sometimes leaving large unpaid bills. In addition, some regulatory agencies feel that residential customers should have lower rates since they cannot pass on electric costs to someone else. For example, one rate schedule charges almost 8 cents/kWh for commercial users during peak season but only a little more than 5 cents/kWh for residential users during the same season.

3.1.6 Small Industrial Rate Schedules

A *small industrial rate schedule* is usually available for small industrial users and large commercial users. The service to these customers often becomes more complex because of the nature of the equipment used in the industry, and their consumption tends to be higher. Consequently, the billing becomes more sophisticated. Usually, the same cost categories occur as in the simpler schedules, but other categories have been added. Some of these are outlined below.

3.1.6.1 Voltage level.

One degree of complexity is introduced according to what *voltage level* the customer needs. If the customer is willing to accept the electricity at transmission voltage levels (usually 50,000 volts or higher) and do the necessary transforming to usable levels on-site, then the utility saves considerable expense and can charge less. If the customer needs the service at a lower voltage, then the utility must install transformers and maintain them. In that case, the cost of service goes up and so does the bill.

The voltage level charge can be handled in the rate schedule in several ways. One is for the utility to offer a percentage discount on the electric bill if the customer owns its own primary transformer and accepts service at a higher voltage than it needs to run its equipment. Another is to increase the energy charge as the voltage level decreases. (This method is shown in the example in Figure 3-8.) Installing their own transformers is often a significant cost-cutting opportunity for industrial users and should be explored. Maintaining transformers is a relatively simple (though potentially dangerous) task, but the customer may also need to install standby transformers to avoid costly shutdowns.

3.1.6.2 Demand billing.

Understanding industrial rate structures means understanding the concept of *demand billing*. Consider Figure 3-8 where energy demands on a utility are plotted against time for two hypothetical companies. Since the instantaneous demand (kW) is plotted over time, the integration of this curve (i.e., the area under the curve) is the total energy (kWh) consumed (see shaded area). Company B and Company A have the same average demand, so the total energy consumed by B equals that of A. Company B's peak demand and its average demand are the same, but Company A has a seasonal peak that is twice as high as its average demand. Because the kWh consumed by each are equal, their bills for consumption will be equal, but this seems unfair. Company B has a very flat demand structure so the utility can gear up for that level of service with high-efficiency equipment. Company A, however, requires the utility to supply about twice the capacity that company B needs but only for one short period of time during the year. This means the utility must maintain and gear up equipment which will only be needed for a short period of time. This is quite expensive, and some mechanism must be used by the utility to recover these additional costs.

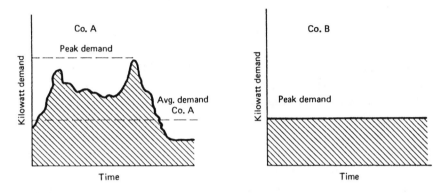

Figure 3-8. Demand Profiles for Two Hypothetical Industrial Firms

To properly charge for this disproportionate use of facilities and to encourage company A to reduce its peak demand, an electric utility will usually charge industrial users for the peak demand incurred during a billing cycle, normally a month. Often a customer can achieve substantial cost reductions simply by reducing peak demand and still consuming the same amount of electricity. A good example of this would be to move the use of an electric furnace from peaking times to nonpeaking times (maybe second or third shifts). This means the same energy could be used at less cost since the demand is reduced. A peak shaving (demand control) example will be discussed in section 3-7.

3.1.6.3 Ratchet clause

Many utility rate structures have a ratchet clause associated with their demand rate. To understand the purpose of the ratchet clauses, one must realize that if the utility must supply power to meet a peak load in July, it must keep that equipment on hand and maintain it for the next peak load which may not occur for another year. To charge for this cost, and to encourage customers to level their demand over the remaining months, many utilities have a ratchet clause.

A ratchet clause usually says that the billed demand for any month is a percentage (usually greater than 50%) of the highest maximum demand of the previous 11 months or the actual demand, whichever is greater. The demand is normally corrected for the power factor. For a company with a large seasonal peaking nature, this can be a real problem. A peak can be set in July during a heavy air conditioning period that the company in effect pays for a full year. The impact of ratchet clauses can be significant, but often a company never realizes this has occurred.

3.1.6.4 Power factor.

Power factor is a complex subject to explain, but it can be a vitally important element in a company's electrical bill. One company the authors worked with had a power factor of 51 percent. With their billing schedule, this meant they were paying a penalty of 56.9 percent on demand billing. With the addition of power factor correction capacitors, this penalty could have been avoided or minimized.

The power factor is important because it imposes costs on a utility that are not recovered with demand and energy charges. Industrial customers are more likely to be charged for a poor power factor. They create greater power factor problems for a utility because of the equipment they use. They are also more likely to be able to correct the problem.

To understand the power factor, you must understand electric currents. The current required by induction motors, transformers, fluorescent lights, induction heating furnaces, resistance welders, etc., is made up of three types of current:

1. *Power-producing current* (working current or current producing real power). This is the current which is converted by the equipment into useful work, such as turning a lathe, making a weld, or pumping water. The unit of measurement of the real power produced from working current is the kilowatt (kW).

2. *Magnetizing current* (wattless or reactive current). This is the current which is required to produce the magnetic flux necessary for the operation of induction devices. Without magnetizing current, energy could not flow through the core of a transformer or across the air gap of an induction motor. The unit of measurement of the reactive power associated with magnetizing current is the kilovar (kVAR) or kilovolt-amperes reactive.

3. *Total current* (current producing apparent power or total power). This is the current that is read on an ammeter in the circuit. It is made up of the vector sum of the magnetizing current and the power-producing current. The unit of measurement of apparent power associated with this total current is the kilovoltampere (kVA). Most alternating current (ac) powered loads require both kilowatts and kilovars to perform useful work.

Power factor is the ratio of real power being used in a circuit, expressed in watts or kilowatts, to the apparent power drawn from the power line, expressed in voltamperes or kilovolt-amperes. The relationship of kW, kVAR, and kVA in an electrical system can be illustrated by scaling vectors to represent the magnitude of each quantity, with the vector for kVAR at a right angle to that for kW (Figure 3-9). When these components are added *vectorially*, the resultant is the kVA vector. The angle between the kW and kVA vectors is known as the *phase angle*. The cosine of this angle is the power factor and equals kW/kVA.

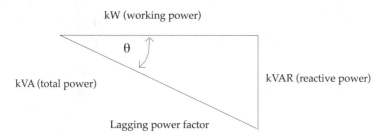

θ = phase angle = measure of net amount of inductive reactance in circuit

$\cos \theta$ = PF = ratio of *real power* to *apparent power*

$$kVA = \frac{kW}{\cos \theta} = \frac{kW}{PF} = \sqrt{(kW)^2 + (kVAR)^2}$$

Figure 3-9. Diagram of AC Component Vectors

Unless some way of billing for a low power factor is incorporated into a rate schedule, a company with a low power factor would be billed the same as a company with a high power factor. Most utilities do build in a power factor penalty for industrial users. However, the way of billing varies widely. Some of the more common ways include:

- Billing demand is measured in kVA instead of kW. A look at the triangle in Figure 3-9 shows that as the power factor is improved, kVA is reduced, providing a motivation for power factor improvement.

- Billing demand is modified by a measure of the power factor. Some utilities will increase billed demand one percent for each one percent the power factor is below a designatd base. Others will modify demand as follows:

$$Billed\ Demand = Actual\ Metered\ Demand \cdot \frac{Base\ Power\ Factor}{Actual\ Power\ Factor}$$

This way, if the actual power factor is lower than the base power factor, the billed demand is increased. If the actual power factor is higher than the base power factor, some utilities will allow the fraction to stay, thereby providing a reward instead of a penalty. Some will run the calculation only if actual power factor is below base power factor.

- The demand or consumption billing schedule is changed according to the power factor. Some utilities will change the schedule for both demand and consumption according to the power factor.

- A charge per kVAR is used. Some companies will charge for each kVAR used above a set minimum. This is direct billing for the power factor.

In addition, since a regular kW meter does not recognize the reactive power, some other measuring instrument must be used to determine the reactive power or the power factor. A kVA meter can be supplied by the utility, or the utility might decide to only periodically check the power factor at a facility. In this case a utility would send a crew to the facility to measure the power factor for a short period of time, and then remove the test meter. Or, many utilities would just install an electronic meter with remote reading agility.

3.1.6.5 The rate schedule.

The previous few sections were necessary in order to be able to present a rate schedule itself in understandable terms. All these complex terms and relationships make it difficult for many managers to understand their bills. You, however, are now ready to analyze a typical rate schedule. Consider Figure 3-10.

Effective in: All territories served

Availability: Power and light service. Alternating current. Service will be rendered at one location at one voltage. No resale, breakdown, auxiliary, or supplementary service permitted.

Rate:
    A. Transmission service (service level 1):
        Customer charge: $637.00/bill/month

        Demand charge applicable to all kW/month of billing demand:
            On-peak season: $10.59/kW
            Off-peak season: $3.84/kW

        Energy charge:
            First two million kWh 3.257¢/kWh
            All kWh over two million 2.915¢/kWh

    B. Distribution service (service level 2):
        Customer charge: $637.00/bill/month

        Demand charge applicable to all kW/month of billing demand:
            On-peak season: $11.99/kW
            Off-peak season: $4.36/kW

        Energy charge:
            First two million kWh 3.297¢/kWh
            All kWh over two million 2.951¢/kWh

    C. Distribution service (service levels 3 and 4):
        Customer charge: $269.00/bill/month

        Demand charge applicable to all kW/month of billing demand:
            On-peak season: $12.22/kW
            Off-peak season: $4.45/kW

Figure 3-10. Typical small industrial rate schedule.
(Courtesy of Oklahoma Gas and Electric Company)

Energy charge:
First two million kWh 3.431¢/kWh
All kWh over two million 3.010¢/kWh

D. Secondary service (service level 5):
Customer charge: $151.00/bill/month

Demand charge applicable to all kW/month of
billing demand:
On-peak season: $13.27/kW
Off-peak season: $4.82/kW

Energy charge:
First two million kWh 3.528¢/kWh
All kWh over two million 3.113¢/kWh

Definition of season:
On-peak season: Revenue months of June-October of any year.
Off-peak season: Revenue months of November of any year
through May of the succeeding year.

Late payment charge: A late payment charge in an amount equal to one
and one-half percent (1-1/2%) of the total balance for services and charges
remaining unpaid on the due date stated on the bill shall be added to
the amount due. The due date shall be twenty (20) days after the bill is
mailed.

Minimum bill: The minimum monthly bill shall be the Customer Charge
plus the applicable Capacity Charge as computed under the above schedule. The Company shall specify a larger minimum monthly bill, calculated
in accordance with the Company's Allowable Expenditure Formula in its
Terms and Conditions of Service on file with and approved by the Commission, when necessary to justify the investment required to provide
service.

Determination of maximum demand: The consumer's Maximum Demand shall be the maximum rate at which energy is used for any period of
fifteen (15) consecutive minutes of the month for which the bill is rendered
as shown by the Company's demand meter. In the event a consumer tak-

Figure 3-10. (*Continued*)

ing service under this rate has a demand meter with an interval greater than 15 minutes, the company shall have a reasonable time to change the metering device.

<u>Determination of billing demand:</u> The Billing Demand upon which the demand charge is based shall be the Maximum Demand as determined above corrected for the power factor, as set forth under the Power Factor Clause, provided that no billing demand shall be considered as less than 65% of the highest on-peak season maximum demand corrected for the power factor previously determined during the 12 months ending with the current month.

<u>Power factor clause:</u> The consumer shall at all times take and use power in such manner that the power factor shall be as nearly 100% as possible, but when the average power factor as determined by continuous measurement of lagging reactive kilovoltampere hours is less than 80%, the Billing Demand shall be determined by multiplying the Maximum Demand, shown by the demand meter for the billing period, by 80 and dividing the product thus obtained by the actual average power factor expressed in per cent. The company may, at its option, use for adjustment the power factor as determined by tests during periods of normal operation of the consumer's equipment instead of the average power factor.

<u>Fuel cost adjustment:</u> The rate as stated above is based on an average cost of $1.60/million Btu for the cost of fuel burned at the company's thermal generating plants. The monthly bill as calculated under the above rate shall be increased or decreased for each kWh consumed by an amont computed in accordance with the following formula:

$$F.A. = A * \frac{(B * C) - D}{10^6} + \frac{P}{S} + \frac{OC}{OS} - Y$$

where

F.A. = fuel cost adjustment factor (expressed in $/kWh) to be applied per kWh consumed

A = weighted average Btu/kWh for net generation from the company's thermal plants during the second calendar month preceding the end of the billing period for which the kWh usage is billed

Figure 3-10. (*Continued*)

B   =   amount by which the average cost of fuel per million Btu
         during the second calendar month preceding the end of the
         billing period for which the kWh usage is billed exceeds
         or is less than $1.60/million Btu; any credits, refunds, or
         allowances on previously purchased fuel, received by the
         company from any source, shall be deducted from the cost
         of fuel before calculating B each month

C   =   ratio (expressed decimally) of the total net generation from
         all the company's thermal plants during the second calendar
         month preceding the end of the billing period for which the
         kWh usage is billed to the total net generation from all the
         company's plants including hydro generation owned by the
         company, or kW produced by hydro generation and pur-
         chased by the company during the same period

D   =   the amount of fuel cost per million Btu embedded in the
         base rate is $2.30

P   =   the capacity and energy cost of electricity purchased by the
         Company, excluding any cost associated with "OC," during
         the second calendar month preceding the current billing
         month, excluding any capacity purchased in said month and
         recovered pursuant to Standard Rate Schedule PCR-1.

S   =   total kWh generated by the company plus total kWh pur-
         chased by the company which are associated with the cost
         included in "P" during the second calendar month preced-
         ing the end of the billing period for which kWh use is billed

OC   =   the difference between the cost of cogenerated power and
         company-generated power (Note that this factor has been
         simplified for purposes of this book.)

OS   =   the company's appropriate Oklahoma retail kWh sales dur-
         ing the twelfth billing month preceding the current billing
         month

Y   =   a factor (expressed in $/kWh) to reflect 90% of the margin
         (profits) from the non-firm off-system sales of electricity to
         other utilities during the 2nd calendar month preceding the
         end of the billing period for which the kWh usage is billed.

Figure 3-10. (*Continued*)

<u>Franchise payment:</u> Pursuant to Order Number 110730 and Rule 54(a) of Order Number 104932 of the Corporation Commission of Oklahoma, franchise taxes or payments (based on a percent of gross revenue) in excess of 2% required by a franchise or other ordinance approved by the qualified electors of a municipality, to be paid by the company to the municipality, will be added pro rata as a percentage of charges for electric service, as a separate item, to the bills of all consumers receiving service from the company within the corporate limits of the municipality exacting the said tax or payment.

<u>Transmission, distribution, or secondary service:</u> For purposes of this rate, the following shall apply:

<u>Transmission service</u> (service level 1), shall mean service at any nominal standard voltage of the company above 50 kV where service is rendered through a direct tap to a company's transmission source.

<u>Distribution service</u> (service levels 2,3, and 4), shall mean service at any nominal standard voltage of the company between 2,000 volts and 50 kV, both inclusive, where service is rendered through a direct tap to a company's distribution line or through a company numbered substation.

<u>Secondary service</u> (service level 5), shall mean service at any nominal standard voltage of the company less than 2,000 volts or at voltages from 2 to 50 kV where service is rendered through a company-owned line transformer. If the company chooses to install its metering equipment on the load side of the consumer's transformers, the kWh billed shall be increased by the amount of the transformer losses calculated as follows:

1% of the total kVA rating of the consumer's transformers * 730 hours

<u>Term:</u> Contracts under this schedule shall be for not less than 1 year, but longer contracts subject also to special minimum guarantees may be necessary in cases warranted by special circumstances or unusually large investments by the company. Such special minimum guarantees shall be calculated in accordance with the company's allowable expenditure formula and its terms and conditions of service on file with and approved by the commission.

Figure 3-10. (*Concluded*)

Let's examine the different components in this rate structure of Figure 3-10.

- *Voltage level.* This utility has chosen to encourage company-owned primary transformers by offering a cheaper rate for both demand and consumption if the company accepts service at a higher voltage level. To analyze what it could save from primary transformer ownership, a company only needs to calculate the dollar savings from accepting service at a higher voltage level and compare that savings to the cost of the necessary transformers and annual maintenance thereof. Transformer losses must be absorbed by the company, and the company must provide a standby transformer or make other arrangements in case of a breakdown.

- *Demand billing.* This utility has chosen to emphasize demand leveling by assessing a rather heavy charge for demand.* Furthermore, the utility has emphasized demand leveling during its summer peaking season.

- *Consumption.* This utility uses a declining block rate for very large users, but this essentially amounts to a flat charge per kilowatt-hour for most consumption levels.

- *Power factor.* The utility has chosen to charge for the power factor by modifying the demand charge. They have decided all customers should aim for a power factor of at least 80 percent and should be penalized for power factors of less than 80 percent. To do this, the peak demand is multiplied by a ratio of the base power factor (80%) to the actual power factor if the actual power factor is below 80%;there is no charge or reward if the power factor is above 80%:

$$Billed\ Demand = Actual\ Metered\ Demand \cdot \frac{Base\ Power\ Factor}{Actual\ Power\ Factor}$$

where the base power factor = .80.

- *Ratchet clause.* The utility has a ratchet clause which says that the billed demand for any month is "65% of the highest on-peak season maximum demand corrected for the power factor" of the previous 12 months or the actual demand corrected for power factor whichever is greater.

---

*Actually, charges are regional. For the Southwest, this is a rather large demand charge. For the Northwest, it would be cheap.

- *Miscellaneous.* Other items appearing in the rate schedule include fuel cost adjustment, late payment charge, and minimum bill. The fuel cost adjustment is based on a formula and can be quite significant. Anytime the cost of energy is calculated, the fuel cost adjustment should be included.

- *Sales tax.* One item not mentioned in the sample schedule is sales tax. Many localities have sales taxes of 6-8% or more, so this can be a significant cost factor. The cost of electrical service should include this charge. One item of interest: Some states have laws stating that *energy used directly in production should not have sales tax charged to it.* This is important to any industry in such a state with energy going to production. Some submetering may be necessary, but the cost savings often justifies this. For example, electricity used in a process furnace should not be taxed, but electricity running the air conditioners in administrative offices would be taxed.

Example 3-1. As an example of rate schedule calculations, let's use the schedule in Figure 3-10 to calculate the September bill for the company whose electric use is shown below:

Month: September 2006
Actual demand: 250 kW
Consumption: 54,000 kWh
Previous high billed demand: 500 kW (July 2006)

Power factor: 75%
Service level: Secondary (PLS, service level 5)
Sales tax: 6%
Fuel adjustment: 1.15¢/kWh (This value is calculated by the utility company according to the formula given in the rate schedule.)

As a first step, the demand should be calculated:

Power factor correction:

billed demand   = (actual metered demand) * (.80/PF)
                   = 250 kW * (.80/.75)
                   = 266.7 kW

minimum billed demand (ratchet clause)
$$= (500 \text{ kW}) * (.65)$$
$$= 325 \text{ kW}$$
billed demand   = max. (266.7 kW, 325 kW)

billed demand   = 325 kW

demand charge (on-peak season)
$$= (325 \text{ kW}) (\$13.27/\text{kW})$$
$$= \underline{\$4312.75}$$

Consumption charge:

(54,000 kWh)($.03528/kWh)   = $1905.12
(54,000 kWh)($.0115/kWh)(fuel adjustment)   = $ 621.00
                        total consumption charge   = $\underline{2526.12}$

Customer charge:

$151.00

Total charge before sales tax:

$4312.75 + $2526.12 + $151.00 = $\underline{6989.87}$

Sales tax:

$6989.87 × (.06) = $\underline{\$ 419.39}$

Total*:

$6989.87 + $ 419.39 = $\underline{7409.26}$

3.1.7 Large Industrial Rate Schedules

Most utilities have very few customers that would qualify for or desire to be on a large industrial rate schedule. Sometimes, however, one or two large industries will utilize a significant portion of a utility's total generating capacity. Their size makes the billing more complex; therefore,

---

*Ignoring franchise payment and late charges.

a well-conceived and well-designed rate schedule is necessary.

Typically a large industrial schedule will include the same components as a small industrial schedule. The difference occurs in the amount charged for each category. The customer charge, if there is one, tends to be higher. The minimum kW of demand tends to be much higher in cost/kW, but all additional kW may be somewhat lower (per kW) than on small industrial schedules. Similarly, the charge per kWh for consumption can be somewhat less. The reason for this is economy of scale; it is cheaper for a utility to deliver a given amount of electrical energy to one large customer than the same amount of energy to many smaller customers.

Figure 3-11 is an example of a large industrial schedule.

STANDARD RATE SCHEDULE   Rate Code No. 530

Large Power and Light (LPL)
(TITLE AND/OR NUMBER)

Availability: Available on an annual basis by written contract to any retail customer. This schedule is not available for resale, standby, breakdown, auxiliary, or supplemental service. It is optional with the customer whether service will be supplied under this rate or any other rate for which he is eligible. Once a rate is selected, however, service will continue to be supplied under that rate for a period of 12 months unless a material and permanent change in the customer's load occurs.

Service will be supplied from an existing transmission facility operating at a standard transmission voltage of 69 kV or higher by means of not more than one transformation to a standard distribution voltage of not less than 2.4 kV. Such transformation may be owned by the company or customer. Service may be supplied by means of an existing primary distribution facility of at least 24 kV when such facilities have sufficient capacity.

Service will be furnished in accordance with the company's rules, regulations, and conditions of service and the rules and regulations of the Oklahoma Corporation Commission.

Net rate: Capacity charge:

$13,750.00:   net per month for the first 2500 kilowatts (kW) or less of
              billing demand

Figure 3-11. Large Industrial Rate Schedule
(Courtesy of Oklahoma Gas and Electric Co.)

$4.20: net per month per kilowatt (kW) required in excess of
        2500 kW of billing demand

$.50:   net per month for each reactive kilovoltampere (kVAR)
        required above 60% of the billing demand

Plus an energy charge:

2.700¢: net per kilowatt-hour (kWh) for the first 1 million kWh
         used per month

2.570¢: net per kilowatt-hour (kWh) for all additional use per
         month

Determination of monthly billing demand: The monthly billing demand
shall be the greater of (a) 2500 kW, (b) the monthly maximum kilowatt
(kW) requirement, or (c) eighty percent (80%) of the highest monthly
maximum kilowatt (kW) requirement established during the previous 11
billing months. The monthly maximum reactive kilovoltampere (kVAR)
required are based on 30-min integration periods as measured by appropriate demand indicating or recording meters.

Determination of minimum monthly bill: The minimum monthly bill
shall consist of the capacity charge. The monthly minimum bill shall be
adjusted according to adjustments to billing and kVAR charges. If the
customer's load is highly fluctuating to the extent that it causes interference with standard quality service to other loads, the minimum monthly
bill will be increased $.50/kVA of transformer capacity necessary to cor-
rect such interference.

Terms of payment: Payment is due within 10 days of the date of mailing
the bill. The due date will be shown on all bills. A late payment charge will
be assessed for bills not paid by the due date. The late payment charge
shall be computed at 1-1/2 % on the amount past due per billing period.

Adjustments to billing:

1. Fuel cost adjustment: The rate as stated above is based on an average
    cost of $2.00/million Btu for the cost of fuel burned at the company's
    thermal generating plants. The monthly bill as calculated under the
    above rate shall be increased or decreased for each kWh consume by
    an amount computed in accordance with the following formula:

Figure 3-11. (*Continued*)

$$FA = A \times \frac{B}{10^6} \times C$$

where   FA=   fuel cost adjustment factor (expressed in dollars per kWh) to be applied per kWh consumed

A=   weighted average Btu/kWh for net generation from the company's thermal plants during the second calendar month preceding the end of the billing period for which the kWh usage is billed

B=   amount by which the average cost of fuel per million Btu during the second calendar month preceding the end of the billing period for which the kWh usage is billed exceeds or is less than $2.00/million Btu; any credits, refunds, or allowances on previously purchased fuel received by the company from any source shall be deducted from the cost of fuel before calculating B each month

C=   ratio (expressed decimally) of the total net generation from all the company's thermal plants during the second calendar month preceding the end of the billing period for which the kWh usage is billed to the total net generation from all the company's plants including hydrogeneration owned by the company, or kWh produced by hydrogeneration and purchased by the company, during the same period

2.   Tax adjustment: If there shall be imposed after the effective date of this rate schedule, by federal, state, or other governmental authority, any tax, other than income tax, payable by the company upon gross revenue, or upon the production, transmission, or sale of electric energy, a proportionate share of such additional tax or taxes shall be added to the monthly bills payable by the customer to reimburse the company for furnishing electric energy to the customer under this rate schedule. Reduction likewise shall be made in bills payable by the customer for any decrease in any such taxes.

Additionally, any occupation taxes, license taxes, franchise taxes,

Figure 3-11. (*Continued*)

and operating permit fees required for engaging in business with any municipality, or for use of its streets and ways, in excess of two percent (2%) of gross revenues from utility business done within such municipality, shall be added to the billing of customers residing within such municipality when voted by the people at a regularly called franchise election. Such adjustment to billing shall be stated as a separate item on the customer's bill.

Figure 3-11. (*Concluded*)

3.1.8 Cogeneration and Buy-Back Rates

Since enactment of the Public Utility Regulatory Policy Act of 1978 (PURPA), there has been significant renewed interest in on-site-generated power. This can be from cogeneration (on-site generation of thermal heat with concurrent, sequential generation of electricity), windmills, solar thermal, solar photovoltaics, or other sources. Generation of this energy for use only on site is often not cost effective due to variability of loads. Resale of excess electricity (when it is available) to the local utility, however, often makes a non-utility electric generation project economically feasible.

PURPA specified that cogenerators that met certain minimum conditions would be designated as Qualifying Facilities (QFs) and would be paid Avoided Costs by the purchasing utilities. To comply with these requirements of PURPA, utilities have developed buy-back rates for this excess electricity. Since the value of this energy may be either less than or greater than the cost to the utility of generating it, buy-back usually requires a separate meter and a separate rate schedule.*

Cogeneration can be an attractive energy cost-saving alternative for facilities that need both electric power and large amounts of steam or hot water. The combined production of electricity and thermal energy can result in fuel savings of 10-30 percent over the separate generation costs. Cogeneration will be discussed again in a later chapter.

3.1.9 Others

Many other rate schedules are being developed as the needs dictate. For example, some utilities have a rate schedule involving interruptible and curtailable loads. An interruptible load is one that can be turned off at

---

*Remember, the time when the industry generates an excess of electricity is probably not a peak time, so the utility really does not need the power as badly.

certain times of the day or year. A utility offers a lower rate as an incentive to companies willing to help decrease the system demand during peaking times of the day or year.

A *curtailable* load is one that the company may be willing to turn off if given sufficient notice. For example, the utility may receive a weather forecast for extreme heat or extreme cold which would result in a severe peaking condition. It may then call its curtailable customers and ask that all the curtailable loads be turned off. Of course, the utility is willing to compensate the customers for this privilege too.

In both cases, the utility compensates its customers for these loads by offering a reduction in the bill. In the case of curtailable loads, the rate reduction occurs every month during the peaking season whether or not the utility actually calls for the turnoff. In the case of interruptible loads, the basic rate is much lower to start with.

3.2   NATURAL GAS

Natural gas rate schedules are similar in structure to electric rate schedules, but they are often much simpler. Natural gas companies also experience a peaking problem. Theirs is likely to occur on very cold winter days and/or when supply disruptions exist. Due to the unpredictable nature of these peak problems, gas utilities normally do not charge for peak demand. Instead, customers are placed into *interruptible priority classes*.

A customer with a high priority will not be curtailed or interrupted unless absolutely necessary. A customer with the lowest priority, however, will be curtailed or interrupted whenever a shortage exists. Normally some gas is supplied to keep customer's pipes from freezing and pilot lights burning. To encourage use of the low-priority schedules, utilities charge significantly less for this gas rate. Most gas utilities have three or four priority levels. Some utilities allow customers to choose their own rate schedule, while others strictly limit the choice.

Figure 3-12 presents a sample rate schedule for four priority levels. Here the industrial customer is limited in choice to priorities 3 and 4.

Some points are demonstrated in this collection of schedules. First, the energy costs decrease as the priority goes down, but the probability of a curtailment or interruption dramatically increases. Second, the winter residential rate has an increasing block component on the block of gas use over 10 Mcf/month. Only very large residential consumers would approach this block, so its intent is to discourage wanton utilization. Like electric rates, fuel cost adjustments do exist in gas rates. Sales taxes also

Residential		**Commercial**	
Priority 1		**Priority 2**	
<u>Winter</u>		<u>Winter</u>	
First 1 ccf/mo	$5.12	First 1 ccf/mo	$6.79
Next 2.9 Mcf/mo	$5.347/Mcf	Next 2.9 Mcf/mo	$5.734/Mcf
Next 7 Mcf/mo	$3.530/Mcf	Next 7 Mcf/mo	$5.386/Mcf
Over 10 Mcf/mo	$3.725/Mcf	Next 90 Mcf/mo	$4.372/Mcf
		Next 1900 Mcf/mo	$4.127/Mcf
<u>Summer</u>		Next 6000 Mcf/mo	$3.808/Mcf
First 1 ccf	$5.12/Mcf	Over 8000 Mcf/mo	$3.762/Mcf
Next 2.9 Mcf/mo	$5.347/Mcf		
Over 3 Mcf/mo	$3.633/Mcf		
		<u>Summer</u>	
		First 1 ccf	$6.79
		Next 2.9 Mcf/mo	$5.734/Mcf
		Next 7 Mcf/mo	$5.386/Mcf
		Next 90 Mcf/mo	$4.372/Mcf
		Next 100 Mcf/mo	$4.127/Mcf
		Next 7800 Mcf/mo	$3.445/Mcf
		Over 8000 Mcf/mo	$3.399/Mcf
Industrial		**Industrial**	
Priority 3		**Priority 4**	
(Second Interruptible)		**(First Interruptible)**	
First 1 ccf	$19.04	First 4000 Mcf/mo or	
Next 2.9 Mcf/mo	$5.490/Mcf	fraction thereof	$12,814.00
Next 7 Mcf/mo	$5.386/Mcf	Next 4000 Mcf/mo	$3.168/Mcf
Next 90 Mcf/mo	$4.372/Mcf	Over 8000 Mcf/mo	$3.122/Mcf
Next 100 Mcf/mo	$4.127/Mcf		
Next 7800 Mcf/mo	$3.445/Mcf		
Over 8000 Mcf/mo	$3.399/Mcf		

Summer periods include the months from May through October.

Winter periods include the months from November through April.

Figure 3-12. Gas Schedules for One Utility
(Courtesy Oklahoma Natural Gas Company)

apply to natural gas bills. Again, some states do not charge sales tax on gas used directly in production.

Natural gas rates differ significantly in different parts of the country. Gas is relatively cheap in the producing areas of Oklahoma, Texas and Louisiana. It is much more expensive in other areas where it must be transported over long distances through transmission pipes. For example, in Florida, gas is almost twice as expensive as shown in the rate structure of Figure 3-12. Gas supplied by Gainesville Regional Utilities (Gainesville, FL) is priced under a flat rate structure (i.e. it does not drop in price with increased use). It costs around $6.00/Mcf for residential use, and almost $5.00/Mcf for commercial use. Interruptible gas service for larger users costs about $4.00/Mcf.

3.3   FUEL OIL AND COAL

Fuel oils are a very popular fuel source in some parts of the country, but they are rarely used in others. Natural gas and fuel oil can generally be used for the same purpose so the availability and price of each generally determines which is used.

Fuel oils are classified as *distillates* or *residuals*. This classification refers to the refining or distillation process. Fuel oils Number 1 and 2 are distillates. No. 1 oil can be used as a domestic heating oil and diesel fuel. No. 2 oil is used by industry and in the home. The distillates are easier to handle and require no heat to maintain a low viscosity; therefore, they can be pumped or poured with ease.

Residual fuel oils include Nos. 4, 5 and 6. Optimum combustion is more difficult to maintain with these oils due to variations in their characteristics that result from different crude oil origins and refining processes. No. 6 or residual bunker C is a very heavy residue left after the other oils have been refined. It has a very high viscosity and must be heated in cold environments to maintain a *pour point* (usually somewhere around 55°F).

The sulfur content of fuel oil normally ranges from .3 to 3.0 percent. Distillates have lower percentages than residuals unless the crude oil has a very high sulfur level. Sulfur content can be very important in meeting environmental standards and thus should be watched carefully.

Billing schedules for fuel oils vary widely among geographical areas of the country. The prices are set by market conditions (supply vs. demand), but within any geographical area they are fairly consistent. Within each fuel oil grade, there is a large number of sulfur grades, so shopping around can sometimes pay off. Basically, the price is simply a flat charge per gallon, so the total cost is the number of gallons used times the price per gallon.

Like fuel oil, coal comes in varying grades and varying sulfur content. It is, in general, less expensive than fuel oil per Btu, but it does require higher capital investments for pollution control, coal receiving and handling equipment, storage, and preparation. Coal is priced on a per ton basis with provisions for or consideration of sulfur content and percent moisture.

Finally, coal does not burn as completely as other fuels. If combustion air is properly controlled, natural gas has almost no unburned combustibles, while fuel oil has only a small amount. Coal, however, is much more difficult to fully combust.

3.4   STEAM AND CHILLED WATER

In some areas of the country, customers can purchase steam and chilled water directly instead of buying the fuel and generating their own. This can occur where there are large-scale cogeneration plants (steam), refuse-fueled plants (steam), or simple economics of scale (steam and/or chilled water). In the case of both steam and chilled water, it is normal to charge for the energy itself (pounds of steam or ton-hours of chilled water) and the demand (pounds of steam per hour or tons of chilled water). A sample hypothetical rate schedule is shown in Figure 3-13.

These rates are often competitive with costs of self-generated steam and chilled water. Purchasing steam and chilled water conserves considerable amounts of capital and maintenance monies. In general, when steam or chilled water is already available, it is worthy of consideration. The primary disadvantage is that the user does not have control of the generating unit. However, that same disadvantage is also true of electricity for most facilities.

3.5   WATER AND WASTEWATER

The energy analyst also frequently looks at water and wastewater use and costs as part of the overall energy management task. These costs are often related to the energy costs at a facility, and are also amenable to cost control. Water use should be examined, and monthly bills should be analyzed similarly to energy bills to see if unusual patterns of consumption are occurring. Water treatment and re-use may be cost effective in areas where water costs are high.

Wastewater charges are usually based on some proportion of the metered water use since the wastewater solids are difficult to meter. This can needlessly result in substantial increases in the utility bill for processes which do not contribute to the wastewater stream (e.g., makeup water for cooling towers and other evaporative devices, irrigation, etc.). A water meter can be installed at the service main to measure the loads not return-

Steam

Steam demand charge:

$1500.00/month for the first 2000 lb/h of demand or any portion thereof

$550.00/month/1000 lb/h for the next 8000 lb/h of demand

$475.00/month/1000 lb/h for all over 10,000 lb/h of demand

Steam consumption charge:

$3.50/1000 lb for the first 100,000 lb of steam per month

$3.00/1000 lb for the next 400,000 lb of steam per month

$2.75/1000 lb for the next 500,000 lb of steam per month

$2.00/1000 lb for the next 1 million lb of steam per month

Negotiable for all over 2 million lb of steam per month

Chilled water

Chilled water demand charge:

$2500.00/month for the first 100 tons of demand or any portion thereof

$15.00/month/ton for the next 400 tons of demand

$12.00/month/ton for the next 500 tons of demand

$10.00/month/ton for the next 500 tons of demand

$9.00/month/ton for all over 1500 tons of demand

(One ton is defined as 12,000 Btu/h, and an hour is defined as any 60 consecutive min.)

Chilled water consumption charge:

$.069/ton • h for the first 10,000 ton • h/month

$.06/ton • h for the next 40,000 ton • h/month

$.055/ton • h for the next 50,000 ton • h/month

$.053/ton • h for the next 100,000 ton • h/month

$.051/ton • h for the next 100,000 ton • h/month

$.049/ton • h for the next 200,000 ton • h/month

$.046/ton • h for the next 500,000 ton • h/month

Base rates: Consumption rates subject only to escalation of charges listed in conditions of service and customer instructions

Figure 3-13. Sample Steam and Chilled Water Rate Schedule

ing water to the sewer system. This can reduce the wastewater charges by up to two-thirds.

3.6   MONTHLY ENERGY BILL ANALYSIS

Once the energy rate structures have been examined, management should now understand how the company is being charged for the energy it uses each month. This is an important piece of the overall process of energy management at a facility. The next step in the examination of energy costs should be to review the bills and determine the average, peak and off-peak costs of energy used during at least the past twelve months.

Energy bills should be broken down into components that can be controlled by the facility. These cost components can be listed individually in tables and then plotted. For example, electricity bills should be categorized by demand costs per kW per month, and energy costs per kWh. The following example illustrates this analysis for an industry in Florida.

Example 3-2. A company in central Florida that fabricates metal products receives electricity from its electric utility at the following general service demand rate structure.

Rate structure:
(Minimum demand of 20 kW/month to qualify for rate)

Customer cost	= $21.00 per month
Energy cost	= $0.04 per kWh
Demand cost	= $6.50 per kW per month
Taxes	= Total of 8%
Fuel adjustment	= A variable amount per kWh each month (which may be a cost or a credit depending on actual fuel costs to the utility).

The electric energy use and costs for that company for a year are summarized below.

The energy analyst must be sure to account for all the taxes, the fuel adjustment costs, the fixed charges, and any other costs so that the true cost of the controllable energy cost components can be determined. In the electric rate structure described above, the quoted costs for a kW of demand and a kWh of energy are not complete until all these additional costs are added. Although the rate structure says that there is a basic charge of $6.50 per kW per month, the actual cost including all taxes is $7.02 per kW per month. The average cost per kWh is most easily obtained by taking the data for the 12-month period and calculating the cost over that period of time. Using the numbers from the table, one can see that this company has an average energy cost of ($42,628.51)/(569,360 kWh) = $0.075 per kWh.

Summary of Energy Usage and Costs

Month	kWh Used (kWh)	kWh Cost ($)	Demand (kW)	Demand Cost ($)	Total Cost ($)
Mar	44960	1581.35	213	1495.26	3076.61
Apr	47920	1859.68	213	1495.26	3354.94
May	56000	2318.11	231	1621.62	3939.73
Jun	56320	2423.28	222	1558.44	3981.72
Jul	45120	1908.16	222	1558.44	3466.60
Aug	54240	2410.49	231	1621.62	4032.11
Sept	50720	2260.88	222	1558.44	3819.32
Oct	52080	2312.19	231	1621.62	3933.81
Nov	44480	1954.01	213	1495.26	3449.27
Dec	38640	1715.60	213	1495.26	3210.86
Jan	36000	1591.01	204	1432.08	3023.09
Feb	42880	1908.37	204	1432.08	3340.45
Totals	569,360	$24,243.13	2,619	$18,385.38	$42,628.51
Monthly Averages	47,447	$2,020.26	218	$1,532.12	$3,552.38

The utility cost data are used initially to analyze potential Energy Management Opportunities (EMOs) and will ultimately influence which EMOs are recommended. For the example above, an EMO that reduces peak demand would save the company $7.02 per kW per month. Therefore, the energy analyst should consider EMOs that would involve using certain equipment during the night shift when the peak load is significantly lower than the first shift peak load. EMOs that save both energy and demand on the first shift would save costs at a rate of $0.075 per kWh. Finally, EMOs that save electrical energy during the off-peak shift should be examined too, but they may not be as advantageous; they would only save at the rate of $0.043 per kWh because they are already using off-peak energy and there would not be any additional demand cost savings.

The energy consumption should be plotted as well as tabulated to show the patterns of consumption pictorially. The graphs often display some unusual feature of energy use, and may thus help highlight periods of very high use. These high-use periods can be further examined to determine whether some piece of equipment or some process was being used much more than normal. The energy auditor should make sure that any discrepancies in energy use are accounted for. Billing errors can also show up on these plots, although such errors are rare in the authors' experience.

Figures 3-14 and 3-15 show graphs of the annual kilowatt-hour and kilowatt billing for the data from the preceding example. An energy auditor examining these graphs should ask a number of questions. Because the months of May through October are warm months in Florida, the kilowatt-hour use during these months would be expected to be higher than during the winter months. However, July shows unexpectedly low usage. In this case the company took a one-week vacation during July, and the plant energy consumption dropped accordingly. In other cases, this kind of discrepancy should be investigated, and the cause determined. The variations between December, January and February again bear some checking. In this example, the plant also experienced shutdowns in December and January. Otherwise, the facility's kilowatt-hour use seems to have a fairly consistent pattern over the twelve-month period.

Kilowatt use also needs some examination. The 18 kW jump from April to May is probably the result of increased air-conditioning use. However, the 9 kW drop from May to June seems odd especially since kilowatt-hour use actually increased over that period. One might expect demand to drop in July commensurate with the drop in energy use, but as long as the plant operated at normal capacity on any day during the month of July, it would be likely to establish about the same peak demand as it did in June. Other causes of large variations for some facilities can be related to meter reading errors, equipment and control system malfunctions, and operational problems.

3.7   ACTIONS TO REDUCE ELECTRIC UTILITY COSTS

Typical actions to reduce kWh consumption involve replacing existing lights with more efficient types; replacing electric heating and cooling

Figure 3-14

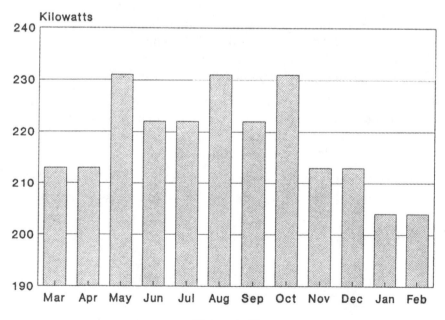

Figure 3-15

equipment with more efficient models; adding insulation to walls and ceilings; replacing motors with high efficiency models and using variable speed drives; recovering heat from air compressors, refrigeration units, or production processes to heat water for direct use or to pre-heat water for steam production; and replacing manufacturing or process equipment by more energy efficient models or processes.

Most of these actions will also result in demand reductions and produce savings through lower kW charges. Other actions that specifically reduce demand involve controlling and scheduling existing loads to reduce the peak kW value recorded on the demand meter. An energy management computer that controls demand is usually better than manual control or time-clock control. If several large motors, chillers, pumps, fans, furnaces or other high kW loads are in use at a facility, then electric costs can almost always be saved through demand limiting or control. All of these areas for savings will be examined in detail in the subsequent chapters.

Example 3-3. As an example of the savings that can be obtained by demand control, consider the use of four large machines at a separately metered production facility where each machine has a demand of 200 kW. The machines could be controlled by a computer which would limit the total demand to 400 kW at any one time. This company has chosen to limit the use of the machines by operational policy which states that no more than two machines should be turned on at any given time.

One morning at 8:00 am a new employee came in and turned on the two idle machines. At 8:30 am the plant foreman noticed that too many machines were running and quickly shut down the extra two machines. What did this employee's mistake cost the facility?

The immediate cost on the month's electric bill has two components. Using the demand rate from Example 3.2, the immediate cost is calculated as:

Demand cost increase   = 400 kW * $7.02/kW
                        = $2808

Energy cost increase   = 400 kW * 0.5 hr * $.043/kWh
                        = $8.60 for the energy.

If the utility rate structure includes a 70% demand ratchet, there would be an additional demand for the next 11 months of (.70 * 800 kW) – 400 kW = 160 kW. This would further increase the cost of the mistake as follows:

Ratchet cost increase	= 11 mo * \$7.02/kW mo * 160 kW
	= \$12,355.20.
Total cost of mistake	= \$2808 + \$8.60 + \$12,355.20
	= \$15,171.80

Under certain conditions a customer may be able to save money by shifting to another rate category [2]. Consider the example of a manufacturing facility which has a meter for the plant area and a separate meter for the office area. The plant is on a demand rate, but the office area is on a non-demand rate since it has a typical demand of around 19 kW for any month. Under the demand rate structure in section 3.6 customers are billed on the demand rate for one full year starting from any month in which their demand exceeds 20 kW for any 30 minute time window. If the office area could establish a very short peak demand of 20 kW or greater for one month it would automatically be shifted to the demand rate, and could likely benefit from the lower cost per kWh on that rate.

3.8   UTILITY INCENTIVES AND REBATES

Many utility rate structures include *incentives* and *rebates* for customers to replace old, energy inefficient equipment with newer, more energy efficient models. Utilities offer such incentives and rebates because it is cheaper for them to save the energy and capacity for new customers than it is to build new power plants or new gas pipelines to supply that additional load. In addition, stringent environmental standards in some areas makes it almost impossible for electric utilities to build and operate new facilities—particularly those burning coal. Helping customers install more energy efficient electrical and gas equipment allows utilities to delay the need for new facilities, and to reduce the emissions and fuel purchases for the units they do operate.

Direct incentives may be in the form of low interest loans that can be paid back monthly with energy savings resulting from the more efficient equipment. Incentives may also be in the form of lower rates for the electricity used to run higher efficiency lights and appliances, and more efficient process equipment. Other incentives include free audits from the utilities and free technical assistance in identifying and installing these energy efficiency improvements.

Indirect incentives also exist, and are often in the form of a special rate for service at a time when the utility is short of capacity, such as a time of day rate or an interruptible rate. The time of day rate offers a lower cost of electricity during the off-peak times, and often also during the off-season times. Interruptible rates allow large use customers to purchase electricity at very low rates with the restriction that their service can be interrupted on short notice. (See Section 3.1.9 for a discussion of interruptible and curtailable loads.)

Rebates are probably the most common method that utilities use to encourage customers to install high efficiency appliances and process equipment. Utilities sometimes offer a rebate tied to the physical device—such as $1.00 for each low-wattage fluorescent lamp used or $10.00 per horsepower for an efficient electric motor. Other rebates are offered for reductions in demand—such as $250 for each kW of demand that is eliminated. Metering or other verification techniques may be needed to insure that the proper kW reduction credit is given to the customer. The load management rate structure shown in Figure 3-7 is a form of rebate for residential customers.

Incentives or rebates can substantially improve the cost effectiveness of customer projects to replace old devices with new, high efficiency equipment. In some cases, the incentives or rebates may be great enough to completely pay for the difference in cost in putting in a high efficiency piece of equipment instead of the standard efficiency model. Subsequent chapters provide additional discussion and examples of utility incentives and rebates, and how they affect equipment replacement decisions.

3.9   ELECTRIC UTILITY COMPETITION AND DEREGULATION

Within a short time of the passage of the national Energy Policy Act of 1992 (EPACT), electric utility interest in DSM programs and levels of rebates and incentives began to decline. EPACT contained a provision that mandated open transmission access—that is, requiring competing utilities to open up their transmission systems to wholesale transactions and wholesale wheeling of power between utilities. EPACT left the issue of retail access—or retail wheeling—in the hands of individual state regulatory agencies. However, EPACT left no doubt that utility deregulation and competition was coming. Within two years of the passage of EPACT, a dozen or more states were actively pursuing retail wheeling experiments or retail wheeling legislation.

Utilities quickly began to restructure their businesses, and began massive cost reductions. The utilities were preparing for becoming the "lowest cost supplier" to their customers, and they feared that some other utility might have lower costs and could eventually capture many of their largest and most lucrative customers. DSM programs were scaled back by most utilities, and were eliminated by others. Rebates and incentives were reduced or eliminated by many utilities, as these were perceived as unnecessary costs and activities in the face of the coming deregulation and retail wheeling. A number of utilities have kept their rebates or incentives, so some of these are still available.

Utility restructuring initially proceeded at a fast pace in those states with high electricity prices, and at a much slower pace in states with low prices. California, New York and Massachusetts had some of the most advanced restructuring plans as of early 1997. California had a restructured competitive market by March 1, 1998. New York had wholesale competition in early 1997 and retail access in early 1998. Massachusetts had retail choice in January 1998. Vermont was not far behind.

However, events of 2000 and 2001 greatly dampened the enthusiasm for restructuring or deregulation, when western state energy price spikes and shortages occurred. Several energy supply companies were found to have created artificial shortages and supply restrictions. As a result, many states delayed, suspended or repealed their legislation and rules on customer choice of their electric supplier. In September 2001, California suspended its retail access policy, and is not expected to have a replacement policy until the time frame of 2008. Even talk of federal "re-regulation" has been common.

However, eventually, many of the same marketing features available for years in deregulated natural gas purchasing will become available for purchasers of electric power. These include the use of brokers, power marketers and risk managers. The brokers will arrange for the purchase of power for customers and the purchase and sale of power for customers who are self-generators. Power marketers will purchase the rights to certain amounts of electric power and will then re-sell it to other purchasers. Risk managers will basically sell futures contracts to help purchasers achieve stable longer-term prices.

This eventual move to deregulation and retail wheeling will have a significant effect on future electricity prices for all large customers. Prices will decline for all large customers, and will decline significantly for many larger facilities. Instead of projecting higher electric costs in the future, these customers should be projecting lower costs. Medium-sized customers should also see cost reductions. The impact on small customers—particularly residential customers—is still unknown in most cases. Some will also see lower prices, but some are likely to actually experience price increases.

In the near term, some energy managers for larger and medium-sized organizations may find themselves more involved with utility programs to provide lower cost electricity than with programs to save energy and demand. Some energy efficiency projects may even have to wait for approval until the final economic analysis can be calculated with the new cost of electricity. However, when the pricing of electricity settles down to its final range in the near future, energy managers will find themselves back in the previous business of finding ways to implement new equipment and processes to save energy and demand.

3.10 SUMMARY

In this chapter we analyzed rate schedules and costs for electricity in detail. We also examined rates and costs for natural gas, coal, fuel oil, steam, and chilled water. A complete understanding of all the rate schedules is vital for an active and successful energy management program.

In the past, few managers have understood all of the components of these rate schedules, and very few have even seen their own rate schedules. The future successful manager will not only be familiar with the terms and the schedules themselves, but he or she will also likely work with utilities and rate commissions toward fair rate-setting policies.

References
1.   Resource: *An Encyclopedia of Utility Industry Terms, Pacific Gas and Electric Company*, San Francisco, CA, 1985.
2.   Capehart, Barney L. and John F. Mahoney, "Minimizing Commercial Electric Bills Through Choice of Rate Structures," *Strategic Planning and Energy Management*, Vol. 9, No. 2, 1989.

Bibliography
Berg, Sanford V., Editor, *Innovative Electric Rates*, Lexington Books, D.C Heath and Company, Lexington, MA, 1983.

Berg, Sanford V. and John Tschirhart, *Natural Monopoly Regulation: Principles and Practices*, Cambridge University Press, 1988.

Bonbright, James C., Albert L. Danielson, and David B. Kamerschen, *Principles of Public Utility Rates*, Public Utility Reports, Inc., Arlington, VA, 1988.

Capehart, Barney, Editor, *Encyclopedia of Energy Engineering and Technology*," Taylor and Francis/CRC Publishing Company, Boca Raton, FL, July, 2007.

Crew, Michael A. and Paul R. Kleindorfer, *The Economics of Public Utility Regulation*, The MIT Press, Cambridge, MA, 1986.

Enholm, Gregory B. and J. Robert Malico, "Electric Utilities Moving into the 21st Century", Public Utility Reports, Arlington, VA, 1994.

Hirsh, Richard F., *Technology and Transformation in the American Electric Utility Industry*, Cambridge University Press, 1989.

Howe, Keith M. and Eugene F. Rasmussen, *Public Utility Economics and Finance*, Prentice-Hall, Inc., Englewood Cliffs, NJ, 1982.

Zajac, Edward E., *Fairness or Efficiency—An Introduction to Public Utility Pricing*, Ballinger Publishing Co., Cambridge, MA, 1978.

Chapter 4

Economic Analysis and
Life Cycle Costing*

4.1 INTRODUCTION

Once an energy management opportunity (EMO) has been identified, the energy manager must determine the cost-effectiveness of the EMO in order to recommend it to management for implementation, and to justify any capital expenditure for the project. If a group of EMOs have been identified, then they should be ranked on some economic basis, with the most cost-effective ones to be implemented first. There are many measures of cost-effectiveness, and sometimes businesses and industries use their own methods or procedures to make the final decisions. The basic elements of cost-effectiveness analysis are discussed in this chapter and some of the common techniques or measures of cost-effectiveness are presented. The emphasis in this chapter is on techniques which use the time value of money, and which calculate the life cycle cost (LCC) of the project, which in our case is the total cost of purchasing and operating a piece of energy consuming equipment or process over its entire lifespan.

The chapter begins by discussing the various types of costs that a project can incur. Section 4.2 discusses "simple" economic analysis methods. Sections 4.3-4.7 develop and discuss material fundamental to the concept of the time value of money and discounted cash flow analysis. Section 4.8 presents decision making based on the time value of money. Sections 4.9 and 4.10 present fundamental life cycle cost concepts, and demonstrate the ease with which they can be applied in evaluating capital investments for energy-related projects. Incorporating taxes and inflation into an economic analysis is discussed in sections 4.11 and 4.12, and potential sources of capital are discussed in section 4.13. A final section on the use of computer software to aid in decision making concludes the chapter.

---

*This chapter was substantially revised and updated by Dr. Camille DeYong, Department of Industrial Engineering and Management, Oklahoma State University.

4.2   COSTS

Organizations incur various types of costs. These are generally classified into two broad categories: expenses, and capital investments. Expenses are the routine, ongoing costs that are necessary for conducting business or operations. Capital investments have four important characteristics. First, they are relatively large, ranging from several thousand to several million dollars, depending on the size of the organization. Second, the benefits of a capital investment are returned over the lifetime of the investment, which is typically several years. Third, capital investments are relatively irreversible. After the initial investment has been made, altering the project significantly, or terminating the project, has substantial (usually negative) implications. Fourth, capital investments can have significant tax implications, depending on the choice of financing methods [1].

4.2.1   Categories of Costs for Capital Investments

The costs incurred for a capital investment can almost always be placed in one of the following categories: acquisition, utilization, and disposal. Acquisition costs are the initial, or first, costs which are necessary to prepare the investment for service. These costs include the purchase price, installation costs, training costs, and charges for engineering work that must be done, permits that must be obtained, or renovations that must be made before the project can be started.

Methods for estimating acquisition costs range from using past experience, to obtaining more precise estimates from vendors, catalogs and databases. The desired accuracy of the estimate depends on where the data will be used. If the data is being collected for determining the feasibility of the project, approximations may be sufficient. If the data is being collected to obtain financing, the estimates must be as accurate as possible.

Utilization costs are those required on a routine basis for operating and maintaining the investment. These include energy, maintenance and repair. Utilization costs can be direct or indirect. Direct costs include labor and materials for routine repair and maintenance. Indirect costs are the costs not directly attributable to the project but necessary for conducting business. Energy costs are usually included in this category. Indirect costs are also often referred to as "overhead." Examples of indirect costs include salaries for staff personnel, janitorial costs, and cleaning supplies. Estimates for utilization costs can be obtained from databases or professional experience.

Our primary interest is in finding energy related costs and performing the economic analysis for evaluating which energy projects represent good investments, and should be implemented. So far we have looked at the general process for conducting an energy audit to determine the energy sources and costs at our facility. It is necessary to have all of the energy bills and to know the rate structure for understanding how our facility is charged for use of those energy sources. Finally, we must have some degree of knowledge of new equipment, processes and technologies that can replace existing ones to save energy and reduce energy costs. The techniques in the following chapters present this information for lighting, air conditioning, motors, boilers and steam systems, insulation and control systems. Using this information the energy manager can determine potential EMOs and can perform the energy analysis to come up with the energy costs and savings of the various alternatives.

Disposal costs are those incurred (or recovered) when the project has reached the end of its useful life. These are costs required to retire or remove the asset. These costs are often referred to as the salvage value if the project has a positive worth at the end of its lifetime. The actual salvage value may depend on many factors, including how well the asset was maintained and the market for used equipment. Estimates for disposal costs/benefits are often obtained from experienced judgments and current values of such used equipment.

4.2.2 Cash Flow Diagrams and Tables

It is often helpful to present costs visually. This can be accomplished with a tool called a cash flow diagram. Alternatively a cash flow table could be used. A cash flow diagram is a pictorial display of the costs and revenues associated with a project. An interest rate or discount rate can also be shown. Costs are represented by arrows pointing down, while revenues are represented by arrows pointing up. The time periods for the costs dictate the horizontal scale for the diagram. Although some costs occur at different points in time, most of the time, an end-of-year approach is sufficient. This chapter will use end-of-year cash flows for all analyses.

Figure 4-1 is a cash flow diagram to look at the costs and benefits of purchasing a heat pump. The costs/benefits associated with the heat pump are:

- The heat pump costs $10,000 initially
- The heat pump saves $2,500 per year in energy costs for 20 years
- The maintenance costs are $500 per year for 20 years
- The estimated salvage value is $500 at the end of 20 years.

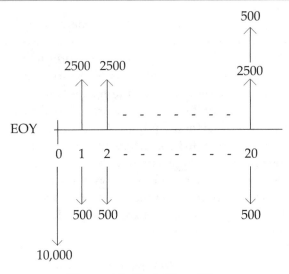

Figure 4-1. Cash Flow Diagram

Costs can also be listed in a tabular format as illustrated in Table 4-1.

Table 4-1. Cash Flow in Tabular Form

End of Year (EOY)	Cash Flow
0	–$10,000
1-19	$2,500 – $500 = $2,000
20	$2,500 – $500 + $500 = $2,500

4.3   SIMPLE PAYBACK PERIOD COST ANALYSIS

One the most commonly used cost analysis methodologies is the Simple payback Period (SPP) analysis. Also called the Payback Period (PBP) analysis, the SPP determines the number of years required to recover an initial investment through project returns. The formula is:

$$\text{SPP} = (\text{Initial cost})/(\text{Annual savings}) \qquad (4\text{-}1)$$

Example 4.1

The heat pump discussed above has an initial cost of $10,000, an energy savings of $2,500 per year, and a maintenance cost of $500 per year. Thus, the net annual savings is $2,000. Therefore, its SPP would be ($10,000)/($2,000/yr) = 5 years.

The advantage of the SPP is its simplicity, and it is easily understood by workers and management. It does provide a rough measure of the worth of a project. The primary disadvantages are: 1) the methodology does not consider the time value of money; and 2) the methodology does not consider any of the costs or benefits of the investment following the payback period. No specific lifetime estimate of the project is required, but it is assumed that the lifetime is longer than the SPP. These limitations mean the SPP tends to favor shorter-lived projects, a bias that is often economically unjustified. However, using the SPP in conjunction with one of the discounted cash flow methodologies discussed later in this chapter can provide a better understanding of an investment's worth.

4.4   ECONOMIC ANALYSIS USING THE TIME VALUE OF MONEY: DISCOUNTED CASH FLOW ANALYSIS

4.4.1 Introduction

Most people understand that money changes value over time. If a person is offered $1,000 today, or $1,000 in one year, almost everyone would choose the $1,000 today. The decision-maker clearly places higher value on money today. This decision occurs for a variety of reasons, including how much the person needs the money, but a primary factor is the recognition that $1,000 today is worth more than $1,000 will be worth in one year.

The change in worth is due to two primary factors: interest (opportunity cost) and inflation. **Interest** is the return earned on money when someone else uses it. If $1,000 were deposited in a bank or invested in a financial instrument (stock, mutual fund, bond, etc.), then in one year, an amount greater than $1,000 would be available for withdrawal. **Inflation** is a decrease in the purchasing power of money. In other words, $1,000 will buy more at a store today than it will in one year. Inflation is discussed in greater detail in Section 4.12 and will not be incorporated into our economic analyses until that time. The effects of both interest and inflation are important to consider in a full economic analysis of energy projects.

Energy management expenditures are typically justified in terms of avoided energy costs. Expenses come at the beginning of a project, while savings/benefits occur later. The sums of money for costs and revenues that need to be compared are paid or received at different points in time. Because money changes value over time, these sums of money, or cash flows, *should not be directly added together unless they occur at the same point in time.* In order to correctly add these cash flows at different times, we need to reduce them to a common basis through the use of the interest rate, also called the discount rate. This method of treatment is called discounted cash flow analysis, and first requires us to understand the mathematics of interest and discounting.

4.4.2 The Mathematics of Interest and Discounting
Two factors affect the calculation of interest: the amount and the timing of the cash flows. The basic formula for calculating interest is:

$$F_n = P + I_n \tag{4-2}$$

where: F_n = a future cash flow of money at the end of the nth year
         P   = a present cash flow of money
         I_n = the interest accumulated over n years
         n   = the number of years between P and F

Interest is stated as a percentage rate that is to be paid for the use of the money for a time period, usually years. Although interest rates can have other compounding periods, this chapter will be limited to yearly interest, and end of year cash flows, for simplicity. A good discussion of interest and cash flows occurring more frequently is available in Thuesen [2]. The term "discount rate" is used to reflect the fact that future sums of money must be discounted by the interest rate to find their initial starting values, or present values. The discount rate is often company-specific. It must be supplied by the company accounting department or corporate accounting level, and is usually known for evaluating investments in new product lines or new facilities. This interest rate is often referred to as the Minimum Attractive Rate of Return (MARR).

4.4.3 Simple Interest
There are two primary types of interest: simple and compound. Simple interest is earned (charged) only on the original principal amount, and paid once, at the end of the time period. The formula is:

   I = $P \times n \times i$, where

   I = the interest accumulated over n years
   P = the original principal amount
   n = the number of interest periods (often measured in years)
   i = the interest rate per period

The general formula for the total amount owed or due at the end of a loan (investment) period of n years when using simple interest is:

$$F_n = P + I$$
$$= P + P \times n \times i$$
$$= P(1 + n \times i) \tag{4-3}$$

Example 4.2

The ABC Corporation wants to borrow $10,000 for 5 years at a simple interest rate of 18%/year. How much interest would be owed on the loan?

$$I = P \times n \times i$$
$$= (\$10,000)(5 \text{ years})(.18/\text{year})$$
$$= \$9,000$$

The total amount owed at the end of the 5 years would be the principal, $10,000 plus the interest, $9,000, for a total of **$19,000**.

4.4.4 Compound interest

Interest which is earned (charged) on the accumulated interest as well as the original principal amount is said to be compounded. In other words, the interest which accumulates at the end of the first interest period is added to the original principal amount to form a new principal amount due at the end of the next period. The power (or penalty) of compound interest is illustrated in Table 4-2.

Table 4-2. Compound Interest

Year	A Amount owed at the beginning of the year	B = i × A Interest owed at the end of the year	C = A + B Total amount owed at the end of the year
1	P $10,000	P ×i $10,000 ×.18 = $1,800	P + P ×i $11,800
2	P + P ×i $11,800	(P + P ×i) ×i $11,800 ×.18 = $2,124	$13,924
3	$13,924	$13,924 ×.18 = $2506	$16,430
4	$16,430	$2957	$19,387
5	$19,387	$3490	**$22,877**

The effect of compound interest is pronounced. When using simple interest, the amount owed at the end of the loan period for the principal amount of $10,000 was $19,000. With compound interest, the amount owed at the end of the same loan period is $22,877. Clearly, a borrower would prefer simple interest, but a lender would prefer compound interest. Since compound interest is used far more commonly in practice than simple interest, the remainder of this chapter uses compound interest.

Formulas for calculating compound interest have been derived for several patterns of cash flows. These formulas are discussed later in this chapter.

4.5   DISCOUNTED CASH FLOWS:
BASICS AND SINGLE SUM ANALYSES

Once we understand the effects of interest or discounting from Section 4.4, we can proceed with our goal of finding a method which correctly allows the addition of two cash flows occurring at different points in time. The general approach used is to reduce cash flows occurring at different times to a common basis through the use of the interest rate or discount rate. This method of treatment is called discounted cash flow analysis, and it is a fundamental approach *that is necessary to use* to correctly account for energy costs and savings in different years.

This section presents the formulas and compound interest factors used to convert a sum of money from its value in one time period to its corresponding value in another time period. In this chapter, interest is assumed to be compounded annually, and cash flows are assumed to occur at the end of the year. The following notation will be used:

i   =   Annual interest rate (or discount rate. Also known as the minimum attractive rate of return or MARR)

n   =   The number of annual interest periods (in our case, the number of years)

P   =   A present value (or present worth)

A   =   A single payment in a series of n equal annual payments

F   =   A future value (or future worth)

As stated in Thuesen [2], four important points apply in the use of discounted cash flow analysis:

1)   The end of one year is the beginning of the next.
2)   P is at the beginning of a year at a time regarded as being the present.

3)   F is at the end of the nth year from a time regarded as being the present.
4)   An A occurs at the end of each year of the period under consideration.

When P and A are involved, the first A of the series occurs one year after P.

When F and A are involved, the last A of the series occurs simultaneously with F.

4.5.1 Single sum, future worth

The first task is to determine how to convert a single sum of money from a present amount to a future amount. This is similar to asking the question "If I borrow $5,000 at 10% interest for five years, how much will I owe at the end of the five years?" In this problem, the present amount, P, and the interest rate are known. The unknown is the future amount, F. The formula for finding F is:

$$F = P(1+i)^n \qquad\qquad (4\text{-}4)$$

The term $(1+i)^n$ is one of the six compound interest factors that are commonly used and tabulated. It is known as the **single sum, future worth factor**, or the **single payment, future worth factor**. The term is also known as the factor $(F/P_{i,n})$, which is read: find F, given P, at i% for n periods (years in our case). It is also often referred to as the "F given P" factor. Both formulas and the compound interest factors will be presented in this chapter, although the factors will be used for most computations.

All compound interest factors are in the same format: $(a/b_{i,n})$, where:

a = unknown quantity—read as "Find a"
b = known quantity—read as "Given b"
i = known interest rate—read as "At i"
n = known number of time periods (usually years)—read as "For n periods of time"

See Table 4-3 for a summary of the formulas, factors, and sample cash flows for various combinations of finding P, F and A.

Although we are only discussing yearly interest and yearly time periods in this chapter, it is important to note that other interest periods can occur, such as quarterly. If this is the case, *the interest period and the time period must match.*

Since the compound interest factors are used extensively in Discounted Cash Flow analysis, numerical tables have been developed for various combinations of i and n. See Table 4-4 for an illustration of the table for i = 10%, with n ranging from 1 to 20 years. This table will be used in the examples in this section. Appendix A contains tables for interest rates other than 10%. Also, any engineering economic analysis text will have tables for a wider range of interest rates.

Table 4-3. Formulas and Factors for P, F, and A

Single Payment   F

Compound Amount:
To Find F
Given P $(F/P,i,n)$ $F = P(1 + i)^n$

Present Worth:
To Find P
Given F $(P/F,i,n)$ $P = F(1 + i)^{-n}$

Uniform Series
A A A A A

To Find F
Given A $(F/A,i,n)$ $F = A\left[\dfrac{(1+i)^n - 1}{i}\right]$

To Find A
Given F $(A/F,i,n)$ $A = F\left[\dfrac{i}{(1+i)^n - 1}\right]$

A A A A A

To Find A
Given P $(A/P,i,n)$ $A = P\left[\dfrac{i(1+i)^n}{(1+i)^n - 1}\right]$

To Find P
Given A $(P/A,i,n)$ $P = A\left[\dfrac{(1+i)^n - 1}{i(1+i)^n}\right]$

Example 4.3

If $5,000 was deposited in an account that paid 10% interest annually, how much would be in the account at the end of five years?
For this problem,

P = $5,000
i   = 10%
n = 5 years
F = ?

Using Equation 4.4,

$$F = 5,000(1 + .10)^5 = \textbf{\$8,053}$$

or using the F/P factor and Table 4-4, with i = 10% and n = 5,

$$F = 5,000(F/P_{10,5}) = 5,000(1.611) = \textbf{\$8,055}$$

The difference in the two answers is due to the rounding of the (F/P) factor in the tables, and is usually not significant.

4.5.2 Single sum, present worth

Using the formula for future worth (equation 4.4), the formula for finding the present worth of a single sum is found by solving for P. Thus, the **single sum, present worth** formula, or **single payment, present worth** formula, is:

$$P = F(1 + i)^{-n} \tag{4.5}$$

The term $(1 + i)^{-n}$ is called the present worth factor, $(P/F_{i,n})$, which is read: find P, given F, at i% for n periods (years).

Example 4.4

An energy manager expects a boiler to last 7 years, and he thinks it will cost about $150,000 to replace at that time. How much money should the company deposit today in an account paying 10% per year in order to have $150,000 available in 7 years?

For this problem:

F = $150,000
i = 10%
n = 7 years
P = ?

Using Equation 4.5,

$$P = \$150,000(1.1)^{-7} = \textbf{\$76,974}$$

Using the present worth factor $(P/F_{10,7})$ and Table 4-4,

$$P = \$150,000(P/F_{10,7}) = \$150,000(0.5132) = \textbf{\$76,980}$$

4.6   DISCOUNTED CASH FLOWS: UNIFORM SERIES

The next type of cash flow pattern to understand is a uniform series of costs or savings. A uniform series is a cash flow pattern with cash flows that are of the same magnitude, occurring at the end of several consecutive periods. We will be using years for our periods. Examples of uniform

Table 4-4. Interest Factors for i = 10%

	Single Sums			Uniform Series			Gradient Series	
n	To Find F Given P (F|P,i%,n)	To Find P Given F (P|F,i%,n)	To Find F Given A (F|A,i%,n)	To Find A Given F (A|F,i%,n)	To Find P Given A (P|A,i%,n)	To Find A Given P (A|P,i%,n)	To Find P Given G (P|G,i%,n)	To Find A Given G (A|G,i%,n)
1	1.1000	0.9091	1.0000	1.0000	0.9091	1.1000	0.0000	0.0000
2	1.2100	0.8264	2.1000	0.4762	1.7355	0.5762	0.8264	0.4762
3	1.3310	0.7513	3.3100	0.3021	2.4869	0.4021	2.3291	0.9366
4	1.4641	0.6830	4.6410	0.2155	3.1699	0.3155	4.3781	1.3812
5	1.6105	0.6209	6.1051	0.1638	3.7908	0.2638	6.8618	1.8101
6	1.7716	0.5645	7.7156	0.1296	4.3553	0.2296	9.6842	2.2236
7	1.9487	0.5132	9.4872	0.1054	4.8684	0.2054	12.7631	2.6216
8	2.1436	0.4665	11.4359	0.0874	5.3349	0.1874	16.0287	3.0045
9	2.3579	0.4241	13.5795	0.0736	5.7590	0.1736	19.4215	3.3724
10	2.5937	0.3855	15.9374	0.0627	6.1446	0.1627	22.8913	3.7255
11	2.8531	0.3505	18.5312	0.0540	5.4951	0.1540	26.3963	4.0641
12	3.1384	0.3186	21.3843	0.0468	6.8137	0.1468	29.9012	4.3884
13	3.4523	0.2897	24.5227	0.0408	7.1034	0.1408	33.3772	4.6988
14	3.7975	0.2633	27.9750	0.0357	7.3667	0.1357	36.8005	4.9955
15	4.1772	0.2394	31.7725	0.0315	7.6061	0.1315	40.1520	5.2789
16	4.5950	0.2176	35.9497	0.0278	7.8237	0.1278	43.4164	5.5493
17	5.0545	0.1978	40.5447	0.0247	8.0216	0.1247	46.5819	5.8071
18	5.5599	0.1799	45.5992	0.0219	8.2014	0.1219	49.6395	6.0526
19	6.1159	0.1635	51.1591	0.0195	8.3649	0.1195	52.5827	6.2861
20	6.7275	0.1486	57.2750	0.0175	8.5136	0.1175	55.4069	6.5081
21	7.4002	0.1351	64.0025	0.0156	8.6487	0.1156	58.1095	6.7189
22	8.1403	0.1228	71.4027	0.0140	8.7715	0.1140	60.6893	6.9189
23	8.9543	0.1117	79.5430	0.0126	8.8832	0.1126	63.1462	7.1085
24	9.8497	0.1015	88.4973	0.0113	8.9847	0.1113	65.4813	7.2881
25	10.8347	0.0923	98.3471	0.0102	9.0770	0.1102	67.6964	7.4580
26	11.9182	0.0839	109.1818	9.159E-03	9.1609	0.1092	69.7940	7.6186
27	13.1100	0.0763	121.0999	8.258E-03	9.2372	0.1083	71.7773	7.7704
28	14.4210	0.0693	134.2099	7.451E-03	9.3066	0.1075	73.6495	7.9137
29	15.8831	0.0630	148.6309	6.728E-03	9.3696	0.1067	75.4146	8.0489
30	17.4494	0.0573	164.4940	6.079E-03	9.4269	0.1061	77.0766	8.1762
36	30.9127	0.0323	299.1268	3.343E-03	9.6765	0.1033	85.1194	8.7965
42	54.7637	0.0183	537.6370	1.860E-03	9.8174	0.1019	90.5047	9.2188
48	97.0172	0.0103	960.1723	1.041E-03	9.8969	0.1010	94.0217	9.5001
54	171.8719	5.818E-03	1.709E+03	5.852E-04	9.9418	0.1006	96.2763	9.6840
60	304.4816	3.284E-03	3.035E+03	3.295E-04	9.9672	0.1003	97.7010	9.8023
66	539.4078	1.854E-03	5.384E+03	1.857E-04	9.9815	0.1002	98.5910	9.8774
72	955.5938	1.046E-03	9.546E+03	1.048E-04	9.9895	0.1001	99.1419	9.9246
120	9.271E+04	1.079E-05	9.271E+05	1.079E-06	9.9999	0.1000	99.9860	9.9987
180	2.823E+07	3.543E-08	2.823E+08	3.543E-09	10.0000	0.1000	99.9999	10.0000
360	7.968E+14	1.255E-15	7.968E+15	1.255E-16	10.0000	0.1000	100.0000	10.0000

series of costs are car payments, house payments, or any other type of regular payment. Most of our energy management projects, or EMOs, produce a uniform series of savings in future energy costs.

Four types of "conversions" will be addressed:

1) Given a present amount, find the equivalent uniform series. This is known as find A, given P, and is denoted by the factor $(A/P_{i,n})$.

2) Given a future amount, find the equivalent uniform series. This is known as find A, given F, and is denoted by the factor $(A/F_{i,n})$.

3) Given a uniform series, find the equivalent present worth. This is known as find P, given A, and is denoted by the factor $(P/A_{i,n})$.

4) Given a uniform series, find the equivalent future worth. This is
    known as find F, given A, and is denoted by the factor $(F/A_{i,n})$.

Formulas for all series conversions are given in Table 4-3. Additional
compound interest factors will be used in the examples, since the series
formulas are more complex than earlier formulas. Examples of each type
of series conversion follow.

Example 4.5

An energy manager has $5,000 available today to purchase a high
efficiency air conditioner with a life of six years. She would like to know
what energy cost savings would be needed each year to justify this project
if the company MARR is 10%.

$$P = \$5,000$$
$$i = 10\%$$
$$n = 6$$
$$A = ?$$

This problem is the first type of conversion: find A, given P. Therefore:

$$A = \$5,000(A/P_{10,6})$$
$$= \$5,000(.2296)$$
$$= \mathbf{\$1,148}$$

Thus, if the air conditioner produces annual energy cost savings of $1148
or greater, the company will earn its MARR at the least. If the savings is
greater than $1148, the actual rate of return will be greater than 10%.

Example 4.6

A heat pump is expected to produce energy cost savings of $1,500
per year over a lifetime of 20 years. What is the equivalent present sum or
present worth for this series of cash flows, if the company MARR is 10%?

$$A = \$1,500$$
$$i = 10\%$$
$$n = 20$$
$$P = ?$$

This is the find P, given A problem.

$$P = \$1,500(P/A\ 10,\ 20) = \$1,500(8.5136) = \mathbf{\$12,770}$$

This $12,770 is the worth in today's dollars of the series of annual sav-
ings over 20 years. It could also be considered as the incremental cost that
could be paid for the heat pump compared to an electric resistance heater,
and still produce a rate of return of 10% on the investment.

Example 4.7

A company needs to begin saving money for the new boiler in example 4.4. The company will make a deposit each year for 7 years to a savings account paying 10% annually. How large should the annual deposits be if they want to have $150,000 in the bank in 7 years?

$$F = \$150,000$$
$$i = 10\%$$
$$n = 7$$
$$A = ?$$

This problem is characterized as a find A, given F problem.

$$A = \$150,000(A/F_{10,7})$$
$$= \$150,000(.1054)$$
$$= \mathbf{\$15,810}$$

Example 4.8

A high efficiency lighting project for a company is saving $10,000 a year in energy costs. If that $10,000 a year is deposited into an energy management savings account paying 10%, how much money will be available in 5 years to use to replace an old chiller with a new, high efficiency model?

$$A = \$10,000$$
$$i = 10\%$$
$$n = 5$$
$$F = ?$$

This is the find F, given A problem.

$$F = \$10,000 \, (F/A_{10,5})$$
$$= \$10,000 \, (6.105)$$
$$= \mathbf{\$61,050}$$

By placing the $10,000 a year energy cost savings into an account that earns 10% interest, the company will have $61,050 to spend on the next project in five years.

4.7 A COST ANALYSIS METHODOLOGY USING DISCOUNTED CASH FLOWS

It is important to have a methodology to follow when performing a discounted cash flow analysis. This chapter recommends the following methodology for performing economic analyses using discounted cash flows:

1. **Define the alternatives.** State the problem, and list all feasible solutions or alternatives which have been selected for economic analysis.

2. **Estimate the relevant costs.** Each alternative from step 1 is defined in terms of its cash flows. Vital information includes the amount, timing, and direction (benefit or cost) of each cash flow.

3. **Analyze the alternatives.** Identify the most cost-effective alternative by analyzing each alternative using the discounted cash flow methodologies described later in this chapter.

4. **Perform sensitivity analyses.** Since the analysis above is generally based on estimated costs, these costs can be varied, depending on the uncertainty of the estimates, to see if the uncertainties have a pronounced effect.

The three pieces of data needed for the above methodology are 1) an estimate of the cash flows, 2) an estimate of the interest rate or discount rate, and 3) an estimate of the life of the project.

4.7.1 Estimation of cash flows

Sources of data for estimates of the cash flows of acquisition costs, utilization costs and disposal costs were discussed in section 4.2.1. Purchase cost estimates are generally obtained from sales personnel, or they can also be based on past experience. Operating and maintenance cost (and savings) estimates can also be based on past experience, from analyses performed by the energy manager, or they can be obtained from vendors, although these are not always accurate. If the costs involve building-related repair and maintenance, the Unit Price Standards Handbook of the U.S. Department of Defense is a good source [3]. This book gives labor-hour estimates by craft, and specifies materials and special equipment needed for each task. Another source for this information is Sack [4]. Salvage value estimates are often based on experience, and communication with vendors.

4.7.2 Interest rate

The interest rate, or discount rate, is a company-specific or organization-specific value. It must be supplied by the company, or corporate, accounting department, and is usually the value the company uses for evaluating investments. The magnitude of the discount rate depends on the source of capital which will be used to finance the project. If money must be borrowed (debt capital), the discount rate is likely to be higher than if company funds (equity capital) are to be used. The discount rate is often known as the Minimum Attractive Rate of Return (MARR).

4.7.3 Lifetime

Estimates for the lifetime of projects are often difficult to obtain, but can usually be found by contacting vendors, or using estimates based on experience. Lifetimes for energy-using equipment can often be found in the ASHRAE Fundamentals Handbook [5]. The lifetime of a project is also referred as the "study period" in economic analysis.

4.7.4 Equivalence of cash flows

Once the concept of converting money from one time period to another is understood, then a method is needed to compare sums of money which are paid or received at different points in time. To accomplish this, cash flows are "moved around" in time using a discount rate, or MARR, and then compared. If two cash flows have the same present value, future value, or annual worth, they are said to be equivalent. In economic analysis, "equivalence" means "the state of being equal in value." Normally, the concept of equivalence is applied to two or more cash flows. The choice of time period is arbitrary. In other words, if two sums of money with the same MARR are found to be equivalent at one point in time, they will be equivalent at any other.

Example 4.9

Is the sum $1000 today equivalent to $1331 three years from today at 10% interest?

Solution:

Compare the values at t = 0 (using present value or present worth):

PV(1) = **$1,000**

PV(2) = $1331 $(P/F_{10,3})$ = $1331(.7513) = **$1,000**

Therefore, they are equivalent at 10%.

Making decisions between cash flow profiles for different projects where the cash flows occur at different times requires using the equivalence of cash flows to make comparisons.

4.8   COST EFFECTIVENESS MEASURES USING DISCOUNTED CASH FLOWS

The goal of this section is to present methods that help energy managers, or other decision makers, determine whether a project is economically feasible or cost effective. Multiple methodologies exist for deciding whether a project is worth pursuing. Section 4.3 discussed the simple payback period. Because this method does not consider the time value of money, SPP should be used in combination with a method that explicitly considers the time value of money.

We cover five economic decision making methods which consider the time value of money in this section: present worth, future worth, annual worth, benefit/cost ratio or savings to investment ratio, and internal rate of return. We present a single example throughout this section so the reader can see the consistency of decisions which are made using these methodologies.

Example 4.10

A single zone heating unit is being used in a small office building. A variable air volume system retrofit can be purchased and installed for a cost of $100,000. The retrofit system is estimated to save 450,000 kilowatt hours per year for its economic life of 10 years. The company uses a MARR of 10%. If the company pays $0.06 per kWh for electricity, and the system will have a salvage value of $500 at the end of its life, should the new system be purchased?

The cash flow table for this example is shown below, in tabular form.

EOY	Cash flow
0	−$100,000
1	450,000 kWh ($0.06/kWh = $27,000
2	$27,000
3	$27,000
4	$27,000
5	$27,000
6	$27,000
7	$27,000
8	$27,000
9	$27,000
10	$27,000 + 500 = $27,500

4.8.1 Present Worth, Future Worth, and Annual Worth

A **present worth** comparison converts all the cash flows to a present worth value; a **future worth** comparison converts all the cash flows to a future value at a common future time (usually at the end of the study period, in this case End Of Year 10); and an **annual worth** analysis converts all the cash flows to a uniform annual series over the study period (in this case, 10 years).

There are three different cash flow patterns in example 4.10: the initial cost, the annual savings, and the salvage value. Each of these is converted using the compound interest factors discussed earlier. Costs carry a negative sign, while benefits carry a positive sign.

The decision rules for present worth, future worth, and annual worth are the same: a positive PW, FW or AW indicates the project is economically feasible and cost effective for a given MARR. It is important to note that the three measures are economically equivalent. This means that if the AW value were converted to either FW or PW, the same numerical value for FW or PW would be obtained. The reader is encouraged to verify this statement using the factors. The ultimate result is that only one of these three measures needs to be calculated. However, different organizations and different people have their own preferences for which of these methods to use. Since all three are in fairly common use, the authors have chosen to discuss each of the three methods.

The first method finds the present worth of the project using the P/A factor to find the equivalent cash flow of a series of $27,000 annual savings for 10 years with a 10% MARR, and the P/F factor to discount the $500 salvage value in year 10 to the present—the start of year 1. The project cost of $100,000 is a present value since it must be paid at the start of the project.

$$PW = -\$100,000 + \$27,000(P/A_{10,\ 10}) + \$500(P/F_{10,10})$$
$$= -\$100,000 + \$27,000(6.1446) + \$500(0.3856)$$
$$= \mathbf{\$66,097}$$

Since the present value of the benefits of the project exceed the present value of the cost by $66,097, this is a highly cost-effective project.

The second method finds the future worth by calculating the equivalent cash flows at the end of year 10. The F/P factor is used to move the $100,000 cost to a value at the end of year 10, and the F/A factor is used to find the equivalent future value of the series of $27,000 annual savings over the project's 10-year life. The $500 salvage value is already based at the end of year 10, so it can be added in directly.

$$FW = -\$100,000(F/P_{10,10}) + \$27,000(F/A_{10,10}) + \$500$$
$$= -\$100,000(2.594) + \$27,000(15.937) + \$500$$
$$= \mathbf{\$171,399}$$

The positive future worth of the project also confirms that it is a cost-effective EMO.

The third method reduces the initial cost and the salvage value to equivalent uniform annual amounts over the life of the project. The A/P factor is used for the cost and the A/F factor is used for the salvage value.

$$\begin{aligned}
\textbf{AW} &= -\$100{,}000(A/P_{10,10}) + \$27{,}000 + \$500(A/F_{10,10}) \\
&= -\$100{,}000(0.1628) + \$27{,}000 + \$500(0.0628) \\
&= \mathbf{\$10{,}751}
\end{aligned}$$

Even after discounting the savings at a 10% MARR, there is still a net savings of \$10,751 per year from implementing this new VAV system.

Since the PW, FW, and AW are all positive, this indicates the project should be pursued, from an economic standpoint.

4.8.2 Benefit/Cost Ratio, or Savings/Investment Ratio

A benefit/cost ratio (BCR), also known as a savings/investment ratio (SIR), calculates the present worth of all benefits, then calculates the present worth of all costs, and takes the ratio of the two sums. The BCR or SIR is another alternative economic decision-making criterion. It is favored by some organizations including parts of the U.S. Government and Military—especially the Army Corps of Engineers. A careful definition of "benefits" and "costs" is important. Benefits are defined to mean all the advantages, less any disadvantages, to the users. Costs are defined to mean all costs, less any savings, that will be incurred by the sponsor. With these definitions, a salvage value would reduce the costs for the sponsor, rather than increase the benefits for the user. (For a complete discussion of Benefit/Cost in the public sector, see Thuesen [2]).

$$BC(i) = \text{Equivalent benefits/Equivalent costs} \qquad (4.6)$$

A benefit/cost ratio greater than one is necessary for the project to be cost-effective. For our example, the only cost is the initial cost, which already occurs at the start of the project, so it is already a present worth. The salvage value of \$500 would serve to reduce the costs. One "benefit" occurs, the annual savings. Note that the ratio is explicitly defined as benefits divided by costs, so the sign on both values in the benefit/cost ratio will be positive.

The calculations required for this example are:

PW of Annual savings	$= \$27{,}000(P/A_{10,10}) = \$27{,}000(6.1446)$
	$= \$165{,}900$
PW of salvage value	$= \$500(P/F_{10,10}) = \$500(0.3856) = \$193$
PW of total benefits	$= \$165{,}900$
PW of total costs	$= \$100{,}000 - \$193 = \$99{,}807$
Benefit/cost ratio	$= \$165{,}900/\$99{,}807 = \mathbf{1.66}$

Since the benefit/cost ratio is greater than one, the project is economically attractive. Some organizations set a low value of the discount rate or

MARR, and then require a minimum BCR or SIR of 1.25 to 1.5 to identify a project as cost effective.

4.8.3 Internal Rate of Return

A final method in this section is to base a project decision on the Internal Rate of Return or IRR of the project. IRR is defined to be that value of the interest rate or discount rate that makes the present worth of the costs of a project equal to the present worth of the benefits of the project. If the computed IRR is greater than the MARR for an organization, the project is cost effective. Many private organizations prefer the IRR method since it produces a rate of return to compare to their MARR that they have already established.

Continuing with the data in Example 4.10, the equation to solve for IRR in this case is:

$$\$100{,}000 = \$27{,}000\ (P/A_{i,10}) + \$500\ (P/F_{i,10})$$

The value of I which makes this equation balance is called the IRR, and can be found using a financial calculator, a computer program, or by trial and error with our standard tables. Since the present worth of the salvage value is going to be quite small, we can start by solving:

$$\$100{,}000 = \$27{,}000\ (P/A_{i,10})$$
$$(P/A_{I,10}) = 100{,}000/(27{,}000) = 3.7037$$

The next step is to look through the tables of P/A factors with $n=10$ to find the value of the P/A factor closest to 3.7037. For $i=25\%$, the P/A factor is 3.5705; for $i=20\%$, the P/A factor is 4.1925. Thus, the value of I, or IRR, is around 24%. This is quite close to the correct answer.

The actual solution for IRR is found using an economic analysis computer program, and is 23.8%. Since the company has a MARR of 10%, the 23.8% rate of return indicates this project is an excellent investment.

4.9   LIFE CYCLE COSTING

The life cycle cost (LCC) for a project or a piece of equipment is its total cost of purchase and operation over its entire service life. This total cost includes the costs of acquisition, operation including energy costs, maintenance and disposal. Most of these costs occur at some future time beyond the purchase date, and must be analyzed using the time value of money. Thus, the discounted cash flow analysis presented in sections 4.4 to 4.8 form a fundamental part of the LCC analysis.

Many businesses and organizations still use economic analysis methods that do not include all costs, and do not use the time value of money. The Simple Payback Period, or SPP, method is commonly used by businesses and other organizations, but it is not a life cycle cost analysis method. Some organizations still make purchase decisions based on lowest initial costs and do not consider the operating and maintenance costs at all. Costs that occur after the project or equipment are purchased and installed are ignored. This omission can create a very inaccurate view of the economic viability of a project in many cases. Use of LCC can lead to more rational purchase decisions, and can often lead businesses to higher profits.

For all federal facilities—and many state government facilities—the direction is very clear. Congress and the President, through executive order, have mandated energy conservation goals for federal buildings, and required that these goals be met using cost-effective measures. The primary criterion mandated for assessing the effectiveness of energy conservation investments in federal buildings is the minimization of life cycle costs. Thus, it is important for all energy managers or energy analysts who deal with federal and state facility projects to understand life cycle costing.

4.10 LCC DECISION MAKING
AMONG MULTIPLE ALTERNATIVES

Most life cycle analyses involve choosing between more than one alternative. Therefore, a simple methodology for making valid economic comparisons between alternatives is necessary. This section uses the present worth method for this comparison. Several other methodologies exist, including future worth, annual worth, internal rate of return, and benefit/cost. Future worth and annual worth are economically equivalent to present worth, and can be interchanged if desired. Internal rate of return and benefit/cost methods require an "incremental" approach to make valid comparisons between projects. See White, Agee and Case for additional information [6]. In addition, multiple solutions for the value of IRR can occur depending on the form of the cash flow. Since present worth is considered the standard against which other methods are judged, present worth is recommended.

The Federal Energy Management Program (FEMP) prescribes the use of a present value based measure of worth in performing LCC analyses. All cash flows must be brought to a single, present-day baseline. In addition, the maximum study period for federal energy conservation projects is 25 years from the date of occupancy. Therefore, the study period should be the lesser of 25 years, the life of the building, or the life of the system. In any case, the study period must be the same for all alternatives under consideration.

The LCC method using present worth, when applied properly, allows the analyst to compare projects with different cash flows occurring at different times. The present worth of the total costs of each alternative is calculated, and the alternative with the lowest LCC is selected. If multiple projects can be selected, then the projects can be ranked by LCC and the project with the next lowest LCC can also be selected.

The present worth analysis in life cycle costing has two important requirements. First, when comparing multiple alternatives, different pieces of equipment may have different lifetimes. In this case, a common "study period" must be chosen so that all alternatives are considered over the same time line. There are four choices for this common time line. They are: shortest life, longest life, least common multiple of the two lives, and an arbitrary choice.

If shortest life is used, a salvage value must be estimated for the longer lived alternatives. If longest life is used, shorter lived alternatives are assumed to be repeatable. If the least common multiple life is used, projects with lives of 3 years and 5 years would be compared over a 15-year study period. Both projects would have to be repeatable. An arbitrary choice results in possibly having to assume both repeatability, and a salvage value. The critical point is that all projects must be compared over the same time horizon.

The second important requirement of the analysis is that the interest rate used (MARR) must be the same for all alternatives.

Example 4.11

An energy efficient air compressor is proposed by a vendor. The compressor will cost $30,000 installed, and will require $1,000 worth of maintenance each year for its life of 10 years. Energy costs will be $6,000 per year. A standard air compressor will cost $25,000 and will require $500 worth of maintenance each year. Its energy costs will be $10,000 per year. If your company uses a MARR of 10%, would you invest in the energy efficient air compressor?

Alternative 1: Energy efficient air compressor
Alternative 2: Standard air compressor

Cash flows:

EOY	Alternative 1	Alternative 2
0	$30,000	$25,000
1	$1,000 + $6,000 = $7,000	$500 + $10,000 = $10,500
2	$7,000	$10,500
3	$7,000	$10,500
4	$7,000	$10,500
5	$7,000	$10,500
6	$7,000	$10,500
7	$7,000	$10,500
8	$7,000	$10,500
9	$7,000	$10,500
10	$7,000	$10,500

(Note: In a Life Cycle Cost Analysis, all cash flows are costs, and the signs are ignored.)

Analysis:

$$LCC(\text{Alt } 1) = \$30,000 + \$7,000(P/A_{10,10})$$
$$= \$30,000 + \$7,000(6.1446)$$
$$= \mathbf{\$73,012}$$

$$LCC(\text{Alt } 2) = \$25,000 + \$10,500(P/A_{10,10})$$
$$= \$25,000 + \$10,500(6.1446)$$
$$= \mathbf{\$89,518}$$

The decision rule for LCC analysis is to choose the alternative with the lowest LCC. Since Alternative 1 has the lowest LCC, it should be chosen.

4.11 TAXES AND DEPRECIATION

Depreciation and taxes can have a significant effect on the life cycle analysis of energy projects, and should be considered for all large-scale projects for organizations that pay taxes. Depreciation is not a "cash flow," but it is considered a business expense by the government, and therefore, lowers taxable income. The effect is that taxes are reduced, and taxes are a cash flow.

Tax laws and regulations are complex and intricate. A detailed treatment of tax considerations as they apply to life cycle analysis is beyond the scope of this chapter, and generally requires the assistance of a tax professional. A high level summary of important considerations follows. The focus is on Federal corporate income taxes, since they apply to most decisions in the U.S., and have general application. The reader is encouraged to consult a tax accountant, or tax lawyer, for specific advice. It should be noted that the principles discussed can apply to tax considerations in other countries, as well.

4.11.1 Depreciation

Depreciation is a recognition that assets decrease in value over time. Depreciation deductions are an attempt to account for this reduction in value. These deductions are called depreciation allowances. To be depreciable, an asset must meet three primary conditions: 1) it must be held by the business for the purpose of producing income, 2) it must wear out, or be consumed in the course of its use, and 3) it must have a life longer than one year.

Multiple methods for calculating depreciation have been allowed in the past. These include straight line, sum of the years digits, declining balance, and the Accelerated Cost Recovery System (ACRS). The depreciation method used depends on when an asset was placed in service. Assets placed in service before 1981 use straight line, sum of the year's digits, or declining balance. Assets placed in service between 1981 and 1986 use the Accelerated Cost Recovery System (ACRS). Assets placed in service after 1986 use the Modified Accelerated Cost Recovery System (MACRS). This chapter will discuss the current (MACRS) depreciation methodology.

MACRS depreciation allowances are a function of 1) the asset's "property class" (a life defined by the government), and 2) the asset's "cost basis" (purchase price plus any costs required to place the asset into service). Note that the salvage value is not used when calculating depreciation using MACRS. Also, the property class "life" may have little relationship to the actual anticipated life of the asset.

Eight property classes are defined for assets which are depreciable under MACRS. Table 4-5 shows the eight classes, and gives some examples of property that falls into each class. Professional tax guidance is recommended to determine the MACRS property class for a specific asset.

Tables for calculating the MACRS allowances are published by the government. These percentages are based on declining balance, switching to straight line, for most property classes. The real property classes are depreciated using straight line. The depreciation percentages for 3-year, 5-year, and 7-year property are shown in Table 4-6.

Table 4-5. MACRS Property Classes

Property class	Example assets
3-year property	Special material handling devices and special tools for manufacturing
5-year property	Automobiles; light and heavy trucks; computers, copiers.
7-year property	Office furniture, fixtures, property that does not fall in other classes
10-year property	Petroleum refining assets, assets used in manufacturing of certain food products
15-year property	Telephone distribution equipment; municipal water and sewage treatment plants
20-year property	Municipal sewers, utility transmission lines and poles
27.5-year residential rental property	Apartment buildings and rental houses
31.5-year non-residential real property	Business buildings

Table 4-6. MACRS Percentages for 3-, 5-, and 7-year Property

Year	3-year	5-year	7-year
1	33.33%	20%	14.29%
2	44.45%	32%	24.49%
3	14.81%	19.2%	17.49%
4	7.41%	11.52%	12.49%
5		11.52%	8.93%
6		5.76%	8.92%
7			8.93%
8			4.46%

Example 4.12

Calculate the depreciation allowance in year 4 for an asset which is 5-year MACRS property, and has a cost basis of $10,000.

Depreciation   = Cost Basis × MACRS %
                = \$10,000 × .1152 = **\$1,152**

4.11.2 Federal Income Taxes

Federal taxes are determined based on a tax rate multiplied by a taxable income. The tax rate depends on the corporation's income range, and varies from 15% to 39% of taxable income (as of December 2010). Taxable income is calculated by subtracting allowable deductions from gross income. Allowable deductions include salaries and wages, materials, interest payments, and depreciation, as well as other "costs of doing business."

Tax paying companies may want to use After Tax Cash Flows (ATCF) when comparing alternatives. The calculation of taxes owed, and the ATCF, requires knowledge of:

- Before Tax Cash Flows (BTCF), the net project cash flows before tax considerations.
- Total loan payments incurred by the project, including a breakdown of principal and interest components of the payments.
- Total bond payments incurred by the project, including a breakdown of the redemption and interest components of the payments; and
- Depreciation deductions attributable to the project.

Once the above information has been determined, the ATCF is calculated in the following way:

Taxable income = BTCF–Loan interest–Bond interest–Depreciation
Taxes = Taxable income (Tax Rate)
ATCF = BTCF–Total loan payments–Total bond payments–Income taxes

At the tax-paying company's choice, all economic evaluation procedures may then be applied to the After Tax Cash Flow.

4.12 INFLATION

One of the fundamental principles supporting a life cycle cost analysis is the recognition that dollars change value over time. There are two primary reasons for this. First is the "opportunity cost," or the interest loss that is incurred when dollars are not invested. The second reason is inflation. Inflation is the term for the loss in purchasing power of a dollar over time, and should be accounted for in any life cycle analysis.

4.12.1 Terminology

Several additional terms must be defined before the term inflation can be explained clearly. They are:

Constant dollars—Constant dollars reflect the purchasing power (not face value) of the cash flow. These dollars are generally stated in relation to purchasing power at some base year, for example, 1990 dollars.

Current dollars—These are the out-of-pocket dollars that will actually change hands at any point in time. In other words, if a person visits a store in ten years, current dollars will be the amount a check is written for. However, the value of the items in the cart if purchased today (1997 dollars) would be the constant dollar amount.

Inflation rate (f)—Rate published by the government; based on the Consumer Price Index.

Real interest rate (j)—The amount of real growth in the earning power of money. This is also known as the inflation-free interest rate.

Market interest rate (i)—Opportunity to earn as reflected by the rates of interest available in finance and business. This rate contains both the inflation effect and the real earning power effect. This is the interest rate quoted by banks and other financial institutions.

The mathematical relationship between market interest rate, the inflation rate, and the real interest rate is:

$$i = f + j + f \times j \tag{4.7}$$

The key to proper analysis under inflation is to match the type of cash flow with the proper interest rate. If cash flows are estimated in current dollars, there are two effects contained in the cash flows: inflation, and real earning power. Therefore, the market interest rate, which includes both effects, is used to discount the cash flows. If cash flows are estimated in constant dollars, they do not contain inflation effects, and the real (inflation-free) interest rate should be used.

Type of Dollars	Interest rate to use
Current (actual)	i = combined (market) interest rate
Constant worth	j = inflation-free interest rate

A current cash flow can be converted to a constant worth cash flow by removing the inflation effect. This is the same principle as discounting any cash flow, and finding the present worth. The equation to be used is:

$$\textbf{Constant \$} = (\text{Current \$}) / (1 + f)^n \tag{4.8}$$

where f is the inflation rate and n is the number of years between the "base year" (usually the present), and the time the money is to be spent. In the same way, constant dollars can be converted to current dollars by multiplying the constant dollars by the inflation effect.

$$\text{Current } \$ = (\text{Constant } \$)\ ((1+f)^n \qquad (4.9)$$

It should be noted that most energy commodities, such as electricity, inflate at a different rate than the overall inflation rate quoted by the government. This rate is called the **escalation rate, k**. In this case, equations 4.7, 4.8 and 4.9 can still be used, with k substituted for the inflation rate, f. Energy escalation rates are quoted by region, end-use sector, and fuel type.

Example 4.14

A company has energy costs of $25,000 a year for the next three years. The cost of energy is subject to escalation, and the energy cost escalation rate is 25%. The company's real discount rate is 4%. Find the present worth (PW) of the energy costs using a) constant dollars and b) current (actual) dollars.

a) Constant dollar analysis

The constant dollar cash flow is $25,000/yr for 3 years. Using the real interest rate of 4% to discount the constant dollar cash flow, we obtain:

$$\begin{aligned}
\text{PW(constant \$)} &= \$25,000(P/F_{4,1}) + \$25,000(P/F_{4,2}) + \$25,000(P/F_{4,3}) \\
&= \$25,000(1.04)^{-1} + \$25,00(1.04)^{-2} + \$25,000(1.04)^{-3} \\
&= \$24,038 + \$23,114 + \$22,225 \\
&= \$69,377
\end{aligned}$$

b) Current (actual) dollar analysis

Two quantities must be calculated before the current (actual) dollar analysis can be performed, 1) the current (actual) dollar cash flow, and 2) the combined interest rate which will be used to discount the current (actual) dollar cash flow.

Current (actual) dollars	$= \text{Constant dollars } (1 + \text{escalation rate})^n$
Current (actual) $ (yr 1)	$= \$25,000(1.25)^1 \qquad = \$31,250$
Current (actual) $ (yr 2)	$= \$25,000(1.25)^2 \qquad = \$39,063$
Current (actual) $ (yr 3)	$= \$25,000(1.25)^3 \qquad = \$48,828$
Combined interest rate	$= .04 + .25 + (.04)(.25) = .30 \text{ or } 30\%$
PW (current/actual $)	$= \$31,250(P/F_{30,1}) + \$39,063(P/F_{30,2})$
	$\quad + \$48,828(P/F_{30,3})$
	$= \$31,250(1.30)^{-1} + \$39,063(1.30)^{-2}$
	$\quad + \$48,828(1.30)^{-3}$
	$= \$24,038 + \$23,114 + \$22,225$
	$= \$69,377$

Note that the PW obtained using either method is the same. This will always be true if the method is properly applied. Thus, it does not matter whether a constant dollar, or current (actual) dollar analysis is used.

4.13 ENERGY FINANCING OPTIONS

Once an energy management opportunity is found to be economically attractive, the next step is to obtain financing to implement the project. Many facilities find it difficult to obtain this financing. In fact, lack of financing is a primary reason EMOs are not implemented. There are, however, a number of methods available for paying for the energy efficiency improvements, and the energy manager should be prepared to present these at the same time he/she presents the EMO. The following is a brief description of some of these options, and their costs.

4.13.1 In-house Capital

Historically, in-house capital, or internal financing, has been the most common way companies finance their EMOs. With this method, the company uses funds available within the organization to finance the project. The cost of using these funds is the return rate that could have been obtained if these funds had been invested elsewhere. If a company does not have sufficient in-house capital available, they may choose to use another financing option.

4.13.2 Utility Rebates/Incentives

Some utility companies offer rebates or other incentives to help their customers reduce the initial cost of an EMO. Typical offerings include:

Rebates—The rebate is the most common utility incentive. The utility reimburses the company for a portion of the cost of implementing an EMO. Rebates may be based on load reduction ($/kW), or based on a fixed dollar amount for each energy-efficient product purchased ($/item).

Recently, many utilities have canceled or reduced their rebate programs. However, due to the newly competitive electricity market, discounts from utilities may become common again.

Direct Utility Assistance—Some utilities pay the installing contractor part or all of the cost of the EMO. Other utilities provide the energy efficiency products or services directly to the customer through utility personnel or through contractors selected by the utility.

Low Interest Loans—Some utilities offer low-interest financing for EMOs. The loan payments may be added to the customer's utility bill.

4.13.3 Debt Financing/Loans

Commercial lending institutions: If internal funds are not available, or are limited to funding only a small percentage of the EMOs, commercial lending is another option. Often, loan payments are less than the monthly savings generated from the EMO. The cost of the financing method is the interest rate the bank charges for lending the money, and must be included in the implementation cost analysis.

Government loans/Bonds: Some states make loans and bonds available to small businesses to help them reduce their energy consumption. For example, the state of Florida has a loan program called the Florida Energy Loan Program (FELP). To qualify for a FELP loan, the company must have 200 employees, or less, and must have made a profit for the last two consecutive years.

4.13.4 Leases

Leasing is a very popular financing option. In 1995, leasing accounted for almost one-third of all equipment utilization. In energy project financing, the monthly lease payment will usually be less than the monthly avoided energy cost, or energy savings; this will result in a positive cash flow.

Capital leases are essentially installment purchases. Little or no initial capital outlay is required to purchase the equipment. The company is considered the owner of the equipment and may take deductions for depreciation and for the interest portion of payments to the lessor. Capital leases are offered by banks, leasing companies, installation contractors, suppliers, and some utilities.

Operating leases are slightly different. In this case, the lessor owns the equipment and leases it to the company for a monthly fee during the contract period. The lessor claims the tax benefits associated with the depreciation of the equipment. At the end of the contract term the company can elect to purchase the equipment at fair market value (or for some pre-determined amount), renegotiate the lease, or have the equipment removed.

4.13.5   Performance Contracts
(Shared Savings, ESCO, Guaranteed Savings)

For companies who do not have funds available in-house, this energy financing option can be quite attractive. The basis for this type of financing is quite simple: the avoided energy cost resulting from the increased energy efficiency of the new equipment pays for the equipment. However, the company must be sure the economic analysis has been done correctly, and must monitor the new equipment to see that it is being used properly.

Performance contracting can be simple if the project involves something like a straight-forward lighting retrofit, where the cost savings can be accurately estimated. In most cases, the cost and risk of the project is largely taken by the contractor, but the user must relinquish a certain amount of control. Also, there are many factors to be considered which will require in-house expertise. Nevertheless, performance contracting is an option that should be considered carefully because of the benefits it can bring if done right.

Shared savings: Shared savings financing offers many of the advantages of operating leases. This option is particularly useful to businesses that do not have the capital to pursue needed energy projects. Like all energy financing options, shared savings is based on using the energy cost savings in the existing energy budget to pay for the project. The shared savings company is usually responsible for the engineering study, energy and economic evaluation, purchase and installation, and maintenance. The customer pays a percentage (50%) of the savings to the shared savings company for a specified period of time (5-10 years).

Performance contract for energy services: This is a variation of a shared savings program. The energy service company (ESCO) contracts with the customer to provide certain energy services, such as providing specific levels of heating, cooling, illumination, and use of equipment for other purposes. The ESCO then takes over the payment of the utility bills that are directly related to the energy services provided.

Insurance guarantee of energy savings: This approach is promoted by some equipment manufacturers and contractors. It consists of an agreement to ensure that the periodic energy cost savings from the energy-efficient equipment will exceed an established minimum dollar value. Often, the minimum guaranteed savings value is set equal to the financing payment value for the same period in order to ensure a positive cash flow during the financing term. The company's risk is minimized with this option, but the cost increases because it includes the cost of an indirect insurance premium.

4.14 LIFE CYCLE COSTING SOFTWARE

The advent of the computer age has significantly helped the practitioner performing life cycle cost analysis. Spreadsheets, such as Microsoft Excel, Lotus 1-2-3, or Quattro Pro, have built in functions which calculates economic values automatically. Users should read the definition of any built-in-function carefully, in order to make certain it is performing the desired calculation.

A common life cycle costing software program is "Building Life Cycle Cost" (BLCC), which is distributed through the National Institute of Standards and Technology. This package was written for governmental agencies who must perform life cycle cost analyses that conform to government standards.

BLCC was designed to provide an economic analysis of capital investments that reduce future costs. It focuses on energy conservation in buildings. Potential users include federal, state, and local governments, as well as private sector users.

BLCC is menu-driven. Data requirements fall under three categories: general assumptions, capital investment data, and operation-related data.

General Assumptions	Capital Investment Data	Operation-related Data
Analysis type	Initial cost	Annual recurring operating and maintenance costs
Base data	Capital replacements	Non-annual recurring operation and maintenance costs
Service data	Residual values	Energy consumption & cost data
Study period	Escalation rates	Discount rate

BLCC provides two report types, one for individual project alternatives, and another for comparative analysis. The individual project report provides an input data listing, a LCC analysis, both detailed and summary, and a yearly cash flow analysis. The comparative analysis provides a listing of all LCCs for all alternatives, with the lowest LCC flagged, as well as comparative economic measures.

The software provides an efficient and effective method to perform the last step in the analysis methodology, a sensitivity analysis. Sensitivity analysis refers to determining how "sensitive" a solution is to changes in the inputs. Since many of the cash flows and interest rates are based on estimates, these estimates can be varied easily to see if minor changes affect the solution.

For more information about software, contact:

BLAST—Blast Support Office, University of Illinois, Dept. of Mechanical and Industrial Engineering, 140 Mechanical Engineering Building, 1206 W. Green Street, Urbana, IL 61801; (800)UI-BLAST or (217)333-3977

ASEAM—ASEAM Coordinator, ACEC Research & Management Foundation, 1015 15th St., NW, Suite 802, Washington, D.C. 20005

BLCC—National Institute of Standards & Technology, Washington, D.C.

4.15 CONCLUSION

This chapter has discussed the fundamental methods needed to perform basic economic evaluations and life cycle cost assessments of capital investment projects. Although the focus was on energy-related projects, the concepts can be applied to any investment. Life cycle cost methodology provides a sound basis to analyze projects and make cost-effective decisions.

References

1. Pratt, D.B., 1994, *Life Cycle costing for Building Energy Analysis*, Student manual, Association of Energy Engineers, Atlanta.
2. Thuesen, G.J. and W.J. Fabrycky, 2000, *Engineering Economy*, 9th Edition, Prentice Hall, Englewood Cliffs, NJ.
3. Ibid.
4. Unit Price Standards Handbook, 1977, Stock No. 008-047-00221-1, Washington, D.C. (also known as NAVFAC P-716, ARMY TB 420-33, and AIR FORCE 85-56).
5. Sack, T.F., 1971, *A Complete Guide to Building and Plant Maintenance*, Prentice Hall, Englewood Cliffs, NJ,
6. 2005 Fundamentals Handbook, Association of Heating, Refrigeration and Air Conditioning Engineers, Atlanta, GA, 2005.
7. White, J.A., K.E. Case, D.B. Pratt, and M.H. Agee, 2001, *Principles of Engineering Economic Analysis*, 4th Edition, John Wiley & Sons, New York, NY.

Bibliography

Capehart, Barney, Editor, *Encyclopedia of Energy Engineering and Technology*, Taylor and Francis/CRC Publishing Company, Boca Raton, FL, July, 2007.

Kennedy, W.J., W.C. Turner and B.L. Capehart, 1994, *Guide to Energy Management*, Fairmont Press, Lilburn, GA

Pratt, D.B., 1997, "Economic Analysis," Chapter 4 in the *Energy Management Handbook*, W.C. Turner, ed., Fairmont Press, Lilburn, GA.

Pratt, D.B., 1994, *Life Cycle Costing for Building Energy Analysis*, Student Manual, Association of Energy Engineers, Atlanta.

Sharp, S. and H. Nguyen, 1995, "Capital Market Imperfections and the Incentive to Lease," *Journal of Financial Economics*, 39, pp. 294

Sullivan, William G, 2012, *Engineering Economy*, 15th Ed., Prentice Hall, Upper Saddle River, NJ.

APPENDIX 4-A: COMPOUND INTEREST FACTORS

Time Value of Money Factors—Discrete Compounding
   i = 10%

	Single Sums			Uniform Series			Gradient Series	
n	To Find F Given P (F|P,i%,n)	To Find P Given F (P|F,i%,n)	To Find F Given A (F|A,i%,n)	To Find A Given F (A|F,i%,n)	To Find P Given A (P|A,i%,n)	To Find A Given P (A|P,i%,n)	To Find P Given G (P|G,i%,n)	To Find A Given G (A|G,i%,n)
1	1.1000	0.9091	1.0000	1.0000	0.9091	1.1000	0.0000	0.0000
2	1.2100	0.8264	2.1000	0.4762	1.7355	0.5762	0.8264	0.4762
3	1.3310	0.7513	3.3100	0.3021	2.4869	0.4021	2.3291	0.9366
4	1.4641	0.6830	4.6410	0.2155	3.1699	0.3155	4.3781	1.3812
5	1.6105	0.6209	6.1051	0.1638	3.7908	0.2638	6.8618	1.8101
6	1.7716	0.5645	7.7156	0.1296	4.3553	0.2296	9.6842	2.2236
7	1.9487	0.5132	9.4872	0.1054	4.8684	0.2054	12.7631	2.6216
8	2.1436	0.4665	11.4359	0.0874	5.3349	0.1874	16.0287	3.0045
9	2.3579	0.4241	13.5795	0.0736	5.7590	0.1736	19.4215	3.3724
10	2.5937	0.3855	15.9374	0.0627	6.1446	0.1627	22.8913	3.7255
11	2.8531	0.3505	18.5312	0.0540	5.4951	0.1540	26.3963	4.0641
12	3.1384	0.3186	21.3843	0.0468	6.8137	0.1468	29.9012	4.3884
13	3.4523	0.2897	24.5227	0.0408	7.1034	0.1408	33.3772	4.6988
14	3.7975	0.2633	27.9750	0.0357	7.3667	0.1357	36.8005	4.9955
15	4.1772	0.2394	31.7725	0.0315	7.6061	0.1315	40.1520	5.2789
16	4.5950	0.2176	35.9497	0.0278	7.8237	0.1278	43.4164	5.5493
17	5.0545	0.1978	40.5447	0.0247	8.0216	0.1247	46.5819	5.8071
18	5.5599	0.1799	45.5992	0.0219	8.2014	0.1219	49.6395	6.0526
19	6.1159	0.1635	51.1591	0.0195	8.3649	0.1195	52.5827	6.2861
20	6.7275	0.1486	57.2750	0.0175	8.5136	0.1175	55.4069	6.5081
21	7.4002	0.1351	64.0025	0.0156	8.6487	0.1156	58.1095	6.7189
22	8.1403	0.1228	71.4027	0.0140	8.7715	0.1140	60.6893	6.9189
23	8.9543	0.1117	79.5430	0.0126	8.8832	0.1126	63.1462	7.1085
24	9.8497	0.1015	88.4973	0.0113	8.9847	0.1113	65.4813	7.2881
25	10.8347	0.0923	98.3471	0.0102	9.0770	0.1102	67.6964	7.4580
26	11.9182	0.0839	109.1818	9.159E-03	9.1609	0.1092	69.7940	7.6186
27	13.1100	0.0763	121.0999	8.258E-03	9.2372	0.1083	71.7773	7.7704
28	14.4210	0.0693	134.2099	7.451E-03	9.3066	0.1075	73.6495	7.9137
29	15.8831	0.0630	148.6309	6.728E-03	9.3696	0.1067	75.4146	8.0489
30	17.4494	0.0573	164.4940	6.079E-03	9.4269	0.1061	77.0766	8.1762
36	30.9127	0.0323	299.1268	3.343E-03	9.6765	0.1033	85.1194	8.7965
42	54.7637	0.0183	537.6370	1.860E-03	9.8174	0.1019	90.5047	9.2188
48	97.0172	0.0103	960.1723	1.041E-03	9.8969	0.1010	94.0217	9.5001
54	171.8719	5.818E-03	1.709E+03	5.852E-04	9.9418	0.1006	96.2763	9.6840
60	304.4816	3.284E-03	3.035E+03	3.295E-04	9.9672	0.1003	97.7010	9.8023
66	539.4078	1.854E-03	5.384E+03	1.857E-04	9.9815	0.1002	98.5910	9.8774
72	955.5938	1.046E-03	9.546E+03	1.048E-04	9.9895	0.1001	99.1419	9.9246
120	9.271E+04	1.079E-05	9.271E+05	1.079E-06	9.9999	0.1000	99.9860	9.9987
180	2.823E+07	3.543E-08	2.823E+08	3.543E-09	10.0000	0.1000	99.9999	10.0000
360	7.968E+14	1.255E-15	7.968E+15	1.255E-16	10.0000	0.1000	100.0000	10.0000

Time Value of Money Factors—Discrete Compounding
   i = 12%

	Single Sums			Uniform Series			Gradient Series	
n	To Find F Given P (F|P,i%,n)	To Find P Given F (P|F,i%,n)	To Find F Given A (F|A,i%,n)	To Find A Given F (A|F,i%,n)	To Find P Given A (P|A,i%,n)	To Find A Given P (A|P,i%,n)	To Find P Given G (P|G,i%,n)	To Find A Given G (A|G,i%,n)
1	1.1200	0.8929	1.0000	1.0000	0.8929	1.1200	0.0000	0.0000
2	1.2544	0.7972	2.1200	0.4717	1.6901	0.5917	0.7972	0.4717
3	1.4049	0.7118	3.3744	0.2963	2.4018	0.4163	2.2208	0.9246
4	1.5735	0.6355	4.7793	0.2092	3.0373	0.3292	4.1273	1.3589
5	1.7623	0.5674	6.3528	0.1574	3.6048	0.2774	6.3970	1.7746
6	1.9738	0.5066	8.1152	0.1232	4.1114	0.2432	8.9302	2.1720
7	2.2107	0.4523	10.0890	0.0991	4.5638	0.2191	11.6443	2.5515
8	2.4760	0.4039	12.2997	0.0813	4.9676	0.2013	14.4714	2.9131
9	2.7731	0.3606	14.7757	0.0677	5.3282	0.1877	17.3563	3.2574
10	3.1058	0.3220	17.5487	0.0570	5.6502	0.1770	20.2541	3.5647
11	3.4785	0.2875	20.6546	0.0484	5.9377	0.1684	23.1288	3.8953
12	3.8960	0.2567	24.1331	0.0414	6.1944	0.1614	25.9523	4.1897
13	4.3635	0.2292	28.0291	0.0357	6.4235	0.1557	28.7024	4.4683
14	4.8871	0.2046	32.3926	0.0309	6.6282	0.1509	31.3624	4.7317
15	5.4736	0.1827	37.2797	0.0268	6.8109	0.1468	33.9202	4.9803
16	6.1304	0.1631	42.7533	0.0234	1.1740	0.1434	36.3670	5.2147
17	6.8660	0.1456	48.8837	0.0205	7.1196	0.1405	38.6973	5.4353
18	7.6900	0.1300	55.7497	0.0179	7.2497	0.1379	40.9080	5.6427
19	8.6128	0.1161	63.4397	0.0158	7.3658	0.1358	42.9979	5.8375
20	9.6463	0.1037	72.0524	0.0139	7.4694	0.1339	44.9676	6.0202
21	10.8038	0.0926	81.6987	0.0122	7.5620	0.1322	46.8188	6.1913
22	12.1003	0.0826	92.5026	0.0108	7.6446	0.1308	48.5543	6.3514
23	13.5523	0.0738	104.6029	9.560E-03	7.7184	0.1296	50.1776	6.5010
24	15.1786	0.0659	118.1552	8.463E-03	7.7843	0.1285	51.6929	6.6406
25	17.0001	0.0588	133.3339	7.500E-03	7.8431	0.1275	53.1046	6.7708
26	19.0401	0.0525	150.3339	6.652E-03	7.8957	0.1267	54.4177	6.8921
27	21.3249	0.0469	169.3740	5.904E-03	7.9426	0.1259	55.6369	7.0049
28	23.8839	0.0419	190.6989	5.244E-03	7.9844	0.1252	56.7674	7.1098
29	26.7499	0.0374	214.5828	4.660E-03	8.0218	0.1247	57.8141	7.2071
30	29.9599	0.0334	241.3327	4.144E-03	8.0552	0.1241	58.7821	7.2974
36	59.1356	0.0169	484.4631	2.064E-03	8.1924	0.1221	63.1970	7.7141
42	116.7231	8.567E-03	964.3595	1.037E-03	8.2619	0.1210	65.8509	7.9704
48	230.3908	4.340E-03	1.912E+03	5.231E-04	8.2972	0.1205	67.4068	8.1241
54	454.7505	2199E-03	3.781E+03	2.645E-04	8.3150	0.1203	68.3022	8.2143
60	897.5969	1114E-03	7.472E+03	1.338E-04	8.3240	0.1201	68.8100	8.2664
66	1.772E+03	5.644E-04	1.476E+04	6.777E-05	8.3286	0.1201	69.0948	8.2961
72	3.497E+03	2.860E-04	2.913E+04	3.432E-05	8.3310	0.1200	69.2530	8.3127
120	8.057E+05	1.241E-06	6.714E+06	1.489E-07	8.3333	0.1200	69.4431	8.3332
180	7.232E+08	1.383E-09	6.026E+09	1.859E-10	8.3333	0.1200	69.4444	8.3333
360	5.230E+17	1.912E-18	4.358E+18	2.295E-19	8.3333	0.1200	69.4444	8.3333

Time Value of Money Factors—Discrete Compounding
  i = 15%

	Single Sums			Uniform Series			Gradient Series	
	To Find F Given P (F|P,i%,n)	To Find P Given F (P|F,i%,n)	To Find F Given A (F|A,i%,n)	To Find A Given F (A|F,i%,n)	To Find P Given A (P|A,i%,n)	To Find A Given P (A|P,i%,n)	To Find P Given G (P|G,i%,n)	To Find A Given G (A|G,i%,n)
n								
1	1.1500	0.8696	1.0000	1.0000	0.8696	1.1500	0.0000	0.0000
2	1.3225	0.7561	2.1500	0.4651	1.6257	0.6151	0.7561	0.4651
3	1.5209	0.6575	3.4725	0.2880	2.2832	0.4380	2.0712	0.9071
4	1.7490	0.5718	4.9934	0.2003	2.8550	0.3503	3.7864	1.3263
5	2.0114	0.4972	6.7424	0.1483	3.3522	0.2983	5.7751	1.7228
6	2.3131	0.4323	8.7537	0.1142	3.7845	0.2642	7.9368	2.0972
7	2.6600	0.3759	11.0668	0.0904	4.1604	0.2404	10.1924	2.4498
8	3.0590	0.3269	13.7268	0.0729	4.4873	0.2229	12.4807	2.7813
9	3.5179	0.2843	16.7858	0.0596	4.7716	0.2096	14.7548	3.0922
10	4.0456	0.2472	20.3037	0.0493	5.0188	0.1993	16.9795	3.3832
11	4.6524	0.2149	24.3493	0.0411	5.2337	0.1911	19.1289	3.6549
12	5.3503	0.1869	29.0017	0.0345	5.4206	0.1845	21.1849	3.9082
13	6.1528	0.1625	34.3519	0.0291	5.5831	0.1791	23.1352	4.1438
14	7.0757	0.1413	40.5047	0.0247	5.7245	0.1747	24.9725	4.3624
15	8.1371	0.1229	47.5804	0.0210	5.8474	0.1710	26.6930	4.5650
16	9.3576	0.1069	55.7175	0.0179	5.9542	0.1679	28.2960	4.7522
17	10.7613	0.0929	65.0751	0.0154	6.0472	0.1654	29.7828	4.9251
18	12.3755	0.0808	75.8364	0.0132	6.1280	0.1632	31.1565	5.0843
19	14.2318	0.0703	88.2118	0.0113	6.1982	0.1613	32.4213	5.2307
20	16.3665	0.0611	102.4436	9.761E-03	6.2593	0.1598	33.5822	5.3651
21	18.8215	0.0531	118.8101	8.417E-03	6.3125	0.1584	34.6448	5.4883
22	21.6447	0.0462	137.6316	7.266E-03	6.3587	0.1573	35.6150	5.6010
23	24.8915	0.0402	159.2764	6.275E-03	6.3988	0.1563	36.4988	5.7040
24	28.6252	0.0349	184.1678	5.430E-03	6.4338	0.1554	37.3023	5.7979
25	32.9190	0.0304	212.7930	4.699E-03	6.4641	0.1547	38.0314	5.8834
26	37.8568	0.0264	245.7120	4.070E-03	6.4906	0.1541	38.6918	5.9612
27	43.5353	0.0230	283.5688	3.526E-03	6.5135	0.1535	39.2890	6.0319
28	50.0656	0.0200	327.1041	3.057E-03	6.5335	0.1531	39.8283	6.0960
29	57.5755	0.0174	377.1697	2.651E-03	6.5509	0.1527	40.3146	6.1541
30	66.2118	0.0151	434.7451	2.300E-03	6.5660	0.1523	40.7526	6.2066
36	153.1519	6.529E-03	1.014E+03	9.859E-04	6.6231	0.1510	42.5872	6.4301
42	354.2495	2.823E.03	2.355E+03	4.246E-04	6.6478	0.1504	43.5286	6.5478
48	819.4007	1.220E-03	5.456E+03	1.833E-04	6.6585	0.1502	43.9997	6.6080
54	1.895E+03	5.276E.04	1.263E+04	7.918E-05	6.6631	0.1501	44.2311	6.6382
60	4.384E+03	2.281E-04	2.922E+04	3.422E-05	6.6651	0.1500	44.3431	6.6530
66	1.014E+04	9.851E-05	6.760E+04	1.479E-05	6.6660	0.1500	44.3967	6.6602
72	2.346E+04	4.263E-05	1.564E+05	6.395E-06	6.6664	0.1500	44.4221	6.6636
120	1.922E+07	5.203E-08	1.281E+08	7.805E-09	6.6667	0.1500	44.4444	6.6667
180	8.426E+10	1.187E-11	5.617E+11	1.780E-12	6.6667	0.1500	44.4444	6.6667
360	7.099E+21	1.409E-22	4.733E+22	2.113E-23	6.6667	0.1500	44.4444	6.6667

Time Value of Money Factors—Discrete Compounding
  i = 20%

	Single Sums			Uniform Series			Gradient Series	
	To Find F Given P	To Find P Given F	To Find F Given A	To Find A Given F	To Find P Given A	To Find A Given P	To Find P Given G	To Find A Given G
n	(F I P,i%,n)	(P I F,i%,n)	(F I A,i%,n)	(A I F,i%,n)	(P I A,i%,n)	(A I P,i%,n)	(P I G,i%,n)	(A I G,i%,n)
1	1.2000	0.8333	1.0000	1.0000	0.8333	1.2000	0.0000	0.0000
2	1.4400	0.6944	2.2000	0.4545	1.5278	0.6545	0.6944	0.4545
3	1.7280	0.5787	3.6400	0.2747	2.1065	0.4747	1.8519	0.8791
4	2.0736	0.4823	5.3680	0.1863	2.5887	0.3863	3.2986	1.2742
5	2.4883	0.4019	7.4416	0.1344	2.9906	0.3344	4.9061	1.6405
6	2.9860	0.3349	9.9299	0.1007	3.3255	0.3007	6.5806	1.9788
7	3.5832	0.2791	12.9159	0.0774	3.6046	0.2774	8.2551	2.2902
8	4.2998	0.2326	16.4991	0.0606	3.8372	0.2606	9.8831	2.5756
9	5.1598	0.1938	20.7989	0.0481	4.0310	0.2481	11.4335	2.8364
10	6.1917	0.1615	25.9587	0.0385	4.1925	0.2385	12.8871	3.0739
11	7.4301	0.1346	32.1504	0.0311	4.3271	0.2311	14.2330	3.2893
12	8.9161	0.1122	39.5805	0.0253	4.4392	0.2253	15.4667	3.4841
13	10.6993	0.0935	48.4966	0.0206	4.5327	0.2206	16.5883	3.6597
14	12.8392	0.0779	59.1959	0.0169	4.6106	0.2169	17.6008	3.8175
15	15.4070	0.0649	72.0351	0.0139	4.6755	0.2139	18.5095	3.9588
16	18.4884	0.0541	87.4421	0.0114	4.7296	0.2114	19.3208	4.0851
17	22.1861	0.0451	105.9306	9.440E-03	4.7746	0.2094	20.0419	4.1976
15	26.6233	0.0376	128.1167	7.805E-03	4.8122	0.2078	20.6805	4.2975
19	31.9480	0.0313	154.7400	6.462E-03	4.8435	0.2065	21.2439	4.3861
20	38.3376	0.0261	186.6880	5.357E-03	4.8696	0.2054	21.7395	4.4643
21	46.0051	0.0217	225.0256	4.444E-03	4.8913	0.2044	22.1742	4.5334
22	55.2061	0.0181	271.0307	3.690E-03	4.9094	0.2037	22.5546	4.5941
23	66.2474	0.0151	326.2369	3.065E-03	4.9245	0.2031	22.8867	4.6475
24	79.4968	0.0126	392.4842	2.548E-03	4.9371	0.2025	23.1760	4.6943
25	95.3962	0.0105	471.9811	2.119E-03	4.9476	0.2021	23.4276	4.7352
26	114.4755	8.735E.03	567.3773	1.762E-03	4.9563	0.2018	23.6460	4.7709
27	137.3706	7.280E.03	681.8528	1.467E-03	4.9636	0.2015	23.8353	4.8020
28	164.8447	6.066E-03	819.2233	1.221E-03	4.9697	0.2012	23.9991	4.8291
29	197.8136	5.055E-03	984.0680	1.016E-03	4.9747	0.2010	24-1406	4.8527
30	237.3763	4.213E-03	1.182E+03	8.461E-04	4.9789	0.2008	24.2628	4.8731
36	708.8019	1.411E-03	3.539E+03	2.826E-04	4.9929	0.2003	24.7108	4.9491
42	2.116E+03	4.725E-04	1.058E+04	9.454E-05	4.9976	0.2001	24.8890	4.9801
48	6.320E+03	1.582E-04	3.159E+04	3.165E-05	4.9992	0.2000	24.9581	4.9924
54	1.887E+04	5.299E-05	9.435E+04	1.060E-05	4.9997	0.2000	24.9844	4.9971
60	5.635E+04	1.775E-05	2.817E+05	3.549E-06	4.9999	0.2000	24.9942	4.9989
66	1.683E+05	5.943E-06	8.413E+05	1.189E-06	5.0000	0.2000	24.9979	4.9996
72	5.024E+05	1.990E-06	2.512E+06	3.951E-07	5.0000	0.2000	24.9992	4.9999
120	3.175E+09	3.150E-10	1.588E+10	6.299E-11	5.0000	0.2000	25.0000	5.0000
180	1.789E+14	5.590E-15	8.945E+14	1.118E-15	5.0000	0.2000	25.0000	5.0000
360	3.201E+28	3.124E-29	1.600E+29	6.249E-30	5.0000	0.2000	25.0000	5.0000

Time Value of Money Factors—Discrete Compounding
 i = 25%

	Single Sums			Uniform Series			Gradient Series	
n	To Find F Given P (F|P,i%,n)	To Find P Given F (P|F,i%,n)	To Find F Given A (F|A,i%,n)	To Find A Given F (A|F,i%,n)	To Find P Given A (P|A,i%,n)	To Find A Given P (A|P,i%,n)	To Find P Given G (P|G,i%,n)	To Find A Given G (A|G,i%,n)
1	1.2500	0.8000	1.0000	1.0000	0.8000	1.2500	0.0000	0.0000
2	1.5625	0.6400	2.2500	0.4444	1.4400	0.6944	0.6400	0.4444
3	1.9531	0.5120	3.8125	0.2623	1.9520	0.5123	1.6640	0.8525
4	2.4414	0.4096	5.7656	0.1734	2.3616	0.4234	2.8928	1.2249
5	3.0518	0.3277	8.2070	0.1218	2.6893	0.3718	4.2035	1.5631
6	3.8147	0.2621	11.2588	0.0888	2.9514	0-3388	5.5142	1.8683
7	4.7684	0.2097	15.0735	0.0663	3.1611	0.3163	6.7725	2.1424
8	5.9605	0.1678	19.8419	0.0504	3.3289	0.3004	7.9469	2.3872
9	7.4506	0.1342	25.8023	0.0388	3.4631	0.2888	9.0207	2.6048
10	9.3132	0.1074	33.2529	0.0301	3.5705	0.2801	9.9870	2.7971
11	11.6415	0.0859	42.5661	0.0235	3.6564	0.2735	10.8460	2.9663
12	14.5519	0.0687	54.2077	0.0184	3.7251	0.2684	11.6020	3.1145
13	18.1899	0.0550	68.7596	0.0145	3.7801	0.2645	12.2617	3.2437
14	22.7374	0.0440	86.9495	0.0115	3.8241	0.2615	12.8334	3.3559
15	28.4217	0.0352	109.6868	0.0091	3.8593	0.2591	13.3260	3.4530
16	35.5271	0.0281	138.1085	0.0072	3.8874	0.2572	13.7482	3.5366
17	44.4089	0.0225	173.6357	5.759E-03	3.9099	0.2558	14.1085	3.6084
18	55.5112	0.0180	218.0446	4.586E-03	3.9279	0.2546	14.4147	3.6698
19	69.3889	0.0144	273.5558	3.656E-03	3.9424	0.2537	14.6741	3.7222
20	86.7362	0.0115	342.9447	2.916E-03	3.9539	0.2529	14.8932	3.7667
21	108.4202	0.0092	429.6809	2.327E-03	3.9631	0.2523	15.0777	3.8045
22	135.5253	0.0074	538.1011	1.858E-03	3.9705	0.2519	15.2326	3.8365
23	169.4066	0.0059	673.6264	1.485E-03	3.9764	0.2515	15.3625	3.8634
24	211.7582	0.0047	843.0329	1.186E-03	3.9811	0.2512	15.4711	3.8861
25	264.6978	0.0038	1054.7912	9.481E-04	3.9849	0.2509	15.5618	3.9052
26	330.8722	3.022E-03	1319.4890	7.579E-04	3.9879	0.2508	15.6373	3.9212
27	413.5903	2.418E-03	1650.3612	6.059E-04	3.9903	0.2506	15.7002	3.9346
28	516.9879	1.934E-03	2063.9515	4.845E-04	3.9923	0.2505	15.7524	3.9457
29	646.2349	1.547E-03	2580.9394	3.875E-04	3.9938	0.2504	15.7957	3.9551
30	807.7936	1.238E-03	3.227E+03	3.099E-04	3.9950	0.2503	15.8316	3.9628
36	3081.4879	3.245E-04	1.232E+04	8.116E-05	3.9987	0.2501	15.9481	3.9883
42	1.175E+04	8.507E-05	4.702E+04	2.127E-05	3.9997	0.2500	15.9843	3.9964
48	4.484E+04	2.230E-05	1.794E+05	5.575E-06	3.9999	0.2500	15.9954	3.9989
54	1.711E+05	5.846E-06	6.842E+05	1.462E-06	4.0000	0.2500	15.9986	3.9997
60	6.525E+05	1.532E-06	2.610E+06	3.831E-07	4.0000	0.2500	15.9996	3.9999
66	2.489E+06	4 017E-07	9.957E+06	1.004E-07	4.0000	0.2500	15.9999	4.0000
72	9.496E+06	1.053E-07	3.798E+07	2.633E-08	4.0000	0.2500	18.0000	4.0000
120	4.258E+11	2.349E-12	1.703E+12	5.871E-13	4.0000	0.2500	16.0000	4.0000
180	2.778E+17	3.599E-18	1.111E+18	8.998E-19	4.0000	0.2500	18.0000	4.0000
380	7.720E+34	1.295E-35	3.088E+35	3.238E-38	4.0000	0.2500	16.0000	4.0000

Time Value of Money Factors—Discrete Compounding
i = 30%

	Single Sums			Uniform Series			Gradient Series	
	To Find F Given P	To Find P Given F	To Find F Given A	To Find A Given F	To Find P Given A	To Find A Given P	To Find P Given G	To Find A Given G
n	(F⏐P,i%,n)	(P⏐F,i%,n)	(F⏐A,i%,n)	(A⏐F,i%,n)	(P⏐A,i%,n)	(A⏐P,i%,n)	(P⏐G,i%,n)	(A⏐G,i%,n)
1	1.3000	0.7692	1.0000	1.0000	0.7692	1.3000	0.0000	0.0000
2	1.6900	0.5917	2.3000	0.4348	1.3609	0.7348	0.5917	0.4348
3	2.1970	0.4552	3.9900	0.2506	1.8161	0.5506	1.5020	0.8271
4	2.8561	0.3501	6.1870	0.1616	2.1662	0.4616	2.5524	1.1783
5	3.7129	02693	9.0431	0.1106	2.4356	0.4106	3.6297	1.4903
6	4.8268	0.2072	12.7560	0.0784	2.6427	0.3784	4.6656	1.7654
7	6.2749	0.1594	17.1128	0.0569	2.8021	0.3569	5.6218	2.0063
8	3.1573	0.1226	23.8577	0.0419	2.9247	0.3419	6.4800	2.2156
9	10.6045	0.0943	32.0150	0.0312	3.0190	0.3312	7.2343	2.3963
10	13.7858	0.0725	42.6195	0.0235	3.0915	0.3235	7.8872	2.5512
11	17.9216	0.0558	56.4053	0.0177	3.1473	0.3177	8.4452	2.6833
12	23.2981	0.0429	74.3270	0.0135	3.1903	0.3135	8.9173	2.7952
13	30.2875	0.0330	97.6250	0.0102	3.2233	0.3102	9.3135	2.8895
14	39.3735	0.0254	127.9125	0.0078	3.2487	0.3078	9.6437	2.9685
15	51.1859	0.0195	167.2863	0.0060	3.2682	0.3060	9.9172	3.0344
16	66.5417	0.0150	218.4722	0.0046	3.2832	0.3046	10.1426	3.0892
17	86.5042	0.0116	235.0139	3.509E-03	3.2948	0.3035	10.3276	3.1345
18	112.4554	0.0089	371.5180	2.692E-03	3.3037	0.3027	10.4788	3.1718
19	146.1920	0.0068	483.9734	2.066E-03	3.3105	0.3021	10.6019	3.2025
20	190.0496	0.0053	630.1655	1.587E-03	3.3158	0.3016	10.7019	3.2275
21	247.0645	0.0040	820.2151	1.219E-03	3.3198	0.3012	10.7828	3.2480
22	321.1839	0.0031	1067.2796	9.370E-04	3.3230	0.3009	10.8482	3.2646
23	417.5391	0.024	1388.4635	7.202E-04	3.3254	0.3007	10.9009	3.2781
24	542.8008	0.0018	1806.0026	5.537E-04	3.3272	0.3006	10.9433	3.2890
25	705.6410	0.0014	2348.8033	4.257E-04	3.3286	0.3004	10.9773	3.2979
26	917.3333	1.090E-03	3054.4443	3.274E-04	3.3297	0.3003	11.0045	3.3050
27	1192.5333	8.386E-04	3971.7776	2.518E-04	3.3305	0.3003	11.0263	3.3107
28	1550.2933	6.450E-04	5164.3109	1.936E-04	3.3312	0.3002	11.0437	3.3153
29	2015.3813	4.962E-04	6714.6042	1.489E-04	3.3317	0.3001	11.0576	3.3189
30	2619.9956	3.817E-04	8.730E+03	1.145E-04	3.3321	0.3001	11.0687	3.3219
36	12646.2186	7.908E-05	4.215E+04	2.372E-05	3.3331	0.3000	11.1007	3.3305
42	6.104E+04	1.638E-05	2.035E+05	4.915E-06	3.3333	0.3000	11.1086	3.3326
48	2.946E+05	3.394E-06	9.821E+05	1.018E-06	3.3333	0.3000	11.1105	3.3332
54	1.422E+06	7.032E-07	4.740E+06	2.110E-07	3.3333	0.3000	11.1110	3.3333
60	6.864E+06	1.457E-07	2.288E+07	4.370E-08	3.3333	0.3000	11.1111	3.3333
66	3.313E+07	3.018E-08	1.104E+08	9.054E-09	3.3333	0.3000	11.1111	3.3333
72	1.599E+08	6.253E-09	5.331E+08	1.876E-09	3.3333	0.3000	11.1111	3.3333
120	4.712E+13	2.122E-14	1.571E+14	6.367E-15	3.3333	0.3000	11.1111	3.3333
180	3.234E+20	3.092E-21	1.078E+21	9.275E-22	3.3333	0.3000	11.1111	3.3333
360	1.046E+41	9.559E-42	3.487E+41	2.868E-42	3.3333	0.3000	11.1111	3.3333

Time Value of Money Factors—Discrete Compounding
  i = 40%

	Single Sums			Uniform Series			Gradient Series	
	To Find F Given P (F|P,i%,n)	To Find P Given F (P|F,i%,n)	To Find F Given A (F|A,i%,n)	To Find A Given F (A|F,i%,n)	To Find P Given A (P|A,i%,n)	To Find A Given P (A|P,i%,n)	To Find P Given G (P|G,i%,n)	To Find A Given G (A|G,i%,n)
n								
1	1.4000	0.7143	1.0000	1.0000	0.7143	1.4000	0.0000	0.0000
2	1.9600	0.5102	2.4000	0.4167	1.2245	0.8167	0.5102	0.4167
3	2.7440	0.3644	4.3600	0.2294	1.5889	0.6294	1.2391	0.7798
4	3.8416	0.2603	7.1040	0.1408	1.8492	0.5408	2.0200	1.0923
5	5.3782	0.1859	10.9456	0.0914	2.0352	0.4914	23637	1.3580
6	7.5295	0.1328	16.3238	0.0613	2.1680	0.4613	3.4278	1.5811
7	10.5414	0.0949	23.8534	0.0419	2.2628	0.4419	3.9970	1.7664
a	14.7579	0.0678	34.3947	0.0291	2.3306	0.4291	4.4713	1.9185
9	20.6610	0.0484	49.1526	0.0203	2.3790	0.4203	4.8585	2.0422
10	28.9255	0.0346	69.8137	0.0143	2.4136	0.4143	5.1696	2.1419
11	40.4957	0.0247	98.7391	0.0101	2.4383	0.4101	5.4166	2.2215
12	56.6939	0.0176	139.2348	0.0072	2.4559	0.4072	5.6106	2.2845
13	79.3715	0.0126	195.9287	0.0051	2.4685	0.4051	5.7618	2.3341
14	111.1201	0.0090	275.3002	0.0036	2.4775	0.4036	5.8788	2.3729
15	155.5681	0.0064	386.4202	0.0026	2.4839	0.4026	5.9688	2.4030
16	217.7953	0.0046	541.9883	0.0018	2.4885	0.4018	6.0376	2.4262
17	304.9135	0.0033	759.7837	1.316E-03	2.4918	0.4013	6.0901	2.4441
18	426.8789	0.0023	1064.6971	9.392E-04	2.4941	0.4009	6.1299	2.4577
19	597.6304	0.0017	1491.5760	6.704E-04	2.4958	0.4007	6.1601	2.4682
20	836.6826	0.0012	2089.2064	4.787E-04	2.4970	0.4005	6.1828	2.4761
21	1171.3556	0.0009	2925.8889	3.418E-04	2.4979	0.4003	6.1998	2.4821
22	1639.8978	0.0006	4097.2445	2.441E-04	2.4985	0.4002	6.2127	2.4866
23	2295.8569	0.0004	5737.1423	1.743E-04	2.4989	0.4002	6.2222	2.4900
24	3214.1997	0.0003	8032.9993	1.245E-04	2.4992	0.4001	6.2294	2.4925
25	4499.8796	0.0002	11247.1990	8.891E-05	2.4994	0.4001	6.2347	2.4944
26	6299.8314	1.587E-04	15747.0785	6.350E-05	2.4996	0.4001	6.2387	2.4959
27	8819.7640	1.134E-04	22046.9099	4.536E-05	2.4997	0.4000	6.2416	2.4969
28	12347.6696	8.099E-05	30866.6739	3.240E-05	2.4998	0.4000	6.2438	2.4977
29	17288.7374	5.785E-05	43214.3435	2.314E-05	2.4999	0.4000	6.2454	2.4983
30	24201.4324	4.132E-05	6.050E+04	1.653E-05	2.4999	0.4000	6.2466	2.4988
36	1.822E+05	5.488E-06	4.556E+05	2.195E-06	2.5000	0.4000	6.2495	2.4998
42	1.372E+06	7.288E-07	3.430E+06	2.915E-07	2.5000	0.4000	6.2499	2.5000
48	1.033E+07	9.680E-08	2.583E+07	3.872E-08	2.5000	0.4000	6.2500	2.5000
54	7.779E+07	1.286E-08	1.945E+08	5.142E-09	2.5000	0.4000	8.2500	2.5000
60	5.857E+08	1.707E-09	1.464E+09	6.829E-10	2.5000	0.4000	6.2500	2.5000
66	4.410E+09	2.268E-10	1.103E+10	9.070E-11	2.5000	0.4000	6.2500	2.5000
72	3.321E+10	3.011E-11	8.302E+10	1.205E-11	2.5000	0.4000	6.2500	2.5000
120	3.431E+17	2.915E-18	8.576E+17	1.166E-18	2.5000	0.4000	6.2500	2.5000
180	2.009E+26	4.977E-27	5.023E+26	1.991E-27	2.5000	0.4000	6.2500	2.5000
360	4.037E+52	2.477E-53	1.009E+53	9.908E-54	2.5000	0.4000	6.2500	2.5000

Time Value of Money Factors—Discrete Compounding
i = 50%

	Single Sums			Uniform Series			Gradient Series	
	To Find F Given P (F|P,i%,n)	To Find P Given F (P|F,i%,n)	To Find F Given A (F|A,i%,n)	To Find A Given F (A|F,i%,n)	To Find P Given A (P|A,i%,n)	To Find A Given P (A|P,i%,n)	To Find P Given G (P|G,i%,n)	To Find A Given G (A|G,i%,n)
n								
1	1.5000	0.6667	1.0000	1.0000	0.6667	1.5000	0.0000	0.0000
2	2.2500	0.4444	2.5000	0.4000	1.1111	0.9000	0.4444	0.4000
3	3.3750	0.2963	4.7500	0.2105	1.4074	0.7105	1.0370	0.7368
4	5.0625	0.1975	8.1250	0.1231	1.6049	0.6231	1.6296	1.0154
5	7.5938	0.1317	13.1875	0.0758	1.7366	0.5758	2.1564	1.2417
6	11.3906	0.0878	20.7813	0.0481	1.8244	0.5481	2.5953	1.4226
7	17.0859	0.0585	32.1719	0.0311	1.8829	0.5311	2.9465	1.5648
8	25.6289	0.0390	49.2578	0.0203	1.9220	0.5203	3.2196	1.6752
9	38.4434	0.0260	74.8867	0.0134	1.9480	0.5134	3.4277	1.7596
10	57.6650	0.0173	113.3301	0.0088	1.9653	0.5088	3.5838	1.8235
11	86.4976	0.0116	170.9951	0.0058	1.9769	0.5058	3.6994	1.8713
12	129.7463	0.0077	257.4927	0.0039	1.9846	0.5039	3.7842	1.9068
13	194.6195	0.0051	387.1390	0.0026	1.9897	0.5026	3.8459	1.9329
14	291.9293	0.0034	581.8585	0.0017	1.9931	0.5017	3.8904	1.9519
15	437.8939	0.0023	873.7878	0.0011	1.9954	0.5011	3.9224	1.9657
16	656.8408	0.0015	1311.6817	0.0008	1.9970	0.5008	3.9452	1.9756
17	985.2613	0.0010	1968.5225	5.080E-04	1.9980	0.5005	3.9614	1.9827
18	1477.8919	0.0007	2953.7838	3.385E-04	1.9986	0.5003	3.9729	1.9878
A	2216.8373	0.0005	4431.6756	2.256E-04	1.9991	0.5002	3.9811	1.9914
20	3325.2567	0.0003	6648.5135	1.504E-04	1.9994	0.5002	3.9868	1.9940
21	4987.8851	0.0002	9973.7702	1.003E-04	1.9996	0.5001	3.9908	1.9958
22	7481.8276	0.0001	14961.6553	6.684E-05	1.9997	0.5001	3.9936	1.9971
23	11222.7415	0.0001	22443.4829	4.456E-05	1.9998	0.5000	3.9955	1.9980
24	16834.1122	0.0001	33666.2244	2.970E-05	1.9999	0.5000	3.9969	1.9986
25	252511683.	0.0000	50500.3366	1.980E-05	1.9999	0.5000	3.9979	1.9990
26	37876.7524	2.640E-05	75751.5049	1.320E-05	1.9999	0.5000	3.9985	1.9993
27	56815.1287	1.760E-05	113628.257	8.801E-06	2.0000	0.5000	3.9990	1.9995
28	85222.6930	1.173E-05	170443.386	5.867E-06	2.0000	0.5000	3.9993	1.9997
29	127834.039	7.823E-06	255666.079	3.911E-08	2.0000	0.5000	3.9995	1.9998
30	191751.059	5.215E-06	3.835E+05	2.608E-06	2.0000	0.5000	3.9997	1.9998
36	2184164.41	4.578E-07	4.368E+06	2.289E-07	2.0000	0.5000	4.0000	2.0000
42	2.488E+07	4.019E-08	4.976E+07	2.010E-08	2.0000	0.5000	4.0000	2.0000
48	2.834E+08	3.529E-09	5.668E+08	1.764E-09	2.0000	0.5000	4.0000	2.0000
54	3.228E+09	3.098E-10	6.456E+09	1.549E-10	2.0000	0.5000	4.0000	2.0000
60	3.677E+10	2.720E-11	7.354E+10	1.360E-11	2.0000	0.5000	4.0000	2.0000
66	4.188E+11	2.388E-12	8.376E+11	1.194E-12	2.0000	0.5000	4.0000	2.0000
72	4.771E+12	2.096E-13	9.541E+12	1.048E-13	2.0000	0.5000	4.0000	2.0000
120	1.352E+21	7.397E-22	2.704E+21	3.698E-22	2.0000	0.5000	4.0000	2.0000
180	4.971E+31	2.012E-32	9.942E+31	1.006E-32	2.0000	0.5000	4.0000	2.0000
380	2.471E+63	4.047E-64	4.942E+63	2.024E-64	2.0000	0.5000	4.0000	2.0000

Time Value of Money Factors—Discrete Compounding
i = 100%

	Single Sums			Uniform Series			Gradient Series	
	To Find F Given P (F | P,i%,n)	To Find P Given F (P | F,i%,n)	To Find F Given A (F | A,i%,n)	To Find A Given F (A | F,i%,n)	To Find P Given A (P | A,i%,n)	To Find A Given P (A | P,i%,n)	To Find P Given G (P | G,i%,n)	To Find A Given G (A | G,i%,n)
n								
1	2.0000	0.5000	1.0000	1.0000	0.5000	2.0000	0.0000	0.0000
2	4.0000	0.2500	3.0000	0.3333	0.7500	1.3333	0.2500	0.3333
3	8.0000	0.1250	7.0000	0.1429	0.8750	1.1429	0.5000	0.5714
4	16.0000	0.0625	15.0000	0.0667	0.9375	1.0667	0.6875	0.7333
5	32.0000	0.0313	31.0000	0.0323	0.9688	1.0323	0.8125	0.8387
6	64.0000	0.0156	63.0000	0.0159	0.9844	1.0159	0.8906	0.9048
7	128.0000	0.0078	127.0000	0.0079	0.9922	1.0079	0.9375	0.9449
8	256.0000	0.0039	255.0000	0.0039	0.9961	1.0039	0.9648	0.9686
9	512.0000	0.0020	511.0000	0.0020	0.9980	1.0020	0.9805	0.9824
10	1024.0000	0.0010	1023.0000	0.0010	0.9990	1.0010	0.9893	0.9902
11	2048.0000	0.0005	2047.0000	0.0005	0.9995	1.0005	0.9941	0.9946
12	4096.0000	0.0002	4095.0000	0.0002	0.9998	1.0002	0.9968	0.9971
13	8192.0000	0.0001	8191.0000	0.0001	0.9999	1.0001	0.9983	0.9984
14	16384.0000	0.0001	16383.0000	0.0001	0.9999	1.0001	0.9991	0.9991
15	32768.0000	0.0000	32767.0000	0.0000	1.0000	1.0000	0.9995	0.9995
16	65536.0000	0.0000	65535.0000	0.0000	1.0000	1.0000	0.9997	0.9998
17	131072.000	0.0000	131071.000	7.629E-06	1.0000	1.0000	0.9999	0.9999
18	262144.000	0.0000	262143.000	3.815E-06	1.0000	1.0000	0.9999	0.9999
19	524288.000	0.0000	524287.000	1.907E-06	1.0000	1.0000	1.0000	1.0000
20	1048576.00	0.0000	1048575.00	9.537E-07	1.0000	1.0000	1.0000	1.0000
21	2097152.00	0.0000	2097151.00	4.768E-07	1.0000	1.0000	1.0000	1.0000
22	4194304.00	0.0000	4194303.00	2.384E-07	1.0000	1.0000	1.0000	1.0000
23	8388608.00	0.0000	8388607.00	1.192E-07	1.0000	1.0000	1.0000	1.0000
24	16777216.0	0.0000	16777215.0	5.960E-08	1.0000	1.0000	1.0000	1.0000
25	33554432.0	0.0000	33554431.0	2.980E-08	1.0000	1.0000	1.0000	1.0000
26	67108864.0	1.490E-08	67108863.0	1.490E-08	1.0000	1.0000	1.0000	1.0000
27	134217728	7.451E-09	134217727	7.451E-09	1.0000	1.0000	1.0000	1.0000
28	268435456	3.725E-09	268435455	3.725E-09	1.0000	1.0000	1.0000	1.0000
29	536870912	1.863E-09	536870911	1.883E-09	1.0000	1.0000	1.0000	1.0000
30	1.07E+09	9.313E-10	1.074E+09	9.313E-10	1.0000	1.0000	1.0000	1.0000
36	68.7E+9	1.455E-11	6.872E+10	1.455E-11	1.0000	1.0000	1.0000	1.0000
42	4.398E+12	2.274E-13	4.398E+12	2.274E-13	1.0000	1.0000	1.0000	1.0000
48	2.815E+14	3.553E-15	2.815E+14	3.553E-15	1.0000	1.0000	1.0000	1.0000
54	1.801E+16	5.551E-17	1.801E+16	5.551E-17	1.0000	1.0000	1.000(	1.0000
60	1.153E+18	8.674E-19	1.153E+18	8.674E-19	1.0000	1.0000	1.0000	1.0000
66	7.379E+19	1.355E-20	7.379E+19	1.355E-20	1.0000	1.0000	1.0000	1.0000
72	4.722E+21	2.118E-22	4.722E+21	2.118E-22	1.0000	1.0000	1.0000	1.0000
120	1.329E+36	7.523E-37	1.329E+36	7.523E-37	1.0000	1.0000	1.0000	1.0000
180	1.532E+54	6.525E-55	1.532E+54	6.525E-55	1.0000	1.0000	1.0000	1.0000
380	2.349E+108	4.255E-109	2.349E+108	4.258E-109	1.0000	1.0000	1.0000	1.0000

Chapter 5

Lighting*

5.0. INTRODUCTION

The lighting system provides many opportunities for cost-effective energy savings with little or no inconvenience. In many cases, lighting can be improved and operation costs can be reduced at the same time. Lighting improvements are excellent investments in most commercial businesses because lighting accounts for a large part of the energy bill—ranging from 30-70% of the total energy cost. Lighting energy use represents only 5-25% of the total energy in industrial facilities, but it is usually cost-effective to address because lighting improvements are often easier to make than many process upgrades.

While there are significant energy-use and power-demand reductions available from lighting retrofits, the minimum lighting level standards of the Illuminating Engineering Society (IES) should be followed to insure worker productivity and safety. Inadequate lighting levels can decrease productivity, and they can also lead to a perception of poor indoor air quality. However, most lighting retrofits will increase productivity, and are most likely to appeal to employees, clients and guests, which can have a positive impact on the facility, business reputation, stock prices[†] and especially sales[¶].

Used as a starting place for an energy management program, lighting can attract immediate employee attention and participation, since everyone has ideas about lighting. Lighting is also seen as a barometer of the attitude of top managers toward energy management: if the office

---

*This chapter was updated by Dr. Eric Woodroof, CEM, CEO of Profitable Green Solutions Company, Santa Barbara, CA.

[†]Wingender, J. and Woodroof, E., (1997) .When Firms Publicize Energy Management Projects: Their Stock Prices Go Up.- How much- 21.33% on Average! Strategic Planning for Energy and the Environment, Summer Issue 1997.

[¶]Profitable Green Strategies Course, www.ProfitableGreenSolutions.com

of the president of a company is an example of efficient lighting, then employees will see that energy management is taken seriously. A lighting retrofit program can be a win-win proposition for the business owner and the employees as it can improve morale, safety, and productivity while reducing life-cycle costs. This chapter provides a brief description of lighting systems, their characteristics, and retrofit options.

5.1   COMPONENTS OF THE LIGHTING SYSTEM

A lighting system consists of light sources (lamps), luminaires (or fixtures), and ballasts. Each component will affect the performance, energy use and annual operating cost of the lighting system. This section discusses each of these components, and provides the basic information on lighting technology needed to successfully accomplish lighting Energy Management Opportunities (EMOs).

5.1.1 Lamp Characteristics

Lamps are rated a number of different ways and each characteristic is a factor to consider in the lamp selection process. The basic ratings include: luminous efficacy (lumens/watt); color temperature (Kelvins); color rendering index (CRI); cost ($); rated life (operation hours); and labor required for relamping. Lamps should carry recognizable name brands and should be purchased from a reputable vendor. Some off-brand lamps, particularly those from some foreign countries, have low light output and short lives.

5.1.1.1 Luminous Efficacy

The luminous efficacy of a lamp is an estimate of the light output (lumens) divided by the electrical power input (watts) under test conditions. Lamps operating outside their design envelope may suffer reduced efficacy. For example, 34-watt energy-saving fluorescent lamps should not be used in environments with temperatures below 60°F; at lower temperatures, they are prone to flickering, their light output is low, and they have short lives. Table 5-1 presents lamp data on many commonly used lamps. More data is available from lamp manufacturers and the IES Lighting Handbook [1].

5.1.1.2 Color Temperature

The color temperature of a lamp describes the appearance of the light generated compared to the perceived color of a blackbody radiator at

Table 5-1. Light Source Characteristics [ref 2, Table 13.4]

	Incandescent, Including Tungsten Halogen	Fluorescent	High-Intensity Discharge			
			Mercury Vapor (Self-Ballasted)	Metal Halide	High-Pressure Sodium (Improved Color)	Low-Pressure Sodium
Wattages (lamp only)	15-1500	15-219	40-1000	175-1000	70-1000	35-180
Life[a] (hr)	750-12,000	7500-24,000	16,000-15,000	1500-15,000	24,000 (10,000)	18,000
Efficacy[a] (lumens/W) lamp only	15-25	55-100	50-60 (20-25)	80-100	75-140 (67-112)	Up to 180
Lumen maintenance	Fair to excellent	Fair to excellent	Very good (good)	Good	Excellent	Excellent
Color rendition	Excellent	Good to excellent	Poor to excellent	Very good	Fair (very good)	Poor
Light direction control	Very good to excellent	Fair	Very good	Very good	Very good	Fair
Source size	Compact	Extended	Compact	Compact	Compact	Extended
Relight time	Immediate	Immediate	3-10 min	10-20 min	Less than 1 min	Immediate
Comparative fixture cost	Low: simple fixtures	Moderate	Higher than incandescent and fluorescent	Generally higher than mercury	High	High
Comparative operating cost	High: short life and low efficiency	Lower than incandescent	Lower than incandescent	Lower than mercury	Lowest of HID types	Low
Auxiliary equipment needed	Not needed	Needed: medium cost	Needed: high cost	Needed: High cost	Needed: High cost	Needed: high cost

[a]Life and efficacy ratings subject to revision. Check manufacturers' data for latest information.

that temperature on the absolute temperature scale (i.e., Kelvin scale). For example, a daylight fluorescent lamp rated at 6300 Kelvins appears bluish, while a warm-white fluorescent lamp rated at 3000 Kelvins appears yellowish. Figure 5-1 shows the color temperatures of commonly used fluorescent lamps.

The energy manager should be sensitive to lamp color when recommending lighting changes and should not recommend changing lamp color unless nearly everyone is in favor of the proposed change. Understanding the color needs of the facility is important too. Some merchandise looks better at a particular color temperature. Example: meat in grocery store looks more appealing under warm lamps which accentuate the reds hues. The same meat in a meat-packing facility can be illuminated with lamps of a higher color temperature because visual appeal is not a factor.

5.1.1.3 Color Rendering Index

The color rendering index (CRI) is a relative indication of how well colors can be distinguished under the light produced by a lamp with a particular color temperature. The index runs from 0 to 1; where a high CRI indicates good color rendering. Many commonly-used lamps have poor CRIs (e.g., the CRI of a typical warm-white fluorescent lamp is 0.42, the CRI of a typical cool white fluorescent lamp is 0.67).

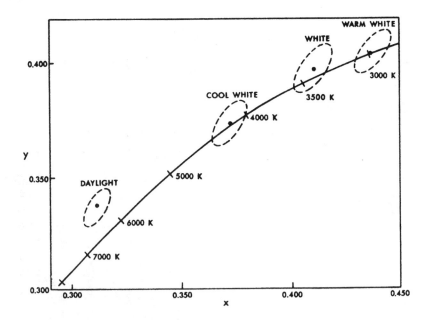

Figure 5-1. Fluorescent Lamp Color Temperatures [ref 1, Figure 5-17]

Rare earth elements in the phosphors of high-efficiency lamps increase light conversion efficacy and color rendition. Light sources such as metal halide lamps have better color rendering abilities than high pressure sodium lamps. The need for accurate color rendering depends on the particular task. Matching colors in a garment factory or photo laboratory will require a much higher CRI lamp than assembling large machine parts.

The color rendering abilities of commonly used lamps are described in Table 5-2.

5.1.2 Lamp Types

Lamps come in a variety of types and have a wide range of characteristics. Choosing the appropriate lamp type depends on the lighting task. Often, the least expensive lamp to buy is not the least expensive to operate. The energy manager should be familiar with the different options available for providing the desired lighting levels.

5.1.2.1 Incandescent Lamps

Incandescent lamps render colors well, are inexpensive to purchase, easily dimmed, small, and controllable which is useful for product display. However, they have relatively short lifespans, low efficacy and are susceptible to failure from heat and vibration. Incandescent lamps rated for long life or rough service have a correspondingly low efficacy. Some energy-saving lamps have a poorly supported filament and should not be used in environments with vibration or other mechanical stresses.

Most incandescent lamps tend to darken with age as tungsten is lost from the filament and deposited on the lamp walls. Figure 5-2 shows most of the commonly used lamps.

A Lamps. These lamps are low cost and commonly used in sizes of 20-1500 Watts. They project light out in all directions. In old industrial plants, look for large A-lamps in pendant fixtures where the lamps are left on most of the time. These are good candidates for replacement with HID lamps.

Reflector (R) Lamps. These lamps are usually more expensive than A-lamps and offer better control of the direction in which light is cast due to a reflective paint on the lamp wall. They have a focal point in back of the lamp, which results in the light from the lamp being dispersed broadly by the reflective surface of the lamp.

Ellipsoidal Reflector (ER) Lamps. These lamps cost about the same as R-lamps, but they are longer and have a focal point in front of the lamp. This

Table 5-2. Color Rendition of Various Lamp Types

	Filament[a]	Clear Mercury	White Mercury	Deluxe White[b] Mercury	Multi-Vapor[b]	Lucalox[a]
			Incandescent, high-intensity discharge lamps			
Efficacy (lm/w)	Low	Medium	Medium	Medium	High	High
Lamp appearance effect on neutral surfaces	Yellowish white	Greenish blue-white	Greenish white	Purplish white	Greenish white	Yellowish
Effect on "atmosphere"	Warm	Very cool, greenish	Moderately cool, greenish	Warm, purplish	Moderately cool, greenish	Warm, yellowish
Colors strengthened	Red Orange Yellow	Yellow Green Blue	Yellow Green Blue	Red Yellow Blue	Yellow Green Blue	Yellow Orange Green
Colors grayed	Blue	Red, orange	Red, orange	Green	Red	Red, blue
Effect on complexions	Ruddiest	Greenish	Very pale	Ruddy	Grayed	Yellowish
Remarks	Good color rendering	Very poor color rendering	Moderate color rendering	Color acceptance similar to CW fluorescent	Color acceptance similar to CW fluorescent	Color acceptance approaches that of WW fluorescent

(Continued)

Table 5-2. Color Rendition of Various Lamp Types (*Continued*)

Fluorescent lamps

	Cool White[b]	Deluxe Cool White[b]	Warm White[a]	Deluxe Warm White[a]	Daylight	White	Soft White/Natural
Efficacy (lm/w)	High	Medium	High	Medium	Medium-high	High	Medium
Lamp appearance effect on neutral surfaces	White	White	Yellowish white	Yellowish white	Bluish white	Pale yellowish white	Purplish white
Effect on "atmosphere"	Neutral to moderately cool	Neutral to moderately cool	Warm	Warm	Very cool	Moderately warm	Warm pinkish
Colors strengthened	Orange Yellow Blue	All nearly equal	Orange Yellow	Red Orange Yellow Green	Green Blue	orange Yellow	Red Orange
Colors grayed	Red	None appreciably	Red, green blue	Blue	Red, orange	Red, green blue	Green, blue
Effect on complexions	Pale pink	Most natural	Sallow	Ruddy	Grayed	Pale	Ruddy pink
Remarks	Blends with natural daylight; good color acceptance	Best overall color rendition; simulates natural daylight	Blends with incandescent light; poor color acceptance	Good color rendition; simulates incandescent light	Usually replaceable with CW	Usually replaceable with CW or WW	Tinted source; usually replaceable with CWX or WWX

[a]Greater preference at lower levels.
[b]Greater preference at higher levels.
Source: General Electric Technical Pamphlet TP-I 19[3]. Reprinted with permission of General Electric Co.

Figure 5-2. Incandescent Bulb Shapes and Designations [ref 1, p. 8-5]

location of the focal point results in the light being more concentrated as it leaves the lamp, and thus the beam is narrower than from an R lamp.

Quartz-Halogen Lamps. These lamps have a short life and low efficacy. They can be a good choice for areas which need lighting on an irregular basis. These lamps should not be cleaned.

Halogen Lamps. These lamps have a higher efficacy and cost than the lamps listed above and are available in many of the same lamp configurations.

5.1.2.2 Fluorescent Lamps

Fluorescent lamps have high efficacy, long life, and low surface luminance; they are cool and are available in a variety of colors. Figure 5-3 shows many of the commonly used fluorescent lamps.

Figure 5-3. Fluorescent Lamp Shapes and Designations
[ref 8, Figure 2-14]

Typical Fluorescent Lamps. Fluorescent lamps are available in standard, high output (HO), and very high output (VHO) configurations. The HO and VHO lamps are useful for low-temperature environments and areas where a lot of light is needed with minimal lamp space.

Energy-saving Lamps. Energy-saving fluorescent lamps which can replace the standard lamps can reduce power demand and energy use by about 15%. They will also decrease light levels about 3-10%. These lamps can only be used with ballasts designed and rated for energy-saving lamps and should not be used in areas in which the temperature falls below 60°F. Fixtures subject to direct discharge from air-conditioning vents are not good candidates for energy-saving fluorescent lamps. Figure 5-4 lists many of the energy-saving lamps available.

Figure 5-4. Energy Saving Fluorescent Lamps [ref 1, Figure 8-115]

Lamp description	Lamp watts	Lamp watts replaced	Lamp current (A)	Lamp volts (V)	Lamp life+ (h)	Nominal length (mm)	Nominal length (in.)	Base (end caps)	Nominal lumens‡§ 3000K RE70	3500K RE70	4100K RE70	3000K RE80	3500K RE80	4100K RE80	5000K RE80
Rapid start															
F17T8	17	—	0.265	70	20,000	610	24	Med. Bipin	1325	1325	1325	1375	1375	1375	—
F25T8	25	—	0.265	100	20,000	914	36	Med. Bipin	2125	2125	2125	2200	2200	2200	—
F3278	32	—	0.265	137	20,000	1219	48	Med. Bipin	2850	2850	2850	2975	2975	2975	2700[b]
F40T8	40	—	0.265	172	20,000	1524	60	Med. Bipin	3600	3600	3600	3725	3725	3725	—
F40T12/U/3	36	40	—	—	12,000	610	24	Med. Bipin	—	—	—	—	—	—	—
F40T12/U/6	34	40	0.45	84	16,000	610	24	Med. Bipin	2800[b]	2800[b]	2800[b]	—	—	—	—
F30T12	25	30	0.453	64	18,000	914	36	Med. Bipin	2090	2090	2090[a]	—	—	—	—
F40T12	34-36‖	40	0.46	73	20,000	1219	48	Med. Bipin	2800	2800	2880	2800	2800	2800	—
F48T12/HO	55	60	—	—	12,000	1219	48	Reces. DC	3850[a]	4075	3850[a]	4400[b]	—	—	—
F96T12/1500	95	110	0.83	126	12,000	2438	96	Reces. DC	8430	8430	8430	8620	8500[a]	8600[c]	—
F96T12/1500	195	215	1.58	137	12,000	2438	96	Reces. DC	—	—	—	—	—	—	—
F48PG17	95	110	1.53	64	12,000	1,219	48	Reces. DC	—	—	—	—	—	—	—
F96PG17	185	215	1.57	144	12,000	2438	96	Reces. DC	—	—	—	—	—	—	—
Preheat start															
F40T12	34	40	0.45	84	15,0000	1219	48	Med. Bipin	—	—	—	—	—	—	—
F90T17	86	90	—	—	9000	1524	60	Mog. Bipin	—	—	—	—	—	—	—
Instant start (Slimline)															
F48T12	30-32	38-40	—	—	9000	1219	48	Single pin	2610	2610	2610	2700[b]	—	—	—
F96T8	40-41	50-51	—	—	7500	2438	96	Single pin	—	—	—	—	—	—	—
F96T8[b]	56	—	0.26	267	15,000	2438	96	Single pin	5800	—	5675	5800	5800	5800	—
F96T12	60	75	044	153	12,000	2438	96	Single pin	5675	5675	5675	5850	5850	5850	—

Nominal Lumens‡§

Lamp description	Lamp watts‡	Lamp watts replaced	4150K CRI 60+ Cool white	3000K CRI 50+ Warm white	3470K CRI 60+ white	6380K CRI 70+ daylight	5000K CRI 90+ C50	4180K CRI 80+ Deluxe cool white	2990K CRI 70+ Deluxe warm white (soft white)	4200K CRI 40+ Lite-white	3200K CRI 80+ Optima 32ND Natural	3570K CRI 80+ C75	7500K CRI 90+ Vita-lite^d	5500K CRI 90+
Rapid start														
F17T8	17	—	—	—	—	—	—	—	—	—	—	—	—	—
F25T8	25	—	—	—	—	—	—	—	—	—	—	—	—	—
F32T8	32	—	—	—	—	—	—	—	—	—	—	—	—	—
F40T8	40	—	—	—	—	—	—	—	—	—	—	—	—	—
F40T12/U/3	36	40	2350^a	2425^a	—	—	—	—	—	2500^a	—	—	—	—
F40T12/U/6	34	40	2480	2530	2550	—	—	—	—	2620	—	—	—	—
F30T12	25	30	1975	2025	—	—	—	—	—	—	—	—	—	—
F40T12	34–36^b	40	2670	2730	2700	2310	2010	1930	1925	2800	2010	—	—	—
F48T12HO	55	60	—	—	—	—	—	—	—	3900^a	—	—	—	—
F96T12HO	95	110	8020	8130	8000	6750^c	—	5750	—	8400	—	—	6000^D	—
F96T12/1500	188	215	13430	—	—	—	—	—	—	13,880	—	—	—	—
F48PG17	95	110	5700^A	—	—	—	—	—	—	—	—	—	—	—
F96PG17	185	215	13,500^A	—	—	—	—	—	—	14,100'	—	—	—	—
Preheat start														
F40T12	34	40	2700	—	—	—	—	—	—	—	—	—	—	—
F90T17	86	90	5765	—	—	—	—	—	—	—	—	—	—	—
Instant start (Slimline)														
F40T12	30.7	40	2475	—	—	—	—	—	2525	—	—	—	—	—
F96T8	40	50	3450^a	—	—	—	—	—	—	—	—	—	—	—
F96T12	60	75	5430	5570	5370	4730	4050	3950	3900	5670	4200	—	—	4015

*The life and light output ratings of fluorescent lamps are based on their use with ballasts that provide proper operating characteristics. Ballasts that do not provide proper electrical values may substantially reduce lamp life, light output or both.

†Rated life under specified test conditions at 3 hours per start. At longer burning intervals per start, longer life can be expected.

‡"RE" indicates "RARE EARTH" type phosphors. This nomenclature has been developed by NEMA to define a system of color rendering information. RE 70 designates a CRI range of 70-79, RE 80 a range of 80-89, and RE 90 ≥ 90.

§At 100 hours. When lamp is made by more than one manufacturer, light output is the average of all manufacturers submitting data.

‖Also in 32 watt cathode cutout but with reduced life.

^a General Electric,

^b Sylvania.

^c Phillip.

^d Duro-Test.

T-Measures for Lamps. The T-measure for a fluorescent lamp is the measure of the diameter of the lamp in eighths of an inch. Thus, a T12 lamp is twelve-eighths of an inch (or 1-1/2 inches) in diameter, while a T10 lamp is 1-1/4 inches in diameter.

T10 Lamps. T10 lamps typically contain phosphors which produce high efficacy and color rendition. They will operate on most ballasts designed for T12 lamps.

T8 Lamps with Electronic Ballasts. T8 lamps produce an efficacy of up to 100 lumens/Watt, the highest efficacy of any fluorescent lamp. They will not generally operate on standard ballasts rated for T12 lamps. They also use less of the toxic materials found in larger fluorescent lamps. T12 fixtures can be retrofitted with T8 lamps by using socket adapters and replacing the old ballasts with the T8 compatible ballasts.

Compact Fluorescent Lamps (CFLs). These twin-tube (TT) and double twin-tube (DTT) lamps are designed to replace many frequently used incandescent bulbs. These lamps can be used to reduce energy use and power demand by over 70%. For example, a 13-Watt fluorescent TT lamp can be used to replace a 60-Watt incandescent lamp. The light produced is similar in appearance to that of an incandescent lamp (i.e., color temperature of 2700 Kelvins for both). Since these lamps produce less heat, space cooling costs are also reduced. Many TT lamps/ballasts have a low power factor of about 0.5-0.6. A high power factor ballast with low harmonic distortion should be specified if a large number of these lamps is used.

Compact fluorescent lamps can be installed as a screw-in or hard-wired conversion kit and have a lifetime of at least 10,000 hours. Frequent cycles of short operation hours will significantly reduce lamp life. One-piece screw-in compact fluorescent lamps have higher life-cycle costs than two-piece screw-in models or the hard-wire models because the useful ballast life ends when the lamp burns out. Using the hard-wire kits will eliminate the possibility of an uninformed maintenance worker throwing away a two-piece screw-in conversion kit when the lamp fails. A number of facilities such as apartment buildings and motels find that screw-in compact fluorescent lamps are often stolen because they are easy to remove and are fairly expensive. Figure 5-5 lists the characteristics of some of the compact fluorescent lamps available.

5.1.2.3 High Intensity Discharge (HID) Lamps

These lamps are relatively expensive initially but offer low life-cycle

Figure 5-5. Compact Fluorescent Lamps [ref 1, Figure 8-1171]

Generic designation NEMA	Lamp watts	Bulb type	Base	Maximum overall length (mm)	(in.)	Rated life[†] (h)	Lamp current (A)	Lamp voltage (V)	Approx. initial lumens[‡]	Lumens per watt	Color[§] temperature and/or CRI
Twin tube 2700K, 3500K, 4100K, 5000K[§] CRI 80 +											
CFT5W/G23	5	T-4	G23	105	4-1/4	7500	0.180	38	250	50	82
CFT7W/G23	7	T-4	G23	135	5-7/16	10,000	0.180	45	400	51	82
CFT9W/G23	9	T-4	G23	167	6-9/16	10,000	0.180	59	600	67	82
CFT13W/GX23	13	T-4	GX23	191	7-1/2	10,000	0.285	60	888	68	82
CFT5W/2G7	5	T-4	2G7	85	3-11/32	10,000	0.180	35	250	50	82
CFT7W/2G7	7	T-4	2G7	115	4-17/32	10,000	0.180	45	400	57	82
CFT7W/2G7[1],‖	9	T-4	2G7	145	5-22/32	10,000	0.180	59	600	67	82
CFT13W/2GX7	13	T-4	2GX7	175	6-29/32	10,000	0.285	59	900	69	82
FT18W/2G11[2]	18	T-5	2G11	229	9	12,000	0.375	60	1250	69	82
FT18W/2G11RS	18	T-5	2G11	267	10-1/2	20,000	0.250	76	1250	69	82
FT24W/2G11	24-27	T-5	2G11	328	12-29/32	12,000	0.340	91	1800	69	82
FT36W/2G11	36-39	T-5	2G11	422	10-5/8	12,000	0.430	111	2900	76	82
FT40W/2G11	40	T-5	2G11	574	22-19/32	20,000	0.270	169	1350	79	82
FT50W/2G11	50	T-5	2G11	574	22-19/32	14,000	0.43	147	4000	80	—
Quad 2700K, 3500K, 4100K, 5000K; CRI 80 +											
CFQ9W/G23	9	T-4	G23-2	111	4-3/8	10,000	0.180	15	575	21	82
CFQ13W/GX23	13	T-4	GX23-2	125	4-29/32	10,000	0.285	15	860	59	82
CFQ10W/G24d	10	T-4	G24d-1	118	45	10,000	0.140	64	600	60	82
CFQ13W/G24d	13	T-4	G24d-1	152	6	10,000	0.170	91	900	69	82
CFQ18W/G24d	18	T-4	G24d-2	175	6-29/32	10,000	0.220	100	1250	69	82
CFQ28W/G24d	26	T-4	G24d-3	196	7-23/32	10,000	0.315	105	1800	69	82
CFQ15W/GX32d	15	T-5	GX32d-1	140	5-1/2	10,000	—	60	900	60	—
CFQ22W/GX32d	20	T-5	GX32d-2	151	5-15/16	10,000	—	53	1200	60	—
CFQ26W/GX32d	27	T-5	GX32d-3	173	6-13/16	10,000	—	54	1600	59	—
CFQ10W/G24q	10	T-4	G24q-1	117	4-5/8	10,000	0.190	64	600	60	82
CFQ13W/G24q	13	T-4	G24q-1	152	6	10,000	0.170	77	900	69	82
CFQ18W/G24q	18	T-4	G24q-2	173	6-13/16	10,000	0.110	80	1250	69	82
CFQ26W/G24q	26	T-4	G24q-3	194	7-5/8	10,000	0.158	80	1800	69	82

(Continued)

Figure 5-5. Compact Fluorescent Lamps [ref 1, Figure 8-1171 (*Continued*)

Typical incandescent lamp substitutes (compact fluorescent lamps, internal ballast)#

Generic designation	Ballast type	Incandescent equivalent (W)	Lamp watts	Bulb type	Base	Related life[+] (h)	Maximum overall length** (mm)	Maximum overall length** (in.)	Lamp current (A)	Lamp voltage (V)	Approx. initial lumens[‡] (Candle-power)	lumens per watt	Color* temperature and/or CRI
7 W[3]	Electronic	25	7	T-4	Med. Screw	10,000	140	5-1/2	0.140	120	400	57	—
11 W[3]	Electronic	40	11	T-4	Mod. Screw	10,000	140	5-1/2	0.170	120	600	55	—
11 W reflector[3]	Electronic	50W .R30	11	P-35	Med. Screw	10,000	148	5-13/16	0.170	120	(315)	—	—
11 W globe[3]	Electronic	30	11	G-32	Med. Screw	10,000	168	6-9/16	0.170	120	450	41	—
15 W[1,2,3]	Electronic	60	15	T-4	Mod. Screw	10,000	172	6-25/32	0.240	120	900	60	81
15 W reflector[3]	Electronic	75W.R40	15	RSB	Med. Screw	10,000	183	7-7/32	0.240	120	(1335)	—	—
15 W globe[3]	Electronic	50	15	G-38	Med. Screw	10,000	188	7-13/32	0.240	120	700	47	82
15 W globe[2]	Magnetic	40-60	15	G-30	Med. Screw	9,000	160	6-5/16	—	120	700	47	—
17 W decorative diffuser[1]	Electronic	60	17	T-24	Med. Screw	10,000	149	5-7/8	0.265	120	950	57	—
18 W decorative diffuser[2]	Electronic	75	18	T-24	Med. Screw	10,000	183	7-7/32	0.240	120	1100	61	—
18 W reflector	Electronic	75W.R40	18	R-40	Mod. Screw	10,000	142	5-19/32	0.240	120	800	44	—
18 W4	Electronic	75	18	T-4	Med. Screw	10,000	175	6-7/8	—	120	1100	61	—
20 W[1,2,3]	Electronic	75	20	T-4	Med. Screw	10,000	203	8	0.265[‡‡]	120	1200	60	—
22 W circular lamp[2]	Magnetic	75	22	T-9	Med. Screw	10,000	130	5-1/8	—	120	1200	54	52
23 W[1,3]	Electronic	90	23	T-4	Med. Screw	10,000	176	6-5/16	0.325	120	1550	67	—
26 W[2]	Electronic	90	26	Y-5	Med. Screw	10,000	203	8	0.430	120	1500	58	—

(*Continued*)

Figure 5-5. Compact Fluorescent Lamps [ref 1, Figure 8-117] (Continued)

Square 2700 K and 3500 K²

Generic designation	Ballast type	Lamp watts	Incandescent equivalent (W)	Bulb type	Base	Related life† (h)	Maximum overall length** (mm)	(in.)	Lamp current (A)	Lamp voltage (V)	Approx. initial lumens‡ (Candle-power)	lumens per watt	Color* temperature and/or CRI
F102D/827/4P		10		2D	GR10Q 4Pl	8000	94	3-23/32	0.18	72	650	65	82
F162D/827/4P		16		2D	GR10Q 4Pl	8000	140	5-1/2	0.195	103	1050	66	82
F212D/827/4P		21		2D	GR10Q 4Pl	8000	140	5-1/2	0.26	101	1350	64	82
F282D/827/4P		28		2D	GR10Q 4Pl	10,000	203	8	0.32	107	2050	73	82
F382D/827/4P§§|||		38		2D	GR10Q 4Pl	10,000	203	8	0.43	110	2850	79	82

Note: Many compact fluorescent lamps come equipped with normal (formerly called low) power-factor ballasts, either integral or as auxiliaries To minimize line losses and to maximize energy savings and lamp efficiency specifiers should require power-factor ballasts and should insist that luminaires, utilizing these lamps, be so equipped.

*These values are the averages of several manufacturers' data.

†At 3 hours per start.

‡Values for reflector lamps are in beam candlepower.

§Not all color temperatures available from each manufacturer.

||Also available in red, blue, green.

#Screw base adapters with integral ballasts are available for retrofitting twin and quad tube lamps into incandescent fixtures.

**Some lamps are available in shorter MOLs.

††HLLDs not yet available for these lamps.

‡‡Lamps are available with higher power factors and lower line currents.

§§Add 2 watts for cathodes when operating on rapid-start circuits.

||||Rapid start life is estimated as 12000 hours @ 3 or more hours per start.

1Philips

2General Electric

3Osram

4Sylvania

costs due to long life and high efficacy. In general, the larger the HID lamp the higher the efficacy. The HID lamp efficacy can be affected by the lamp position, and some of these lamps have a significant color shift and loss of efficacy near the end of their rated life. Figure 5-6 shows some of the commonly used HID lamps.

Mercury Vapor Lamps. Mercury vapor lamps were the first HID lamps. They can offer good color rendering and low-to-moderate efficacy. Self-ballasted mercury vapor lamps are a direct replacement for large incandescent lamps but have only 30-50% of the efficacy of typical mercury vapor lamps. Mercury vapor lamps are good candidates for replacement with more efficient light sources such as metal halide, high-pressure sodium, and low-pressure sodium.

Metal Halide lamps. Many of these lamps produce bright white light. They are used in applications which require good color rendering and high lighting levels such as sports facilities. They do not have as long a life as most other HID lamps. Some metal halide lamps can be installed in fixtures designed to operate mercury vapor lamps.

High Pressure Sodium (HPS) Lamps. HPS lamps offer high efficacy, long life, and a relatively small light source which is easily controlled. The light produced does not render colors well but is useful for tasks in which color rendering is not critical. These lamps are usually the best choice for warehouses, factories, exterior floodlighting, and streetlighting.

5.1.2.4 Low-Pressure Sodium (LPS) Lamps

LPS lamps offer the highest efficacy of any light source (i.e., up to 180 lumens/Watt). They have a long life but are fairly large compared to HID lamps. Their size requires a larger, more complex reflector design to efficiently utilize the light produced. The color of the light is nearly

A   T   PAR   E   R   BT   R

Figure 5-6. HID Lamp Shapes and Designations [ref 8, Figure 2-16]

a monochromatic yellow under which very little color discrimination is possible. LPS lamps are common in European street lighting systems.

LPS is useful for interior security lighting. The distinctive yellow light tells law enforcement personnel that the premises should be unoccupied. Thieves and vandals can be mistaken for employees working late when better color-rendering light sources are used for night-time security.

5.1.3 Ballasts

Light from discharge-type lamps (e.g., fluorescent, mercury vapor, metal halide, high-pressure sodium, and low-pressure sodium) is produced indirectly by a cathode exciting a gas in which an electrical arc forms which then emits light. In a fluorescent lamp, mercury vapor emits ultra-violet radiation which strikes the sides of the lamp wall where phosphors convert it to visible light. A ballast is required to start and operate all discharge lamps.

Each lamp/ballast combination should have test results which state the ballast factor (BF). The lumen rating for a lamp is based on a particular lamp/ballast combination with a ballast factor of 1.0. The ballast factor indicates the light output of a particular system relative to a standard test ballast on which the lamp lumen ratings are based. For example, if a four-foot F32T8 has a rated lamp light output of 3000 lumens, and it is used with a ballast which has a ballast factor of 1.1 for the lamp/ballast combination, the system produces about 3300 lumens.

The ballast label usually gives the electric current drawn for particular lamp types, states whether the ballast does not contain polychlorinated biphenyls (PCB), and shows a wiring diagram. Many older ballasts, especially those manufactured before 1975, utilize PCB oil which contains some of the most toxic chemicals known. Leaking PCB-laden ballasts should be handled as a hazardous waste material during disposal. Ballasts which do not contain PCB can currently be landfilled in most states.

National standards require that all new ballasts have a minimum efficiency and contain no PCBs. Ballasts are usually labeled low-heat (LH), very-low heat (VLH), super-low heat (SLH), energy-saving, or electronic. Not all electronic ballasts on the market will operate at higher efficiency than magnetic ballasts. Claims of "full light output" can be distracting. Use the ballast factor to determine the actual efficiency of the ballast. Many of these new ballasts have much longer operating lives than the ballasts they replace. Some low-efficiency replacement ballasts are still available but should not be used.

Unlike magnetic core-and-coil ballasts, some electronic ballasts can be dimmed, but they also can produce significant levels of harmonic distortion. When recommending electronic ballasts, specify ballasts with a total harmonic distortion (THD) less than 20%.

Strobing and flickering lamps strain the ballast. Strobing occurs as the ballast attempts to restart and operate the lamp(s). The problem is sometimes due to a loose or corroded connection, but most often the lamp has come to the end of its useful life. As lamps age, their light output decreases and they become more difficult to start and operate. If checking the connections and lamps does not reveal the cause of the flickering, then ballast replacement may be necessary.

5.1.4 Luminaires (Fixtures): Lenses, Diffusers, and Reflectors

The luminaire is the complete lighting fixture. It consists of a housing socket, the light source (lamps) and the components which distribute the light such as the lens, the diffuser and the reflector.

The coefficient of utilization (CU) of a lighting fixture is the ratio of the light falling on the working surface to the light produced by the lamp in that fixture. Light is absorbed and converted to heat by surfaces, by lamp-to-lamp interactions, and by the lenses. Tables of common CU values can be found in the IES Lighting Handbook [1].

5.1.4.1 Types of Fixtures

Some common types of lighting fixtures are listed below. Retrofit options are noted where appropriate.

Jars and Globes: These fixtures typically have low CU values. Although they generally use incandescent lamps, compact fluorescent lamps can also be used in jar and globe fixtures.

Wall Surface-Mounted: These fixtures have moderate CU values.

Pendant: These fixtures utilize a variety of lamp types. They often use bare lamps, but are sometimes fitted with globes, lenses, or diffusers. Large incandescent lamps (e.g., 500-1500 Watt) are frequently used, but more efficient sources are available. These fixtures often hang from the ceilings of production and warehouse areas. They can sometimes be lowered to increase the light levels at the work surfaces if the resulting light distribution is acceptable.

Track and Recessed: These fixtures (sometimes referred to as downlights) baffle some light when A-lamps or deeply-recessed R-lamps are used.

Most of the baffled light is converted to heat. Ellipsoidal reflector lamps (ER) are designed to cast more light out of recessed fixtures. If an adjustable-depth can-type fixture without a specular reflector looks bright inside, there is potential for using an ER lamp. However, a protruding ER lamp is not a good application.

Although there are some excellent downlights available which have been designed for CFLs, in general a compact fluorescent screw-in lamp should not be used in a track or recessed fixture unless lighting levels can be reduced. A CFL casts nearly all of its light out to the sides instead of downward where it is needed in a downlight.

Area Lights: These fixtures are commonly known as barn lights. They cast light in nearly all directions. Much of the light is lost to the sky and overhead trees, or it trespasses off the property. Cutoff luminaires which use reflectors to direct the light where it is needed can produce the same light levels for the area of interest with about a third of the power.

Streetlights: Some types such as barnlights and cobraheads cast light upward and away from the area of interest. Cutoff luminaires offer a good alternative here too.

Figure 5-7 demonstrates the application of well-controlled lighting fixtures to reduce outdoor lighting costs an poorly applied fixtures.

Exit Signs: These older signs typically use two 20-Watt or two 25-Watt incandescent lamps which can be replaced by a 2- to 5-Watt LED (light emitting diode) lamp. This will reduce operating costs by more than 90%. These LED exit lamps can last up to 90,000 hours when run continuously.

Floodlights: Look for low-efficacy sources such as mercury vapor lamps. High pressure sodium floodlights offer excellent savings.

5.1.4.2 Lenses, Diffusers and Reflectors

Adding lenses or reflectors to a lighting fixture changes the light distribution pattern (i.e., photometrics). Replacing the lens or diffuser can be helpful in situations where the system was poorly designed or the use of the space has changed. For example, many office areas which were lighted appropriately for traditional paperwork now have glare on video terminals because people are using computers instead of typewriters or pens. Polarized lenses can reduce the glare but they also reduce the amount of light leaving the fixture. Figure 5-8 shows how relationships can change when the workstation is altered.

The top two illustrations indicate effective ways of achieving desired lighting for the area involved (indicated by the dotted line). The bottom two approaches waste energy and the "light trespass" effect of the "spill light" can irritate neighbors.

Figure 5-7. Effective Outdoor Lighting [ref 7, Figure 11]

Prevailing relationships between workers, their tasks, and their environments are altered significantly when traditional "white paper tasks" (on left) are changed to VDT-based tasks (on right).

Figure 5-8. Prevailing Lighting Relationships [ref 7, Figure 7]

Glare from the specular surfaces reduces visibility. This veiling glare can be reduced with careful fixture positioning and proper selection of diffusers. Glare on video display terminals can often be alleviated by tipping the monitor down or putting a wedge under the back. Figure 5-9 shows how glare can reach the eyes of a worker.

Most fixtures have built-in reflectors. A fixture with a baked white enamel finish has a reflectivity of about 0.88. The practice of removing half the lamps and adding a reflector to a fixture will reduce the amount of light leaving a fixture. Therefore, this is only practical when the area is over-lighted.

The performance of lenses, diffusers, and reflectors depends on environmental factors such as dirt accumulation, oxidation, glare, discoloration due to UV light exposure, and vandalism. Surface finishes react with airborne chemicals and lose some light transmission capability as they age. Choosing materials appropriate for the work environment and cleaning the fixtures regularly will prolong their useful lives. Outdoor lighting presents some special problems. Vandals often delight in shooting at exterior lighting fixtures. Polycarbonate lenses resist damage due to firearms and other projectiles. However, they also discolor from prolonged exposure to UV light produced by the lamp or the sun. Acrylic lenses typically last longer

Veiling reflections commonly occur when the light source is directly above and in front of the viewer, in the "offending zone."

Figure 5-9. Veiling Reflections [ref 7, Figure 3]

than even UV-inhibited polycarbonate lenses. Table 5-3 gives a comparison of the performance of acrylic and polycarbonate lenses over time.

5.2   DETERMINING LIGHTING NEEDS

A variety of techniques are available for estimating the lighting levels in a given space. The average illuminance method described in the Illuminating Engineering Society IES Lighting Handbook [1] incorporates the major variables affecting light utilization: amount of light produced by lamps, amount of light exiting the fixture, mounting height and spacing of fixture, fixture photometrics, lumen dirt depreciation, lamp lumen depreciation, ballast factor, and room surface finish characteristics. A worksheet to calculate lighting levels is also found in the IES Lighting Handbook.[1]

The IES has developed standards for appropriate lighting levels for typical applications. Lighting levels are generally expressed in terms of illuminance, which is measured in footcandles. Table 5-4 shows the illuminance category for a number of commercial and industrial applications.

Once the appropriate illuminance category has been identified, Table 5-5 can be used to determine the range of illuminance values needed

Table 5-3. Comparison of Acrylic and Polycarbonate Lenses
[ref 7, Table 4]

	Acrylic	High-Impact Acrylic	Polycarbonate
Light Transmission	92%	90%	88%
Aging-Light Stability	10-15 yrs	10-15 yrs	3-4 yrs
Impact Strength	1	10X	30X but degrades rapidly
Haze	Under 3%	Under 3%	Under 3% but degrades
Scratch Resistance	Very Good	Good	Good
Burning Character (U.L. Class)	Slow	Slow	Self-Ext.
Smoke Generation	Slight	Slight	High
Resistance to Heat	200°F	180°F	250°F
Type Smoke	Nontoxic	Nontoxic	Toxic
Relative Cost	1X	2X	4X

Table 5-4. Lighting Recommendations for Specific Tasks

Area-activity	Illuminance category
Bakeries	D
Classrooms	D to E
Conference rooms	D
Drafting rooms	E to F
Hotel lobbies	C to D
Home kitchens	D to E
Inspection, simple	D
Inspection, difficult	F
Inspection, exacting	H
Machine shops	D to H
Material handling	C to D
Storage, inactive	B
Storage, rough, bulky items	C
Storage, small items	D
	Footcandles
Building entrances	1-5
Bulletin boards, bright surroundings, dark surfaces	100
Bulletin boards, dark surroundings, bright surfaces	20
Boiler areas	2-5
Parking areas	1-2

to achieve desirable lighting levels. In using these tables, the IES recommends that the lower values be used for occupants whose age is under 40 and/or where the room reflectance is greater than 70% and that the higher values be used for occupants more than 55 years old and/or where the room reflectance is less than 30%. For occupants between 40 and 55 years of age and where the reflectance is 30-70% or where a young occupant is combined with low reflectance or an older person is in a high-reflectance environment, the intermediate values should be used. In addition, the need for speed and accuracy influences the amount of light needed, with higher speed and accuracy demanding more light.

Lighting levels can sometimes be reduced if there is sufficient contrast between the work and its surroundings. However, too much contrast between the work and the ambient environment will fatigue the eyes. For example, occupants of dark paneled offices who work with pencil on brightly-illuminated white paper will often experience eye fatigue.

The illuminance values considered acceptable have changed through the years in response to emerging technology. Recommended light levels were relatively low in the incandescent lamp era due to restrictions resulting from heat production. From 1945-1960 more efficient light sources were developed and energy costs plummeted, leading to a tripling of the recommended values [3]. Rising energy and power demand costs in the 1970's spurred the industry to reduce the recommended lighting levels to the minimum necessary to provide adequate illumination.

5.3   MAINTAINING THE LIGHTING SYSTEM

In addition to a proper choice of light sources, ballasts, and luminaires, the efficiency of a lighting system depends on maintenance policies. Maintenance includes both cleaning and relamping.

5.3.1 Luminaire Maintenance

The performance of lenses, diffusers, and reflectors depends on environmental factors such as dirt accumulation, oxidation, vandalism, and degradation due to ultra-violet (UV) light exposure. Typical fluorescent lamp performance under various temperature conditions is shown in Figure 5-10.

Lamps, fixtures, reflectors, lenses, and diffusers collect dust and insects. Dust accumulation on lighting fixtures and on surfaces adjacent to lighting fixtures reduces light utilization by up to 40 percent and increases heat production. Periodic cleaning of the fixtures will maintain higher

Table 5-5. Illuminance Categories and Illuminance Values

Type of activity	Illuminance category	Ranges of illuminances		Reference work plane
		Lux	Footcandles	
Public spaces with dark surroundings	A	20-30-50	2-3-5	
Simple orientation for short temporary visits	B	50-75-100	5-7.5-10	General lighting throughout spaces
Working spaces where visual tasks are only occasionally performed	C	100-150-200	10-15-20	
Performance of visual tasks of high contrast or large size	D	200-300-500	20-30-50	
Performance of visual tasks of medium contrast or small size	E	500-750-1000	50-75-100	Illuminance on task
Performance of visual tasks of low contrast or very small size	F	1000-1500-2000	100-150-200	
Performance of visual tasks of low contrast and very small size over a prolonged period	G	2000-3000-5000	200-300-500	
Performance of very prolonged and exacting visual tasks	H	5000-7500-10000	500-750-1000	Illuminance on task, obtained by a combination of general and local (supplementary lighting)
Performance of very special visual tasks of extremely low contrast and small size	I	10000-15000-20000	1000-1500-2000	

Figure 5-10. Lamp Performance vs. Temperature

and more uniform light levels. All lamps should be cool before cleaning. Gloves should be worn when cleaning any mirror-like reflective part of a luminaire. Quartz lamps should not be cleaned.

Outdoor lighting has some special maintenance problems. Poorly-sealed gaskets allow insects to clog the lenses of outdoor lighting fixtures. Dead insects can completely block out light. Overgrown vegetation can also reduce lighting levels from outside fixtures. Regular trimming of shrubs and trees will help to fully utilize these light sources. Small trees planted under or near lighting fixtures can quickly grow to block the light source.

5.3.2 Establishing the Lighting System Maintenance Schedule

Establishing a good maintenance schedule for a lighting system takes three steps. First, you must determine the maintenance character-istics of the luminaires in your facility. Table 5-6 lists the maintenance categories for a variety of luminaires.

The next step is to determine what dirt conditions the luminaires are likely to experience. Table 5-7 shows dirt conditions for representative areas. Once you know both the maintenance category from Table 5-6 and the appropriate dirt conditions for the facility from Table 5-7, then use the graphs in Figure 5-11 to set the luminaire maintenance schedule. These graphs show the effect that dirt accumulation has upon lighting levels

Table 5-6. Maintenance Categories for Luminaire Types

Maintenance category	Top enclosure		Bottom enclosure	
I	1.	None	1.	None
II	1.	None	1.	None
	2.	Transparent with 15% or more uplight through apertures	2.	Louvers or baffles
	3.	Translucent with 15% or more uplight through apertures		
	4.	Opaque with 15% or more uplight through apertures		
III	1.	Transparent with less than 15% uplight through apertures	1.	None
	2.	Translucent with less than 15% uplight through apertures		
	3.	Opaque with less than 15% uplight through apertures		
IV	1.	Transparent unapertured	1.	None
	2.	Translucent unapertured	2.	Louvers
	3.	Opaque unapertured		
V	1.	Transparent unapertured	1.	Transparent unapertured
	2.	Translucent unapertured	2.	Translucent unapertured
	3.	Opaque unapertured		
VI	1.	None	1.	Transparent unapertured
	2.	Transparent unapertured	2.	Translucent unapertured
	3.	Translucent unapertured	3.	Opaque unapertured
	4.	Opaque unapertured		

Table 5-7. Degrees of Dirt Conditions

	Very clean	Clean	Medium	Dirty	Very dirty
Generated dirt	None	Very little	Noticeable but not heavy	Accumulates rapidly	Constant accumulation
Ambient dirt	None (or none enters area)	Some (almost none enters)	Some enters area	Large amount enters area	Almost none excluded
Removal or filtration	Excellent	Better than average	Poorer than average	Only fans or blowers if any	None
Adhesion	None	Slight	Enough to be visible after some months	High—probably due to oil, humidity, or static	High
Examples	High-grade offices, not near production; laboratories; clean rooms	Offices in older buildings or near production; light assembly; inspection	Mill offices; paper processing; light machining	Heat treating; high-speed printing; rubber processing	Similar to "Dirty" but luminaires within immediate area of contamination

Source: Courtesy of the Illuminating Engineering Society of North America.

over a period of months.

These graphs are for average conditions. Actual conditions may warrant a more frequent cleaning schedule than the graphs indicate. Using open fixtures and diffusers often reduces the potential for light loss since there are fewer surfaces for settling, and particles and dust can be more easily removed from these fixtures. However, an environment with potentially explosive dust; such as flour, corn, coal, etc., has to have sealed explosion-proof fixtures. Typical lamp lumen depreciation, luminaire dirt depreciation, and room surface dirt depreciation curves for fluorescent lamps are shown in Figure 5-12.

5.3.3 Relamping strategies

The usual strategy for replacing lamps in many facilities is to wait until a lamp burns out and then replace it (called spot relamping). This relamping strategy is not necessarily the best one for a facility to follow because it does not consider such factors as labor costs or lumen depreciation. It is often more economical to replace all of the fluorescent and HID lamps in a facility at one time (called group relamping). Spot relamping is more labor intensive and results in less efficient lighting than group relamping. However, spot relamping can be more practical for lamps with a short life such as incandescent lamps.

The reason that spot relamping results in a loss of lighting efficiency is that the amount of light that comes from a lamp declines with the age of the lamp. Performance curves, such as those in Figures 5-13, 5-14, and 5-15 show how the light output is reduced as a function of ordinary usage. Note that the life of a lamp is measured in hours of use rather than installed hours. In addition to degraded performance of individual lights, the total lighting system performance is decreased as individual lamps burn out. Typical mortality curves, showing the percent of lamps in service as a function of time, are given in Figures 5-16, 5-17, and 5-18.

As these figures show, lamp mortality at 85-100% of rated life is about twenty-five times that of lamps aged 0-70% of rated life. Therefore, fixtures should be group relamped when lamps are between 70-80% of the rated lamp life.

The average useful lamp life is shown in Figure 5-19.

Since most F40T12 four-foot fluorescent lamps are rated for 20,000 hours, they should be replaced after 16,000 hours of operation. Lamp life ratings indicate the point at which half of the lamps are likely to have failed. Fluorescent lamp lifetimes are rated at three hours per start, HID lamps are rated at five hours per start. Operating the lamps for shorter periods will reduce lamp life.

Figure 5-11. Luminaire Dirt Depreciation Factors

[Fraction of full light output is the scale on the y-axis.] (Courtesy of the Illuminating Engineering Society [1])

**Figure 5-12. Lumen Maintenance Curve
For Fluorescent Lighting Systems [ref 8, Figure 2-20]**

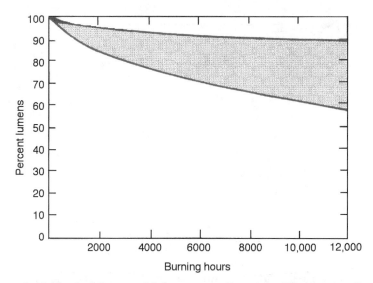

Figure 5-13. Typical Lumen Maintenance Curve for Fluorescent Lamps
(From General Electric Technical Pamphlet TP-105.
Courtesy of General Electric Co.)

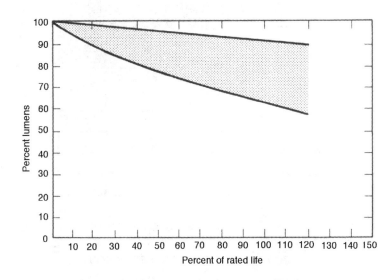

Figure 5-14. Typical Lumen Maintenance Curve for Filament Lamps
(From General Electric Technical Pamphlet TP-105.
Courtesy of General Electric Co.)

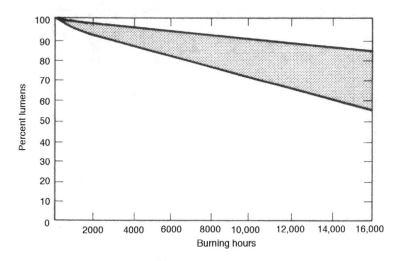

Figure 5-15. Lumen Maintenance Curve for HID Lamps
(From General Electric Technical Pamphlet TP-105.
Courtesy of General Electric Co.)

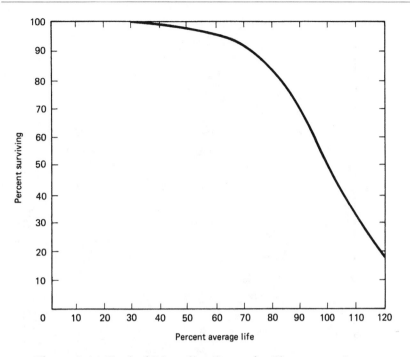

Figure 5-16. Typical Mortality Curve for Fluorescent Lamps
(From General Electric Technical Pamphlet TP-105.
Courtesy of General Electric Co.)

Figure 5-17. Typical Mortality Curve for Filament Lamps
(From General Electric Technical Pamphlet TP-105.
Courtesy of General Electric Co.)

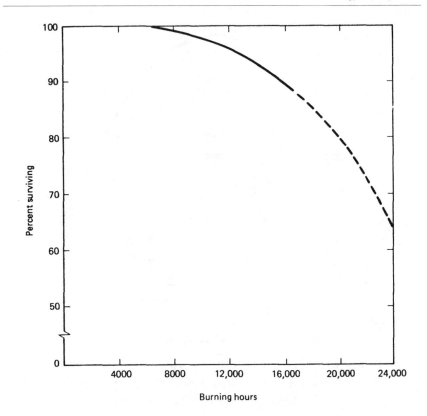

Figure 5-18. Typical Mortality Curve for HID (Mercury Vapor) Lamps
(From General Electric Technical Pamphlet TP-105.
Courtesy of General Electric Co.)

Group relamping can:

1. **Reduce Labor Costs:** Spot relamping can require up to 30 minutes
    to move furnishings or equipment, set up, replace the lamp, and put
    equipment away (e.g., ladder, lamps, tools). In group relamping, a
    fixture can usually be relamped and cleaned in 5 minutes. The loss-
    of-use cost of replacing a few lamps before they have burned out
    is generally less than the increased cost of the labor to individually
    replace burned out lamps.

2. **Reduce Lamp Costs:** Purchasing a large number of lamps at one
    time allows for high volume discounts. Fewer purchases result in
    less time spent ordering, receiving, and stocking lamps.

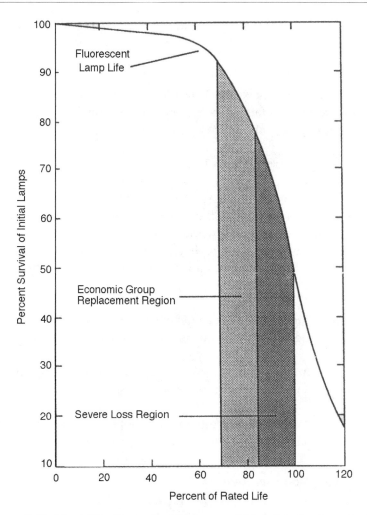

Figure 5-19. Mortality Curve for Fluorescent Lighting Systems [ref 8, Figure 2-20]

3. **Allow Lamp Maintenance to be Scheduled:** Relamping can be scheduled for slow periods for the maintenance staff. Scheduled relamping also allows a regular schedule to be set for regular inspection and cleaning of lamps and fixtures.

4. **Maintain Higher and More Uniform Lighting Levels:** The light output of a lamp decreases with age. Group relamping insures that all lamps have a high light output. Fewer lamp failures, less flickering, and reduced swirling and color shift produce a safer, more comfort-

able work environment. If the lighting system was over-designed to allow for loss of light levels, group relamping may allow for some delamping.

5.   **Reduce Inventory Needs:** Fewer lamps must be stored in inventory since fewer spot failures occur in group relamped fixtures when the relamping interval is set correctly.

6.   **Insure the Correct Lamp Use:** Spot relamping often results in the installation of a variety of lamp types with inconsistent light output levels, lifespans, and colors. This can occur when the inventory of spare lamps runs out and the person who purchases a quick replacement either does not know the correct lamp type, cannot locate the correct type, or does not realize that the cheapest lamp may not be the best value. Group relamping provides an opportunity to install the newest energy-efficient lamps.

7.   **Extend Ballast Life:** Ballasts have to work harder to start and operate strobing lamps which are near the end of their life.

8.   **Reduce Interruptions in Work Area:** Group relamping prevents most of the unplanned lamp replacements.

General Electric [4] has provided the following cost formulas for determining relamping costs:

Spot Replacement Costs

$$C = L + S \qquad (5\text{-}1)$$

Group Relamping Costs

$$C = \frac{L + G}{I} \qquad (5\text{-}2)$$

where: C   = total replacement cost per lamp
          L   = net price per lamp
          S   = spot replacement labor cost per lamp
          G   = group replacement labor cost per lamp
          I   = group relamping interval (% of rated lamp life)

Example 5-1: An office building contains a number of small (400 ft^2) rooms, each of which has four two-lamp fluorescent fixtures. Every time a maintenance person changes lamps, they must bring a ladder into the room and clear away furniture. It takes the person 15 minutes to replace one lamp. It takes 25 minutes to replace all the lamps in a room and clean the luminaires if all the work is done at one time. The lamps cost $0.85 each, and labor costs are $10/hour. The lamps are used about 2000 hours/year. The average lamp life is 20,000 hours. Determine whether group relamping with I = 0.8 is preferable to spot relamping for this building.

Solution: Find the cost for spot relamping and group relamping using equations 5-1 and 5-2, and select the method with lowest cost.

$$
\begin{aligned}
L &= \$0.85 \text{ per lamp} \\
S &= \text{(time to replace lamp in hours)} \times \text{(cost of labor per hour)} \\
  &= 15/60 \times \$10 = \$2.50 \\
G &= \text{(time to replace lamp in hours)} \times \text{(cost of labor per hour)} \\
  &= 3/60 \times \$10 = \$0.50. \\
I &= 0.8
\end{aligned}
$$

and thus

$$
\begin{aligned}
C_{spot} &= 0.85 + 2.50 \\
&= \underline{\$3.35/lamp} \\
C_{group} &= (0.85 + 0.50)/0.80 \\
&= \underline{\$1.69/lamp}
\end{aligned}
$$

Therefore, the decision should be made to use group relamping. The reduced labor cost more than offsets the cost of the lost lamp life; in fact, the savings from group relamping is almost twice the full cost of each lamp.

5.4   THE LIGHTING SURVEY

To perform an objective evaluation of the lighting system, the energy auditor must gather the following data: lighting needs/objectives, hours/days that lighting is required, lighting levels, type of lamps, age of lamps, age of the lighting installation, ambient environment of lighting fixtures (e.g., dust exposure, air temperature, etc.), room surface characteristics, type of ballasts, condition of fixtures, and relamping practices. Building plans are not useful unless the facility was built as planned and few modifications have been made.

Forms are useful for recording the specific lighting data needed. Chapter Two, Figure 2-5 provided a sample lighting data collection form. Figure 5-20 shows a sample data collection form for recording the lighting system condition, while Figure 5-21 illustrates a sample form for recording lighting needs.

Two basic surveys should be conducted to look for savings opportunities: one to see how the facility operates while in production and another to determine the lighting practices when the facility is dormant or shut down for the night.

5.4.1 Interviews with Employees

Interviews with the managers and workers help the energy auditor to evaluate the relamping and maintenance practices, determine problems with the lighting system, ascertain the employees' satisfaction level, find out when light is needed, and uncover the potential for cost savings. The first question should be—Are you happy with your lighting? Major retrofits such as fixture replacement and color changes affect everyone in the work environment, so the opinions of all participants should be considered.

5.4.2 Measuring Light Levels

A light meter is needed to measure illuminance levels. An inexpensive digital light meter is practical and rugged for screening lighting levels and determining relative values. A self-calibrating digital light meter (photometer) is very useful to get fast, accurate, and repeatable measurements. Using a lightmeter, the auditor should record light intensity readings for each area and task of the facility.

Taking notes on the types of tasks performed in each area will help the auditor select alternative lighting technologies that might be more energy efficient. Other items to note are the areas that may be infrequently used and are potential candidates for controlling the lighting with occupancy sensors, or the areas where daylighting may be feasible.

5.5   REGULATORY/SAFETY ISSUES

The lighting industry is encountering increasing safety and environmental concerns. Some of the materials used in lighting fixtures are, or will soon be, labeled hazardous for disposal.

For example, older ballasts and capacitors may contain polychlorinated biphenyl (PCB) oil which should be sent to a facility certified for handling hazardous wastes. Federal law 40CFR Part 761 requires proper

Location _____   Light type (HPS, FL, etc.) _____

Lamps _____

Watts   Condition of system*

Lighting level (fc)	No.	Each	Total	% burning	Luminaires	Reflective surfaces

*From Table 5-7

Figure 5-20. Lighting System Condition Form

Location	Hours when light is needed	Importance of color rendition	Task lighting possible?	Light levels required (fc)
————	————	————	————	————
————	————	————	————	————
————	————	————	————	————
————	————	————	————	————
————	————	————	————	————

Figure 5-21. Lighting Needs Form

disposal of leaking ballasts containing PCB oil. Fluorescent lamps contain mercury vapor (Hg), antimony (Sb), cadmium (Cd) and other toxic chemicals. The new T8 lamps (e.g., 1" diameter) use less phosphor material and mercury gas than the conventional T12 lamps (e.g., 1-1/2" diameter). Using T8 lamps should reduce disposal costs and environmental impacts. New T5 lamps (e.g. 5/8" diameter) are even less of a problem.

5.5.1 Lighting Safety Issues

Lamps are fragile and break easily. Broken fluorescent lamps are difficult to transport and recycle. Areas subject to vibration or other mechanical stress should be illuminated with durable lamps or with fixtures which have adequate containment for broken lamps within the fixture housing. Delamping inexpensive fluorescent lighting fixtures can also be hazardous if the lamp pins come in contact with the fixture housing.

Insulation placed on top of lighting fixtures recessed in the ceiling may pose a fire hazard unless the fixture is rated for insulation contact. The insulation can increase the operating temperature of the lamps and ballasts of fluorescent or high intensity discharge (HID) fixtures, which will reduce the lifespan of all the system components.

High intensity discharge lamps have arc tubes which operate at high temperatures. Pieces of hot arc tube can fall from the fixture if the lamp wall is fractured. Some manufacturers recommend using lenses or fixture housings capable of containing incendiary materials.

5.5.2 Energy Policy Act of 1992

As part of the Federal Energy Policy Act of 1992, Congress placed restrictions on the production and sale of inefficient fluorescent and incandescent lamps [10]. These restrictions will require facility managers to take a hard look at the emerging high-efficiency lighting technologies when many of the lamps currently in use become difficult or impossible to find.

Standard cool-white and warm-white F40T12 lamps, the fluorescent lamps most often seen in commercial lighting fixtures, are specifically targeted by the Act. The manufacture and sale of low-efficacy F96T12 lamps and many reflector-type and PAR-type incandescent lamps are also restricted.

Table 5-8 from the Energy Policy Act of 1992 lists the minimum lamp efficacy and color rendering index requirements for fluorescent and incandescent lamps. Manufacturers must comply with these restrictions between 18 and 36 months after the date of the Act—which was October 24, 1992.

While lamps can be exempted from these restrictions for specialty applications, this law means that facility managers will be able to reduce life-cycle operation costs for lighting more easily. Since there are some gaps in suitable lamp replacements, particularly for incandescent downlights, the law will also spur research and development of new lighting technology. The 2005 Energy Policy Act further regulates and mandates minimum efficiency regulations, incentives and research dollars for lamps, ballasts, fixtures, LEDs, transformers, electrical equipment, motors and appliances. Changes/impacts specific to lighting are documented in Table 5-9.

In addition, the 2005 Energy Policy Act created tax incentives for installing energy efficient equipment. Significant tax rebates are available if the lighting project saves energy. More information about tax incentives (and utility incentives) can be found at:

www.dsireusa.org
www.energytaxincentives.org
http://www.efficientbuildings.org
http://www.lightingtaxdeduction.org

5.6   IDENTIFYING POTENTIAL EMOS

Lighting is used primarily for workplace illumination, for safety, and for decoration. In each of these uses, the same three questions can be asked: (1) How much light is needed? (2) How must the light be controlled? (3) How can lighting be provided most efficiently? When examin-

Table 5-8. Criteria for Manufacturing Lamps

Fluorescent Lamps

Lamp Type	Nominal Lamp Wattage	Minimum CRI	Minimum Average Lamp Efficacy (LPW)	Effective Date (Months)
4-foot medium bi-pin	>35W	69	75.0	36
	<35W	45	75.0	36
2-foot U-shaped	>35W	69	68.0	36
	<35W	45	64.0	36
8-foot slimline	>65W	69	80.0	18
	<65W	45	80.0	18
8-foot high output	>100W	69	80.0	18
	<100W	45	80.0	18

Incandescent Reflector Lamps

Nominal Lamp Wattage	Minimum Average Lamp Efficacy (LPW)	Effective Date (Months)
40-50	10.5	36
51-66	11.0	36
67-85	12.5	36
86-115...	14.0	36
116-155...	14.5	36
156-205	15.0	36

ing an existing system of lighting, the answers to these questions can be used to decrease lighting cost and improve lighting efficiency.

Lighting improvements provide cost savings in a number of ways: reduced energy use and power demand; reduced heat production; lower life-cycle lamp costs; reduced need for maintenance; and increases in safety and productivity. In some instances, improving lighting quality

Table 5-9
2005 Energy Policy Act—Minimum Energy Efficiency Regulations

Product	Effective Date	Description
Medium screw-based compact fluorescent lamps (CFLs)	CFLs manufactured on or after January 1, 2006 must meet new requirements	All CFLs used must meet selected specifications in the 2001 ENERGY STAR® Standard, including: minimum initial efficacy levels, minimum lumen maintenance at 1000 hours and 40% of rated life, rapid cycle stress test, and minimum lamp life of 6000 hours (see ENERGY STAR Sell Sheet, pc 85281).
Electromagnetic fluorescent ballasts for T12 Lamps	**OEM Impact** Cannot manufacture EM ballasts for new fixtures after January 1, 2009 Cannot sell EM ballasts for use in new fixtures after October 1, 2009	Elimination of electromagnetic ballasts that operate: • One 4-foot T12 lamp • Two 4-foot T12 lamps • Two 8-foot T12 slim line lamps
	Distributor Impact EM replacement ballasts can be sold until July 20, 2010.	Electromagnetic ballasts marked "For Replacement Only," containing not more than 10 ballasts per package can be sold under a limited time exemption.
	EXEMPT Dimming ballasts Sign ballasts Residential ballasts	These ballasts do not need to meet minimum efficiency regulations.
Mercury vapor ballasts	Effective January 1, 2008. Can no longer be manufactured or imported.	Bans the sale of mercury lamp ballasts essentially banning the sale of new mercury vapor fixtures.
Illuminated EXIT signs Pedestrian cross-walk modules Traffic signal heads	Effective January 1, 2006. Can no longer be used in new installations.	Must meet ENERGY STAR® efficiency levels (input power $<= 5$ watts). Essentially forces the use of LEDs.
Torchiere fixtures	Effective January 1, 2006	Torchieres shall not be capable of operating with lamps that total more than 190 watts.
Transformers	Effective January 1, 2007	Low voltage dry-type transformers must meet NEMA TP-1 standards.

increases worker productivity, and can result in greater profitability for a facility since the benefits of even a small change in worker productivity can vastly outweigh the relamping and maintenance costs. Figure 5-22 presents the results of a study of productivity in an office building when the lighting levels were first cut in half, then restored.

In examining the lighting system, the energy auditor should ask three questions: first, whether the light is needed; second, whether the correct amount of light is being used; and third, what is the most cost-effective lighting technology to use to supply the correct amount and quality of light. To locate the energy management opportunities, the auditor should specifically:

• Identify and characterize the visual tasks, and determine the contrast of the work to the surrounding surfaces.

Productivity

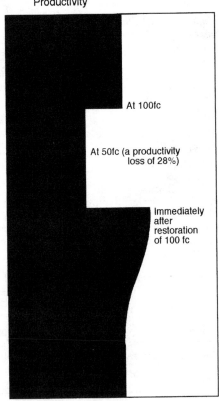

At 100fc

At 50fc (a productivity loss of 28%)

Immediately after restoration of 100 fc

Figure 5-22. Productivity Loss from Light Reduction [ref 6, Figure 2]

- Look for the potential to use daylighting and task-specific lighting to displace high ambient, artificial lighting levels.
- Determine the appropriate lighting levels and the quality of light needed.

- Select alternative lighting systems to meet the needs, and analyze the cost-effectiveness of each alternative.

- Select the best alternative to implement.

The remainder of this section provides some specific recommendations for areas that can result in cost-effective improvements in lighting systems. The advice of a qualified lighting consultant should be solicited before undertaking any major lighting retrofit to ensure the task is completed with the best available technology and the lowest life-cycle costs.

5.6.1 Delamping

Major savings can be obtained by removing some of the lamps that are producing excessive levels of illumination. The lighting levels in many facilities have been over-designed to allow for poor maintenance and relamping practices. The first place to check for excessive light levels is in corridors and at work stations. The range of footcandles for each task was given in Table 5-4. Other places where lighting should be examined carefully include warehouses and storage areas.

There is often a good potential for delamping fixtures when planned group relamping is practiced. For example, a fluorescent light fixture with four new F40T12 40-watt lamps may provide over twice the actual illumination specified by IES standards for a particular task. In such a case, half of the lamps can often be removed while still providing the area with sufficient light levels and light distribution patterns. Hallways with 4-lamp fixtures frequently offer a good opportunity for removing half the lamps while maintaining adequate light levels.

If two lamps are removed from a four-lamp fixture, it is usually better to remove either the inboard or outboard set. Which set depends on the fixture design. Measuring the light levels after removing each set of lamps in turn will reveal which set should be removed. The best light distribution is typically achieved when the lamps are centered in the fixture. There are low-cost kits available for repositioning lamps. Although some reflector sales people advertise their product as capable of maintaining the same fixture light output with half the lamps, much of the increase in light levels is due to centering the lamps in the fixture.

There is usually one ballast for each two fluorescent lamps. The ballast for rapid-start lamps will continue to consume some power even when the lamps are removed. Disconnecting the ballast and capping the power leads will eliminate this power draw and provide a readily accessible replacement ballast already mounted in the fixture. The ballasts in instant-start (IS) fixtures automatically disconnect when the lamps are removed. Leaving burned-out lamps in IS fixtures reduces ballast life.

Example 5-2: The packing and shipping area in a plastic jar production facility is lighted with 50 fluorescent fixtures that each have four F40T12 40-Watt lamps. It operates two shifts per day for 250 days a year. Light level measurements show that the average illumination level is about 110 footcandles. Is delamping warranted, and if so, how much can be saved if electricity costs 8 cents per kWh?

Solution: The IES illumination level standard (Tables 5-4 and 5-5) for packing and shipping—which are material handling tasks—is 50 footcandles as an upper limit for tasks involving large items. Thus, half of the lamps can be removed if the resulting light distribution pattern is still acceptable. Assume that each light fixture has two ballasts, and that each ballast serves two lamps. The two lamps and ballast removed will save 80 Watts for the two lamps, and an additional 15%—or 12 Watts—for the ballast. The total energy cost savings from this delamping can be calculated as follows:

Cost savings = (92 Watts/fixture) × (50 fixtures) × (16 hours/day) ×
                (250 days/year) × (1 kWh/1000 Wh) × ($.08/kWh)
             = $1472/year

5.6.2 Task Lighting

Ambient lighting levels can be reduced when adequate task lighting is supplied for the work. Ambient lighting levels of 25 footcandles or less are frequently sufficient if the individual work areas have sufficient light from task-dedicated lighting fixtures. Figure 5-23 demonstrates the placement of supplementary luminaires.

Figure 5-23. Supplementary Luminaires [ref 1, Figure 9-11]. Examples of placement of supplementary luminaires: a. Luminaire located to prevent veiling reflections and reflected glare; reflected light does not coincide with angle of view. b. Reflected light coincides with angle of view. c. Low-angle lighting to emphasize surface irregularities. d. Large-area surface source and pattern are reflected toward the eye. e. Transillumination from diffuse source.

5.6.3 Relamping

Replacing existing lamps, ballasts and luminaires with newer, more energy-efficient models offers the potential for significant savings on lighting system costs in many facilities. Lamp replacement is often a simple procedure if the new lamp works with the existing ballast and fixture. For example, replacing existing F40T12 40-watt lamps with F40T12 34-watt lamps in appropriate areas offers an easy and cost-effective lighting system improvement. Replacing existing F40T12 40-watt lamps with F32T8 32-watt lamps is a little more complicated as it requires a change of ballasts and lamp sockets in addition to the lamp change. This may still be very cost-effective, but it costs more for equipment and labor, and must be analyzed to see if it is truly cost-effective for a particular facility.

As another example, a 2'×4' lighting troffer (fixture) using four F40T12, 34-watt energy-saving lamps can be retrofitted with three F40T10 40-watt high-efficacy lamps or three F32T8 lamps and an electronic ballast. These alternatives will maintain or increase light levels and reduce energy use by about 14% and 35%, respectively.

There are many potential lamp substitutions possible, and it is important to know what kind of substitutions are reasonable. Table 5-10 shows some of the lamp substitutions which can produce lower life-cycle costs. Many other lamps are also suitable for replacing commonly used lamps.

Table 5-11 shows more possible lamp substitutions including some compact fluorescent lamps.

Table 5-10. Lamp Substitutions [ref 5, Table 5-8]

Present lamp	Substitute	Light level	Energy saved (w/%)
60-W In	30-W RI	100%	30/50
(1000 h)	50-W RI	200%	10/16
60-W In	50-W RI	200 + %	10/16
(2500 h)	55-W PAR/FL	200 + %	5/8
75-W In	55-W PAR/FL	150%	20/27
100-W In	75-W IR	125%	25/25
(750 h)	75-W PAR/FL	200 + %	25/25
	75-W ER	200 + %	25/25
100-W In	75-W PAR/FL	100 + %	25/25
(2500 h)			
150-W PAR/FL	100-W PAR/FL	70%	50/33
150-W R/FL	100-W PAR/FL	150%	50/33
200-W In	150-W PAR/FL	200 + %	50/33
(750 h)	(2000 h)		
30-W F	30-W EEL	87%	5/16
40-W F	40-W EEL	89%	6/15
96-W F	96-W EEL	91%	15/20
96-W F/HO	96-W EEL/HO	91%	15/14
96-W F/SHO	96-W EEL/SHO	90%	30/14
175-W MD	100-W HPS	104%	75/42
400-W MD	200-W HPS	96%	200/50
300-W In	150-W HPS	250%	120/50
750-W In	150-W HPS	104%	570/80
1000-W In	200-W HPS	93%	770/80

[a]Abbreviations:

EEL	energy-efficient fluorescent light (such as Watt-Mizer, Super-Saver, etc.)
ER	elliptical reflector (shape and inside coating of lamp)
F	fluorescent
FL	floodlight
HO	high output: 1000-ma filament
HPS	high-pressure sodium
h	hour (mean life expectancy)
In	incandescent
MD	mercury deluxe: mercury vapor corrected to improve color
PAR	parabolic aluminized reflector (see ER)
RI	reflective coated incandescent
SHO	superhigh output: 1500-ma filament
W	Watt

Table 5-11. Lamp Substitutions (including TT)
[ref 7, Table 2]

Standard Lamp	Replacement Lamp	Wattage Savings[2]	Comparative Light Output of Replacement Lamp[3]	Value of Energy Savings over Life of Replacement Lamp at $0.08/kWh	Other Benefits[4]
60W Incandescent	55W Reduced-Wattage Incandescent	5	=	$0.40	
	13W TT Compact Fluorescent with Ballast Adapter	44.5	+	$35.60	x
75W Incandescent	70W Reduced-Wattage Incandescent	5	=	$0.40	
	22W Circline Fluorescent	45	=	$43.20	
	18W Compact Fluorescent	57	=	$34.20	x
100W Incandescent	95W Reduced-Wattage Incandescent	5	=	$0.40	
	44W Circline Fluorescent	56	=	$33.60	x
75W PAR-38 Spot or Flood Incandescent	65W PAR-38 Spot or Flood Incandescent	10	=	$1.60	
	45W Incandescent (Halogen)	30	=	$4.80	
150W R-40 Flood Incandescent[5]	75W ER-30 Incandescent[5]	75	=	$12.00	
	120W ER-40 Incandescent[5]	30	++	$4.80	x
150W PAR-38 Spot or Flood Incandescent	90W PAR-38 Spot or Flood Incandescent (Halogen)	60	=	$9.60	
	120W PAR-38 Incandescent	30	=	$4.80	
300W R-40 Flood Incandescent[5]	120W ER-40 Incandescent[5]	180	=	$28.80	

(Continued)

Table 5-11. Lamp Substitutions (including TT) (Continued)

Standard Lamp	Replacement Lamp	Wattage Savings[2]	Comparative Light Output of Replacement Lamp[3]	Value of Energy Savings over Life of Replacement Lamp at $0.08/kWh	Other Benefits[4]
500W Incandescent	450W Self-Ballasted Mercury Vapor[6]	50	•	$64.00	
1,000W Incandescent	750W Self-Ballasted Mercury Vapor[6]	250	–	$320.00	
F-40 Fluorescent	F-40 Reduced-Wattage, High-Efficiency Fluorescent	7	=	$11.20	
	F-40 Reduced-Wattage, High-Efficiency Cathode-Disconnect Fluorescent	9.5	=	$15.20	
	F-40 Reduced-Wattage, High-Efficiency, Color-improved Fluorescent	7	•	$11.20	x
	F-40 Reduced-Wattage, High-Efficiency, Color-improved Cathode-Disconnect Fluorescent	9.5	•	$15.20	x
	F-40 High-Brightness Fluorescent	0	+	$0.00	x
F-40 Fluorescent (U-Shape)	F-40 Reduced-Wattage, High-Efficiency Fluorescent (U-Shape)	7	=	$11.20	
F-96 Fluorescent	F-96 Reduced-Wattage, High-Efficiency Fluorescent	17.5	•	$16.80	
F-96 HO Fluorescent	F-96 HO Reduced-Wattage, High-Efficiency Fluorescent	21	•	$20.20	
F-96 1,500 MA Fluorescent	F-96 1,500 MA Reduced-Wattage, High-Efficiency Fluorescent	25	•	$20.00	

175W Mercury Vapor	150W Retrofit High-Pressure Sodium	40	+ +	$ 38.40	x
250W Mercury Vapor	215W Retrofit High-Pressure Sodium	65	+ +	$62.40	x
400W Mercury Vapor	325 Retrofit Metal Halide	70	+ +	$112.00	x
	400W Retrofit Metal Halide	0	+ +	$ 0.00	x
	360W Retrofit High-Pressure Sodium	60	+ +	$76.80	x
1,000W Mercury Vapor	880W Retrofit High-Pressure Sodium	160	+ +	$204.80	x
	950W Retrofit Metal Halide	50	+ +	$48.00	x

NOTES

1. This table does not indicate all possible lamp replacement options and, in some cases, replacing the ballast and lamp, or relying on a new fixture, ballast and lamp will provide better overall performance and energy management than the replacement shown. All numbers reported in the table are approximations, and in certain cases assumptions are made about the types of fixtures and other conditions involved. Consult manufacturers for accurate data relative to direct replacements possible for a given installation as well as any ballast operating temperature or other restrictions which may apply.

2. Wattage savings include ballast losses, where applicable, assuming use of a standard ballast. Actual ballast losses to be experienced depend on the specific type of ballast involved and operating conditions which affect its performance. In those cases where wattage savings exceed the difference in lamp wattage (if any), operation of the replacement lamp also has the effect of reducing ballast losses.

3. Symbols used indicate the following; + + (substantially more) + (more) = (about the same) • (less) – (substantially less). Consult manufacturers for accurate information relative to conditions unique to the lamps and installations involved.

4. Other benefits typically provided by retrofit lamps include: maintenance costs due to longer lamp life; improved productivity, safety / security, quality control, etc., due to higher light output; ability to reduce the number of lamps installed systemwide due to higher output of retrofit lamps, and improved color rendition.

5. When installed in a stack-baffled downlight.

6. For high voltages only.

Example 5-3: Calculate the annual savings from replacing 40-watt F40T12/Workshop lamps with 34-Watt energy-saving lamps in two hundred (200) 4-lamp fixtures which are operated continuously. Assume the following:

The F40T12/Workshop lamps cost $1.00 each and last for 12,000 hours. The 34-Watt F40T12 lamps cost $1.50 each and last for 20,000 hours. Electric energy costs $0.05 per kWh. The demand charge is $5.50 per kW per month. The facility is not air conditioned.

Solution:
Annual Energy Savings (ES):

$$ES = \text{(\# of fixtures)} \times \text{(\# of lamps/fixture)} \times \text{(wattage of low-efficiency lamps–wattage of high-efficiency lamps)} \times \text{(annual operating hours)}$$

$$= \text{(200 fixtures)} \times \text{(4 lamps/fixture)} \times (40 - 34) \text{ Watts/lamp} \times \text{(8760 hours/yr)}$$

$$= 800 \text{ lamps} \times 6 \text{ Watts/lamp} \times 1 \text{ kW}/1000 \text{ Watts} \times 8760 \text{ hr/yr}$$

$$= 42{,}048 \text{ kWh/yr}$$

Energy Cost Savings (ECS):

$$ECS = ES \times \text{(cost of electricity)}$$

$$= 42{,}048 \text{ kWh/yr} \times \$0.05/\text{kWh}$$

$$= \$2102/\text{yr}$$

Demand Reduction (DR):

$$DR = \text{(\# of lamps)} \times \text{(wattage reduction)}$$

$$= \text{(800 lamps)} \times \text{(6 Watts)} \times 1 \text{ kilowatt}/1000 \text{ Watts}$$

$$= 4.8 \text{ kW}$$

Annual Demand Cost Savings (ADCS):

$$ADCS = DR \times \text{Demand Charge}$$

$$= \quad 4.8 \text{ kW} \times \$5.50/\text{kW}/\text{month} \times 12 \text{ months}/\text{yr}$$

$$= \quad \$317/\text{yr}$$

Annual Lamp Cost Savings (ALCS):

The Workshop lights only have a lifetime of 12,000 hours, where the replacement lamps have a lifetime of 20,000 hours. Each of these costs must be annualized to determine the actual cost savings. The total number of lamp hours used in one year is found from multiplying the number of lamps times the hours of use in one year.

Total annual use = 800 × 8760 hr = 7,008,000 lamp hours

1.   Workshop light cost:

To compute the number of Workshop lights needed for one year, divide the total annual lamp hours needed by the life of one Workshop light.

Number of lights needed = 7,008,000/12,000 = 584 lamps

The annual cost is:

584 lamps/yr × $1.00/lamp = $584

2.   34 Watt light cost:

Number of lights needed = 7,008,000/20,000 = 350 lamps

The annual cost is:

350 lamps/yr × $1.50/lamp = $525

ALCS = $584 - $525 = $59 per year

The total annual savings from this relamping EMO is the sum of all the savings calculated above.

Total annual cost savings = $2102 + $317 + $59 = $2478

Example 5-4: Upgrade T12 Lighting System to T8
A hospital had 415 T12 fluorescent fixtures, which operate 24 hours/day, year round. The lamps and ballasts were replaced with T8 lamps and electronic ballasts, which saved about 30% of the energy, and provided higher quality light. Although the T8 lamps cost a little more (resulting in additional lamp replacement costs), the energy savings quickly recovered the expense. In addition, because the T8 system produces less heat, air conditioning requirements during summer months will be reduced. Conversely, heating requirements during winter months will be increased. The air conditioner COP = 2.6 and the A/C runs at full load 2000 hours/year. The heater efficiency is 80%, and the heater operates at full load 1500 hours per year. Fuel costs $4.00/MCF.

DATA
- Number of fixtures ....415 fixtures
- Number of lamps per fixture .... 2 lamps/fix
- Annual Operating Hours ....8,760 hours
- kWh cost .... $.05/kWh
- kW cost..$120/kW-year
- Retrofit cost per fixture $45/fixture

Existing Fixture Data
- Input watts per 2 lamp T12 fixture ....86 watts/fix
- Lamp life...20,000 hours
- Lamp cost...$1.50/lamp

Proposed Fixture Data
- Input watts per 2 T8 fixture..60 watts/fix
- Lamp life...20,000 hours
- Lamp...$3.00/lamp
- T8 ballast.. $16
- Total number of ballasts needed...415 ballasts
- Total number of lamps needed... 830 lamps

Calculations
kW Savings
= (# fixtures) [(Present input watts/fixture) – (Proposed input watts/fixture)]
= (415)[(86 watts/T12 fixture)-(60 watts/T8 fixture)]
= 10.8 kW

kWh Savings
= (kW savings)(Annual Operating Hours)
= (10.8 kW)(8,760 hours/year)
= 94,608 kWh/year

Air Conditioning Savings
= (kW savings)(Air Conditioning Hours/year)(1/Air Conditioner's COP)
= (10.8 kW)(2000 hours)(1/2.6)
= 8,308 kWh/year

Additional Gas Cost
= (kW savings)(Heating Hours/year)(.003413 MCF/kWh)(1/Heating Efficiency)(Gas Cost)
= (10.8 kW)(1,500 hours/year)(.003413 MCF/kWh)(1/0.8)($4.00/MCF)
= $ 276/year

Additional Lamp Replacement Cost
= [(# fixtures)(# lamps/fixture)][((annual operational hours/proposed lamp life)(proposed lamp cost)) − ((annual hours operation/present lamp life)(present lamp cost))]
= [(415 fixtures)(2 lamps/fixture)][((8,760 hours/20,000 hours)($ 3.00/ T8 lamp)) − ((8,760 hours/ 20,000 hours)($ 1.50/T12 lamp))]
= $ 545/year

Total Annual Dollar Savings
= (kW Savings)(kW charge)+[(kWh savings)+(Air Conditioning savings)](kWh cost) − (Additional gas cost) − (lamp replacement cost)
= (10.8 kW)($ 120/kW year)+[(94,608 kWh)+(8,308 kWh)]($ 0.05/kWh) − ($ 276/year) − ($ 545/year)
= $ 5,621/year

Implementation Cost
= (# fixtures) (Retrofit cost per fixture)
= (415 fixtures) ($ 45/fixture)
= $ 18,675

Simple Payback
= (Implementation Cost)/(Total Annual Dollar Savings)
= ($ 18,675)/($ 5,621/year)
= 3.3 years

Example 5-5:
Replace Incandescent Lighting with Compact Fluorescent Lamps
A power plant has 111 incandescent fixtures which operate 24 hours/day, year round. The incandescent lamps were replaced with compact fluorescent lamps, which saved over 70% of the energy, and last over ten times as long. Because the lamp life is so much longer, there is a maintenance relamping labor savings. Air conditioning savings or heating costs were not included because these fixtures are located in a high-bay building which is not heated or air-conditioned.

DATA
*   Number of fixtures 111
*   Yearly operating hours .... 8,760 hrs/yr
*   kWh cost $0.05/kWh
*   kW cost... $120/kW-year
*   Labor cost .... $ 20/hour
*   Maintenance relamping cost
    (5 minutes per fixture) .... $ 1.70/lamp
*   Retrofit labor cost... $ 15/fixture

Existing fixture data
*   Number of lamps per fixture
*   Wattage of each lamp.. 150 watts
*   Lamp type (Incandescent).. 200 A19
*   Cost of incandescent bulb .... $ 1.93/lamp
*   Average incandescent lamp .... 750 hours

Proposed fixture data
*   Lamp type (CFL)..T-9
*   Number of lamps per fixture
*   Input watts per fixture.. 30 watts
*   Cost of CFL ballast, lamp and socket extender $30
*   Cost of CFL 30 watt lamp... $10/lamp
*   Average CFL lamp life .... 10,000 hours

Calculations
Watts Saved Per Fixture
=   (Present input watts/fixture) - (Proposed input watts/fixture)
=   (150 watts/fixture) - (30 watts/fixture)
=   120 watts saved/fixture

kW Savings
= (# fixtures)(watts saved/fixture)(1 kW/1000 watts)
= (111 fixtures)(120 watts/fixtures)(1/1000)
= 13.32 kW

kWh Savings
= (Demand savings)(annual operating hours)
= (13.32 kW)(8,760 hours/year)
= 116,683 kWh/year

Lamp Replacement Cost
= [(Number of Fixtures)(cost per CFL Lamp)(operating hours/lamp life)] - [(Number of existing incandescent bulbs)(cost per bulb)(operating hours/lamp life)]
= [(111 Fixtures)($10/CFL lamp)(8,760 hours/10,000 hours)] – [(111 bulbs)($1.93/type "A" lamp)(8,760 hours/750 hours)]
= $ – 1,530/year§   §Negative cost indicates savings.

Maintenance Relamping Labor Savings
= [(# fixtures)(maintenance relamping cost per fixture)] [((annual hours operation/present lamp life))-((annual hours operation/proposed lamp life))]
= [(111 fixtures)($1.70/fixture)][((8,760/750))-((8,760/10,000))]
= $ 2,039/year

Total Annual Dollar Savings
= (kWh savings)(kWh cost)+ (kW savings)(kW cost) – (lamp replacement cost) + (maintenance relamping labor savings)
= (116,683 kWh)($.05/kWh)+(13.32)($120/kW year) – (–1,530/year) + (2,039/year)
= $ 11,002/year

Total Implementation Cost
= [(# fixtures)(cost/CFL ballast and lamp)] + (retrofit labor cost)]
= (111 fixtures)($45/fixture)
= $ 4,995

Simple Payback
= (Total Implementation Cost)/(Total Annual Dollar Savings)
= ($ 4,995)/($11,002/year)
= 0.5 years

Note that for air-conditioned facilities, the operating costs associated with lower wattage or more efficient lamps will be reduced because they produce less heat. Less air conditioning will be needed with this lower amount of heat produced. This will be discussed further in the air conditioning chapter in Section 6.4.1.5.

Lighting options should be compared to the existing system in a recently relamped and cleaned condition. A reasonable estimate of the lighting levels in the relamped system can be made from measuring lighting levels and noting the age/operation hours of the lamps. Some lamp sales representatives demonstrate how their lamp is superior to the lamps in use by relamping a fixture for comparison with adjacent fixtures. The results can be misleading due to light losses caused by lamp lumen depreciation (i.e., old lamps) and lumen dirt depreciation (i.e., the relamped fixture is usually wiped clean during lamp installation). Make sure both fixtures are clean and contain new lamps before comparing lamp alternatives.

5.6.4 Ballasts

Ballasts are an important part of a lighting system, and each ballast uses from five to twenty percent of the power of the lamp it is associated with. Furthermore, the ballast draws some power even if the lamp has been removed. Therefore, when a lamp is removed from a fixture, the ballast should usually be removed too. The ballast can be stored for future use, saving additional replacement costs. Ballasts should also be replaced if they overheat or smoke.

When older coil and core ballasts in a lighting fixture fail, replacement with an electronic ballast should be considered. New, electronic ballasts are much more efficient than the older magnetic ballasts, and offer desirable features such as dimming capabilities. When T8 fluorescent lamps are used, an electronic ballast is usually specified, too. A T5 must have an electronic ballast to perform well.

5.6.5 Lighting Control Technologies

Areas which are seldom occupied do not need constant light, yet lights are frequently left on in such places. Lights should not be left on in warehouses and storage areas unless the lights serve some function—illuminating storage areas for assistance in finding a product or reading labels, security, or other identified functions. Occupancy sensors have benefits to offer in these cases.

Fluorescent lamps should be turned off if they will not be used for five minutes or more. HID lamps should be turned off if they will not be used for about thirty minutes and quick restart time is not critical. HID lamps take up to fifteen minutes to regain full light output after restarting.

There are a number of cost-effective control EMOs that can be used to turn off lights that are not needed, or to utilize daylighting to supplement artificial lighting. These control technologies are discussed below, and additional discussion of controls is covered in chapter nine. Example 9-3 illustrates the savings from an application of occupancy sensors.

Switches: Many types of switches are available for controlling lighting. The simplest is the standard wall-mounted snap switch. Switches should be installed in the areas in which the fixtures are controlled. Rewiring to reduce the number of fixtures controlled by a single switch increases the ability of occupants to control the amount of lighting that is used. Installing switches next to one another frequently results in all the available lighting operating at once because people tend to turn on all the switches at once. If switches are installed next to each other, installing the switch upside-down that controls the least-needed lights will reduce the chance of that switch being turned on accidentally.

Other types of switches control lighting fixture operation on the basis of lighting levels, time, motion, or infrared radiation. Exterior lighting should be controlled by a light-sensitive switch. Photocells operate the lighting between dusk and dawn. They are available in various sensitivities. It is best to use photocells which turn the fixture on when they fail; this provides a good signal that replacement is necessary. The fail-off type can remain undetected and leave a facility without security lights.

Photocell input can also be used as a basis for controlling interior lighting. Some energy management control systems (EMCS) can use photocell input data for automatically adjusting indoor lighting levels to maintain a constant value when dimmable ballasts are used.

Timers: Timers can be used to control outdoor lighting but some are subject to inaccuracy due to seasonal changes in day-length, daylight savings time changes, clock slippage, power outages, and manual over-ride. Adjustments should be made to simple timers about four times per year to prevent unnecessary operation of equipment. Timers can be used in conjunction with photosensors to reduce lighting costs if the lighting can be turned off before dawn.

Occupancy Sensors: Occupancy sensors can also be used to reduce unnecessary lighting use. Infrared sensors are directional and useful for active areas; ultrasonic sensors are fairly non-directional. The sensor's coverage of the area must be complete or nuisance cutoffs will occur and the occupants will remove the sensors.

Example 5-6: The prisoner holding cells in a courthouse utilize two fixtures with two F40T12 lamps each to illuminate an 8'×12' area. The cells are usually occupied a maximum of 45 minutes per day but they are illuminated about 12 hours per day. How much can you save by installing occupancy sensors in these cells? Calculate the simple payback period and return on investment if the sensors cost $70 installed and have a ten year life? Electricity costs 8 cents per kWh.

Solution: Assume that the occupancy sensor has a delay built into its operation, and that the lights will operate 1 hour a day, five days a week, 50 weeks a year. Further assume that each light fixture with its ballast consumes 80 Watts for the two lamps, and an additional 15%—or 12 Watts—for the ballast. Thus, the total savings from the occupancy sensor is found from:

Cost Savings = (2 fixtures/cell) × (92 Watts/fixture) × (1 kW/1000 Wh)
                 × (11 hours/day) ×
                 (5 days/week) × (50 weeks/year) × ($.08/kWh)

              = $40.48/year per cell

SPP   = Installed cost/annual savings

         = $70/$40.48/year

         = 1.73 years

ROR   = Solution for i to:

           $40.48 * (P/A, i, 10) = $70

ROR   = 57.2%

Dimmers: Dimmers are good for areas which require low ambient lighting levels most of the time with an occasional need for bright lighting. Solid-state dimmers operate by reducing the voltage supplied to the lamps. This reduces energy use and extends lamp life. However, fluorescent and HID lamps cannot be dimmed without dimmable ballasts. Rheostat dimmers are not recommended for any application because they produce considerable heat and do not save energy.

5.6.6 Other Lighting EMOs

5.6.6.1 Exterior Lighting

Exterior lighting is another area in which lighting energy is often wasted. In motels, for example, peripheral lighting is often left on both day and night. Such waste can be easily corrected with a timer or with a light switch turned on and off by a photocell. Each of the perimeter and outside lights should be carefully considered to see when it should be on, how much light is needed for the intended function, and whether more efficient lighting sources would work as well as those now being used.

5.6.6.2 Daylighting

Windows and skylights are often used to add light in a given area. One problem is that windows admit radiant heat as well as light, and it may be more expensive to remove the heat than to supply the light. In that case, the windows should be treated with exterior-mounted solar screens, louvers, or a reflective film with a low shading coefficient and a high percentage transmission of visible light. Daylighting is discussed in more detail in Chapter 13.

5.6.6.3 Environmental Factors

An area can appear to be dark if the walls, floors, or ceilings are painted (or otherwise covered) in dark colors or are greasy or dirty. Using light colors for paint and flooring, or cleaning these surfaces more often can make the existing light more effective and thereby save money.

5.6.7 Selecting Lights for a New Facility

Any time a new facility is built, or an existing facility is expanded, there is a significant opportunity to save on energy costs by selecting and installing cost-effective, energy-efficient lighting systems at the time of construction. It is almost always cheaper to install correct equipment the first time than to retrofit existing equipment. Greater first costs may produce significantly lower operating costs, and provide cost-effective savings for the facility. Unfortunately, many design decisions are made on the basis of first cost rather than life-cycle costs which include operation and maintenance costs. Utilizing life-cycle costing can assure the lowest lighting costs throughout the life of the lighting system.

Example 5-7. Gator Plastics Company is experiencing such a growth in demand for their products that they are planning on adding a new production room. Their industrial engineer (IE) is responsible for selecting the lighting system to be installed. The IE has identified two alternative lighting systems. Alternative One uses 40 fluorescent light fixtures with four 34-Watt lamps in each fixture, together with a four-lamp ballast that consumes a total of 20 watts. This system costs $2000 to purchase and $2000 to install. Alternative Two uses 40 fixtures with three 40-Watt T10 lamps, together with one three-lamp electronic ballast that consumes a total of 15 Watts. This system costs $2400 to purchase and $2000 to install. If each lighting system lasts six years, the lights are used 2000 hours per year, electricity costs $0.08 per kWh, and the company's investment rate is 8%, which alternative should the IE select? Calculate the three standard economic evaluation measures for each alternative—SPP, ROR and B/C ratio.

Solution:
Alternative One: The operating cost is found as follows:
Annual cost = (40 fixtures) × (156 Watts/fixture) × (2000 hours)
                       × (1 kWh/1000 Wh) × ($0.08/kWh)
              = $998.40

Alternative Two:
Annual cost = (40 fixtures) × (135 Watts/fixture) × (2000 hours)
                       × (1 kWh/1000 Wh) × ($0.08/kWh)
              = $864.00

Economic evaluation: Alternative Two costs $400 more than Alternative One, but it saves ($998.40 - $864.00) = $134.40 per year in operating costs. To determine if this additional cost is a good investment, we need to calculate the three standard economic performance measures.

SPP = Initial cost/Annual cost savings
      = $400/$134.40 = <u>3.0 years</u>

The ROR is found by solving:

($134.40) (P/A, i, 6) = $400
ROR =   24.6%

Using the methods from Chapter 4 to find the present worth of the annual savings gives:
B/C = $134.40 × (P/A, 8%, 6)/$400

$$= \quad \$134.40 \times 4.623 / \$400 = \underline{1.55}$$

Thus, for many companies the added cost of alternative two would be considered a good investment.

5.7   LIGHTING CHECKLIST

Lighting and maintenance costs can be reduced with a concurrent improvement in worker productivity, safety, and comfort. Figure 5-24 presents a checklist of energy-saving guidelines for lighting developed by the Illuminating Engineering Society.

5.8   NEW TECHNOLOGIES AND APPROACHES

Technology is advancing at an unprecedented rate and these changes affect how organizations are structured, their flexibility, marketing, operations, etc. We are living in a digital world with more computer screens, technology and active lighting systems. Because humans are mostly visual communicators, lighting is evolving to fulfill the needs of our rapidly changing society. For these reasons, it is recommended to consult with a lighting designer prior to a retrofit/redesign—there are many variables and the impact on human productivity can completely obliterate or enhance the financials of a project. However, this section outlines some new technologies, which can be valuable in your design approach.

5.8.1 It Is a Small World

Smaller fixtures are allowing designers to put lighting fixtures closer to the task (where the light is needed). This breakthrough changes the design because instead of the conventional "blast more light through an efficient fixture, lens and then through the air", we can have a target lit from a close range (or even "back-lit") with much less light. Over the next decade, thin-film "light panels" will be hitting the market with new applications for creating "soft ambient light"… even from walls!

5.8.2 LEDs (Light Emitting Diodes)

This should provide the next lighting revolution within 5 years. Many lighting manufacturers are actively developing LED white light systems that could replace fluorescent lamps. Today, this opportunity already exists in signs, retail displays, medical/dental operating rooms

Figure 5-24. Checklist of Energy Saving Guidelines for Lighting Ideas [ref 1, Figure 4-6] [ref 6, figure 2]

Lighting Needs

- Visual tasks: specification — Identify specific visual tasks and locations to determine recommended illuminances for tasks and for surrounding areas.

- Safety and esthetics — Review lighting requirements for given applications to satisfy safety and esthetic criteria.

- Overlighted application — In existing spaces, identify applications where maintained illumination is greater than recommended. Reduce energy by adjusting illuminance to meet recommended levels.

- Groupings: similar visual tasks — Group visual tasks having the same illuminance requirements, and avoid widely separated workstations.

- Task lighting — Illuminate work surfaces with luminaires properly located in or on furniture; provide lower ambient levels.

- Luminance ratios — Use wall washing and lighting of decorative objects to balance brightnesses.

Space Design and Utilization

- Space plan — When possible, arrange for occupants working after hours to work in close proximity to one another.

- Room surfaces — Use light colors for walls, floors, ceilings and furniture to increase utilization of light, and reduce connected lighting power to achieve required illuminances. Avoid glossy finishes on room and work surfaces to limit reflected glare.

- Space utilization: branch circuit wiring — Use modular branch circuit wiring to allow for flexibility in moving, relocating or adding luminaires to suit changing space configurations.

- Space utilization: occupancy — Light building for occupied periods only, and when required for security or cleaning purposes (see chapter 31, Lighting Controls).

(Continued)

Figure 5-24. Checklist of Energy Saving Lighting Ideas (*Continued*)

Daylighting

- Daylight compensation — If daylighting can be used to replace some electric lighting near fenestration during substantial periods of the day, lighting in those areas should be circuited so that it may be controlled manually or automatically by switching or dimming.

- Daylight sensing — Daylight sensors and dimming systems can reduce electric lighting energy.

- Daylight control — Maximize the effectiveness of existing fenestration-shading controls (interior and exterior) or replace with proper devices or shielding media.

- Space utilization — Use daylighting in transition zones, in lounge and recreational areas, and for functions where the variation in color, intensity and direction may be desirable. Consider applications where daylight can be utilized as ambient lighting, supplemented by local task lights.

Lighting Sources: Lamps and Ballasts

- Source efficacy — Install lamps with the highest efficacies to provide the desired light source color and distribution requirements.

- Fluorescent lamps — Use T8 fluorescent and high-wattage compact fluorescent systems for improved source efficacy and color quality.

- Ballasts — Use electronic or energy efficient ballasts with fluorescent lamps.

- HID — Use high-efficacy metal halide and high-pressure sodium light sources for exterior floodlighting.

- Incandescent — Where incandescent sources are necessary, use reflector halogen lamps for increased efficacy.

- Compact fluorescent — Use compact fluorescent lamps, where possible, to replace incandescent sources.

- Lamp wattage reduced-wattage lamps — In existing spaces, use reduced-wattage lamps where illuminance is too high but luminaire locations must be maintained for uniformity. *Caution:* These lamps are not recommended where the ambient space temperature may fall below 16°C (60°F).

- Control compatibility — If a control system is used, check compatibility of lamps and ballasts with the control device.

- System change — Substitute metal halide and high-pressure sodium systems for existing mercury vapor lighting systems.

Luminaires
- Maintained efficiency

  Select luminaires which do not collect dirt rapidly and which can be easily cleaned.

- Improved maintenance

  Improved maintenance procedures may enable a lighting system with reduced wattage to provide adequate illumination throughout system or component life.

- Luminaire efficiency:

  Check luminaire effectiveness for task lighting and for overall efficiency; if ineffective or inefficient,

- replacement or relocation

  consider replacement or relocation.

- Heat removal

  When luminaire temperatures exceed optimal system operating temperatures, consider using heat removal luminaires to improve lamp performance and reduce heat gain to the space. The decrease in lamp temperature may, however, actually increase power consumption.

- Maintained efficiency

  Select a lamp replacement schedule for all light sources, to more accurately predict light loss factors and possibly decrease the number of luminaires required.

Lighting Controls
- Switching; local control

  Install switches for local and convenient control of lighting by occupants. This should be in combination with a building-wide system to turn lights off when the building is unoccupied.

- Selective switching

  Install selective switching of luminaires according to groupings of working tasks and different working hours.

- Low-voltage switching systems

  Use low-voltage switching systems to obtain maximum switching capability.

- Master control system

  Use a programmable low-voltage master switching system for the entire building to turn lights on and off automatically as needed, with overrides at individual areas.

- Multipurpose spaces

  Install multicircuit switching or preset dimming controls to provide flexibility when spaces are used for multiple purposes and require different ranges of illuminance for various activities. Clearly label the control cover plates.

- "Tuning" illuminance

  Use switching and dimming systems as a means of adjusting illuminance for variable lighting requirements.

- Scheduling

  Operate lighting according to a predetermined schedule.

(Continued)

Figure 5-24. Checklist of Energy Saving Lighting Ideas (*Conclusion*)

- Occupant/motion sensors — Use occupant/motion sensors for unpredictable patterns of occupancy.

- Lumen maintenance — Fluorescent dimming systems may be utilized to maintain illuminance throughout lamp life, thereby saving energy by compensating for lamp-lumen depreciation and other light loss factors.

- Ballast switching — Use multilevel ballasts and local inboard-outboard lamp switching where a reduction in illuminances is sometimes desired.

Operation and Maintenance

- Education — Analyze lighting used during working and building cleaning periods, and institute an education program to have personnel turn off incandescent lamps promptly when the space is not in use, fluorescent lamps if the space will not be used for 5 min or longer, and HID lamps (mercury, metal halide, high-pressure sodium) if the space will not be used for 30 min or longer.

- Parking — Restrict parking after hours to specific lots so lighting can be reduced to minimum security requirements in unused parking areas.

- Custodial service — Schedule routine building cleaning during occupied hours.

- Reduced illuminance — Reduce illuminance during building cleaning periods if building is not otherwise occupied.

- Cleaning schedules — Adjust cleaning schedules to minimize time of operation, by concentrating cleaning activities in fewer spaces at the same time and by turning off lights in unoccupied areas.

- Program evaluation — Evaluate the present lighting maintenance program, and revise it as necessary to provide the most efficient use of the lighting system.

- Cleaning and maintenance — Clean luminaires and replace lamps on a regular maintenance schedule to ensure proper illuminance levels are maintained.

- Regular system checks — Check to see if all components are in good working condition. Transmitting or diffusing media should be examined, and badly discolored or deteriorated media replaced to improve efficiency.

- Renovation of luminaires — Replace outdated or damaged luminaires with modern ones which have good cleaning capabilities and which use lamps with higher efficacy and good lumen maintenance characteristics.

- Area maintenance — Trim trees and bushes that may be obstructing outdoor luminaire distribution and creating unwanted shadows.

and other specialty lighting applications. Why? Beyond being low watt, close-range light sources, LEDs can offer multi-colored light without a lot of heat or maintenance. For a retailer, they can light a window display for products with different colors, creating different moods and attracting different buyers. The same principle applies for signs—more flexibility. In operating rooms, smaller lights closer to the patient with less heat being generated... that means happier patients and doctors! In addition, art museums have applications due to reduced "light wear and tear" on paintings. In airports, runways, taxiways and other signaling/marking applications, LEDs are reducing lamp replacement and labor costs.

5.8.3 Fluorescent Technology Update
Generations

Fluorescents continue to expand their applications as they get more efficient and thinner, thereby letting more light out of the fixture. T-8s (and in some cases T-5 lamps) have become the standard retrofit from T-12 technologies. At this time, there are no less than 3 generations of T-8 fluorescent lamps:

- 1st Generation T-8 Lamps with 2800 lumens

- 2nd Generation T-8 Lamps with 2950-3000 lumens

- 3rd Generation T-8 Lamps with 3100 lumens

Today's T-8 electronic ballasts also allow more configurations, such as ballast factor adjustments to provide the optimal amount of light for each system and each dollar invested.

Indirect Approaches and Successes

Due to the digital office environment, indirect lighting has proven popular as well as efficient. There are several case studies showing how indirect lighting has saved 50% of the lighting energy while improving the visual environment, flexibility and performance of a workspace or classroom.* Figure 5-25 below is a picture of an up-down fixture that allows light to be directed up only; down only; or both up and down. This fixture can be used in either direct light mode, indirect light mode, or both when the room needs more light. You can configure this fixture to optimize the

_____

*Several examples include: Patagonia, Google, GE, Home Depot, etc. Other examples can be downloaded from the Resource Vault. at www.ProfitableGreenSolutions.com.

Figure 5-25. Up-Down Light Fixture

lighting for audio visual presentations or for general illumination in a classroom or office.

Technology Integration

Many fixtures now offer integrated occupancy sensors (and also photocells for daylight measurement and dimming). This offers a new level of automation for the energy management program. Another opportunity is to give "power to the people" by integrating lighting and environmental comfort controls to the occupant via a web interface.

5.9   SUMMARY

The lighting system in a facility is an important area to examine and to improve in terms of energy efficiency and quality of light. This chapter has discussed the lighting system, described the components of the system, and provided suggestions for ways to improve the system. Lighting technology is changing at a rapid pace, and new lamps and ballasts are being developed and marketed almost daily. Major energy savings opportunities exist in most older lighting systems, and additional cost-effective savings is often possible in relatively new systems since technology is continually improving in this area.

References

1. *IES Lighting Handbook, Application and Reference Volume, Ninth Edition.* Illuminating Engineering Society of North America, New York, 2000.
2. Harrold, Rita M., Lighting, Chapter 13, in *Energy Management Handbook, Second Edition*, Senior Editor W.C. Turner, Fairmont Press/Prentice Hall, Atlanta, GA, 1993.
3. Pansky, Stanley H., *Lighting Standards, Tracing the development of the lighting standard from 1939 to the Present*, Lighting Design & Application, Illuminating Engineering Society, February, 1985.
4. *Lighting Maintenance*, General Electric Technical Pamphlets TP-119 & TP-105, General Electric Co., Cleveland OH, Feb. 1969.
5. *General Electric Company Technical Pamphlet TP-105*, General Electric Company, Cleveland, OH.
6. *Getting the Most from Your Lighting Dollar, 2nd Edition* Pamphlet, National Lighting Bureau.
7. *Profiting from Lighting Modernization* Pamphlet, National Lighting Bureau, 1987.
8. *Electric Utility Guide To Marketing Efficient Lighting*, Western Area Power Administration & American Public Power Administration, 1990.
9. Griffin, C.W., *Energy Conservation in Building: Techniques for Economical Design*, The Construction Specifications Institute, 1974.
10. *Energy Policy Act of 1992*, P.L. 102-486 (H.R. 776), U.S. Congress, October 24, 1992.

Bibliography

Advanced Lighting Guidelines: 1993, Electric Power Research Institute, EPRI TR-1010225, May, 1993.

Capehart, Barney, Editor, *Encyclopedia of Energy Engineering and Technology*," Taylor and Francis/CRC Publishing Company, Boca Raton, FL, July, 2007.

Compact Fluorescent Lamps, Electric Power Research Institute, Palo Alto, CA, 1991.

Helms, Ronald N. and M. Clay Belcher, *Lighting for Energy Efficient Luminous Environments*, Prentice-Hall, Englewood Cliffs, NJ, 1991.

Hollister, D.D., Lawrence Berkeley Laboratory, Applied Sciences Division, "Overview of Advances in Light Sources." University of California, 1986.

Lighting Equipment and Accessories Directory, Illuminating Engineering Society of North America, March, 1993.

Lighting Handbook for Utilities, Electric Power Research Institute, Palo Alto, CA,

1993.

Lindsey, Jack L., *Applied Illumination Engineering,* Fairmont Press, Atlanta, GA, 1991.

Retrofit Lighting Technologies, Electric Power Research Institute, Palo Alto, CA, 1991.

Sorcar, P.C., *Rapid Lighting Design and Cost Estimating.* McGraw-Hill, Inc., 1979.

Thumann, Albert, *Lighting Efficiency Applications, Second Edition,* Fairmont Press, Atlanta, GA, 1992.

Verderber, R.R. and O. Morse, Lawrence Berkeley Laboratory, Applied Science Division. "Performance of Electronic Ballasts and Other New Lighting Equipment." University of California, 1988.

Verderber, R.R. and O. Morse, Lawrence Berkeley Laboratory, Applied Science Division. "Cost Effective Lighting," University of California, 1987.

Chapter 6

Heating, Ventilating, and Air Conditioning

6.0 INTRODUCTION

The heating, ventilating, and air conditioning (HVAC) system for a facility is the system of motors, ducts, fans, controls, and heat exchange units which delivers heated or cooled air to various parts of the facility. The purpose of the HVAC system is to add or remove heat and moisture and remove undesirable air components from the facility in order to maintain the desired environmental conditions for people, products or equipment. Providing acceptable indoor air quality is a critical function of the HVAC system, and air movement to remove odors, dust, pollen, etc. is necessary for comfort and health. It may also be necessary to air-condition an area to protect products or to meet unusual requirements such as those in a laboratory or a clean room.

The HVAC system is responsible for a significant portion of the energy use and energy cost in most residential and commercial buildings. Because many industrial facilities do not have heated or cooled production areas, HVAC energy use does not account for as great a portion of the total energy use for these facilities. However, a number of manufacturing plants are fully heated and air conditioned, and almost all industrial facilities have office areas that are heated and cooled. Thus, looking for ways to save on the energy costs of operating a facility's HVAC system is an important part of any energy management program.

Many facilities have HVAC systems that were designed and installed during periods of low energy costs; these are often relatively expensive to operate because energy efficiency was not a consideration in the initial selection of the system. In addition, many HVAC systems are designed to meet extreme load conditions of very hot or very cold weather; they are then poorly matched to the average conditions that are experienced most of the time. Thus, improving the operation of the HVAC system provides many opportunities to save energy and reduce costs. In this chapter we describe how an HVAC system works, discuss the major components of HVAC systems, analyze heating and cooling loads and ventilation requirements, and give methods for improving the energy efficiency of existing HVAC systems.

6.1   HOW AN HVAC SYSTEM WORKS

Air in a facility absorbs heat from lights, people, industrial processes, and the sun, and air conditioning removes the excess heat in order to provide a comfortable working environment. The air conditioning system also removes excess humidity. In periods of cold weather, the heating system adds heat if the working environment is too cold for worker comfort. During the heating season, moisture may be added to increase the humidity. The HVAC system also provides ventilation and air movement even when no heating or cooling load is present.

HVAC systems vary depending on the fluid that is used as a heat exchange medium (usually water or air), on the particular requirements for the system, and on the type of system that was in style when the building was originally built. All heating systems have certain components in common: a source of heat, some means for transferring the heat from the point of generation to the point of use, and a control system. The source of heat is usually a boiler, a furnace, or the sun. For cooling systems, the source of cold temperature is usually a chiller, although cold air can be supplied either as the exhaust air from a cold area or as cool air brought into the facility during periods when the outside air is at a lower temperature than the inside air. The heat or cold is usually transferred from a furnace, a boiler, or a chiller to the air, and this air is distributed to the points of use.

An alternative system may distribute heated or chilled water to the points of use where the water heats or cools the air to be blown into the room. The control system may be as simple as a thermostat that turns on a furnace when it senses room temperature below a preset level, or it may be very elaborate, controlling air volume, humidity, and temperature through monitoring inputs from many sensors and actuating valves, motors, and dampers.

6.1.1 Dual-Duct System

One way to understand how HVAC systems work is to first learn how one system works and then to see other systems as variations of that system. One of the more widely found HVAC systems is the dual-duct system, which is illustrated in Figure 6-1. Outside air is introduced through dampers (see Figure 6-1) which filter and control the amount of incoming air. This outside air is mixed with return air; the amount which

Figure 6-1. Dual-duct HVAC system.

is mixed is controlled by a return air damper. The air is then blown by a supply fan into both a hot duct and a cold duct. The air in the hot duct passes through heating coils and is sent at a preset temperature to the rooms where it is used. The air in the cool duct goes through conditioning coils and thence to the rooms; its temperature is set when it leaves the cooling coils. At each room, the cold and hot air go into a mixing box controlled by a thermostat; from the mixing box, the air with the desired temperature and moisture level enters the room. Later the air, with the heat, the cold, and the contaminants from the room, is removed through a return air grille and exhausted to the outside and/or returned as part of the intake air.

The dual-duct system has the advantage that it can accommodate widely differing demands for heating and cooling in different zones of a building by changing the ratio of hot to cold air in each zone. Its control system is also easily understood and relatively maintenance-free. The dual duct system has three main disadvantages. First, it requires much ductwork with attendant cost and space usage. Second, when the hot and cold ducts are next to each other, unproductive heat transfer takes place, and energy use and costs increase. Third, the use of energy to heat and cool simultaneously makes the dual-duct system a relatively inefficient system with respect to the energy required to perform the HVAC function.

Most of the original dual-duct systems in facilities have been modified so that the energy efficiency of the system is much better. Often this means that one of the ducts has been shut down so that heating and cooling are not provided at the same time for the entire system. This is an acceptable solution for some facilities with fairly uniform loads in all areas, and where moisture control is not a problem. Some of the systems have been modified to become variable air volume systems which are discussed in Section 6.1.4.1. Other systems use one duct for supplying heat or cooling to the core area of the facility, and perimeter fan-coil units are used to supply heat or cooling to the areas of the facility that are likely to need additional conditioned air.

6.1.2 Single-Duct, Terminal Reheat System Operation

A second type of system is a single-duct, terminal reheat system, and is illustrated in Figure 6-2. In this system, outside air enters through dampers, is mixed with return air in a mixing box or plenum, and is forced by a supply fan through a cooling unit. The air that has been cooled passes through a single supply duct to mixing boxes which contain a heating unit of some type—typically a hot water coil, and the air is then sent into a room. The return air system is similar to that described for the dual-duct system. When the source of heat for the reheat coil is a boiler, a common design fault is to have pumps continuously running water from the boiler

Figure 6-2. Terminal reheat HVAC system.

through the heating coil system. This uses electricity constantly for the pumps but avoids the thermal shock that might occur if cold water were injected into a warm boiler. A better alternative is to install a mixing valve at the boiler so that the pumps can be shut down when heating is not needed.

Many original terminal reheat systems used electric resistance heaters for the source of reheat. When electric energy costs greatly increased, most of these systems were modified to eliminate many of the terminal reheat units, and to use hot water coils and a gas or oil-fired boiler to supply the heat actually needed. Some of these terminal reheat systems were also modified to become variable air volume systems (see Section 6.1.4.1). Reheat systems are being used more now since ventilation standards have changed, and more outside air must now be brought into the facility.

6.1.3 System components

The typical components found in HVAC systems include dampers, grilles, filters, coils, fans, ductwork, and a control system. Each of these contributes to the operation of the HVAC system as follows:

Dampers. A damper controls a flow of air. If the damper is open, the air can flow unimpeded; if it is closed, the flow is reduced to 5-10% of open-damper flow, with the percentage dependent on the construction and maintenance of the damper. Dampers are usually used to regulate the flow of outside air into a system or to control the flow from one part of the system to another part, as in the case of a return air damper (A in Figure 6-2). In Figure 6-2, if the return air damper is closed and the outside air damper is open, all the heat (or cooling) in the return air is lost to the surrounding atmosphere. If the return air damper is open and the outside damper closed, then all the air is recirculated. Most HVAC systems operate somewhere between these two extremes, since some outside air must be supplied to buildings to meet health and safety code requirements.

Grilles. A screen—or grille—is usually placed upstream from a damper to catch bugs, lint, and debris before they go into the air distribution system. When a grille gets plugged or is blocked by objects such as furniture or stored material, this part of the air distribution system does not work. The authors saw a school where the temperature was consistently too hot for comfort although the air conditioning control system and other components seemed to be functioning properly. The grilles, however, were completely plugged with lint. When the grilles were cleaned, the problem disappeared.

Fans. Fans provide the power to move air through the air distribution system. A typical fan has three main parts: a motor, belts or a chain to transmit power from the motor to the fan blades (many small fans are direct drive), and the blades with their housing. If any one of these parts fails or is not connected properly, the fan will not move air, and this part of the air conditioning system will not work.

Heat Exchange Surfaces. The air that circulates through the HVAC system must usually be heated or cooled in order to be useful. This heating or cooling takes place when air is forced around a coil or finned surface containing hot or cold fluid. If these heat exchange surfaces are fouled with dirt, grease, or other materials with poor heat conduction properties, heat exchange will be inefficient, and more heat or cooling energy will have to be used in order to heat or cool air to the desired temperature.

Ductwork. Ductwork directs and conducts air from the heat exchange surfaces to the rooms where the hot or cold air is desired, and it conducts the exhaust air from these rooms back to the mixing plenum and to the outside. This function is impaired if the ducts leak or if loose insulation or other obstructions slow the airflow within the ducts.

HVAC Control System. An HVAC control system transforms the operating instructions for desired environmental conditions into the air temperatures and ventilation volumes desired in the working environment. The control system has the task of regulating the HVAC system so that these instructions are met as nearly as possible.

The control system accomplishes its function through a system of sensors, actuators, and communication links. The sensors send appropriate electrical or pneumatic signals when some temperature, pressure, or humidity threshold has been crossed. These signals are sent through a communication network, generally using either an electrical or pneumatic system. Signals can also be multiplexed through electrical supply lines, or radio transmission can be used. Upon reaching their destinations, the control signals are then translated into additional pneumatic or electrical signals that are used to open or close dampers, to regulate fans, and to initiate or stop the source of heating or cooling.

Dampers can be opened or closed to regulate the amount of incoming air, the amount of exhaust air that is mixed with fresh air, and the amount of air that is introduced into an area. Fans can be turned on or off to increase air coming into a room or being exhausted from it, or their speed can be regulated so that the amount of air coming into a room is no more than is needed for proper ventilation and temperature control. Boilers or chillers can be turned on to provide heat or cooling for the heat exchangers. Boiler controls are discussed in more detail in Chapter Eight.

6.1.4 Other HVAC System Types

Section 6.1.1 described the working of a dual-duct system, and Section 6.1.2 described the working of a terminal reheat system. Other common HVAC system types are variable air volume (VAV), fan coil, unit ventilator, induction, steam, and hot water systems. These systems are described briefly below because an energy auditor will invariably run into each of them at some time. For more detail, see References 6 and 7.

6.1.4.1 VAV systems.

In a variable air volume (VAV) system, tempered (heated or cooled) air is forced into a room at a rate dependent on the amount of heating or cooling desired. If the air volume needed is relatively constant, then a simple approach is to use dampers in VAV boxes installed in each duct opening where the air enters the room. If the air volume needed at different times changes significantly, the fan motor must have an adjustable speed drive so that the volume of air moved can be carefully controlled. If less heating or cooling is desired, less hot air is blown into the room. The advantage of this system is that only the amount of air needed is used, and, since power requirements vary as the cube of air volume moved, less air volume means less electrical consumption.

Some of the disadvantages of this system include the complexity and difficulty of maintaining the controls, and the need to use and control high-velocity airstreams. However, new control technologies have improved these problems, and this system is probably the most widely used HVAC system for installation in new buildings. Increased ventilation requirements for health and safety have resulted in more VAV systems needing fan motors with adjustable speed drives. Previously, VAV boxes in rooms could be shut down far enough to provide very low air flow rates; now this cannot be done, so the air flow must be more controllable to meet the required ventilation air flows. In addition, if moisture control is needed in a facility, it is necessary to have reheat to bring the temperature of the overcooled air back up to a comfortable temperature. Thus, even with the use of latest technology VAV systems it is still usually necessary to have reheat available.

6.1.4.2 Fan coil systems.

The fan coil system provides heat or cooling by using a fan to move room air across heating or cooling coils and back into the room. No outside air is introduced for ventilation, and no ductwork for outside air is needed. Air is usually distributed directly from the unit, and no ductwork is needed for supply air or return air. Control of the conditioned air is provided by varying the amount of heating or cooling fluid circulated through the coil and/or the fan speed.

6.1.4.3 Unit ventilators.

In this system, air is brought in directly from the outside and heated or cooled as it enters the room. A window air conditioner falls into this category as do many of the individual packaged room units in motels. The advantage of this system is that each room can be individually and easily controlled; the disadvantages are that installation costs are high and the occupants do not usually control temperatures so as to minimize energy consumption.

6.1.4.4 Induction units.

In these units, high-pressure supply air flows through nozzles and induces additional room air flow into the unit. This secondary air flows over heating or cooling coils and back into the room. This system provides both ventilation and heating or cooling at relatively low capital and energy costs. It also gives good local control of temperature. Its disadvantages are that its controls are complex and that each unit must be maintained regularly to keep it free of lint and dust.

6.1.4.5 Steam units.

In these systems, heat is produced from steam that condenses in radiators and is transferred either by fans or by natural convection. The condensate is then returned to the boiler where the steam was generated. The advantages of such systems include low initial and maintenance expenses for a multiroom installation. Disadvantages may include the need to operate the boiler when only a small part of the boiler design capacity is needed—for example, when only one or two rooms need heat and the boiler was designed to meet the needs of an entire building.

6.1.4.6 Water systems.

Water systems range from complex high-temperature units to the more familiar two-pipe units found in many old apartment buildings. In a typical water system, hot or cold water is pumped through coils and heats or cools air that is drawn around the coil by natural convection or by fans. In a two-pipe system, water enters the radiator through one pipe and leaves by another. In this system, complex valving is necessary to be able to change the system from heating to cooling, and the system operators must be skilled. In a four-pipe system, two pipes take and remove hot water and two take and remove chilled water, with the relative amounts of each depending on the amount of heating or cooling desired. The four-pipe system involves more plumbing than the two-pipe system but avoids the necessity for changing from hot to cold water throughout the system.

The main advantage of water systems is that they move a large amount of heating and cooling energy in return for a small amount of pumping energy; the amount of distribution energy per unit of heating or cooling is significantly less than that of an air system. The main disadvantage is the large amount of plumbing involved. Piping is expensive to buy and to install, and leaks in piping can cause far more expensive consequences than leaks in air-duct systems.

6.1.4.7 Heat pump systems.

A heat pump system is an HVAC system which uses the vapor-compression refrigeration cycle in a reverse mode. A heat pump system can move heat either to the inside or to the outside, so it can provide heating or cooling as the need arises. Single compressor systems up to about 25 to 30 tons are the most common. If a facility is going to be air conditioned, then the heat pump system is often a low cost system to provide the heating needed. In the moderate climate areas of the South and West, air-to-air heat pumps are very effective for heating. Use of water-to-air or ground-source heat pumps greatly expands the area where these systems are cost-effective.

6.2   PRODUCTION OF HOT AND COLD FLUIDS FOR HVAC SYSTEMS

6.2.1 Hot Fluids

Hot air, hot water and steam are produced using furnaces or boilers which are called primary conversion units. These furnaces and boilers can burn a fossil fuel such as natural gas, oil or coal, or use electricity to provide the primary heat which is then transferred into air or water. Direct production of hot air is accomplished by a furnace which takes the heat of combustion of fossil fuels or electric resistance heat, and transfers it to moving air. This hot air is then distributed by ductwork or by direct supply from the furnace to areas where it is needed.

Hot water is produced directly by a boiler which takes the heat from combustion of fossil fuels or electric resistance heat, and transfers it into moving water which is then distributed by pipes to areas where it is needed. A boiler might also be used to add more heat to the water to produce steam which is then distributed to its area of need. The combustion process and the operation of boilers and steam distribution systems are described in detail in Chapters Seven and Eight.

6.2.2 Cold Fluids

Cold air, cold water and other cold fluids such as glycol are produced by refrigeration units or by chillers, which are the primary conversion units. Refrigeration units or chillers commonly use either a vapor-compression cycle or an absorption cycle to provide the primary source of cooling which is then used to cool air, water or other fluids to be distributed to areas in which they are needed.

6.2.2.1 The basic vapor-compression cycle.

Room air conditioners and electrically powered central air conditioners with capacities up to 20-30 tons (or up to 100 tons with multiple compressors) operate using the basic vapor-compression cycle which is illustrated in Figure 6-3. There are four main components in a refrigeration unit using the vapor-compression cycle: the compressor, the condenser, the expansion valve and the evaporator. There is also a working fluid which provides a material that experiences a phase change from liquid to gas and back in order to move heat from one component of the system to another. The working fluid was historically a chlorofluorocarbon or CFC, but these CFCs are being phased out because of the damage they cause to the ozone layer. Hydrofluorocarbons or HFCs, or Hydrogen-CFCs or

Figure 6-3. Vapor compression refrigeration cycle.

HCFCs are already in use in some systems, and are expected to serve as replacements for CFCs until totally ozone-safe working fluids are developed. The replacement of CFCs is discussed in more detail in Section 6.2.3.

We can start the description of the vapor-compression cycle operation at any point in the diagram, so let's start at the compressor. As the working fluid enters the compressor, it is in the state of a low temperature and low pressure vapor. After compression, the fluid becomes a high pressure, superheated vapor. This vapor then travels to the condenser which is a heat transfer coil that has outside air blowing through it. As the heat from the vapor is transferred to the outside air, the vapor cools and condenses to a liquid. This liquid then travels to the expansion valve where both its pressure and temperature are reduced. Next, the low pressure, low temperature liquid travels to the evaporator, which is another heat transfer coil. Air from inside the conditioned space is blown through the evaporator coil, and heat from this air is absorbed by the working fluid as it continues to expand to a low pressure and low temperature vapor as it passes through the evaporator. The inside air has now been cooled as a result of some of its heat content being absorbed by the evaporator, and this cool air can be distributed to the area where it is needed. Finally, the cycle repeats as this low pressure and low temperature vapor from the evaporator returns to the compressor.

Rapid heat transfer from the inside air to the evaporator coil, and from the condenser to the outside air is critical to the proper operation and energy efficiency of a refrigeration unit using the vapor-compression cycle. These coils must be kept clean to allow rapid heat transfer. If the coils are dirty or air flow is partially blocked because of physical obstructions or shrubbery, the refrigeration unit will not work as effectively or as efficiently as it should. Duct leakage is also a common reason for poor cooling or low air flow. Proper operation also requires the correct amount of working fluid in the system. If leaks have allowed some of the fluid to escape, then the system should be recharged to its rated level.

6.2.2.2 Chillers.

A typical chiller provides cold water or some cold fluid such as glycol which is supplied to areas where secondary units such as fan-coil units are used to provide the cooling that is needed at each location. Chillers have capacities that vary widely, from a few hundred tons to several thousand tons. The majority of chillers use either the vapor-compression cycle or the absorption cycle as the basic cooling mechanism, and have secondary fluid loops that reject the unwanted heat to the outside air or water, and provide the cold fluid to the areas where it is needed. The schematic diagram illustrated in Figure 6-4 is typical of a water chiller that is water cooled.

In Figure 6-4 the condenser cooling water is usually supplied by a closed loop that goes to a cooling tower. The cooling tower is an evaporative cooler that transfers the heat from the water to the outside air through the process of evaporation as the water is sprayed or falls through the air. If lake water or ground water were used in an open loop, the water would simply be supplied from one location and returned to a different location in the lake or in the ground.

The chilled water produced by the evaporator is circulated in another secondary closed loop to the parts of the facility where it will be used to provide air conditioning or process cooling. Individual fan coil units can be used in rooms, or centralized air handling units can be used to take a larger quantity of cooled air and distribute it to various parts of the facility. Part of the chilled water may be used to circulate through production machines such as plastic injection molding machines, welders, or metal treatment baths. Chilled water or other chilled fluids may be used to provide refrigeration or freezing capability for various types of food processing, such as meat packing or orange juice processing.

There are three types of mechanical compressors used in chillers. Small compressors used in chillers with capacities up to about 50 tons are almost always reciprocating compressors; they may also be used in chill-

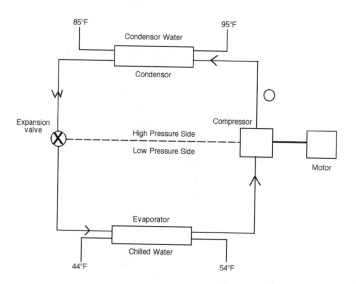

Figure 6-4. Diagram of a typical chiller.

ers up to around 250 tons. Rotary screw compressors are used in chillers as small as 40-50 tons, and in large units up to and somewhat over 1000 tons. Centrifugal compressors are used in chillers as small as 75 tons, and in very large units up to and over 5000 tons.

6.2.3 Replacement of CFCs with alternative working fluids

Releases of many working fluids or refrigerants commonly used in HVAC systems can cause damage to the ozone layer in the earth's stratosphere [9]. The refrigerants degrade to release chlorine molecules which then proceed to break down ozone. Ozone, a molecule containing three oxygen atoms, is effective in filtering out ultra-violet (UV) radiation. The hole in the ozone layer over the Antarctic has allowed a large increase in the amount of UV radiation reaching the earth's surface. Sheep in southern Argentina received sunburns for the first time in recorded history when the ozone hole drifted northward. More importantly, the increase in UV radiation may damage marine crustaceans which form a vital link near the bottom of the food chain in the world's oceans.

The production of chlorinated fluorocarbons (CFC) used in refrigeration and air-conditioning systems ceased on January 1, 1996. Several of the most common refrigerants (i.e., R-11 and R-12) are no longer manufactured. They will be available from recyclers for a short period of time but probably at high cost. Several alternative refrigerants are suitable substitutes.

There are problems associated with most of the alternatives in many applications. The energy efficiency and capacity of the systems may be less than that of systems using CFC's. This will require more energy use and its associated pollution and production of greenhouse gases. Other problems include: incompatibility with commonly used lubricants and gasket materials, toxicity, and difficulty in retrofitting existing systems (e.g., replacement of sensors, motors, impellers, gears, etc.).

The current substitutes are R-123 and R-245ca for R-11 and R-134a for R-12. Another short-term remedy is to use halogenated chlorofluorocarbons (HCFC) such as the commonly used R-22 or a mixture of R-22 with other compounds (i.e., near azeotropic compounds). HCFC compounds have less ozone depletion potential (ODP) than CFC compounds (i.e., ODP of R-11 is 1.0 by definition). For example, R-22 has an ODP of 0.05. HCFC production is slated to be phased out by 2030, but there is pressure to accelerate the timetable.

Large industrial refrigeration needs can be met with existing compounds such as ammonia or water/lithium bromide. These are well-developed technologies which predate the common use of CFC compounds for refrigeration.

Many facilities will replace their existing equipment with new systems containing refrigerants with low ODP and then recover the existing refrigerants for use in their other equipment.

6.3   ENERGY EFFICIENCY RATINGS FOR HVAC SYSTEM COMPONENTS

6.3.1 Boilers and furnaces

The efficiency of furnaces and boilers is specified in terms of the ratio of the output energy supplied to the input energy provided. This efficiency is shown in equation 6-1 below:

$$\text{Efficiency (\%)} = [\text{heat output}/\text{heat input}] \times 100 \qquad (6\text{-}1)$$

Efficiency specifications differ depending on where the heat output is measured. Combustion efficiency will be the highest efficiency number since it measures heat output at the furnace or boiler. Combustion efficiency can be measured with a stack gas analyzer, or it can be measured by determining the temperature and flow rate of the air or water from the furnace or boiler, and then calculating the heat output.

Furnace efficiencies can range from 65% to 85% for most standard furnaces, and up to 98% for pulse-combustion, condensing furnaces. Boiler efficiencies range from about 65% for older, smaller boilers to 85% for newer, larger models. Furnace and boiler efficiencies should be checked periodically, and tuned up to keep the efficiencies at their upper levels and reduce energy costs.

6.3.2 Air conditioners

The efficiencies of air conditioners are usually measured in terms of their Energy Efficiency Ratios or EERs, or their Seasonal Energy Efficiency Ratios or SEERs. In either case, the EER or SEER is specified as:

$$\text{EER or SEER} = \text{Btu of cooling}/(\text{watt-hours of electric energy input}) \quad (6\text{-}2)$$

The EER value is measured at a single temperature for the outside air, while the SEER involves a weighted average of the EERs over a typical season with a range of outside temperatures. SEERs usually range from 0.5 to 1.0 units higher than the corresponding EERs. Air conditioning units with capacities of five tons or less are rated with SEERs, while units over ten tons are rated with EERs. Current levels of SEERs for air conditioners can reach 18 or greater, but most units have SEERs around 12-14. Federal appliance efficiency standards and ASHRAE standards increased further with standards set in the 1992 National Energy Policy Act [10]. In January, 2006, minimum standards for air conditioners SEER went to 13.

Example 6-1: A five ton air conditioner has an average electric load of 6 kW. What is its SEER?

Solution: Using Equation 6-2 gives:

SEER = (5 tons) (12,000 Btu/hr/ton)/(6 kW)(1000 W/kW)

= 60,000 Btu/6,000 Wh

= 10 Btu/Wh

6.3.3 Chillers

Chiller efficiency is usually measured in terms of a Coefficient of Performance (COP) which is expressed as:

COP = heat absorbed by the evaporator/[(heat rejected
        by the condenser) – (heat absorbed
        by the evaporator)]   (6-3)

Chiller efficiencies depend on what type of compressor is used in the chiller, and whether air or water cooling is used. COPs may be as low as 2.5 for small chillers, and up to 7.0 for large, water-cooled, centrifugal or screw compressor chillers. Absorption chillers have COPs that range from 0.4 to 1.2.

Chiller efficiencies may also be expressed as EERs, where

EER = COP × 3.412 Btu/Wh   (6-4)

Example 6-2: A 100-ton chiller has a COP of 3.5. What is its electrical load?

Solution: Using Equation 6-2 and rearranging it gives:
electrical load = cooling capacity/EER
= (100 tons)(12,000 Btu/hr/ton)/(3.5)(3412 Btu/kWh)
= 100.5 kW

Chiller efficiencies are greatly affected by the amount of load on the system. A more useful measure of the efficiency of a chiller is to look at its IPLV (integrated part load value) or its NPLV(non-standard part load value).

6.4   HEATING, COOLING, AND VENTILATING LOADS

One of the easier ways to reduce costs in HVAC systems is to reduce the amount of energy that must be added to or extracted from an area to bring the area to the desired temperature range. Two major strategies for accomplishing this are available: (1) reduce the heating or cooling load; (2) change the targeted temperature range. The amount of *cooling* needed in an area can be reduced by reducing the amount of heat brought into the area through the walls, by reducing the number of people present, by reducing the number of heat sources such as lighting, or by modifying the energy consumption characteristics of the industrial processes. The amount of *heating* needed can be reduced by increasing the amount of machinery located indoors, by capturing some of the heat from lights, by reducing infiltration of cold outside air, or by insulating roofs or ceilings so that less heat escapes. In any of these examples, the cooling or heating load is being changed.

The second strategy is to change the temperature range that is considered to be desirable. This means changing the temperature limit above which air cooling occurs—the upper set point, and the temperature limit below which heating occurs—the lower set point. If heating does not start until the temperature is 55°F or lower, less heat will be used than if the threshold temperature for heating is 65°F. Similarly, an upper control limit of 85°F is more economical than a limit of 75°F. Changing these temperatures has the effect of changing the heating or cooling load imposed upon the HVAC system, although none of the heat sources are changed.

6.4.1 Heating and Cooling Load Calculations

The heating and cooling loads in a building occur because of (1) heat given off by people; (2) radiant energy from the sun that enters through windows, is absorbed by furniture, walls, and equipment, within the building, and is later radiated as heat within the building; (3) heat conducted through the building envelope (walls, roofs, floors, and windows) to or from the environment around the building; (4) waste heat given off by processes and machinery within the building; (5) heat given off by lighting; and (6) heat or cooling lost to ventilation or infiltration air. In this section we emphasize managing the energy costs of an existing building by examining those aspects of the heating and cooling load that can be changed by moderate or low expense or by scheduling. Reference 1 can be used for design purposes because it contains a sufficiently detailed methodology for calculating these heat and cooling losses and their interactions.

<u>6.4.1.1 Heating and cooling load: people.</u>
    People give off heat, and the amount they give off depends on the type of work they are doing, the temperature of their surroundings, and whether they are men or women. Table 6-1 gives representative values for the heat given off under various conditions. If no cooling or heating takes place during nonworking hours, the figures in Table 6-1 can be used directly. If cooling or heating takes place when the work force is not present, the later reradiation of heat given off by people and absorbed by equipment and surroundings must be taken into account as described in the ASHRAE Fundamentals Handbook [1].

Table 6-1. Rates of Heat Gain from People

Activity	Total heat gain for male adults (Btu/h)
Seated at rest	400
Seated, writing	480
Seated, typing	640
Standing, light work or slow walking	800
Light bench work	880
Normal walking, light machine work	1040
Heavy work, heavy machine work, lifting	1600

Note: Heat gain from adult females is assumed to be 85% of that for adult males.
Source: From *1993 Fundamentals Handbook*, ©1993. Reprinted with permission from the American Society of Heating, Refrigerating, and Air Conditioning Engineers, Atlanta, Ga.

    People-generated heat can be managed in several ways. The first management technique is that of scheduling: decreasing the number of people in an air-conditioned area during the time of peak energy consumption will decrease the amount of heat that must be removed and thus will decrease this component of the peak demand. If the work of people can be scheduled when the outside temperature is lower than the inside temperature, it is often possible to remove people-generated heat by introducing colder outside air rather than by using mechanical refrigeration. Another technique is to remove people. This is accomplished by replacing people by automated equipment such as automated storage and retrieval systems. Removing the people decreases the cooling load and may completely eliminate the necessity for cooling (or heating) an area.

6.4.1.2 Heating and cooling load: solar radiation.

The cooling load due to solar radiation through windows can be calculated by

$$q = \Sigma \, (A \times SC \times MSHG \times CLF) \qquad (6\text{-}5)$$

where q = cooling load (Btu/hr)
         A = window area (square feet)
         SC = shading coefficient
         $MSHG$ = maximum solar heat gain (Btu/hr/ft^2)
         CLF = cooling load factor

The total cooling load is found by summing the cooling loads for all surfaces of the building envelope. In Equation 6-5, the window area A must be measured. The shading coefficient SC depends on the kind of glass, the indoor shading (roller blinds or venetian blinds), and the average outside wind velocity. For single-pane glass, typical values range from 0.53 for 1/2-inch heat-absorbing glass to 1.00 for 1/8-inch clear glass; a white roller shade inside a clear 3/8-inch glass pane gives a shading coefficient of 0.25. Additional values of the shading coefficient can be obtained from glass vendors and from Reference 1. Changing glass types and installing shades are energy management measures that can make a significant change in the shading coefficient. Adding sun screens can reduce the shading coefficient to about 0.2, and using reflective film on windows can provide a shading coefficient of around 0.4.

The maximum solar heat gain and the cooling load factors, however, are fixed by the location and design of the structure. The maximum solar heat gain (MSHG) depends on the month, the hour, the latitude, and the direction the surface is facing. For 40° north latitude (Denver and Salt Lake City), these factors are shown in Table 6-2. The cooling load factor (CLF) measures the fraction of energy absorbed at a given time that is radiated as heat at a later time. This factor, like the shading coefficient, depends on the interior shading. Representative values for this factor are given in Table 6-3 for a building with medium-weight construction (approximately 70 lb of building material per square foot of floor area). Additional values for these factors can be found in Reference 1.

Table 6-2. Maximum Solar Heat Gain Factors
(Btu/h•ft² for 40° North Latitude)

	Surface orientation					
	N	NE/NW	E/W	SE/SW	S	Horizontal
Jan.	20	20	154	241	254	133
Feb.	24	50	186	246	241	180
March	29	93	218	236	206	223
April	34	140	224	203	154	252
May	37	165	220	175	113	265
June	48	172	216	161	95	267
July	38	163	216	170	109	262
Aug.	35	135	216	196	149	247
Sept.	30	87	203	226	200	215
Oct.	25	49	180	238	234	177
Nov.	20	20	151	237	250	132
Dec.	18	18	135	232	253	113

Source: From Reference 1. Reprinted with permission from the American Society of Heating, Refrigerating, and Air Conditioning Engineers, Atlanta, Ga.

Equation 6-5 shows what factors can be altered to manage the solar component of the cooling load. For buildings in a hot climate, it is often cost effective to cover all outside windows and thus reduce the area A through which solar radiation enters the building. If, however, heating is more of an expense than cooling, it may be possible to increase the area and use passive solar heating.* The shading coefficient (SC) can be changed by installing shades, heat-absorbing glass, or reflective coatings. The maximum solar heat gain (MSHG) can be used to help decide where to locate people and equipment within a building relative to the location of windows. More heat will come in if the windows are facing south than

---

*For more detail on the possibilities of this approach, see Reference 2.

Table 6-3. Cooling Load Factors for Glass

Solar time	Direction window is facing								
	N	NE	E	SE	S	SW	W	NW	Hor.
Without Interior Shading									
2 a.m.	.20	.06	.06	.08	.11	.13	.13	.12	.14
4	.16	.05	.05	.06	.08	.10	.10	.09	.11
6	.34	.21	.18	.14	.08	.09	.09	.09	.11
8	.46	.44	.44	.38	.14	.12	.10	.11	.24
10	.59	.40	.51	.54	.31	.15	.12	.14	.43
12	.70	.33	.39	.51	.52	.23	.14	.17	.59
2 p.m.	.75	.30	.32	.40	.58	.44	.29	.21	.67
4	.74	.26	.26	.33	.47	.58	.50	.42	.62
6	.79	.21	.21	.25	.36	.53	.55	.53	.47
8	.50	.15	.15	.18	.25	.33	.33	.32	.32
10	.36	.11	.11	.14	.18	.24	.23	.22	.24
12	.27	.08	.08	.10	.14	.18	.17	.16	.18
With Interior Shading									
2 a.m.	.07	.02	.02	.03	.04	.05	.05	.04	.05
4	.06	.02	.02	.02	.03	.04	.04	.03	.04
6	.73	.56	.47	.30	.09	.07	.06	.07	.12
8	.65	.74	.80	.74	.22	.14	.11	.14	.44
10	.80	.37	.62	.79	.58	.19	.15	.19	.72
12	.89	.27	.27	.49	.83	.38	.17	.21	.85
2 p.m	.86	.24	.22	.28	.68	.75	.53	.30	.81
4	.75	.20	.17	.22	.35	.81	.82	.73	.58
6	.91	.12	.11	.13	.19	.45	.61	.69	.25
8	.18	.05	.05	.07	.09	.12	.12	.12	.12
10	.13	.04	.04	.05	.07	.08	.08	.08	.08
12	.09	.03	.03	.04	.05	.06	.06	.06	.06

Note: Solar time = local standard time (LST) + 4 min × (LST meridian − local longitude) + correction for the month (from − 13.9 min in February to + 15.4 min in October).
Source: From Reference 1. Reprinted with permission from the American Society of Heating, Refrigerating, and Air Conditioning Engineers, Atlanta, Ga.

if the windows are facing north. If a close visual contact with the outside is desired but the sun heat is not, both conditions can be met with windows facing the north. The cooling load factor is reduced by shading, since less heat is absorbed by furniture and walls to be reradiated later.

6.4.1.3 Heating and cooling load: conduction

In addition to inside heating caused by radiation of solar energy absorbed by inside materials, the sun heat combines with the outside temperatures to create heating due to conduction through the walls and the roof. Although writing and solving the heat conduction equations are beyond the scope of this text, several observations can be made that are relevant to energy management. First, the amount of heat gain or loss through a wall depends on the thermal conductance—or U value—of the wall. Adding insulation to walls or roofs can significantly reduce the unwanted heat gain or heat loss through a wall and can be very cost effective if the main component of the cooling load is from conduction rather than from inside sources. Thermal conductance, insulation, and the calculation of heat transfer through walls and roofs are discussed in detail in Chapter 11.

Second, the construction of the wall and roof is also an important factor, with the material used to build the structure having a potential thermal storage effect that can be utilized. The so-called *flywheel effect* describes this: In the same way that a flywheel at motion tends to remain in motion, a warm wall tends to remain warm and to radiate heat after the sun is down. The amount of heat and the time the wall continues to radiate depend on the construction of the wall in the same sense that the speed and running time of a released flywheel depend on the construction of the wheel. If an existing building has massive walls, it may be possible to schedule work hours so as to take advantage of the flywheel effect and obtain free heating or cooling. These points are discussed in Reference 3.

Heating or cooling loads due to conduction through a wall can be approximated using Heating Degree Day (HDD) or Cooling Degree Day (CDD) data. Definitions of HDDs and CDDs and calculation methods were given in Section 2.1.1.2. The method involves the use of the heat flow equation 11-9 given in Chapter 11. In this equation, A is the area of the wall, U is the thermal conductance of the wall, Q is the heat flow through the wall, in Btu/yr and the temperature difference ΔT is replaced by the number of heating or cooling degree days, HDD or CDD.

$$Q = U \times A \times (DD/year) \times 24\,h/day \qquad (6\text{-}6)$$

Example 6-3: A wall has an area of 100 ft² and has a thermal conductivity of 0.25 Btu/ft²•h•°F. If there are 3000 degree days in the heating season, what is the total amount of heat lost through the wall?

Solution: The heat lost through the wall is found using equation 6-6 as:

$$Q = (0.25 \text{ Btu/ft}^2 \bullet h \bullet °F) \times 100 \text{ ft}^2 \times (3000 \text{ °F days/year})$$
$$\times 24 \text{ h/day}$$
$$= \underline{1,800,000 \text{ Btu/year}}$$

The degree day method has many potential problems that limit its value in finding the total heating needs of a building or facility. First, it only provides an approximate value for the heat flow due to conduction. It does not take into account the moisture in the air; internal heat loads that might provide less than or greater than a 5°F inside temperature gain; solar heat gain; or many other factors.

6.4.1.4 Heating and cooling load: equipment.

The fourth major source of heating is from equipment. The energy consumption in Btu per hour from computers, ovens, industrial processes such as solder pots that use much heat, and many other types of equipment can either be read directly from nameplates or can be approximated from gas or electricity usage by assuming that every kWh of electricity contributes 3412 Btu of heat at the point of use and that each Mcf (thousand cubic feet) of gas has an energy content of 1 million Btu.

The efficiency for a single-phase motor is usually 50-60%, and the efficiency for a three-phase motor is usually 60-95%. The energy used by other kinds of equipment can be found in Reference 1, from equipment vendors, and from gas and electric utilities. Decreasing or rescheduling the amount of equipment using electricity or gas at a given time has a twofold effect. First, the actual energy used at that time is decreased. Second, the heat introduced into the space is also reduced. If this heat must be removed by cooling, the cooling load is thus reduced by turning off this equipment. If this heat is desirable, Equations 6-9, 6-10 and 6-11 can be used to indicate how much waste heat can be made available by scheduling or reducing the number of electric motors.

The amount of heat given off by a motor can also be determined from its nameplate rating in kW if there is one, or it can be estimated from the nameplate voltage and current used by the motor. If it is a single-phase motor, the energy consumed per hour by the motor is given by

$$kWh/h = kW \times use\ factor\ (fraction\ of\ time\ that$$
$$the\ motor\ is\ in\ use) \qquad (6\text{-}7)$$

or by the calculation involving the voltage, current and power factor

$$kWh/h = voltage \times current \times power\ factor \times use\ factor \qquad (6\text{-}8)$$

The amount of heat that is given off in the space being analyzed depends on the job the motor is doing. In most cases, all of the energy used by the motor shows up as heat added to the space. This may seem odd at first, but when the physical factors are considered, the result is quite clear. First, the motor is not 100% efficient, so some of the motor energy is lost directly to heat through losses in the armature and field windings, losses in the motor core and friction losses in bearings. This heat makes the motor feel hot to the touch, and directly heats up the space.

The remaining motor energy is then used for some application. However, in most cases, that application is such that the remaining motor energy is also converted to heat somewhere in the space. Consider a typical application where a motor is being used to drive a conveyor belt in a production room. The useful energy from the motor is used to overcome the friction of the rollers and belt, and the inertia of the items being moved on the conveyor, and thus is converted completely to heat. Unless the conveyor belt extends outside the space being considered, all of the motor's useful energy becomes heat inside that space. Thus, all of the energy supplied to the motor eventually becomes heat somewhere in the space being considered.

It is useful to consider the few instances where the energy from a motor is not converted completely to heat in the immediate space. This can only occur if some of the energy of the motor is stored in the product being made and leaves the space as embedded energy in the product. For example, if a motor is being used to compress a spring that is then used in a device which is further assembled, then a small part of the motor's energy is transferred to the spring in the form of potential energy. Some other examples might be energy in compressed gases, or energy in frozen foods. Although there are some cases where this energy storage and removal are involved, it is usually a very small part of the overall energy used by the motor, and for all practical considerations, all of the energy used by a motor ends up as heat in the immediate space.

The heat given off by a single phase motor—in Btu per hour—can be found as follows

$$Btu/h = kW \times \text{use factor} \times 3412 \, Btu/kWh \qquad (6\text{-}9)$$

or

$$Btu/h = \text{voltage} \times \text{current} \times (\text{power factor}) \times (3412$$
$$Btu/kWh) \times \text{use factor}/(1000 \, wh/kWh) \qquad (6\text{-}10)$$

For a three-phase motor, the energy converted to heat per hour is given by equation 6-9 since the calculation assumes the kW rating of the motor is known. If the kW rating is not listed on the nameplate for the motor, then the voltage, current and power factor must be used, along with the factor 1.732 for proper determination of the power in a three phase motor.

$$Btu/h = \text{voltage} \times \text{current per phase} \times (\text{power factor})$$
$$\times 1.732 \times 3.412 \times \text{use factor} \qquad (6\text{-}11)$$

6.4.1.5 Heating and cooling load: lighting.

Heat generated from lighting is another example in which all of the energy used is generally converted to heat in the immediate space. A very small part of the energy supplied to lights appears in the form of visible light. Incandescent lamps convert about 2-3% of the energy they use into visible light. The remaining 97% is immediately converted to heat which enters the surrounding space. Fluorescent lights are more efficient, and they convert around 10% of their input energy into visible light. Even so, the remaining 90% is converted directly into heat from the lamps and from the ballasts.

Next consider what happens to the visible light once it is produced. The photons of light strike surfaces such as floors, walls, desks, machines and people where some of the energy is absorbed and becomes heat. Some of the light energy is reflected, but it then strikes the same surfaces and more energy is converted to heat. Unless some of the light escapes outside the area of interest, all of the energy in the light eventually becomes heat. Thus, except in some relatively rare instances, all of the energy supplied to lights in an area quickly becomes heat in that same area.

In some cases, heating systems have been designed and sized to utilize the heat from lights as a significant source of heat. This is not an efficient heating strategy, but it may need to be recognized and dealt with to improve the energy efficiency of this kind of system. In other cases, light fixtures may be ventilated to the outside, or to another unheated area. Here, the heat from the lights does not contribute much of a load to the space when air conditioning is considered. However, during the heating season, this is a significant loss of heat that increases the energy that must be supplied by the regular heating system. If the ventilating ducts in the lighting fixtures can be easily closed during the heating season and reopened during the cooling season, this is an energy efficient operation strategy.

To calculate the amount of heat that is added to a space from lighting, remember that each kWh of electricity is equivalent to 3412 Btu. Thus the heat produced each hour is found by

$$Btu/h = 3412 \, Btu/kWh \times K \, kW \qquad\qquad (6\text{-}12)$$

where K is the number of kW of lighting load. Note that K must include the power consumption for any ballasts that are connected to the lights. To obtain the total heat produced from lighting, Equation 6-12 is calculated for each hour and the total obtained.

Some of the ways of reducing the amount of lighting energy needed to illuminate a space to a prescribed level include replacing lamps with more efficient lamps, cleaning the luminaries, and painting adjacent surfaces. These and other measures were discussed in detail in Chapter 5. The air conditioning savings from reducing lighting energy is illustrated in the example below.

Example 6-4: A common relamping EMO was described in Example 5-3 in Chapter Five. In that example, 200 40-Watt lamps were replaced with 34-watt lamps in a facility that was not air conditioned. If the facility had been air conditioned, there would be an additional savings depending on the number of hours that the air conditioning was needed. Calculate this additional savings assuming the air conditioner has a COP of 2.8.

Solution: Example 5-3 calculated that the energy saved from the relamping was 42,048 kWh/yr. As discussed earlier, all of this lighting energy becomes heat that must be removed by the air conditioner. If the facility is one that has a number of heat-generating machines and processes which results in the facility having to be cooled 24 hours per day, all of the lighting energy saved translates to air conditioning savings. The heat reduction from the lighting energy savings can be found using Equation 6-12 re-written as:

$$Btu \quad = 3412 \, Btu/kWh \times K \, kW \times h$$

$$= 3412 \text{ Btu}/\text{kWh} \times 42,048 \text{ kWh}$$

$$= 143.5 \times 10^6 \text{ Btu}$$

The electric energy savings from the air conditioner can now be found by dividing this Btu quantity by the EER of the air conditioner.

$$\text{A}/\text{C energy savings} \quad = 143.5 \times 10^6 \text{ Btu}/(2.8 \times 3.412 \text{ Btu}/\text{Wh})$$

$$= 15.02 \times 10^6 \text{ Wh}$$

$$= 15,020 \text{ kWh}$$

A more direct way to get this same result is just to divide the lighting energy savings by the air conditioner COP.

$$\text{A}/\text{C energy savings} \quad = 42,048 \text{ kWh}/2.8$$

$$= 15,020 \text{ kWh}$$

This increased savings from considering air conditioning makes a significant change in the cost-effectiveness of the relamping program. In this case, it increases the savings by over one-third.

6.4.1.6 Heating and cooling load: air.

The sixth major category of heating or cooling load comes from energy used to heat, cool, or humidify air. This part of the heating or cooling load can be reduced by weatherstripping, caulking, and tightening windows; by installing loading dock shelters; by replacing broken windows; and by other measures designed to reduce or eliminate air leakage to or from the outside. Air infiltration can occur through the envelope at many places, and there are infiltration-reducing techniques unique to each place. Such techniques range from caulking cracks and openings to sealing loading docks. It can also prove worthwhile to prevent airflow from conditioned areas of a plant to unconditioned areas (or vice versa) by installing airflow barriers indoors such as a plastic curtain.

Infiltration or exfiltration generally involves air moving through openings. The annual amount of energy lost through a hole or crack can be estimated from

$$\text{Btu/year} = V \times 1440 \times .075 \times .24 \times (\text{HDD} + \text{CDD}) \tag{6-13}$$

where V = volume of air entering or leaving, in cubic feet per minute
     1440 = number of minutes per day
     .075 = pounds of dry air per cubic foot
     .24 = specific heat of air, in Btu per pound per degree F
     HDD = heating degree days per year, in days × degrees F
     CDD = cooling degree days per year, in days × degrees F

This formula is a modification of that found in Reference 1. The impact of water vapor is ignored in Equation 6-13 because in many climatic regions exhausting it is often a benefit for cooling and a disbenefit for heating, and the total effect is usually negligible. In climatic areas where moisture is a problem for heating or cooling, the effect of water vapor may need to be considered.

6.4.2 Ventilation Requirements for Health

In addition to satisfying the need for a comfortable working environment, the HVAC system must also provide ventilation to remove noxious substances from the air. Many ventilation requirements are specified by state and local health and safety codes. Table 6-4 shows some of the minimum outdoor air requirements specified by ASHRAE Standard 62-2004 [11]. These requirements can often be met without having to heat or cool excessive amounts of outside air by installing special equipment such as self-ventilating hoods, or by isolating contaminant areas from other parts of a plant. Newer ventilation requirements are given in ASHRAE 62-2007 and 2010.

Getting rid of cigarette smoke is another problem. Where people are not smoking, the recommended ventilation standards range from 15-30 cfm/person; in smoking lounges, the recommended standard is 60 cfm/person. In this situation, the total amount of ventilation air is reduced by having separate areas for smokers; a separation which is enforced by law in many states. In Table 6-4, some applications involving air contaminants such as those from dry cleaners, and smoking areas may require additional ventilation, or at least special equipment to remove the smoke or contaminants.

6.4.3 Ventilation Standards for Comfort

One of the main reasons for using air conditioning is to keep people comfortable. The definition of a comfortable condition changes with clothing styles, the number and quantity of local drafts, and the price of energy for heating, cooling, and humidifying. Comfort is generally defined,

Table 6-4. Outdoor Air Requirements for Ventilation
(From ASHRAE Standard 62-2004)

Applications	Outdoor air requirements cfm per person)
Offices	
Office space	20
Reception area	15
Conference room	20
Data entry area	20
Dry cleaners, Laundries	
Commercial laundry	25
Commercial dry cleaner	30
Smoking lounge	60

however, by a temperature range that varies depending on the relative humidity of the air. Three measures take advantage of the comfort zone: lowering the minimum temperature at which heating is initiated, raising the maximum temperature at which cooling is initiated, and changing the humidity in the air to more nearly conform to that of the outside environment. If outside air can be used without either humidifying or dehumidifying, considerable energy can be saved. The exact amount depends on the HVAC system, but note that evaporating 1 pound of water uses about 1040 Btu. The method of cooling air so that condensation takes place and then reheating the dried air to a comfortable temperature is also costly. An alternative approach in some cases is the use of heat pipes which is discussed in Section 6.6. Another method for saving energy is to reduce the amount of excess ventilation air to that required by the given environment; the cube law for fan horsepower (Equation 6-14) can be used to estimate possible savings.

Note that management bears the burden of selling the employees on energy-saving measures or of introducing the changes so gradually that they are not noticed. Thus, one good method of energy management that relates to comfort is to allow temperatures to change slowly. Several studies have shown that people are relatively insensitive to slow changes in air temperature [4,5]. This insensitivity can be used to advantage by turning the air conditioning down or off 30 minutes to an hour before quitting time and letting the temperature drift. Ventilation must still be provided, though.

6.5   IMPROVING THE OPERATION OF THE HVAC SYSTEM

6.5.1 Basic Operating Rules

The objective of learning how an HVAC system operates is to manage that system more efficiently. HVAC system management can be improved by careful attention to the following operating rules.

Operating Rule 1. Heat to the lowest temperature possible, and cool to the highest temperature possible. Set the hot and cold air temperatures on the hot and cold sides of a dual-duct system so that one zone is receiving only hot air and one zone is receiving only cold air. The hot temperature is thus set so that the system meets the heating needs of the coldest room and cooling needs of the warmest zone. It automatically meets the temperature needs of all the other zones.

Operating Rule 2. Avoid heating or cooling when heating or cooling is not needed. For example, heating or cooling for people is not needed when people are not in a building at times such as weekends or at night. At those times, a building temperature can be allowed to drift with the only constraint being the safe temperature of building components or other material contained within the building. Avoid heating or cooling warehouses unless they contain people or materials sensitive to heat or to cold.

Operating Rule 3. Learn how your control system is supposed to work and then maintain it properly. A consistent problem that plagues buildings is a control system that does not work. For example, a control system is not functioning as intended if return air dampers are blocked open, and the HVAC system heats or cools all the air used for ventilation. People seldom understand the way a two-thermostat system is supposed to work, and they turn the wrong thermostat, causing heating at night but none when the heating is actually needed.

Operating Rule 4. To insure that the minimum required amount of ventilation air is being used, adjust the ventilation system by altering the control system settings or by changing pulleys on fans or their drive motors, or by using variable speed drives. One very useful relationship is

$$hp_A / hp_B = (cfm_A / cfm_B)^3 \qquad\qquad (6\text{-}14)$$

This is the cube law for fan horsepower [8]. It is useful in calculating the energy consumption to be saved from reducing the ventilation requirements.

Example 6-5: ACE Industries presently has a 5-hp ventilating fan that draws warm air from a production area. The motor recently failed and they think they can replace it with a smaller motor. They have determined that they can reduce the amount of ventilation air by one third. What size motor is needed now?

Solution: Use Equation 6-14, and note that the ratio of the new to old cfm rate is 2/3. Thus, the new hp needed is:

New hp = $(2/3)^3 \times 5$ hp = 0.3×5 hp = 1.5 hp

Since 1 hp is equivalent to .746 kW at 100% efficiency, the power savings achieved by a reduction of H hp is given by

$$\text{Power saved (kW)} = (H \text{ hp} \times .746 \text{ kW/hp})/\text{EFF} \qquad (6\text{-}15)$$

where EFF is the motor efficiency, usually between .70 and .90. This expression can only be used if the old and new motors have the same efficiencies. If this motor is running constantly, the action of reducing the fan horsepower reduces both the demand and the energy part of the electric bill; otherwise it may affect the demand but must be multiplied by the hours of use to determine the amount and cost of energy saved.

Example 6-6: What is the electrical load reduction for the smaller fan motor in Example 6-5 if the 5-hp motor had an efficiency of 84%? The new 1.5-hp motor has an efficiency of 85.2%.

Solution: We cannot use Equation 6-15 since the two motors do not have the same efficiency. We must calculate the electrical load from each motor and determine the difference.

Old load = (5 hp) × (0.746 kW/hp)/(0.84) = 4.44 kW

New load = (1.5 hp) × (0.746 kW/hp)/(0.852) = 1.31 kW

Electric load reduction = 3.13 kW

However, since the two motor efficiencies are approximately equal, we can make an approximation by using Equation 6-15:

Motor load reduction = (3.5 hp) × (0.746 kW/hp)/(0.84) = 3.11 kW

This is very close to the correct value.

Operating Rule 5. If you do not need heating, cooling or ventilation, turn off the HVAC system. Unnecessary conditioning and ventilation cost money. Find out when the conditioning and ventilation is needed and arrange to have the HVAC system running only at those times.

6.5.2 Inspecting the HVAC system

First, the auditor should determine whether the room is being cooled or heated more than is necessary. Ideally, a person in a jacket should be comfortable in the winter, and a person in shirt sleeves should be comfortable in the summer. Excessive heating or cooling is another unnecessary energy cost.

Next, the energy auditor must inspect the HVAC system thoroughly to determine whether it is operating properly. Every HVAC system has heat transfer surfaces to enable heating or cooling to take place, every HVAC system has some means of transporting its working fluid to the point of use from the point where heating or cooling is supplied, and every HVAC system has a set of controls which govern its operation. Inspecting each one of these component areas is necessary in a complete energy audit. Since the most common systems are those in which air is the working fluid, the rest of this discussion will be confined to air-cooled systems.

6.5.2.1 Heat transfer surfaces

A heat transfer surface is the surface where a hot or cold fluid gives up or receives heat from air that is passing around it. Typically, heat transfer surfaces are designed so that a hot or cold liquid flows through pipes surrounded by fins. The fins are used to increase the heat transfer rate. These heat transfer surfaces must be examined periodically and be maintained to continue to work properly. The points to be inspected in this part of the HVAC system are the fluid flow lines and the heat transfer surfaces. Do relevant gauges show that fluid is flowing? Are "hot water" pipes actually hot? Can you hear the sound of fluid flowing? When you examine the heat transfer surfaces, are the fins and coils clean, or are they fouled with dirt or grease or dust? Fins must be clean to function effectively and efficiently.

6.5.2.2 Air transportation system

The air transportation system moves air from the outside, mixes outside air with return air, and removes used air either to the outside or to the supply fans for use as return air. The main components of this system are the ductwork, dampers, and filters and the fans, blowers, and associated motors. The ductwork can have insulation hanging loose. The ducts can leak air through untaped seams, or they can be crushed by adjacent piping. Loose insulation can be detected by removing one duct panel and examining the inside of the duct with a flashlight; untaped seams and crushed ducts can be detected with a quick visual inspection. Dampers and filters can also be inspected visually. The dampers should be clean and should close, and their mechanical linkages should be connected to the control actuators. Filters should be installed and reasonably clean; return air grilles should also be inspected to see whether they need to be cleaned. Filters should be cleaned or changed at periodic intervals.

The HVAC fans and blowers should be examined carefully. Fans and blowers should be operational, and the belts should be aligned correctly so that the fan pulley and the motor pulley are in a straight line. Be particularly thorough in examining fan belts and motor connections; the authors have observed instances where motors had been installed and were running but were not connected to the fans they were supposed to drive. Motors should also be inspected to see that they are properly connected with balanced voltages to all three legs of a three-phase system. Motors should also be free from excess bearing noise. Fans or blowers should be reasonably clean, since accumulated dust detracts from their efficiency. The fan or blower should rotate in the correct direction, and the fan shaft should not bind.

6.5.2.3 The control system

The HVAC control system detects pressures or temperatures and compares these with preset values. Depending on the result, the control system sends electrical or pneumatic signals to open or close dampers, open or close valves, and turn furnaces, chillers, and blower motors on or off. Clearly it is important that the control system function properly; otherwise, the HVAC system does not work as intended, and will not be energy efficient.

The first step in inspecting the control system is to examine the thermostats. A reliable industrial thermometer can be used to calibrate each thermostat thermometer; if the temperature difference between the two temperature readings is significant, the thermostat should be checked by a vendor. Next the thermostat set point temperature should be raised and lowered to see if the heat or cold comes on or shuts off; this procedure tests the entire HVAC control system.

Gauges should be checked to see whether they are connected and whether they are reading within the correct operating ranges. The compressor that supplies compressed air to the control system should be inspected to see that it is working properly and not leaking oil or water into the control system; if either oil or water gets into the controls, a complete replacement of the control system is often necessary. An air dryer is almost always a necessity on the air supply system to keep the controls working properly.

6.6   HEAT PIPES

In many areas of the country, removing humidity is the chief energy cost of air conditioning. This is because moisture is condensed out of the air by the cooling coil and the colder the air when it passes over the cooling coil, the more moisture is removed. Therefore, in areas where the humidity is high, the air must be cooled much lower than the desired temperature in order to remove the moisture and then heated back to the desired temperature. This means that energy is required to overcool the air; additional energy is required to reheat the air; and the equipment must be oversized in order to overcool the air which increases the power demand of the air conditioning system.

The older, energy-inefficient air-conditioners were designed with very cold cooling coils; newer, high-efficiency models are often designed with warmer coils. These coils are larger and require less energy to operate, and one of the reasons why they save energy is because they do not remove as much moisture. Although this is not a problem in some parts of the country, in areas with high humidity the energy-efficient air-conditioners often leave the conditioned air uncomfortably humid. This problem is generally solved by lowering the temperature setting of the thermostat which uses additional cooling energy and may negate the savings from the energy efficient model.

One energy savings solution is the heat pipe [12]. Heat pipes are relatively new on the commercial air conditioning scene. Although they were first developed near the turn of the last century, their commercial use only recently became feasible as a result of research on a NASA contract. A heat pipe is a metal tube that is filled with an evaporative fluid and then sealed at both ends. When heat is applied to one end of the tube, the fluid inside evaporates and the vapor moves to the other, cooler end by some sort of capillary action through an inside wick. The vapor then condenses in the cooler end and returns to the warmer end by gravity. The heat pipe is activated by the temperature difference between the two ends and does not use energy to operate. See Figure 6-5.

When used in an air conditioning system, one end of the heat pipe is placed in the return air system and is heated by the warm return air.

Figure 6-5. Diagram of heat pipe.

That heat is "free" because no additional energy is expended to generate the heat. Because the heat pipe absorbs heat from the return air, that air is cooler when it goes to the cooling coil so the cooling coil can work at a lower temperature. That means that the cooling coil can remove more moisture from the air; it also means that the cooling load on the compressor is lower. The other end of the heat pipe is placed in the supply, or conditioned, air system. The heat from the first section in the return air stream is transferred to this section which is used to reheat the chilled air and lower the relative humidity before the air is distributed throughout the system. The process of reheating the air causes that end of the heat pipe to become chilled. See Figure 6-6.

Heat pipe
heat exchanger

Figure 6-6.
A high-efficiency air conditioner/dehumidifier using heat pipes.
Courtesy of Florida Solar Energy Center.

Heat pipes are also useful for areas which need large amounts of outside air for ventilation. Fresh air brings in both heat and humidity which means that larger air conditioning systems are usually needed. The addition of heat pipes to the air conditioning system allows the addition of up to 20% fresh air without increasing the size of the system.

Heat pipes are suitable for a number of air conditioning situations. Industries which require a low humidity level or humidity control include: electronic component production, assembly and storage; film drying, processing and storage; drug, chemical and paper manufacturing and storage; printing; candy and chocolate processing and storage. Other examples of industries or areas which would benefit from heat pipes are hospital operating rooms, libraries, grocery stores, telephone exchanges and relay stations, clean rooms, underground silos, and places with indoor swimming pool or spa facilities.

6.7   THERMAL STORAGE

Thermal energy storage offers one of the most promising technologies for effective electrical peak load management in buildings and other facilities [13]. Heating and cooling energy needs provided by electrically powered equipment usually correspond to the time of a facility's peak demand, and contribute to increased electric costs that occur from the demand charge. A thermal energy storage system may produce chilled water or ice off-peak for later use in cooling a facility, or produce heated air or water off-peak for later use in heating a facility.

Cool storage systems operate by producing and storing chilled water or ice during the evening and night when electric rates are low, and drawing on that stored water or ice during the day when the cooling load is greatest. The high-demand electrically powered chillers are shifted to operation during off-peak periods, and are especially cost-effective for facilities that have time-of-day rates which offer low-cost energy during the night. The storage of chilled water or ice also allows the facility to operate with smaller-sized chillers since the peak cooling load is handled with the combination of the small chiller and the cool storage. For new buildings or facilities, this reduction in the size of the chillers often pays most of the cost for the cool storage system. Other cooling system operating cost savings also result from this approach because of the downsizing of fans, pumps and ducts due to the lower temperatures of distributed air because of the very low temperatures of the stored water or ice.

Thermal energy storage systems typically increase the overall system energy consumption because of the storage losses, but they also significantly decrease the costs associated with peak load or peak demand charges. The increase in energy for cooling with storage is often moderated due to the higher efficiency of the chiller system operating at lower night time temperatures.

Many electric utilities offer rebates or incentives for facilities to install thermal storage systems. The incentive may be in the form of low rates for off-peak energy use, or in the form of a direct rebate based on the number of kW moved off-peak.

6.8   SUMMARY

In this chapter, we have explained the functions and the components of HVAC systems. Once the functions and components are understood, the energy manager can find ways to reduce the heating and cooling loads and thus reduce this element of energy costs. When the manager also understands how each of the HVAC system components works, he or she is then prepared to improve the operation of the physical system. By using the operating rules presented, and by adding additional rules unique to your own system, this understanding can be translated into improved operating policies for both HVAC equipment and for the people affected by the HVAC system.

References
1. 2005 *ASHRAE Handbook of Fundamentals*, American Society of Heating, Refrigerating, and Air Conditioning Engineers, Atlanta, Ga., 2005.
2. Donald Rapp, *Solar Energy*, Prentice-Hall, Englewood Cliffs, N.J., 1981.
3. W.J. Kennedy, Jr., and Richard E. Turley, "Scheduling and the Energy Economics of Wall Design," in *Proceedings of the 1981 Spring Annual Conference & World Productivity Conference*, Detroit, Michigan, Institute of Industrial Engineers, Atlanta, Ga., 1981.
4. Larry G. Berglund and Richard R. Gonzalez, "Human Response to Temperature Drift," *ASHRAE Journal*, Aug. 1978, pp. 38-41.
5. L.G. Berglund, and R.R. Gonzalez, "Application of Acceptable Temperature Drifts to Built Environments as a Mode of Energy Conservation," *ASHRAE Transactions*, Vol. 84, Part 1, 1978.
6. National Electric Contractors Association and National Electrical Manufacturers Association, *Total Energy Management: A Practical Handbook on Energy Conservation and Management*, Third Edition, National Electrical Contractors Association, Washington, D.C., 1986.
7. 2004 *ASHRAE Handbook for Systems and Equipment*, American Society of Heating, Refrigerating,, and Air Conditioning Engineers, Atlanta, Ga., 2004.

8. 2003 *ASHRAE Handbook for HVAC Applications*, American Society of Heating, Refrigerating, and Air Conditioning Engineers, Atlanta, Ga., 2003.
9. *CFC's: Time of Transition*, American Society of Heating, Refrigerating and Air Conditioning Engineers, Atlanta, GA, 1989.
10. Energy Policy Act of 1992, P.L. 102-486 (HR 776), U.S. Congress, October 24, 1992.
11. ANSI/ASHRAE Standard 62-2001, Ventilation for Acceptable Indoor Air Quality, American Society of Heating, Refrigerating, and Air Conditioning Engineers, Atlanta, Ga., 2001.
12. *Commercial High Efficiency Dehumidification Systems Using Heat Pipes, A Technical Manual*, Heat Pipe Technology Company, Alachua, FL, 1993.
13. "Thermal Storage Forum," *ASHRAE Journal*, April, 1990, pp 26-66, American Society of Heating, Refrigerating and Air Conditioning Engineers, Atlanta, GA.

Bibliography

Frank E. Beaty, *Sourcebook of HVAC Details*, McGraw-Hill Book Company, New York, NY, 1987.

Capehart, Barney, Editor, *Encyclopedia of Energy Engineering and Technology*," Taylor and Francis/CRC Publishing Company, Boca Raton, FL, July, 2007.

Robert Chatenver, *Air Conditioning and Refrigeration for the Professional*, Wiley, New York, NY, 1988.

Harold R. Colen, *HVAC System Evaluation*, R.S. Means Company, Kingston, MA, 1990.

Roger W. Haines, *HVAC System Design Handbook*, TAB Professional and Reference Books, Blue Ridge Summit, PA, 1988.

HVAC Systems—Applications, Sheet Metal and Air Conditioning Contractors National Association, Vienna, VA, 1987.

Industrial Ventilation, Committee on Industrial Ventilation, American Conference of Governmental Industrial Hygienists, Lansing MI, 1979.

John I. Levenhagen, *HVAC Controls and Systems*, McGraw-Hill Book Company, New York, NY, 1993.

Raymond K. Schneider, *HVAC Control Systems, Second Edition*, Wiley, New York, NY, 1988.

Variable Air Volume Systems Manual, 1980 ed., Honeywell, Inc., Minneapolis, 1980.

Chapter 7

Understanding and Managing Boilers

The largest use of energy in many facilities is in boilers. Boilers generate steam which is then used to provide space heat, process heat, mechanical power, and possibly electricity. Since their energy use is so significant, boilers are a good place to start in looking for ways to reduce energy costs. Chapter 7 discusses boilers—what they are, what they do, their inputs and outputs, and how to locate and evaluate energy management measures associated with boilers. With this as a background, Chapter 8 presents a more detailed discussion of steam distribution systems—their components, controls, and energy management.

7.0 INTRODUCTION

The purpose of a boiler is to change the energy in a fuel into a more useful form, generally steam. In this chapter, we explain what a boiler is, how it operates, how it is instrumented, and some of the more common ways to save energy in its operation and use. The next chapter explains the components of a steam distribution system, their instrumentation, and the calculation of savings from tested energy management measures.

Many commercial and industrial facilities use fossil fuel-fired boilers to produce steam or hot water for space heating or for process heating. These boilers are usually the major consumers of the fuel used in the facility, so anyone involved in energy management needs to know how the boiler and steam distribution systems work, what operating parameters of a boiler are important to energy consumption and steam performance, and how the performance of a boiler can be improved.

7.1 How boilers work

A boiler is either a fire tube boiler—with tubes containing flame and surrounded by water—or a water tube boiler—with tubes containing water surrounded by flame. The basic operation of a fire tube boiler is shown in Figure 7-1. Fuel and air are combined in a burner, go through tubes and heat up water, and leave through a flue. The water is brought into the boiler, surrounds the tubes containing the fire from the burner, rises to the top under pressure, and leaves the boiler for use in industrial processes or to generate electricity.

Figure 7-1. Fire Tube Boiler, courtesy of Trams for Bath, City of Bath, United Kingdom

Another example of a fire tube boiler is shown in Figures 7-2 and 7-3, used with permission of Cleaver-Brooks. The one shown here is a packaged boiler, shipped from the factory as a complete assembly, with burner, control systems, operating and safety controls all piped and/or wired into the assembly. In the boiler shown, flame goes from left to right, is diverted at the right end for a second pass and goes from right to left, then is diverted first to the right, finally to the left, where it exits the boiler. Successive passes take more heat from the flame.

In a large commercial water-tube boiler, shown in Figure 7-4, water starts in a cylindrical drum at the bottom of the boiler chamber. Water travels upward by convection through the tubes that make up the wall of the combustion chamber and is converted to steam as it is heated by the flame in the boiler. At the top of the chamber, the steam is collected in a cylindrical drum. From this drum, the steam travels

FRONT HEAD
WITH
MOTOR,
CONTROLS,
LINKAGE
AND
BURNER

GAS
TRAIN

OIL PIPING

Figure 7-2. Fire Tube Boiler, Assembled

FOURTH PASS

THIRD PASS

FIRST PASS

SECOND PASS

Figure 7-3. Fire Tube Boiler, Cutaway (photographs courtesy of Cleaver-Brooks)

through economizers and ultimately is used, lost, or returns to the boiler. An example of a packaged industrial water tube boiler is given in Figures 7-5 and 7-6.

A third kind of boiler is the *fluidized bed* boiler. In a fluidized bed boiler, such as the one shown in Figure 7-7, the fuel is burned on a bed of limestone fragments that are kept suspended by air forced into the bed from below. There are many advantages to such a system. First, many

Figure 7-4.. Water tube boiler, courtesy of Trams for Bath, City of Bath, United Kingdom

different fuels can be burned-waste plastic from retorts, peat, municipal solid waste, rice hulls, and many more. High sulfur coal can be burned without the need of flue gas scrubbers. The temperature is low enough that lower amounts of nitrogen oxides are produced than with other combustion methods. Finally, less slag is produced on the water-cooled furnace walls. More detail is given in Reference 7.

7.2   Boiler components

In addition to the tubes and the boiler chamber areas between the tubes, each boiler has a burner and a flue. Most boilers also have water treatment, economizer tubes, and some environmental protection in the flue.

Figure 7-5. Industrial Water
Tube Boiler

Figure 7-6. Industrial Water Tube Boiler, Cutaway (photographs courtesy of Cleaver-Brooks)

Figure 7-7. Fluidized Bed Boiler (Reproduced with the permission of Ahlstom Power, Inc., Windsor, CT from Combustion Fossil Power (Copyright 1991))

The kind of burner depends upon the fuel. Figure 7-8 shows some of the major components of one kind of burner and gives an idea of how it works. The fuel comes in from the left, is mixed with air in the swirl vanes, and burns in the turbulence zone. Much more detail is given at *www.marineengineering.org.uk*. In boilers fired by coal or residential trash and yard waste, the fuel may be moved into the fire zone by grates or inserted into a fluidized bed from above. These and other boilers are covered in detail in Reference 1.

The flue is the stack that carries off combustion products and excess combustion air. Heat in the flue gas is lost to the air, along with heat lost by radiation. Flue gas usually goes through an economizer—a heat exchanger where some of the heat from the flue gas is used to heat the makeup water going to the boiler. Heating the makeup water with an economizer can increase the efficiency of the boiler from 2 to 4%, according to Cleaver Brooks (Reference 2).

In addition to an economizer, another auxiliary that is present in most boilers is some kind of water treatment unit. Water treatment is necessary to control the acidity in the boiler water and to remove dis-

Figure 7-8. Typical Boiler Burner, courtesy of Marine Engineering, United Kingdom

solved oxygen and carbon dioxide. Both objectives can be accomplished with the use of chemicals inserted into the boiler feed water through a chemical feeder; a mechanical deaerator system with a surge tank for excess feed water is often used as well. The purpose of water treatment is to keep the acidity of the water at the right levels to reduce corrosion and scale buildup in the piping system. The purposes of the surge tank are to recover condensate that is hot and has already been chemically treated and to keep surges in condensate from damaging the boiler.

As water passes through boiler tubes, becomes steam, and is returned as condensate, it picks up dissolved and particulate solids. If these are not eliminated, they degrade the boiler operation. To reduce the concentration of these solids and other undesirable sludges, a certain amount of water is removed from the boiler each day, and this is called *blowdown*. This water needs to be cooled and possibly treated chemically before being drained into a sewage system. The cooling is accomplished by a blowdown heat recovery unit that uses the waste heat in the blowdown to heat incoming boiler water.

Environmental protection equipment associated with a boiler may also be necessary and depends upon boiler design, the fuel, and the regulatory climate of the particular boiler. Such equipment can include wet scrubbers, electrostatic precipitators, cyclone separators, or fiber filters. Some boiler manufacturers recirculate flue gas to get rid of nitrogen oxides (NO_x), and most modern boiler burners are designed with NO_x and SO_x (sulfur oxides) in mind.

7.3   BOILER CONTROLS AND GAUGES (FROM REFERENCE 2).

Boiler controls and gauges need to be operating correctly in order for a boiler to be operated safely. Parameters that need to be monitored are the following:

- Flame. The flame is controlled by the amount of air and fuel that are entering the burner. The flame can be visually inspected or can be automatically monitored. Starting a boiler generally requires a set sequence of operations involving fuel pressure and temperature, the air registers, the blower motor, the pilot flame, and steam temperature and pressure—all under the control of a programmable logic controller. The controller monitors each of these and will stop the startup sequence if any parameter exceeds limits. During the operating phase, the flame is continually monitored, and alarms are sounded and/or the boiler is shut down if potentially dangerous conditions are detected.

- Temperature. The steam pressure is directly related to the steam temperature, and both of these are monitored. If the temperature or pressure goes above a preset limit, the control system is designed to shut down the boiler before damage to the boiler or an explosion occurs.

- Low water. If the water level in the boiler gets too low, excess heating of boiler tubes and consequent damage can take place—in fact, many accidents have been caused by low boiler water. To prevent such problems, the water level is shown on a gauge and is limited by a low water cutoff that shuts down the boiler when the water level gets too low.

- Others. Depending on the boiler type and size, there may also be switches to control various boiler functions in the case of low gas pressure, high gas pressure, low oil temperature, high oil temperature, low oil pressure, and others.

7.4   BOILER FUELS

A boiler is almost always designed for a particular fuel. The best situation is probably one where a waste stream can be used as fuel—for example, plastic waste from retorts, excess refinery gas that would otherwise be flared, or hog fuel—bark that is removed from trees in the manufacture of plywood. In looking over a manufacturing facility, this kind of waste stream should be examined carefully. In most situations, however, the fuel available is natural gas, fuel oil of some sort, or coal. In Section 7.4.1, conventional fuels are discussed together with their advantages and disadvantages; this discussion forms a base for Section 7.4.2 on less conventional fuels, including various types of industrial waste.

7.4.1 Characteristics of Common Industrial Fuels

A comparison of industrial fuels must examine the following characteristics of each fuel: (1) cost per Btu as a raw material; (2) availability in any kind of weather and any international political climate; (3) complexity of the on-site equipment need to transport and burn the fuel; (4) problems associated with the storage of the fuel; (5) emissions caused by combustion; and (6) historical success of the technology for boilers using this fuel. Consider coal, fuel oil, and natural gas in the light of these characteristics.

7.4.1.1 Coal

The fuel of choice for most large boilers is coal. Coal is abundant and available, and it is possible to purchase coal on long-term contracts to keep price fluctuations down. The technology to burn coal is well-known, and it can be delivered wherever a freight train can go. Disadvantages of coal include the need to control sulfur oxide and nitrogen oxide emissions as well as particulates. The material handling equipment is generally more complex than for either fuel oil or natural gas. Coal is classified by the amount of specific materials it contains, as in Table 7-1.

The cost of coal as delivered ranges from \$35 to \$70 per ton, about \$1.40 to \$2.80 per million Btu. It is readily available, but it is subject to labor strikes and occasionally to weather-caused transportation problems. At the boiler plant, the coal must be removed from its trucks, railroad cars, or slurry pipeline and transported to the boiler, often with intermediate processing. Burning coal usually involves complicated mechanical equipment such as spreading stokers or atomizing burners. The technology of this equipment is, however, well-developed, and the

Table 7-1. Analysis of a Sample Coal

Component	Weight %
% Ash	2.8 %
% H_2O	3.5 %
Moisture and ash-free	
% Volatile material	42.4%
% C	76.2
% H	6.2
% O	1.8
% N	2.6
% S	1.8
Low heating value (see note)	12,700 Btu/lb

Note: The *high heating value* includes the amount of heat necessary to heat and vaporize all the water in the coal and created in the combustion of the coal. If this heat is subtracted, the result is the *low heating value*.

procedures for maintaining the equipment are well-known. Coal storage does not present a major problem, although it (1) requires space, (2) requires material handling, (3) can be subject to spontaneous combustion, and, (4) can occasionally freeze. These problems are not new, and their solutions are well known.

Since the Clean Air Act of 1977, the amount of NO_x and SO_x (nitrogen and sulfur oxides) in the flue gas emissions have been regulated by law, and so the amount of these oxides in the flue gas for a given coal is a matter of some concern. Most coal-fired boilers operate more efficiently at high (2100 - 2500°F) temperatures than at lower temperatures. Unfortunately, more NO_x forms at these temperatures, and the residence time in the boiler is not sufficient for it to dissociate into O_2 and N_2. SO_x is also a problem, depending on the sulfur content of the coal being burned. The reduction of NO_x and SO_x requires expensive flue gas treatment and has made low-sulfur fuels and low temperature combustion processes (such as fluidized-bed combustion) more competitive economically. The technology for mining and burning coal has evolved over many centuries; both the problems and their solutions are well-known.

7.4.1.2 Fuel Oil

This classification includes everything from No. 1, a distillate, such as the kerosene used for home heating, to No. 6, a heavy residual oil.

Prices for No. 6 fluctuate around \$40-\$85 for a 42-gallon barrel. The availability of this fuel is not affected by weather, but the world political climate can change both the supply and the cost dramatically as was shown by the Arab Oil Embargo of 1973, the Gulf War of 1991, and the Iraq War of 2003. The offloading and storage of fuel oil is usually uncomplicated, but its viscosity increases in cold weather, and, depending on the grade, it may be necessary to provide auxiliary heating before it will flow. Methods for solving the viscosity problem are well known. As with coal, NO_x and SO_x emissions can present a problem.

7.4.1.4 Natural Gas

Natural gas many advantages: it is clean; it has no particulate emissions and its exhaust can therefore be used in gas turbines; it is easy to transport with pipelines; and it is easy to burn. Because of its convenience and because its price was kept low by government regulation for a long time, it has been the preferred fuel for many applications. The decontrol of natural gas has, however, allowed the price of gas to increase and to fluctuate, and has made other sources of energy more attractive than before. NO_x and SO_x do not seem to cause the same problems when gas is the fuel as they do for coal and fuel oil, and gas remains the fuel of choice for many users. The technology is well-known and less complicated that that for coal or fuel oil.

7.4.1.4 Unconventional Fuels

Companies are continually searching for fuels less expensive than coal, fuel oil, and gas. A natural source for an inexpensive fuel is manufacturing waste, and this can be almost anything with hydrocarbons. Waste materials currently being used as fuels include pulp mill liquor, sawdust, food processing waste, municipal garbage [7], coal wash water, coffee grounds, cardboard, hog fuel (wet bark from plywood operations), and bagasse (sugar cane after the liquid has been extracted). At least one company disposes of waste plastic for a fee and uses it as fuel in a fluidized bed operation. Using industrial waste as fuel can simplify the refuse disposal problem for a company as well as provide it with an inexpensive source of heat. Some sample analyses are given in Table 7-2, with natural gas included for comparison purposes.

Still, there are some problems associated with burning any new fuel. The technology for dealing with coal, gas, or fuel oil is well-known. Using a new fuel, however, raises the following questions:

- How high in the combustion chamber should the new fuel be in-

Table 7-2. Examples of Composition of Non-traditional Fuels

Fuel	Sulfur (S)	Hydro-gen (H)	Carbon (C)	Compo-nent Oxygen (O)	Moisture (H$_2$O)	Ash	Heating Value (Btu/lb)
Pine bark (dry basis)	0.1%	5.6%	53.4%	37.9%	(50%)	2.9%	9,030
Natural gas	—	23.3	74.72	1.22	—	(0.76 %N$_2$)	22,904
Fuel oil No. 6	12.0	10.5	85.7	0.92	2.0	0.08	18,270
Coke breeze	0.6	0.3	80.0	0.5	7.3	11.0	11,670
Bagasse	—	2.8	23.4	20.0	52.0	1.7	4,000
Municipal garbage (metals removed)	0.1 0.4	3.4-6.3	23.4-42.8	15.4-31.3	19.7-31.3	9.4-26.8	3100-6500

Source: References 1 and 6.

jected into the boiler? (This is critical in burning municipal waste.)
- What kind of problems will the ash or residue create?
- What modifications are needed to burners?
- How will the new fuel be transported to and within the facility?
- What storage problems can be expected?
- How regular will the supply be?

It pays to examine the operation of a successful facility using the proposed fuel before making a significant new investment, and it is essential to have the best possible engineering advice.

Before leaving this section, it should be noted that other factors must be taken into account in the decision to use an alternate fuel. First is the need to have some kind of backup boiler if the waste fuel is not available. This problem is particularly severe when one company uses the waste that is generated by another company; it was also responsible for some of the early difficulties experienced in the attempt to use municipal refuse as a fuel. A second major factor is the political climate. It is necessary to determine what government agencies must give their approval before a particular plan can be put into effect. In the case of

municipal refuse, political problems have probably delayed more projects than technical difficulties, especially where intermediate storage has
been seen as a problem.

7.4.2 Cost Comparison Example
7.4.2.1 Present situation

To illustrate some of the factors involved in a typical choice of
boiler fuels, consider the following example. A company is using process
steam at rates that sometimes reach 300,000 lb/h. The company presently has a gas-fired boiler capable of meeting its needs. This and other
local companies are sending significant amounts of similar combustible
wastes to a local landfill. A company engineer has suggested that these
wastes might be used as a replacement source of fuel. His preliminary
study has indicated that this usage of waste as fuel will be acceptable to
EPA and to other local and federal authorities, that the other companies
will buy into this solution, and that there will be no negative environmental aspects to such usage. The study showed three alternatives to
be viable:

(1)   Continue buying gas and sending the waste to the local landfill;

(2)   Construct two boilers, one for waste and capable of efficient op-
      eration from 90,000 to 200,000 lb/h, and one burning coal with an
      efficient operating range of 30,000 to 100,000 lb/h;

(3)   Construct a single waste-fired boiler with an efficient capacity
      of 210,000 to 300,000 lb/h, and charge $15.00/T for burning ac-
      ceptable industrial wastes, estimated at 30,000 T/yr from nearby
      companies.

The first step is to determine the details and costs of each alternative. Alternative 1, the present system, uses purchased gas and has
costs of trash hauling and landfill fees added to the usual operation and
maintenance of the boiler. Gas presently costs $5.00/million Btu. Pres-
ent gas costs are $2,500,000/yr. This represents about 50×10^{10} usable
Btu/yr. The company presently produces 40,000 tons of combustible
waste per year. This waste has been analyzed and found to contain 16%
ash by weight and to have a heating value of 6390 Btu/lb as fired. It
is estimated that a waste-fired boiler of the type contemplated would
have an efficiency of 75%. The usable heat content of this waste is
therefore 6390 Btu/lb x 2000 lb/T x 40,000 T x .75 = 38.3×10^{10} Btu/
yr, an amount which would need to be supplemented by some other

energy source to meet the needs of the plant. This waste is presently transported to a landfill at a cost of $1.25/T and then landfilled at $2.50/T tipping fee. Waste hauling costs for this company are not expected to increase, but landfill costs are expected to increase 30%/yr for the next 5 years and 10%/yr thereafter. These same rates will hold for any ash that is landfilled.

Alternative 2, the two boiler combination, avoids the gas cost and all of the cost of waste haulage and waste landfilling. This alternative, however, causes the company to incur the initial capital cost of the boilers and higher operating and maintenance costs than under the present system. In addition, there is the cost of hauling and landfilling the ash. The coal to be used has a heating value of 12,780 Btu/lb and an ash content of 9.6%. The coal boiler efficiency is estimated as 82%, giving the effective heating value of the coal as 21 million Btu/T. The amount of coal needed is calculated from $(50 - 38.3) \times 10^{10}$ Btu/(12,780 Btu/lb $\times$ 2000 lb/T $\times$ .82 efficiency) = 5580 T/yr. Coal costs are projected to be $55.00/T for the near future. Ash comes from the waste and from the coal; the amount from the waste is 40,000 $\times$ .16 = 6400 T/yr; the amount from the coal is 5580 T $\times$ .096 = 536 T/yr.

Alternative 3, the large waste-fired boiler, avoids the gas cost and all of the waste haulage and landfilling expense but incurs a larger capital cost. This alternative will help pay for itself with the revenue generated from industrial customers in addition to the company costs it avoids. This revenue is estimated as 30,000 T/yr $\times$ $15.00/T, or $450,000/yr. This alternative, however, has ash haulage and landfilling costs. Since the total amount of waste burned per year is 70,000 T, the ash to be disposed of is 70,000 T $\times$ .16 = 11,200 T/yr.

These costs are summarized in Table 7.3. In addition to these costs, a complete analysis would require the depreciation schedule for each item of capital equipment, the required after-tax rate of return, and any cost inflation that is anticipated. With all of these data, the alternatives could be easily analyzed in one or more spreadsheets. To complete the analysis it would be necessary to do a number of sensitivity analyses testing the sensitivity of the results to a range of different likely cost scenarios.

7.5   THE HEAT BALANCE FOR BOILERS

The fundamental tool for analyzing efficiency improvements to boilers and steam distribution systems is the heat balance. A heat balance determines the heat coming into a system from all sources and the

Table 7-3. Costs of Waste-burning Boiler Alternatives

	Present System	Two Boilers	One Big Boiler
FIRST COST	None	$12,500,000	$14,000,000
ANNUAL COSTS			
Gas	$2,500,000	$0	$0
Coal	$0	$306,900	$0
Boiler Maintenance	$50,000	$300,000	$250,000
Waste Transportation	$50,000 (40,000 T × $1.25)	$0	$0
Waste Landfilling (First year)	$100,000 (40,000 T × $2.50)	$0	$0
Ash Transportation	$0	$8,700 (6940 T × $1.25)	$14,000 (11,200 T × $1.25)
Ash landfilling (first year)	$0	$17,350 (6940 T × $2.50)	$28,000 (11,200 T × $2.50)
ANNUAL REVENUES Waste from other companies			$450,000 (30,000 T × $15.00/T)

heat leaving the system. The input and the output are then balanced so that all of the heat is accounted for. Using the heat balance requires understanding the heat content of steam—a concept embodied in the term *enthalpy*—and understanding the basic principles of heat transfer. These are explained in more detail in Chapter 8. The results, however, are important here and are shown in Figures 7-9 and 7-10. These figures are based on the data and concepts of Table 7-4.

With the data from Table 7-4 it is possible to estimate the amount of each heat flow in the boiler and to thereby choose the best place to look for energy savings. This is the procedure:

Step One: First calculate the heat gains or inputs to the boiler system. Sources of this heat energy are fuel, combustion air, makeup water, and returned condensate.

Heat Gain from Fuel. The amount of heat energy available from the fuel is the fuel heating value multiplied by the amount of fuel used per hour.

Table 7-4. Terms Used in Boiler Energy Balance

Data	Description	Data needed
Flue gas	Exhaust gas with combustion products	Temperature (°F) Volume (cfm or lb/hr)
Flue surface Losses	Losses from the heated surface of the flue	Area (ft^2) Surface temperature (°F) Ambient temperature Around flue(°F)
Steam usage Pressure (psia)	Amount of steam actually used	Amount (lb/hr)
Condensate return	Used steam returned as condensate	Amount (lb/hr) Temperature (°F)
Boiler surface losses	Losses from the heated surface of the boiler	Area (ft^2) Surface temperature (°F) Ambient temperature Around boiler (°F)
Makeup water	Water added to make up for steam losses blowdown, and unreturned condensate	Water temperature (°F) Amount (lb/hr)
Boiler blowdown	Water exhausted from boiler to control Chemistry and remove particulate matter	Temperature (°F) Amount (lb/hr)
Combustion air	Air introduced to boiler to provide oxygen for combustion	Temperature (°F)
Fuel	Source of boiler heat	Amount (tons/day) Heat content (Btu/lb)

In this example, the energy input is [(101.3 tons/day)/(24 h/day)] x 2000 lb/ton x 14,610 Btu/lb, or 123.3 million Btu/h.

Heat Gain from Combustion Air. As shown in Figure 7-9, the temperature of the incoming combustion air is 80°F. Since the enthalpy of the flue gas was calculated relative to the temperature of the entering combustion air, the relative enthalpy of the combustion air is zero. This convention avoids counting the enthalpy of combustion air in both the input and the output computations.

Heat Gain from Makeup Feedwater. The makeup water enters the boiler at the ambient temperature of 80°F and thus does not represent a heat gain or a heat loss.

Heat Gain from Condensate Return. Condensate returned to the boiler accounts for a substantial amount of heat energy input. This input is 64,000 lb/h × (150°F - 80°F) Btu/lb, or 4,480,000 Btu/h.

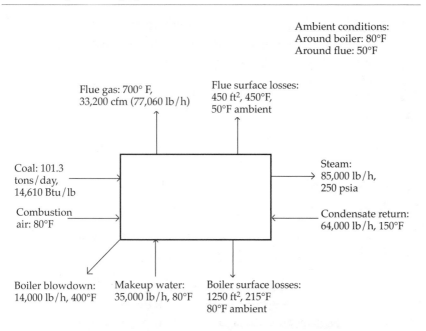

Ambient conditions:
Around boiler: 80°F
Around flue: 50°F

Flue gas: 700° F,
33,200 cfm (77,060 lb/h)

Flue surface losses:
450 ft², 450°F,
50°F ambient

Coal: 101.3
tons/day,
14,610 Btu/lb

Combustion
air: 80°F

Steam:
85,000 lb/h,
250 psia

Condensate return:
64,000 lb/h, 150°F

Boiler blowdown:
14,000 lb/h, 400°F

Makeup water:
35,000 lb/h, 80°F

Boiler surface losses:
1250 ft², 215°F
80°F ambient

Figure 7-9. Mass Balance for a Typical Boiler

Step Two: Now, calculate the heat losses as follows:

Radiative and Convective Losses. Much of the energy lost from boilers is radiated to the environment or transferred to the environment via convection. Useful formulas for these heat losses are given in Reference 4:

$$\text{radiative loss} = A \times .1714 \times 10^{-8} \times (T_S^4 - T_R^4)$$

and

$$\text{convective loss} = A \times .18 \times (T_S - T_R)^{4/3}$$

where

$$A = \text{surface area in ft}^2$$

$$.1714 \times 10\text{-}8 = \text{Stefan-Boltzmann constant in Btu}/(h \times ft^2 \times °R4)$$

$$T_S, T_R = \text{surface and room temperatures, respectively, in} \\ °R (°F + 460)$$

Using these equations, radiative and convective losses from the boiler are

radiative loss $= 1250 \times 0.1714 \times 10^{-8} \times [(215 + 460)^4 - (80 + 460)^4]$
                 $= 214.25 \times 10^{-8} \times (2.076 \times 10^{11} - 8.503 \times 10^{10})$
                 $= 262{,}600$ Btu/h
convective loss $= 1250 \times .18 \times (215 - 80)^{4/3}$
                 $= 155{,}800$ Btu/h

From the flue,
radiative loss $= 450 \times 0.1714 \times 10^{-8} \times [(450 + 460)^4 - (50 + 460)^4]$
                 $= 476{,}700$ Btu/h
convective loss $= 450 \times .18 \times (450 - 50)^{4/3}$
                 $= 239{,}700$ Btu/h

Therefore, the total radiative and convective losses are approximately 1,134,800 Btu/h.

Heat Loss in Steam. The heat energy contained in steam is called *enthalpy* and is made up of latent heat, or the heat required to vaporize water, plus sensible heat, energy required to heat pure steam—generally proportional to the temperature difference through which the steam was heated. Enthalpy tables are given in Tables 7-5 and 7-6. For 250 psia steam, the enthalpy is 1201.1 Btu/lb. Thus the heat carried from the boiler in the steam is 85,000 × 1201.1, or 102,090,000 Btu/h.

Heat Loss in Boiler Blowdown. The enthalpy of water at 400°F is 375.1 Btu/lb, hence the blowdown of 14,000 lb/h represents a loss of 5,251,000 Btu/h.

Heat-Loss in Flue Gas: There are three ways (at least) to estimate the heat loss in flue gas.

(1)   Use the composition of the coal and the chemical reactions in the combustion to estimate the amounts of each flue gas component. Then determine the enthalpy of each component, and add them all up. This requires a good knowledge of chemistry and a lot of hard work.

(2)   Use a combustion efficiency curve such as Figure 7-11. Use of this curve requires a measurement of the percent flue gas oxygen which is readily available. Suppose for this problem that the flue gas is 6 percent oxygen. The temperature rise in the flue gas is 700°F – 80°F, or 620°F.

Table 7-5. Properties of Saturated Steam and Saturated Water (pressure)

Pressure psia	Steam Temp F	Specific Volume (ft³/lb)	Enthalpy, Btu/lb		
			Saturated Liquid	Latent Heat	Steam
0.0886	32.018	3302.4	0.00	1075.5	1075.5
0.10	35.023	2945.5	3.03	1073.8	1076.8
0.15	45.453	2004.7	13.50	1067.9	1081.4
0.20	53.160	1526.3	21.22	1063.5	1084.7
0.30	64.484	1039.7	32.54	1057.1	1089.7
0.40	72.869	792.1	40.92	1052.4	1093.3
0.5	79.586	641.5	47.62	1048.6	1096.3
0.6	85.218	540.1	53.25	1045.5	1098.7
0.7	90.09	466.94	58.10	1042.7	1100.8
0.8	94.38	411.69	62.39	1040.3	1102.6
0.9	98.24	368.43	66.24	1038.1	1104.3
1.0	101.74	333.60	69.77	1036.1	1105.8
2.0	126.07	173.76	94.03	1022.1	1116.2
3.0	141.47	118.73	109.42	1013.2	1122.6
4.0	152.96	90.64	120.92	1006.4	1127.3
5.0	162.24	73,53	130.20	1000.9	1131.1
6.0	170.05	61.98	138.03	996.2	1134.2
7.0	176.84	53.65	144.83	992.1	1136.9
8.0	182.86	47.35	150.87	988.5	1139.3
9.0	188.27	42.40	156.30	985.1	1141.4
10	193.21	38.42	161.26	982.1	1143.3
14.696	212.00	26.80	180.17	970.3	1150.5
15	213.03	26.29	181.21	969.7	1150.9
20	227.96	20.087	196.27	960.1	1156.3
30	250.34	13.744	218.9	945.2	1164.1
40	267.25	10.497	236.1	933.6	1169.8
50	281.02	8.514	250.2	923.1)	1174.1
60	292.71	7.174	262.2	915.4	1177.6
70	302.93	6.205	272.7	907.8	1180.6
80	312.04	5.471	282.1	900.9	1183.1
90	320.28	4.895	290.7	894.6	1185.3
100	327.82	4.431	298.5	888.6	1187.2
120	341.27	3.728	312.6	877.8	1190.4
140	353.04	3.219	325.0	868.0	1193.0
160	363.55	2.834	336.1	859.0	1195.1
180	373.08	2.531	346.2	850.7	1196.9
200	381.80	2.287	355.5	842.8	1198.3

(Continued)

Table 7-5. (Continued)

Pressure psia	Steam Temp F	Specific Volume (ft³/lb)	Enthalpy, Btu/lb		
			Saturated Liquid	Latent Heat	Steam
250	400.97	1.8432	376.1	825.0	1201.1
300	417.35	1.5427	394.0	808.9	1202.0
350	431.73	1.3255	409.8	794.2	1204.0
400	444.60	1.1610	424.2	780.4	1204.6
450	456.28	1.0318	437.3	767.5	1204.8
500	467.01	0.9276	449.5	755.1	1204.7
550	476.94	0.8418	460.9	743.3	1204.3
600	486.20	0.7698	471.7	732.0	1203.7
700	503.08	0.6556	491.6	710.2	1201.8
800	518.21	0.5690	509.8	689.6	1199.4
900	531.95	0.5009	526.7	669.7	1196.4
1000	544.58	0.4460	542.6	650.4	1192.9
1100	556.28	0.4006	557.5	631.5	1189.1
1200	567.19	0.3625	571.9	613.0	1184.8
1300	577.42	0.3299	585.6	594.6	1180.2
1400	587.07	0.3018	598.8	576.5	1175.3
1500	596.20	0.2772	611.7	558.4	1170.1
2000	635.80	0.1883	672.1	466.2	1138.3
2500	668.11	0.1307	731.7	361.6	1093.3
3000	695.33	0.0850	801.8	218.4	1020.3
3208.2	705.47	0.0508	906.0	0	906.0

Table 7-6. Properties of Saturated Steam and Saturated Water (temperature)

Temp F	Volume, ft³/lb		Enthalpy, Btu/lb		
	Pressure psia	Specific Volume (ft³/lb)	Saturated Liquid	Latent Heat	Steam
32	0.08859	3305	−0.02	1075.5	1075.5
35	0.09991	2948	3.00	1073.8	1076.8
40	0.12163	2446	8.03	1071.0	1079.0
45	1.14744	2037.8	13.04	1068.1	1081.2
50	0.17796	1704.8	18.05	1065.3	1083.4
60	0.2561	1207.6	28.06	1059.7	1087.7

(Continued)

Table 7-6. (*Continued*)

Temp F	Volume, ft³/lb		Enthalpy, Btu/lb		
	Pressure psia	Specific Volume (ft³/lb)	Saturated Liquid	Latent Heat	Steam
70	0.3629	868.4	38.05	1054.0	1092.1
80	0.5068	633.3	48.04	1048.4	1096.4
90	0.6981	468.1	58.02	1042.7	1100.8
100	0.9492	350.4	68.00	1037.1	1105.1
110	1.2750	265.4	77.98	1031.4	1109.3
120	1.6927	203.26	87.97	1025.0	1113.6
130	2.2230	157.33	97.96	1019.8	1117.8
140	2.8892	123.00	107.95	1014.0	1122.0
150	3.718	97.07	117.95	1008.2	1126.1
160	4.741	77.29	127.96	1002.2	1130.2
170	5.993	62.06	137.97	996.2	1134.2
180	7.511	50.22	148.00	990.2	1138.2
190	93.40	40.96	158.04	984.1	1142.1
200	11.526	33.64	168.09	977.9	1146.0
210	14.123	27. 82	178.15	971.6	1149.7
212	14.696	26.80	180.17	970.3	1150.5
220	17.186	23.15	188.23	965.2	1153.4
230	20.779	19.381	198.33	958.7	1157.1
240	24.968	16.321	208.45	952.1	1160.6
250	29.825	13.819	218.59	945.4	1164.0
260	35.427	11.762	228.76	938.6	1167.4
270	41.856	10.060	239.95	931.7	1170.6
280	49.200	8.644	249.17	924.6	1173.8
290	57.550	7.460	259.4	917.4	1176.8
300	67.005	6.466	269.7	910.0	1179.7
310	77.67	5.626	280.0	902.5	1182.5
320	89.64	4.914	290.4	894.8	1185.2
340	117.99	3.788	311.3	878.8	1190.1
360	153.01	2.557	332.3	862.1	1194.4
380	195.73	2.335	353.6	844.5	1198.0
400	247.26	1.8630	375.1	825.9	1201.0
420	308.78	1.4997	396.9	806.2	1203.1
440	381.54	1.2169	419.0	785.4	1204.4
460	466.9	0.9942	441.5	763.2	1204.8
480	566.2	0.8172	464.5	739.6	1204.1

(Continued)

Table 7-6. (*Continued*)

Temp F	Volume, ft³/lb		Enthalpy, Btu/lb		
	Pressure psia	Specific Volume (ft³/lb)	Saturated Liquid	Latent Heat	Steam
500	680.9	0.6749	487.9	714.3	1202.2
520	812.5	0.5596	512.0	687.0	1199.0
540	962.8	0.4651	536.8	657.5	1194.3
560	1133.4	0.3871	562.4	625.3	1187.7
380	1326.2	0.3222	589.1	589.9	1179.0
600	1543.2	0.2675	617.1	550.6	1167.7
620	1786.9	0.2208	646.9	506.3	1153.2
640	2059.9	0.1802	679.1	454.6	1133.7
660	2365.7	0.1443	714.9	392.1	1107.0
680	2708.6	0.1112	758.5	310.1	1068.5
700	3094.3	0.0752	822.4	172.7	995.2
705.5	3208.2	0.0508	906.0	0	906.0

The 6 percent oxygen intersects the combustion efficiency curve at approximately 35% excess air; at a temperature rise of 600°F, this corresponds to a boiler efficiency of about 77%. The largest loss in the boiler is up the stack, so the flue gas heat loss is approximately equal to the energy input from the fuel × (100% − 77%), or 123,300,000 × .23, or 28,360,000 Btu/h

(3)   Sum the energy inputs and the energy outputs, and assume that the flue gas losses make up the difference. In this case,

Input enthalpies:

Coal	123,300,000 Btu/h
Makeup water	0 Btu/h
Condensate return	4,480,000 Btu/h
Input air	0 Btu/h
	127,780,000 Btu/h

Output enthalpies:

Steam	102,100,000 Btu/h
Flue gas	To be determined
Blowdown	5, 251,000 Btu/h
Surface losses	1, 134,000 Btu/h
	108,500,000 + Flue gas

Since the heat enthalpy coming in is equal to the heat enthalpy leaving, the flue gas accounts for (127,780,000 – 108,500,000) or about 19,300,000 Btu/h. A very careful chemical analysis gives 21,910,000 Btu/h. The difference between these estimates and the previous figure is probably accounted for by the particular boiler efficiency curve—a boiler efficiency curve for this particular boiler would probably give an answer much closer to 20,000,000 Btu/h.

It is convenient to summarize the input and output heat energy flows as shown in Figure 7-10.

The heat balance of Figure 7-10 serves several objectives. First, it is a clear way of presenting the energy input and output of a system, whether it is a boiler or an industrial process. This diagram can thus be used to communicate energy flows, whether the communication is part of the data-checking process or part of a presentation being made in defense of a proposal. Second, preparing the diagram necessitates examining the system in detail in order to find and quantify the energy inputs and outputs. Many of these may not have been evaluated before, and their magnitudes are probably different from what they are thought to be. Third, the balance immediately and visibly suggests areas to examine in order to improve the system operation. It also clearly indicates the level of emphasis to place on the various energy management possibilities. Some of the possibilities are discussed in the next chapter.

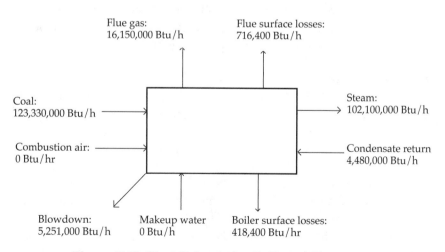

Flue gas:
16,150,000 Btu/h

Flue surface losses:
716,400 Btu/h

Coal:
123,330,000 Btu/h

Steam:
102,100,000 Btu/h

Combustion air:
0 Btu/hr

Condensate return
4,480,000 Btu/h

Blowdown:
5,251,000 Btu/h

Makeup water
0 Btu/h

Boiler surface losses:
418,400 Btu/hr

Figure 7-10. Heat Balance for Boiler of Figure 7-9

7.5.1 A Brief Note on Boiler Chemistry

The chemistry of combustion reactions is complicated, and the turbulent flow of gases within the combustion chamber complicates the problem even more. To give a flavor of the computations, the combustion of one ft^3 of methane (CH_4) gives these results:

Heat value: 23.879 Btu
     less   2.359 Btu used to vaporize the water vapor
                     produced in combustion
Air required for
     combustion: 9.528 ft^3
Flue products under complete combustion:
                CO_2 1.0 ft^3
                H_2O 2.0 ft^3
                N_2 7.528 ft^3
(N_2 makes up approximately 79% of the combustion air)

When similar computations are performed for all the components of a fuel, approximate values can be obtained for the heating value of a fuel and for the volume and weight of combustion products. Niessen gives a excellent presentation of these computations.[6]

7.6   BOILER EFFICIENCY IMPROVEMENTS

Some of the most important boiler efficiency improvements are (1) oxygen/air optimization, (2) load management—matching the boiler capacity to the steam demand, and, (3) maintaining the boiler system.

7.6.1 Combustion Air Optimization.

If the amount of air provided is not enough for complete combustion, some of the potential heat value in the fuel is lost. This happens, for example, when enough air is present to create CO but not CO_2. Soot can also be formed, and the boiler can give off excessive smoke. If conditions are right, the boiler can explode. If, however, too much combustion air is used, the excess heated air goes out the stack, taking the heat with it. Optimizing the amount of excess air thus becomes an area where energy management can be profitable.

The minimum amount of air to use for complete combustion depends upon a number of factors—the design of the boiler, the load on the boiler, the condition of the system, applicable local regulations, and more. The exact amount is difficult to calculate, but the consequences of too little oxygen—smoke, carbon monoxide, possible boiler problems—are worse than the consequences of too much oxygen—higher stack temperatures, lower boiler efficiency. Consequently most boilers are designed for a specified amount of excess air. The boiler design is also for a given load, and the load on a boiler can change from hour to hour, so again the margin of safety demands some excess oxygen.

But the more excess oxygen, the lower the boiler efficiency and the higher the fuel cost. Determining whether to reduce the amount of excess air then becomes a worthwhile question for investigation. For a given fuel and boiler, combustion efficiency curves such as Figure 7-11 are available and helpful. The left axis at the bottom gives the percent of oxygen in the flue gas. This amount can be determined by Orsat analysis or from a permanently installed monitor. [Reference 7 has an excellent discussion of boilers and excess air.] The flue gas oxygen is related to the percent excess air by the curve shown. The lines in the graph at the top of the figure relate percent air and combustion efficiency for a given stack temperature rise. Since this temperature rise can be measured, it is possible to determine the combustion efficiency of the present system. In Figure 7-11, a reading of 9% flue gas oxygen percent combined with a stack temperature rise of 500°F gives excess air of about 75% and an overall combustion efficiency of about 76.4%. This is seen from the middle arrow sequence. If it were possible to reduce the percent of excess air to 10% without changing the stack temperature rise, the outer arrow sequence shows that the efficiency could increase to about 81.6%. This represents a decrease in fuel usage and costs of about 5% and is therefore significant.

In the above example, the combustion efficiency could be further increased by adding heat recovery units to lower the stack temperature rise and reclaim heat that would otherwise be wasted. The new stack temperature, however, must be well above the dew point of flue gas constituents. For example, hydrochloric acid and sulfuric acid can both be present in vapor form, and the condensation of either will cause corrosion of the metal parts of the flue. (Hydrochloric acid is formed by the combustion of materials that contain chlorine, such as plastics; sulfuric acid is formed by the combustion of materials that contain sulfur, including many boiler fuels.)

7.6.2 Improving Load Management

Another significant source of cost savings can be load management. Any boiler will be more efficient at one load than at others. Depending

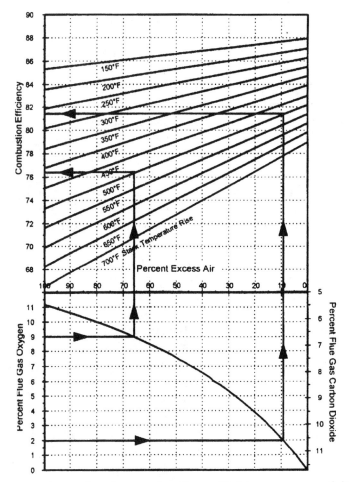

Figure 7-11. Example of a combustion efficiency curve and its use, (From S.A. Parker; R.B. Scollon; and R.D. Smith, "Boilers and Fired Systems," Chapter 5 in *Energy Management Handbook, Fourth Edition,* Wayne C. Turner, Senior Editor, Fairmont Press, 2001, reproduced with permission).

upon the total load, it may be better to use one boiler at full load than two at partial loads. As an example, suppose that your company has boilers 1, 2, and 3, with the loads and efficiencies shown in Table 7-5, and that you are determining the optimum way to operate these boilers under low, medium, and high load conditions of 80,000 Btu/h, 130,000 Btu/h, and 300,000 Btu/h, respectively.

First, notice that there is a difference between boilers. At a load of

Table 7-5. Boiler Efficiencies for Example

Boiler No.	Steam load $(10^3\ Btu/h)$	Combustion Efficiency (%)	Output $(10^6\ Btu/h)$	Input $(10^6\ Btu/h)$
1	170	85.0	197.7	232.6
	120	88.0	139.5	158.6
	85	65.0	98.8	152.1
2	190	84.0	220.1	263.0
	130	82.0	151.2	184.3
	80	75.0	93.0	124.0
3	330	88.0	383.7	436.0
	235	83.0	273.3	329.2
	170	60.5	197.7	326.7

80,000 Btu/h, Boiler #2 uses less input Btu/h than #1 and therefore less fuel; #3 is even worse. At a load of 130,000 Btu/h, #1 has an efficiency between 85.0% and 85.7% and is the best choice. At 300,000 Btu/h, if #3 is used, assuming 87.0% combustion efficiency gives $(300/330) \times 380 \times 10^6$, or 345.5×10^6 Btu/h input; if #1 and #2 are used at loads of 170,000 and 130,000 respectively, the total input is 350×10^6 Btu/h. This would appear to be a significant savings, depending upon the number of hours each demand rate would hold and upon the difficulty of changing back and forth from single to multiple boiler operation.

7.6.3 Maintenance

Proper maintenance is necessary for the safe and efficient operation of any boiler. Boilers with fire and with steam under pressure are, however, inherently dangerous, and SAFETY PRECAUTIONS MUST BE KNOWN AND OBSERVED AT ALL TIMES. Make sure that all safety interlocks are functioning before assigning or doing any work on a boiler. Nothing in the following section should be construed as superseding any manufacturers' instructions or local safety and environmental regulations. Many of the points presented here are available in references 8 and 9 and in more detail at *www.abma.com* (accessed May 5, 2004), the web site of the American Boiler Manufacturers Association.

7.6.3.1 Water treatment

Impurities in boiler water and returned condensate can cause scale on the insides of pipes, thereby inhibiting heat and material flow, and they can cause corrosion. Impurities can be minimized by proper water treatment and by blowdown—adjustment of the concentration of impurities in boiler water by allowing some water with concentrated impurities in the steam drum—blowdown—to be replaced by new water. The blowdown water can be used as a source of heat for incoming water or where other heat is desired. Continuous water treatment is necessary to maintain the acidity (Ph) at an acceptable level; blowdown is necessary to keep the sludge concentration in the steam drum at acceptable levels. It is important to determine the Ph and the impurities in the water either continuously or often each day and to have specific measures for keeping both within prescribed limits.

7.6.3.2 Air leaks

Another maintenance action that should be taken regularly is a search for air leaks. Ambient air that leaks into the boiler causes a reduction in boiler efficiency as explained earlier; heated air that leaks out is wasted heat. Leaks frequently occur around seals, where refractory brick has come loose, at the ends of tube bundles, and at access doors and where ash is removed. A negative pressure inside the boiler can also cause air infiltration.

To check for leaks, infrared units are available which give comparisons of surface temperatures and incidentally show where insulation is damaged. It is also possible to use chemical smoke. Other measures are generally given in the operating manuals for packaged boilers or can be obtained from references 7 or 9.

7.6.3.3 General rules

It is critically important to have all gauges working and accurate. Safety interlocks should be checked daily. Other boiler components that should be checked on a regular basis include valves, ash-handling units, pumps, burners, and all fans. It has consistently been proven worthwhile to insure that each of these is operating correctly and safely.

References
1.   The Babcock and Wilcox Company: *Steam, Its Generation and Use*, 1992
2.   Cleaver-Brooks Division of Aqua-Chem, Inc., Boiler Room Guide, publication CB-7853, 1997.
3.   Philip S. Schmidt, "Steam and Condensate Systems" Chapter 6 in *Energy Management Handbook, Fourth Edition*, Wayne C. Turner, Senior Editor,

Fairmont Press, 2001.

4. Frank Kreith, *Principles of Heat Transfer, Fourth Edition,* Harper and Row, New York, NY 1986.

5 Federal Energy Administration and Environmental Protection Agency Guidelines for Industrial Boiler Performance Improvement, Publication EPA-600/8-77-003a, 1977.

6. Walter R. Niessen, *Combustion and Incineration Processes, Second Edition,* Marcel Dekker, 1995

7. S.A. Parker; R.B. Scollon; and R.D. Smith, "Boilers and Fired Systems," Chapter 5 in *Energy Management Handbook, Fourth Edition,* Wayne C. Turner, Senior Editor, Fairmont Press, 2001.

8. W.J. Kennedy, "Energy Systems Maintenance" Chapter 14 in *Energy Management Handbook, Fourth Edition,* Wayne C. Turner, Senior Editor, Fairmont Press, 2001.

9. W.H. Axtman, Boiler Fuel Management and Energy Conservation, American Boiler Manufacturers Association, available in some form at *www.abma. com.*

Bibliography

Capehart, Barney, Editor, *Encyclopedia of Energy Engineering and Technology,*" Taylor and Francis/CRC Publishing Company, Boca Raton, FL, July, 2007.

Chapter 8

Steam Distribution Systems

Chapter 7 introduced boilers and their energy management. Chapter 8 discusses the system that uses steam and returns the condensate to the boiler, namely the steam distribution system.

8.1   INTRODUCTION

Once the boiler has produced steam, the steam goes to processes where the heat of the steam is used as an energy source. As the energy in the steam is converted for use, the steam becomes cooler, and some of it condenses as water. This condensate can then be returned to the boiler to again be converted into steam. To understand this process and to realize some of the many opportunities for energy cost savings in the steam distribution system, it is first necessary to understand the concept of enthalpy, i.e. the measurement of the heat energy contained in the steam. Using enthalpy, it is possible to understand economies available in proper management of the steam distribution system components, in particular steam lines, the hot water system, steam traps, and tracer lines. Cogeneration is also relevant and is included here.

8.1.1 Enthalpy

At atmospheric pressure and 212°F it takes about 970 Btu to convert 1 lb of water to steam. This energy, called latent heat, is given up whenever the steam condenses. Steam also carries sensible heat, energy proportional to the temperature difference through which the steam was heated. Steam under pressure contains additional energy due to the mechanical work done on the vapor. (The latent heat is usually lower at higher pressures.) The sum of the latent heat, the sensible heat, and the mechanical work is called enthalpy and, when expressed in Btu/lb, specific enthalpy. (This definition of enthalpy does not take into account

313

the internal kinetic energy of the steam, but this can usually be neglected in energy management. [1]) When water is present with the steam, as in most steam distribution systems, the steam is said to be saturated, and the pressure increases as the temperature increases. The enthalpy and specific volume are found in Tables 7-5 and 7-6 for various pressure and temperature conditions.

These tables have many uses. For example, suppose that a steam leak is estimated to be losing 150 lb of steam per hour from a 1200 psi line to an ambient temperature of 70°F. From Table 7-5, the enthalpy at 1200 psi is 1184.8 Btu/lb and, from Table 7-6, the enthalpy at 70°F is 38 Btu/lb. (From 70 to 212°F, the water is all liquid, and the only heat is sensible. Raising the temperature of 1 lb of water 1 takes 1 Btu.) Thus the total change in enthalpy is 1185 – 38, or 1147 Btu/lb. A loss of 150 lb/hr represents a loss of 150 lb/hr x 1147 Btu/lb, or 172,000 Btu/hr. If fuel costs $8.00 per million Btu, this represents a cost of $1.38/h or about $33/day on days when the facility is running constantly. The amount of flash steam released to the atmosphere by this leak is the volume of steam at 212°F multiplied by the estimated pounds of steam lost. Using Table 7-6 yields 26.80 ft³/lb x 150 lb/hr, or 4020 ft³/hr.

In some applications, steam is present without water. Such steam is said to be superheated, and its enthalpy and specific volume at various temperatures and pressures are given in a separate table not included in this text.

Example 8-1: In one part of your plant, you are currently heating 1000 gallons of water per hour from 60°F to 150°F with gas. A the same time in this area you have noticed that you are venting 100 psia steam to the air. (The tables show pressure in psia—absolute pressure in pounds per square inch—which is 14.7 lb/in² greater than gauge pressure—psig.) How many pounds per hour of 100 psia steam would be required to meet the water heating needs with a 90% efficient steam coil, if the steam leaves the coil at 70°F?

Solution: The amount of heat needed is 1000 gallons × 8.345 lb/gallon × (150°F - 60°F) = 751,000 btu/h. Each pound of 100 psia steam enters with 1187.2 btu, leaves with 1092.1 and gives up 1187.2 – 1092.1 = 95.1 Btu. So 751,000/(95.1 × .90) = 8,775 lb/h would be needed.

Example 8-2: Use of flash steam. Given 500 lb/h of condensate at 200 psia drops to 50 psia through a pressure reducing valve. The condensate must quickly go to the boiling temperature at the lower pressure, with the surplus heat causing the lower pressure condensate to turn to steam. This is "flash steam." In this situation, how much low pressure steam would be available if the steam were allowed to flash off?

Answer:	Enthalpy at 200 psia	= 1198.3 Btu/lb
	Enthalpy at 50 psia	= 1174.1 Btu/lb
	Heat available for flashing	= 24.2 Btu/lb
	Enthalpy of low pressure vapor —enthalpy of low-pressure liquid	= 923.9 Btu/lb
	Proportion evaporated	= 24.2/923.9 = 2.6%
	Flash steam available	= 500 × 0.026 = 13 lb/h

8.2   STEAM DISTRIBUTION SYSTEM COMPONENTS

Figure 8-1 shows components of a typical steam distribution system. When the steam comes from the boiler, its pressure and temperature are 100 psig (pounds per square inch as shown by the gauge, i.e. the difference between its actual pressure and atmospheric pressure) and 337.9, where its enthalpy is 1190 Btu/lb. As the steam proceeds through piping, it gives up some of its heat, and some of it condenses and is drawn off by steam traps and returned to the boiler. At each of the processes where it is used, steam is gives up some of its heat, steam traps separate the condensate, and the condensate returns through the steam distribution system to the boiler. Toward the end of the system, the steam has dropped in pressure to 50.3 psig and cooled to 297.97°F. At this point its enthalpy is 1179.7 Btu/lb, indicating that 1190 – 1179.7, or 10.3 Btu/lb have been used in the system. If the boiler is producing 2000 lb/h, then the steam energy used is 20,600 Btu/h.

8.2.1 Steam Lines

Lines that carry steam give off heat, and the amount of heat they give off depends upon the steam temperature, the ambient temperature around the lines, and the amount of insulation surrounding them. Chapter 11 discusses insulation in detail and shows how to calculate the amount of insulation that minimizes the cost of insulation together with the cost of lost heat.

Occasionally steam lines develop leaks, and these can create opportunities for cost savings. As steam lines give off heat, some of the steam within them condenses, and the condensate must be removed, generally through steam traps.

Figure 8-1. A Steam Distribution System (From [1], with permission)

8.2.1.1 Steam Leaks

Steam leaks can be a major waste of energy associated with steam lines if large amounts of steam are lost; condensate leaks represent lost heat and loss of treated water. In many environments, steam leaks can be detected by their hissing; in noisy environments, it may be necessary to use an industrial stethoscope or an ultrasonic leak detector. Evidence of possible condensate leaks includes pools of hot water, dripping pipes, and rust spots on pipes. While looking for leaks, put your hand close to the pipes. (But don't do this with very high pressure steam. Steam leaks in high pressure steam lines can have enough force to seriously injure you.) If a pipe is too hot to touch, it probably should be insulated. The costs of steam leaks can be estimated by the following equation:

$$\text{Cost/yr} = (\$/\text{Btu of steam}) \times (\text{lb steam lost/h}) \times (\text{Btu/lb}) \times (\text{operating hours/yr})$$

The pounds of steam lost can either be measured directly, estimated on the basis of experience, or calculated with a formula. Grashof's formula [2] gives the number of pounds of steam lost per hour through an orifice of area A as:

$$\text{Lb/h} = 0.70 \times 0.0165 \times 3600 \times A \times P^{97}$$

where

0.70 = coefficient of discharge for hole
   (for a perfectly round hole, this coefficient is 1)
0.0165 = a constant in Grashof's formula
   = number of seconds in an hour
A = area of hole in square inches
P = pressure inside steam line in psia

Using this formula and the steam tables in Chapter 7, it is possible to develop tables giving steam and Btu losses for any given leak size for a plant whose steam pressure is known. When the cost per million Btu is also known, it is then possible to estimate the cost of steam leaks directly from the size of the leaks. Table 8-1 gives an example of such calculations for a 100 psig system, assuming that the energy cost is $6.00/million Btu.

A crude but frequently effective way of evaluating steam leaks was given by Waterland in an early version of the Energy Management Handbook [3]. His method is based on an arbitrary rating system and

Table 8-1. Heat and dollar losses from each leak of 100 psig steam

Hole diameter	Steam loss (lb/h)	Heat loss (Btu/h)	Dollar loss ($/year)
1/8	51	60,670	531
1/4	203	241,490	2,115
3/8	457	543,650	4,762
1/2	812	965,960	8,462

works as follows: Tour a defined plant area at a weather condition or time of day when leaks are quite prominent, making a note of each steam leak and rating it as a wisp, moderate leak, or severe leak. Assign a value of 25 lb/h for each wisp, 100 lb/h for a moderate leak, and 500 to 1000 lb/h for the severe leaks. When more than 20 leaks are evaluated, the total leakage determined in this way will usually be within 25% of the actual steam loss.

8.2.1.2 Steam lines—insulation
    Adding insulation to hot steam pipes can result in significant savings in energy as well as removing a possible burn hazard from a work environment. Insulation for pipes is discussed in detail in Section 11.1.5.2.

8.2.2 Steam traps
    Steam traps and the condensate return system separate condensate from the steam distribution system and thereby perform three important functions:

• Prevent water hammer
• Return condensate to the boiler
• Improve the quality of the steam for further processes

Steam traps also help in the removal of air and dissolved gases from the steam thus removing two insulators and sources of inefficiency.
    The sizing and selection of steam traps and other fluid control equipment and their installation in a steam distribution system are presented in detail in references 4 and 5. Two of the most common trap types are shown in Figure 8-2 and 8-3. Figure 8-2 shows an inverted bucket steam trap. In this trap, steam enters the bucket from the left, causing it to float

in the condensate that surrounds it. There is a small hole in the top of the bucket, and air escapes through this hole. Since the air is at a high pressure, it pushes condensate into the bucket, the bucket loses buoyancy, and the air and some condensate escape. When steam goes through the vent hole, it is condensed, and condensate again escapes.

In the float and thermostatic steam trap, shown in Figure 8-3, the outlet valve is initially closed. Condensate flows into the trap and the float begins to rise. When enough condensate is in the trap, the outlet is uncovered, and condensate is released. The thermostat operates to remove air and dissolved gases from the system.

The third common type of steam trap is the thermostatic steam trap, one of which is shown in Figure 8-4. The bellows contain a mixture of alcohol and distilled water with a boiling point slightly less than the temperature of the steam. When the condensate is cool, the bellows contract, and the condensate goes out through the valve in the bottom. As the steam heats up the bellows, the alcohol vaporizes and expands the bellows, closing the valve and preventing the trap from leaking steam. Other steam trap types include liquid expansion ther-

Figure 8-2. Inverted bucket steam trap (Courtesy of Armstrong Machine Works)

Figure 8-3. Float and thermostat steam trap (Courtesy of Armstrong Machine Works)

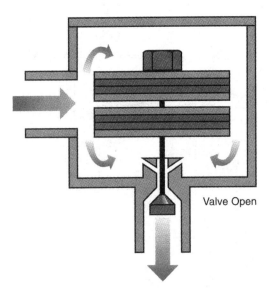

Figure 8-4. Thermostatic steam trap (From [4], with permission of Spirax Sarco, Inc.)

mostatic traps, thermodynamic traps bimetallic traps, and hybrid trap types. These are discussed in Reference 4.

8.2.3 Steam trap problems and maintenance

Steam traps can fail open or fail closed. If a trap fails open, the effect and the cost are the same as a steam leak. If it fails closed, it is not separating condensate from steam, and water hammer, corrosion, and structural failure (caused by the weight of condensate) can result. If it fails closed, it can also be subject to freezing, depending on the climate.

There are three traditional methods for testing steam traps. The first is to put the blade of a long screwdriver on the trap and the handle to an ear, and listen for the clicks as the trap opens and closes. This works, but it requires a trained ear and a steam trap that is accessible. This can be refined with the use of an ultrasonic detector that masks out background noise or with a stethoscope like that used in medicine. A second method is a visual inspection at an upstream blowdown valve where the valve can be manually opened and blowdown removed. This leaves pure steam, and the trap should be shut. A third method uses infrared testing to show steam leaks, and this method can show steam blowing through open steam traps that are difficult to reach. These methods all require training and personnel.

A new method for monitoring steam traps on a continuous basis has been developed by Armstrong under the trade name of Steam Eye™. With this system, traps are tested continuously and the results are reported by a radio transmitter. By having a temperature probe in each steam trap, such a unit can signal if the trap is operating normally, if steam is blowing through the trap without any condensing, or if the trap is flooded with condensate and inoperative. The results of monitoring are displayed in three files: monitored traps, non-critical trap faults, and new critical trap faults. Since this monitoring system saves much of the time needed to determine the location and nature of steam trap problems and is probably more accurate, it appears to be a significant advance over the time when traps were examined individually by maintenance personnel. More information on this system can be found at the Armstrong company web site: *www.armstrong-intl.com* (accessed May 12, 2004).

Another patented method for automatic monitoring of steam traps is the Spira-tek Leak Detector System, by Spirax Sarco [4]. This consists of a detection chamber connected to the condensate line on the inlet side of a trap. When the condensate and gas are in balance, the trap is working. On the other hand, when the amount of gas increases, this is an indication that the trap has failed. These detectors can also be installed for remote reading by the plant energy management system.

Example 8.3: A recent inspection of your facility has revealed four steam traps with 1/2" orifices that are blowing steam into the condensate return. The steam pressure for the four traps was 150 psia, and the condensate return is gravity fed at atmospheric pressure. At $6.00 per million Btu, how much are these steam losses costing, assuming continuous operation of the plant?

Solution: If steam is blowing from a trap, its enthalpy in the steam line is 1194.0 Btu/lb, and its enthalpy of the saturated liquid at ambient temperature, 48.04 Btu/lb. The amount of heat lost from the steam lines per pound of steam is then 1146.0 Btu/lb. To estimate the amount of steam loss, Grashof's formula gives 4212 lb/hr. The manufacturers, however, give 1087 lb/hr (interpolating in the tables in [4]), a figure which takes into account the reduction in area caused by condensate. Using the manufacturer's figure gives 1146 × 1087 × 4 traps × $6.00/million Btu = $29.90 per hour in fuel costs. This is really quite a lot of money, and the amount is probably conservative.

8.2.4 Water treatment

Scaling has an adverse effect on heat transfer. The more scale buildup, the less heat is transmitted through pipe walls. This scaling can be prevented by proper water treatment, making water treatment one of the essential elements in boiler management. The amount of water treatment needed depends on the hardness of the water and the quantity of water used. Because the condensate is pure water, returning it to the boiler saves money on additional water treatment.

8.3   TRACER LINES

A tracer line is a small steam line, usually copper or steel, wrapped around pipes, pumps, and instruments used with hot liquids or condensate; its purpose is to maintain the temperature of the liquid. Tracer lines are usually attached between the line or pump and the insulation as shown in Figure 8-5. Tracer lines range in diameter from 3/8" to 2," with 1/2" the most common. These lines can keep water and other products from freezing even in very cold weather, but the price of maintaining a high temperature in cold weather is the necessity to manage the condensate in the tracer lines. Sizing and maintaining of tracer lines are discussed in detail in References 4 and 5.

8.4   WASTE HEAT RECOVERY

Waste heat is heat that goes into the atmosphere to some other heat sink without providing any appreciable benefit to the user. Examples with waste heat include flue gases, boiler blowdown exhausted to the

Figure 8-5. Examples of tracer lines (From [5], with permission of Spirax Sarco, Inc.)

air, heated air exhausted directly to the environment by vents, and heat lost from pipes that pass directly through unheated spaces. (Note that when these pipes provide protection against freezing, the heat radiated from the pipes is not waste heat.) Waste heat can be used to generate steam, as a source of power for turbines, or to provide heat for incoming fluid streams. Any of these uses can improve the efficiency of energy utilization in a plant, whether the source of the waste heat is a boiler, an industrial process, or an HVAC system. Benefits from waste heat recovery often include fuel savings and lower capital cost of heating and cooling equipment. Other benefits can include increased production capacity and, under some circumstances, revenue from the sales of recovered heat or energy [7].

8.4.1 Analyzing the Potential for Waste Heat Recovery

Waste heat sources and their uses can be conveniently categorized by the temperature at which the heat is exhausted, as shown in Table 8-2. [7]

The use of waste heat to power a turbine or pump should be considered if turbine work or pumping is needed and if enough energy is available to justify the cost involved. If the economics justify the production of electricity, cogeneration, discussed in Section 8.6, may be a possibility. The use of waste heat to heat a fluid stream should be considered if:

(1)   the waste heat source is close enough to the fluid stream that the fluid temperature will still be high enough to be useful even after taking into account all heat lost in transporting the fluid from source to stream,

(2)   using waste heat from the source will not create problems at the source, and

(3)   the transfer of heat from the source to the stream is technically feasible.

The first step in analyzing an industrial process for possible waste heat recovery is the collection of data sufficient to describe the process with a heat balance as was described in Section 7.5. The next step is to find all point sources of heat use or exhaust, to determine the annual Btu and mean temperature for each, and to show this information on an input-output diagram of the facility. Another way to summarize these

Table 8-2. Sources of Waste Heat

Temperature	Source	Use
High (1200+)	Exhausts from direct-fired industrial processes: 　　Cement kiln (dry), 11 50-1350°F 　　Steel heating furnaces, 1700-1900°F 　　Glass melting furnaces, 1800-2800°F 　　Solid waste incinerators, 1200-1800°F 　　Fume incinerators, 1200-2600°F	Cogeneration
Medium (450-1200)	Exhausts: 　　Steam boiler, 450-900°F 　　Gas turbine, 700-1000°F 　　Reciprocating engine, 450-1100°F 　Heat-treating furnaces, 800-1200°F 　Drying and baking ovens, 450-1100°F 　Annealing furnace cooling systems, 800-1200°F	Steam generation
Low (90-450)	Process steam condensate, 130-190°F Cooling water from: 　　Furnace doors, 90-130°F 　　Bearings, 90-190T 　　Welding machines, 90-190°F 　　Air compressors, 80-120°F 　　Internal combustion engines, 150-250°F Hot-processed liquids, 90-450°F Hot-processed solids, 200-450°F	Supplemental Heating

data is shown in Figure 8-6. The most promising candidates for heat recovery are then examined in detail. Table 8 -3 shows the kind of input needed. If at all possible, the waste heat from a process should be used to improve the efficiency of that same process at its heat source. This practice avoids transportation losses and helps keep processes independent from each other.

8.4.2 The Economics of Waste Heat Recovery

The benefits from waste heat recovery can be substantial; therefore the benefits included in the economic analysis must be as complete as possible. Table 8-4 lists a number of benefits; this list should be augmented by the particular benefits of each specific project. The analysis must also

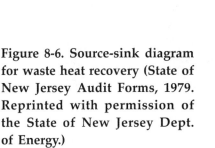

Figure 8-6. Source-sink diagram for waste heat recovery (State of New Jersey Audit Forms, 1979. Reprinted with permission of the State of New Jersey Dept. of Energy.)

include complete details of the costs involved and the amount by which these costs are reduced by any tax benefits. Table 8-5 gives such a list.

8.4.3 Waste Heat Recovery Equipment

The factors that determine which equipment to select for waste heat recovery are (1) the fluid temperature at the source, (2) the intended use for the waste heat, and (3) the distance the heated fluid (if any) must be transported. The most common equipment types and their uses are shown in Table 8-6.

8.4.3.1 Recuperators

A recuperator is a heat transfer device that passes gas to be heated through tubes that are surrounded by a gas that contains excess heat. The heat is transferred from the hot gas to the tubes and through the tube walls to the cool gas inside the tubes. Examples of two types of recuperators are shown in Figures 8-7 and 8-8.

8.4.3.2 Heat wheels

A heat wheel is a large porous wheel that rotates between two adjacent ducts, one of which contains a hot gas and the other a cooler gas. In carrying heat from the hot gas to the cool gas, the heat wheel either captures some of the escaping heat in the hot gas, if that is desired, or cools the hot incoming gas, if that is desired. If the escaping gas is contaminated, the wheel can be constructed with a purge section so that the contaminants are flushed from the wheel after they have given up a

Table 8-3. Data needed to do an in-depth analysis of a particular waste heat recovery measure

Source data

      Location:_____

      Fluid used as a heat source: _____

      Flow rate: _____lb/h

      Present temperature of fluid:_____°F

      Present pressure of fluid:_____psig

      Dew point of fluid:_____°F

      Minimum allowable temperature:_____°F

Sink data

      Location:_____

      Fluid to be heated: _____

      Flow rate: _____lb/h

      Present temperature of fluid: _____°F

      Maximum allowable temperature of fluid: _____°F

      Present pressure of fluid: _____psig

      Maximum containable pressure of fluid: _____psig

Heat transport path data

      Horizontal distance:_____ft

      Vertical distance:_____ft

      Mass pumped: _____lb/h

      Pumping cost: _____$/h

      Estimated heat loss per foot: _____Btu/h

      Heat transported: _____Btu/h

      Temperature drop: _____°F

      Pressure drop: _____psig

General data

      Building location: _____

      Heating cost, for planning purposes: $_____/Btu

      Electricity cost, for planning purposes: $ -/kWh

      Production:_____(amount and units)/year

      Production cost savings: $_____unit output

Table 8-4. Economic benefits from waste heat recovery

Benefits	How Quantified
Reduced fuel costs	$/Btu × Btu/h × operating h/year
Increased production capacity	Marginal profit/unit × increase in capacity × no. h increased capacity will be used each year
Reduced overhaul costs	Cost per overhaul/mean time between overhauls
Reduced regular maintenance costs	Cost of maintenance per hour × no. hours of maintenance per operating hour
Lower capital costs of furnaces and heating and cooling equipment	Vendors' quotations
Sales of energy to utilities	Firm contracts with utilities, or regulations of state public service commission
Reduced pollution abatement equipment costs	Vendors' quotations

significant part of their heat. These wheels, illustrated in Figures 8-9 and 8-10, have high heat transfer efficiencies and can be used to significantly reduce either heating or cooling loads. It should be noted, moreover, that they can be designed to transfer latent heat, i.e. the heat in water vapor in the gas stream, as well as the usual sensible heat.

8.4.3.3 Air preheaters/economizers

In an air preheater or economizer, hot gas flowing through a series of closed channels transfers heat to cooler gas in adjacent channels. This kind of equipment allows the use of hot flue gas to preheat combustion air and reduces the amount of heat that must be supplied by fuel.

8.4.3.4 Run-around coils

A run-around coil heat exchanger consists of two heat exchanger coils connected by piping; a pump is also usually required. The heat is picked up by the heat exchange fluid in one coil. The fluid is then pumped to the other coil where the heat is removed and used. This heat transfer method makes waste heat recovery possible when the source and sink are somewhat separated.

Table 8-5. Costs and cost reductions to be considered in waste heat recovery (Taken, in part, from Reference 6)

Cost	Source of information
Capital cost of equipment	Vendors
Energy tax credits	Internal Revenue Service
Capital investment tax credits	Internal Revenue Service
State and local tax credits	State and local tax information sources
Increased property taxes	State and local tax information sources
Engineering design costs	Local consulting engineers (plan on 2-5% of total project costs)
Installation costs	Time and equipment cost estimates
Cost of production downtime during installation	In-house production, time, and cost estimates
Operation and maintenance costs of new equipment	Vendors

Table 8-6. Waste heat recovery equipment [7]

Source Fluid	Use	Equipment
Exhaust gases over 450°F	**Preheat combustion air:**	
	Boilers	Air preheaters
	Furnaces	Recuperators
	Ovens	Recuperators
	Gas turbines	Regenerators
Exhaust gases under 450°F	**Preheat boiler feedwater or**	**Economizers makeup water**
Condenser exhaust gases	Preheat liquid or solid	Heat exchangers feedstocks
Any waste heat source	Generate steam, power, or	Waste heat boilers electricity
Any waste heat source	Transfer heat	Run-around coils, heat wheels
Any waste heat source	Cooling	Absorption chillers

Figure 8-7. Radiation recuperator. (From [7]. Courtesy of the U.S. Department of Commerce.)

Figure 8-8. Convective-type recuperator. (From [7], Courtesy of the U.S. Department of Commerce.)

Figure 8-9. Simple heat wheel (Used with permission of the American Society of Heating, Refrigerating, and Air Conditioning Engineers, Inc.)

Figure 8-10. Heat wheel with purge section. (From [7], Courtesy of the U.S. Department of Commerce.)

8.4.3.5 Finned-tube heat exchangers

This type of heat exchanger is a tube surrounded by perpendicular fins. The fins help to transfer heat from the tube to the surrounding air by enlarging the heat transfer surface area. If the surrounding fluid has a higher temperature than the tube, the transfer works in the opposite way and transfers heat from the fluid to the material inside the tube. This type of heat exchanger is very common and is often used on boilers to recover some of the heat that would otherwise be lost in the stack gas. It is also used in baseboard heating and in automobile radiators.

8.4.3.6 Heat pipe heat exchangers

A heat pipe is a gas-to-gas heat exchanger and was described and illustrated in Section 6.6 of Chapter 6. As discussed there, heat pipes have been applied to moisture removal in air conditioning systems. They can also be used as air preheaters for a boiler or furnace, as heat reclaimers from waste streams, and as heat exchangers in drying and curing ovens. Heat pipes have a variety of applications in industrial waste heat recovery because of their high efficiency and compact size.

8.4.3.7 Waste heat boilers

If gas leaving some industrial process is sufficiently hot to vaporize water or some other working fluid, it may be possible to use a waste heat boiler. This type of boiler uses waste heat to produce vapor or steam which can then be used directly in the industrial process or can be run through a turbine, pump, or generator to generate electricity or shaft horsepower. Water has been the customary fluid for this purpose, but working fluids with lower boiling points are becoming more common.

8.5   IMPROVING THE HOT WATER
DISTRIBUTION SYSTEM

The hot water system is often a part of the overall boiler and steam distribution system where steam is used to produce the lower temperature hot water needed for process use, cleaning and bathroom use. This system can also provide a number of energy management opportunities. In performing an energy audit of this system, first see if more hot water is heated than is needed; second, determine whether the hot water can be heated to a lower temperature without adverse effects; and, third, assess the need for insulation.

Need for hot water. To estimate the need for hot water, look at the industrial cleaning uses first and then examine the personnel needs for washing. The amount and temperature of hot water needed in industrial processes will be defined by each process, and the auditor should make sure that neither the amount nor the required temperature are overstated. Each process should be examined to see whether cold water can be substituted for hot. In a laundry, for example, the unavailability of hot water can lead to the discovery that cold water works as well as hot, without any change in procedures except for a possible change in detergents.

The generally accepted figures for warm water for personal washing needs are 2-3 gal/day per person in offices, 20-25 gal/day per person in homes, and 40 gal/day in hospitals. If your facility exceeds these amounts, it may be worthwhile to valve off (carefully!) one or more hot water tanks and see if any complaints result. Another necessary step in this audit is to check hot water faucets for leaks. Hot water leaks are expensive; a continuous leak of 1 gal/hour of water at 155°F represents a loss of 7 million Btu/year, equivalent to about 7000 ft^3 of natural gas.

Temperature. To find out whether the temperature of the hot water can be lowered, first determine the present temperature. If the water is used for bathrooms, its delivered temperature can be reduced to 105°F by adjusting the thermostat on the hot water heater. Be sure to check the hot water thermostat. In one case, both the minimum and the maximum temperatures were set at 180°F, and one employee had complained about scalding. If the present temperature is too high, lower the hot water thermostat to the desired level, possibly in 5 degree increments to avoid complaints. Also consider eliminating the use of hot water completely. Most of us have used washrooms in gasoline stations or rest areas where only cold water was provided without any ill effects.

If the water is used in a kitchen, it must be hot enough to meet local health codes—usually at least 140°F. In many cases, this temperature is achieved by heating the water to an excessive temperature at the hot water tank; this increases the heat loss between the hot water tank and the kitchen. This loss can be avoided by having water temperatures of 105-110°F in the kitchen with a booster hot water unit in the kitchen for the dishwasher. This unit should be on a timer so that water is heated only when needed.

Insulation. Were hot water is needed, it may be desirable to insulate the pipes, hot water tanks, and valves. If a pipe or tank feels hot to the touch, its temperature is probably 120°F or higher, and it is probably losing a significant amount of heat to the surrounding air. Valves and flanges have large heat transfer surfaces, and custom-made insulation for them may also be a worthwhile investment.

8.6   COGENERATION

Cogeneration is the process of sequentially producing both electricity and steam from a single fuel source. A cogeneration facility uses some of the thermal energy that a plant producing only electric power would otherwise reject to the environment. Thus cogeneration can produce a given amount of electric power and thermal energy for 10 to 30% less fuel than a plant which produces the same amount of electricity alone. The fuel saving for a particular plant is highly dependent on both the cogeneration technology used and the quantity of thermal energy needed. For many facilities, cogeneration offers a way to provide both low-cost electric power and the large amounts to thermal energy needed for process use [8]. For example, a citrus producing plant in Florida installed a cogeneration plant to provide low-cost electricity and the large quantities of steam needed to pasteurize orange juice and sterilize the bottles it is packed in.

Cogeneration has become more attractive because any cogenerated electricity that contributes to the total peak deliverable capacity of an electric utility enables the utility to avoid building an equal amount of new capacity, and the utility is obligated to buy such electricity from the cogenerator. (There are usually strict conditions on this purchase: The company must use some of the electricity it produces, the cogenerated electricity must be controlled and interconnected in such a way that it will not damage the utility network, and it must not force a net loss of revenues upon the utility). Cogeneration also offers additional benefits to the utility, since any environmental pollution from the electric power production is the responsibility of the cogenerator rather than of the utility.

Cogeneration offers a number of benefits to the facility, particularly if it is in an industry that has a source of waste fuel. Wood chips, black liquor, bagasse, hogged fuel (tree waste in a plywood mill that has been chopped up or "hogged"), garbage, and waste heat are all sources of fuel that would replace the need for a facility to purchase expensive oil, gas, or coal. High-efficiency cogeneration helps industries and businesses compete in national and international markers and helps keep more jobs in these industries and businesses. A facility can reduce its energy cost by lowering its need to buy expensive power from the electric utility. It may also get some income from selling excess power to the utility. It is also possible that it may get a more favorable electric billing schedule.

8.6.1 Cogeneration Technologies

There are four major technologies available for cogenerating electric energy and thermal energy. Three of these are called topping cycles, where fuel is burned to generate electrical energy, and then the remaining heat or steam is used in an industrial process. The fourth technology is the bottoming cycle, where fuel is burned to produce industrial process heat, with the remaining heat used to generate electrical energy. These technological distinctions are important since they have a significant effect on how the efficiencies and environmental impacts are defined and evaluated.

In a steam turbine topping cycle application, fuel such as coal, gas, or oil is burned in a boiler to produce high temperature and pressure steam, and that steam powers a turbine that drives an electrical generator. Only a portion of the energy in the steam is converted to electrical energy, and the remaining thermal energy is available for an industrial process such as drying or cooking.

In a gas turbine topping cycle application, gas or light oil is burned in a version of a jet engine, and the mechanical shaft power is used to drive an electrical generator. Then the waste heat from the gas turbine is captured and used directly for drying or other purposes or is sent to a waste heat boiler where it produces steam for an industrial process.

In a diesel engine topping cycle application, gas or light oil is burned in a diesel engine—similar to a truck or marine diesel engine—and the mechanical shaft power is used to drive an electrical generator. Then the waste heat from the combustion exhaust, the jacket cooling water, and the oil cooling system can be used directly, or the waste heat from the exhaust can be sent to a waste heat boiler to produce steam for an industrial process.

In a bottoming cycle operation, fuel such as gas, oil or coal is burned to produce high temperatures for an industrial process such as drying cement in a kiln or melting glass in a furnace. The waste heat from the industrial process is then captured and sent to a waste heat boiler where it produces steam to drive a turbine generator for electrical energy production.

Many industries use the topping cycle for cogenerating electric energy and process heat or steam. Other industries use bottoming cycles with processes where large amounts of heat are generated in chemical reactions, and then this heat is captured and used in waste heat boilers and for power generation—with additional fuel supplied. A particular example is in the phosphate industry where sulfur is burned to produce sulfur dioxide which is then reacted with water to produce sulfuric acid. The sulfuric acid is then used as a principal component of the process which yields the final product which is phosphoric acid. In this process, the sulfur is burned, and tremendous amounts of heat are produced. The heat is really a waste product, but when it is captured and sequentially used it has substantial value in its ability to reduce the amount of oil, gas or coal that is then needed to produce high temperature steam for generating electric energy.

8.6.2 Efficiency of Cogeneration

Whenever industrial steam or heat is generated, there is an opportunity for simultaneously generating high-efficiency electric power. If industrial steam is generated at a pressure and temperature above that required for the end use and the steam is then brought down to the desired pressure and temperature through a turbine generator, electric power can be produced with about half the fuel needed by a new centralized power plant. If industrial heat is produced for a production process, the remaining heat is a waste product, and any use of that heat represents a large overall efficiency improvement and a reduced need for added fossil fuel.

Cogeneration, however, is a broad technical term, and there are substantial differences in the efficiency of different cogeneration technologies and different cogeneration facilities. Some facilities cogenerate in a highly efficient manner; others cogenerate only incidentally [9]. If, after generating electricity, a topping cycle cogenerator uses all the remaining thermal energy for a production process, then the plant cogenerates as efficiently as it can. Conversely, if a substantial part of the remaining heat is exhausted to a cooling tower or cooling air, then the efficiency may be quite low. For a cycle using waste heat, efficiency is somewhat difficult to define, since it uses energy that would otherwise be lost to the environment. However, any fuel use that is displaced by the available waste heat is an energy savings.

The best cogenerators are more than twice as efficient as new coal-fired power plants. For example, performance measures for generating electric power are associated with net heat rates; typical heat rates are 10,000 – 10,500 Btu/kWh for large, modern coal-fired central station power plants and 4300 – 4700 Btu/kWh for highly-efficient, steam-topping cogeneration plants. Industrial size gas turbine topping cycles average around 5500 Btu/kWh and diesel engine topping cycles about 6500 Btu/kWh. Waste added from an industrial process may reduce the heat rate of an associated generating unit to the range of 5000 Btu/kWh.

8.6.3 Conditions for Successful Application

Cogeneration is not a new idea. In process industries where a great deal of heat is needed, many of the facilities were designed to incorporate cogeneration into the process. For example, one of the largest producers of magnesium takes the magnesium from magnesium chloride by electrolysis. The magnesium chloride comes from salt brines that have been dried. Drying the salt brines takes a large amount of heat; the electrolysis uses large quantities of electricity. The only way that this process can be carried out economically is to generate the electricity and use the waste heat to drive off the water from the brines. Another similar use of cogeneration appears in refineries.

In addition to industrial process use, cogeneration has also been used for district heating, where steam or hot water is supplied to a district by a utility whose primary product is electricity. Such use became less prevalent after World War II, when fuel was cheap and the economies of returned condensate were more attractive than those of wide-area distribution of steam and hot water. More recently, cogeneration is being used in large apartment buildings and in shopping centers as a less expensive source of both heating and electricity.

Cogeneration offers the promise of cheap electricity, with the electrical price under the control of the plant owner rather than of a utility company. But this promise may well be an illusion unless at least five conditions are met.

Condition 1. The need for heat must coincide with the need for electricity unless there are facilities for substantial heat or electrical energy storage. For example, one major university uses 400°F water for building heat and has an average electrical demand of 1.15 MW with peak demands of 2.3 MW. Although cogeneration seemed feasible in this situation, the highest electrical usage came from air conditioning in the summer at the same time that the demand for heat was at a minimum. If the university decided to use air conditioning based on a central chilled water facility with absorbing chillers, there might still be an economic opportunity for cogeneration. Hotels, motels, and hospitals are other good candidates because they have a coincident need for electricity and hot water. Most industries also fit this condition.

Condition 2. Unless the economics are overwhelming, retrofitting for cogeneration should not be considered. Bottoming cycle equipment, in particular, is usually very heavy and can require substantial renovating of building foundations. To define these and other installation problems, it is a good idea to get a detailed feasibility study from a reputable engineering firm with experience in bottoming cycles. Furthermore, a facility designed to produce electricity in addition to the process heat now provided (or vice versa) will necessarily be larger than the present facility. This increase in scale is usually expensive. Other problems can also arise. The authors have had experience with a lumber products operation where the boiler was operated completely on hogged fuel—chopped tree wastes. Changing this unit into a cogeneration facility would have required more energy than was available from the hogged fuel, necessitating either the purchase of gas or the development of some new fuel source. These mill operators were not enthusiastic about cogeneration. Alternatively, the authors have also seen many facilities where the boilers were greatly oversized for the existing loads,. Such oversized boilers might well be economic to run at full capacity and cogenerate electricity while still providing all the stem the facility needs.

Condition 3. If the economic justification for a cogeneration project depends on the sale of electricity, this sale must be based on legally binding documents, and the conditions under which the electricity can be sold, particularly to utilities, must be spelled out, and their economic consequence evaluated. An electric utility is under no obligation to buy electricity in any way that would result in a net loss. In particular, an electric company is not obligated to pay premium rates for electricity delivered when the company has a surplus, and the company is not obligated to tie into a less controlled electric source having a phase or frequency difference that could harm the electrical system. The cost and constraints imposed by such control should be considered and carefully evaluated—they can be great enough to discourage a sale of electricity in favor of using all the electricity in the cogenerating plant. Many facilities that cogenerate find that supplying their own loads provides significant savings without having to sell power to the utility.

Condition 4. The source of fuel for the cogeneration system should be considered carefully. When electricity is sold to a utility, the price is based in part on the reliability of the cogenerating source, and this places greater burdens on the cogenerator. If the source of fuel is some type of industrial waste, this may have to be supplemented with purchased fuel at an unacceptable cost. The lumber operation described above illustrates this problem. For those facilities supplying their own loads, the use of waste of alternative fuels provides an advantage and can lead to the generation of very low-cost electric power compared to utility purchase.

Condition 5. The cogenerator must be prepared to purchase or provide backup heat and/or electricity for occasions when the cogeneration unit is out of service. This is an important economic consideration, since many utilities charge substantial rates to provide backup power. However, the reliability of most industrial cogeneration facilities is well over 90%, and this equipment is maintained well and repaired quickly.

Cogeneration is similar to other projects in that there are always a large number of considerations that influence the economics of the particular project. Overall analysis of the cost effectiveness of cogeneration is complex, and a specialized consultant should always be employed to help evaluate the various factors involved. However, the cogeneration and small power production area is growing rapidly, and this technology offers the potential for attractive savings in many facilities.

8.7   SUMMARY

In this chapter we have presented the concept of enthalpy and have provided examples of its application in the analysis of steam distribution systems. We have also described the equipment and potential of waste heat utilization and improvements to the hot water distribution system and have provided some guidelines for maintaining and improving these systems. In addition, the general concept of cogeneration has been presented, along with some broad recommendations for its cost-effective application.

References
1.   Philip S. Schmidt, "Steam and Condensate Systems" Chapter 6 in *Energy Management Handbook, Fourth Edition*, Wayne C. Turner, Senior Editor, Fairmont Press, 2001.
2.   Theodore Baumeister, Eugene A. Avallone, and Theodore Baumeister III, *Marks' Standard Handbook for Mechanical Engineers*, McGraw-Hill, 1978.
3.   Alfred F. Waterland, "Energy Auditing: A Systematic Search for Energy Savings Opportunities" in Chapter 3, *Energy Management Handbook*, Wayne C. Turner, Senior Editor, New York, 1982.
4.   Spirax Sarco, Inc., *Design of Fluid Systems: Hookups*, Spirex Sarco, Inc., Blythewood, SC 29016, 2001.
5.   Spirax Sarco, Inc., *Design of Fluid Systems: Steam Utilization*, Spirex Sarco, Inc., Blythewood, SC 29016, 2001.
6.   The Babcock and Wilcox Company: *Steam, Its Generation and Use*, 40th Edition, 1992

7.   Kenneth G. Kreider and Michael B. McNeil, *Waste Heat Management Guidebook*, National Bureau of Standards Handbook 121, U.S. Department of Commerce, 1977.
8.   Barney L. Capehart and Lynne C. Capehart, "Efficiency in Industrial Cogeneration: The Regulatory Role," Public Utilities *Fortnightly*, March 15, 1990.
9.   Lynne C. Capehart and Barney L. Capehart, "Efficiency Trends for Industrial Cogeneration," Public Utilities *Fortnightly*, April 1, 1991.

Bibliography

Capehart, Barney, Editor, *Encyclopedia of Energy Engineering and Technology*," Taylor and Francis/CRC Publishing Company, Boca Raton, FL, July, 2007.

Chapter 9

Control Systems and Computers

9.0 INTRODUCTION

Energy use can be controlled in order to reduce costs and maximize profits. The controls can be as simple as manually turning off a switch, but often automated controls ranging from simple clocks to sophisticated computers are required. Our view is that the control should be as simple and reliable as possible. Consequently, this chapter starts with manual controls and proceeds through timers, programmable controllers, and digital computers.

As one moves through this hierarchy of controls, each level of automation and complexity requires additional expenditure of capital. That is, the automated controls are more expensive, but they do more. Because choosing the proper type of control is often a difficult task, we will explore this decision process.

Computers can also help the energy manager in the analysis of proposed and present energy systems. Some excellent large-scale computer simulation programs have been written that enable the energy analyst to try alternative scenarios of energy equipment and controls, so in the last part of this chapter we discuss these computer programs and their use. BLAST 3.0 and DOE-2.1D are the two described in depth, but several others are mentioned.

9.1 WHY CONTROLS ARE NEEDED

Every piece of energy-consuming equipment has some form of control system associated with it. Lights have on-off wall switches or panel switches, and some have timers and dimmer controls. Motors have on-off switches, and some have variable speed controls. Air conditioners have thermostats and fan switches; they sometimes have night setback controls or timers. Large air conditioning systems have extensive controls consisting of several thermostats, valve and pump controls, motor speed controls, and possibly scheduling controls to optimize the operation of all of the components. Water and space heaters have thermostats and pump controls or fan motor controls. Large heating systems have modulating controls on the boilers and adjustable speed drives on pumps and variable air volume fans.

These controls are necessary for the basic safety of the equipment and the operators, as well as for the proper operation of the equipment and systems. Our interest is in the energy consumption and energy efficiency of this equipment and these systems, and the controls have a significant impact on both these areas. Controls allow unneeded equipment to be turned off, and allow equipment and systems to be operated in a manner that reduces energy costs. This may include reductions in the electric power and energy requirements of equipment, as well as the power and energy requirements associated with other forms of energy such as oil, gas and purchased steam.

9.2   TYPES OF CONTROLS

In this section, we present the different types of controls in order of increasing complexity and cost. In each subsection, the control discussed can perform the functions covered in that subsection as well as all those functions covered in the preceding subsections. For example, the functions discussed in the second subsection on timers can be performed very well by a timer or any of the succeeding types of controls (programmable controllers, microprocessors, and large computers) but not by a manual system.

9.2.1 Manual Systems

Manual control systems can be used to turn equipment off when it is not needed. Turning equipment off when not in use can lead to dramatic savings. For example, lights are often left on at night, but they should normally be turned off whenever possible. (Often a small series of lights is left on for security purposes.)

One of the best opportunities for manual control exists in the area of exhaust and makeup air fans. These fans are often located at the top of a high ceiling, and they are frequently left on unnecessarily because their running is undetectable without close scrutiny. The savings for turning off exhaust fans is twofold. First, electricity is no longer required to run the fan motor, and, second, conditioned air is no longer being exhausted. Consider the following example.

Example 9-1: Suppose that a fan is exhausting air at a rate of 10,000 ft³/ min from a welding area. The fan is run by a 5-hp motor and is needed for two shifts (8:00 a.m. to 12:00 midnight) 5 days/ week. Previously, the fan has been left running all night and on weekends. If the space is not air-conditioned and is heated to 65°F by a gas furnace that is 80% efficient, and the efficiency of the motor is 84%, what is the savings for turning the fan off at night and on the weekends? Gas costs $5.00/million Btu and electricity, $.08/kWh. (There will be no demand savings since peaking does not occur at night). Assume the outside temperature averages 30°F for the hours the fan can be shut off. (This would have been determined through weather data analyses, as discussed in Chapter 2.)

Solution:

    1. Electricity savings:

    The electric energy savings from turning the motor off during nights and weekends is found by multiplying the motor load in kW times the number of hours saved (number of hours the motor is not running). This energy savings in kWh is then multiplied by the energy cost to get the dollar cost savings.

Electric energy savings =

$$5\,\text{hp} \times \frac{.746\,\text{kW}}{\text{hp}} \times \frac{1}{.84} \times \left[\frac{5\,\text{days}}{\text{week}} \times \frac{8\,\text{h}}{\text{day}} \times \frac{2\,\text{days}}{\text{week}} \times \frac{24\,\text{h}}{\text{day}}\right] \times \frac{52\,\text{weeks}}{\text{year}} \times \frac{\$.08}{\text{kWh}}$$

$$= \$1{,}625.57$$

    2. Heating savings: (see equation 6-13 in Chapter Six)

Heating cost savings =

$$\frac{10{,}000\,\text{ft}^3}{\text{min}} \times \frac{60\,\text{min}}{\text{h}} \times \left[\frac{5\,\text{days}}{\text{week}} \times \frac{8\,\text{h}}{\text{day}} \times \frac{2\,\text{days}}{\text{week}} \times \frac{24\,\text{h}}{\text{day}}\right] \times \frac{52\,\text{weeks}}{\text{year}}$$

$$\times \frac{.075}{\text{ft}^3} \times \left[\frac{.24\,\text{Btu}}{\text{lb}\,°\text{F}} \times \left(65°\text{F} - 30°\text{F}\right) \times \frac{\$5.00}{10^6\,\text{Btu}} \times \frac{1}{0.8}\right]$$

$$= \$10{,}810.80$$

where the density of air $= .075 \text{ lb/ft}^3$ and the specific heat of air $= .24 \text{ Btu/lb}°\text{F}$.

3. Total annual savings $= \underline{\$12{,}436.37}$

In another example, a large office complex made a detailed study of building utilization and found that only a few tenants worked at nights or on the weekends. By making provisions for these few, the office complex was able to reduce lighting and space conditioning, saving about a third of its annual energy bill. For most industrial plants and many office buildings, the use of night setback to lower heating temperatures offers significant savings with little or no capital expenditure.

The calculation procedure for determining this savings is relatively simple and involves heat loss calculations during the hours of setback. Bin weather data on outside temperatures and inside thermostat settings are required. The heat losses are calculated for the old thermostat setting and again for the revised setting. The difference is the heating savings in Btu. To simplify this procedure, or at least to give an approximation, a nomograph is given in Figure 9-1. The following example shows how to perform the calculation.

Example 9-2: A manufacturing company of $100{,}000 \text{ ft}^2$ is located in an area where heating demands are 4000 degree days. The company keeps its thermostats set at $70°\text{F}$ all the time even though it works only one shift. Presently, the company figures it consumes $240 \times 10^3 \text{ Btu/ft}^2$ of gas for heating. (Normally, this can be estimated from gas bills.) If the company pays $\$4.50/10^6$ Btu for its natural gas and the heaters are 75% efficient, what would the savings be for turning the thermostats back to $55°\text{F}$ at night when the building is not occupied?

Solution: To use the nomograph, follow the heavy black lines in Figure 9-1. The savings are approximately $125 \times 10^3 \text{ Btu/ft}^2$. Total savings then are found by first determining the actual fuel savings and then finding the dollar savings.

Fuel savings in Btu $= (125 \times 10^3 \text{ Btu/ft}^2) (100{,}000 \text{ ft}^2) \times 1/.75$
                                       $= \underline{16{,}700 \times 10^6 \text{ Btu}}$

Savings in dollars $= (16{,}700 \times 10^6 \text{ Btu}) (\$4.50/10^6 \text{ Btu})$
                                       $= \underline{\$75{,}150/\text{year}}$

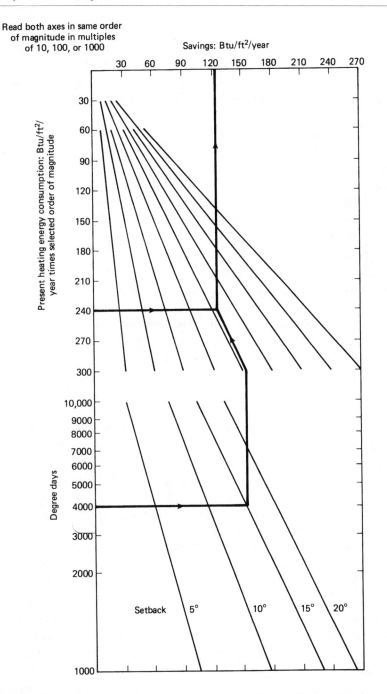

Figure 9-1. Estimation of savings via thermostat night setback. (*Identifying Retrofit Projects for Buildings*, FEA/D-76-467, Sept. 1976.)

The savings for thermostat setback can be substantial, as shown in the example. In warm climates similar savings can result from turning off the air conditioners at night and on weekends, but the dollar amount is usually less since heating demands peak at night, while cooling demands peak during the day. Of course, the energy manager needs to be careful to ensure that the night setback does not cause heating plant problems or cause process problems, e.g., changing the tolerances on large metal parts or affecting the hardening rate of thermocure resins. This depends on the particular equipment and controls used as well as the thermal load of the building itself.

Night setback can also be applied to process areas. For example, large furnaces such as brick kilns should be turned off when possible, but often the preheat time required and/or the thermal wear on the furnace walls makes this impossible. Many times, however, the thermostat can be adjusted downward significantly without causing these problems. Trial and error may be required to determine the optimum setting.

9.2.2 Basic Automatic Controls—Timers and Dimmers

The next step in level of control complexity is the use of automatic controls such as timers and dimmers. Timers can range from very simple clocks to fairly complicated central time clocks with multiple channels for controlling numerous pieces of equipment on different time schedules. Automatic timer controls can range from simple thermostats each with a built-in time clock (costing somewhere around $100 each) to a central time clock that overrides all the thermostats. An installed single-channel central time clock will cost around $1000, but it can control numerous thermostats if all are on the same schedule. Different setback schedules require multiple channels, increasing the cost somewhat.

Some companies have utilized time clocks to duty cycle equipment such as exhaust fans. For example, a large open manufacturing area will likely have several exhaust fans. If there are six fans, then a central time clock could turn one fan off each 10 minutes and rotate so that each fan is off 10 minutes of each hour, but no more than one fan is ever off at the same time. This saves on electrical consumption (kWh) to run the fan, electrical demand (kW, since one fan is off at any time), and heating (since less conditioned air is exhausted). General ventilation over a wide area is maintained. Of course, care must be taken to ensure that no ventilation problems develop.

The use of timers allows a company to start-stop equipment at exactly the correct time. It is not necessary to wait for maintenance people to make their rounds, turning off equipment and adjusting thermostats.

However, although timers don't forget to do their job, they do suffer from other problems. For example, power outages may require timers to be reset unless a battery backup is used. Also, arrival or termination of daylight savings time requires all timers to be set up an hour in the spring and back an hour in the fall. Finally, the clocks must be maintained and replaced as they wear out.

The authors had an opportunity to audit a plant that had sophisticated time clock controls on its equipment, but management was not maintaining the clocks. The 7-day time clocks allowed for night and weekend setbacks. The audit was done on a Thursday, but the time clocks read Saturday. Consequently, the thermostats were on night setback, and the employees were cold. To remedy this, maintenance had purchased several additional portable heaters. If they had come in on a Saturday, when the clock read Monday, the plant would have been nice and warm. In this case, the poorly maintained clocks were costing the company a great deal of money. Timers and any other type of control system must be maintained.

Another type of control that has some attractive savings potential is a light dimmer. Dimmers can be automatically controlled depending on time, and on natural lighting levels if photocell sensors are used. It is important to be sure the dimming system chosen actually reduces electrical consumption and is not simply a rheostat (variable resistor) that consumes the same amount of energy regardless of the amount of light delivered. Supermarkets can often use relatively sophisticated dimming systems. For example, supermarkets might:

1. Use photocells to detect natural light and dim the window lights as appropriate.

2. Use photocells that turn parking and security lights off at dawn and on at dusk.

3. Use photocells to determine dusk so that interior lighting can be reduced. (Studies have shown that people coming from a dark street to a brightly lit room are actually uncomfortable. Lower lighting levels are preferred.)

As with timers, photocells and dimmers must be maintained. They sometimes fail, and if undetected, the failure can cause other more severe problems. A regular maintenance schedule of checking photocells and dimmers should be used. Photosensors were also discussed in Chapter 5.

9.2.3 Programmable Controllers

A programmable controller is a control device that has logic potential but is not powerful enough to be called a computer. As might be expected, it fills a need for systems requiring more than a timer but less than a computer. It can do all that timers can do and considerably more but at a cost significantly less than that of a computer. Fixed logic devices —such as timers—are primarily useful in buildings under 50,000 ft^2 with maybe a dozen control points. (A control point is a switch, a thermostat, or a control actuator.) Computerized systems would be applicable in buildings of 100,000 ft^2 with more with 100 control points. The middle group would be suitable for programmable controllers. Some control system selection guidelines are given in reference 1.

A programmable controller adds logic capability to control systems. Demand shedding is a prime example. When the controller senses that the electrical demand is approaching a critical programmed level, the unit then shuts off equipment and/or lights to keep the demand from passing that critical level. As shown in Chapter 3, demand can be a large part of an electrical bill, so the savings can be significant.

Another example is excess air control for a boiler or any larger combustion unit. By sensing CO_2 or O_2 and perhaps CO levels in exhaust, the controller can adjust the combustion air intake to yield optimum combustion efficiency. As shown in Chapter 7, this can be a real money saver. Continuous control through the use of a programmable controller allows the air intake to be adjustable according to the heating demand on the unit.

Programmable controllers can also be used to control outside air for heating, ventilating and air conditioning systems. In air conditioning, if the outside air is more comfortable than the inside air, the outside air should be used rather than returning the inside air. In fact, sometimes the air conditioning units can be turned off completely and outside air used for cooling. Programmable controllers can sense the difference between outside and inside enthalpy and determine the optimum damper setting. The same controller can shut off outside air completely for early morning start-up and nighttime operation during heating seasons. Outside air control is discussed further in the next section.

9.2.4 Computerized Systems

Most computerized energy management control systems (EMCSs) sold today are microprocessor-based. These capabilities run from a few control points up to several thousand, with the larger ones often performing fire-safety functions, equipment maintenance status monitoring and report generation as well as energy management [2]. The technology is changing very rapidly and there are many vendors in the field, each introducing new equipment. A potential user should consult several vendors and be well prepared to discuss the facility's needs. Some general approaches to the selection and specification of computer control systems for energy management are given in references 3 and 4.

EMCS users must recognize the need for feedback. The computer may have sent a signal to turn off a load, but was the load actually turned off? A sensor is required to feed the control status back to the computer. Also, it is helpful if the computer maintains a record of when control was exercised. With these records, histograms can be developed periodically to show how frequently any given load is being shed.

Additional control options are available to the EMCS user, but mostly they are some combination of the techniques discussed in the previous sections. Figure 9-2 is useful in summarizing these techniques. In Figure 9-2(a), the original electrical demand profile is shown before control is applied. Note that the area under these curves is the integration of demand over time and thus is the kWh consumption. In Figure 9-2(b), demand control is applied. Here, a peak demand is determined, and loads are shed once that peak is approached. Shedding requires predetermining what loads can be shed and in what priority. For example, display lighting would likely be shed before office lighting.

Many systems also "remember" loads previously shed. They will rotate among shedable loads, and will obey preset maximum shed times. As Figure 9-2(b) shows, shed loads must sometimes be recovered and sometimes not. For example, shedding refrigeration saves demand, but sooner or later the unit must catch up. Shedding lights, on the other hand, saves energy because lighting cannot be recovered. As shown, then, some consumption is shifted but usually not as much as was shed. Demand savings remain the predominant goal.

In Figure 9-2(c), a fixed start-stop schedule is utilized. Now units are turned on and off at exactly the same time each day. No longer are personnel required to make rounds, turning equipment on and off. In Figure 9-2(d), an optimized start-stop schedule is employed. The precise time of need is determined each day, and the equipment is turned on at that time. For example, if the outside and inside temperatures are both warm, the heating units do not have to be turned on as early as they would be if the respective temperatures were quite cold.

Figure 9-2(e) shows what happens when the use of outside air is optimally controlled. In this case, the plant requires air conditioning. The fans—but maybe not the compressors—are turned on early in the

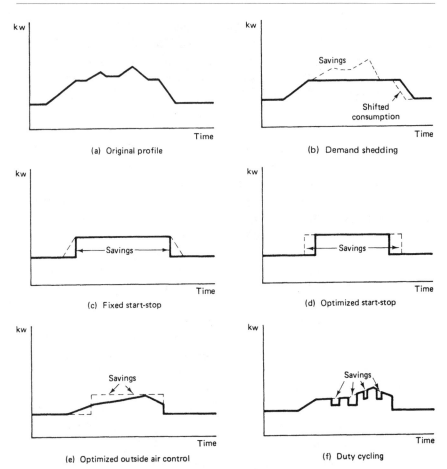

Figure 9-2. EMCS control techniques. (From Dick Foley, "Reducing Waste Energy with Load Controls," *Industrial Engineering*, July 1979, p. 24. Extracted with permission from Institute of Industrial Engineers, Inc., 25 Technology Park/Atlanta, Norcross, GA 30092, ©1979.)

morning to draw in cool outside air for precooling. As the daytime temperatures warm, less outside air is used, and the compressors have to run longer; at the peak air conditioning time a minimum of outside air is used. Toward the end of the operating time, it may again become profitable to utilize outside air.

Duty cycling is depicted in Figure 9-2(f). Here loads are selected to be turned off on a predetermined schedule. For example, in our earlier discussion exhaust fans were turned off 10 minutes out of each hour. Duty cycling will produce a savings in kWh since the units are operating less. If these schedules are determined so that some piece of equipment is being cycled off at any time that peaks are likely to occur, a demand savings will also result.

All the techniques shown in Figure 9-2 deal with electrical consumption. Other techniques cannot be easily demonstrated in such a figure because of their nature and/or because they affect fuel consumption rather than electricity. Examples include the following:

1. Light dimming, as discussed in Section 9.1.2.

2. Combustion air control for furnaces. This affects fuel consumption.

3. Night setback for heating. This normally affects fuel consumption instead of electricity.

4. Surge protection. If power outages occur, EMCSs can be programmed to start turning loads off to prevent an extremely large surge of power once the service is reconnected.

5. Temperature reset. Here the temperature of supply air or water is modified to meet actual demand. For a heating system, the supply air temperature may be reduced by 10 to 20°F when heating demands are small. This could save substantial fuel.

There are several different generic forms of computer EMCS configurations available. In a centrally controlled system, control is vested in one central unit—a microprocessor or microcomputer. The control points are accessed directly (a *star network*) or through common wiring (a *common data bus network*). These two configurations are shown in Figure 9-3. In the star network, control is more direct but installation is considerably more expensive. So, its use is limited to facilities with few control points. The common bus design allows for common use of wiring, so its installation cost is less for facilities with a large number of control points. Some of the current activities in the design of EMCSs involve the movement to standardized communication methods called protocols [5]. Standardized systems should be cheaper and easier to maintain and expand.

Instead of using a centrally controlled EMCS, many newer EMCS systems use a distributed configuration where remote processing units using microprocessors or microcomputers perform the actual control functions. A central unit is still used, but primarily for coordination and report generation. This is illustrated in Figure 9-4 using a star network.

Points being controlled

Star network

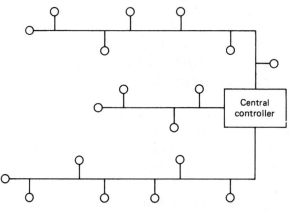

Common bus network

Figure 9-3. Centrally controlled EMCS star network and common bus network.

The remote control units vary in capability from a simple transfer function to complete control. In a true distributed control system, the remote controllers can function with or without the central unit—at least for a period of time. Intelligent remote controllers are only slightly more expensive than unintelligent ones. Most designers predict that future EMCSs will have more distributed control as the cost of remote controllers is reduced. Of course, systems can be hybrids so that some star networking is used along with common bus designs. Also, some remote controllers may be intelligent and others not, and some points may be directly controlled by the central controller even in a distributed control basic design.

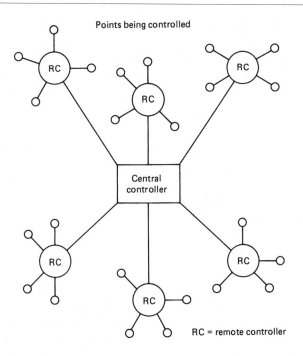

Figure 9-4. Distributed control EMCS star network.

An important factor to consider in the selection and cost of EMCSs is the type of control needed at the points being controlled. The simplest and cheapest is a digital point which is a simple on-off control. Examples include switches for fans, lights, and motors. Analog controls, on the other hand, are more complex and therefore more expensive. Analog controls are needed when signals of varying intensity are required. For example, control of outside air for air conditioning systems requires an analog signal to adjust the position of a supply vent damper. As the outside air cools, the EMCS will continue to open the damper, allowing more outside air to enter. A significant number of analog points will run up the cost of an EMCS rather rapidly, but the savings should also increase. Careful studies should be conducted in order to determine the optimum number of control points.

EMCSs sometimes fail to provide the energy cost savings that have been predicted. These reasons usually fall into one of the categories shown in Table 9-1. The prudent energy manager will consider these potential pitfalls and plan accordingly.

Table 9-1. Why EMCS units fail to produce the desired results.

1. *Simple things not done first.* The EMCS can do many things, but it should not be asked to do the things that should have already been done. For example, the EMCS cannot turn a standard lighting system into an energy-efficient one.

2. *Simple alternatives overlooked.* An EMCS may not be necessary. Manual or time control may suffice.

3. *Requirements not carefully defined.* This is the most important reason. The buyer must define the requirements before choosing the system. Considerable planning is necessary.

4. *Inadequate buyer commitment/inadequate seller backing.* All too often, buyers seem to expect the EMCS units to install, program, and maintain themselves. Sometimes the seller misrepresents the amount of work necessary to get the EMCS operational. The energy manager should insure that the seller will back the product and provide the necessary technical aid.

5. *Poor vendor assessment.* The energy manager should screen the vendors carefully. Ask for reference letters and check with other energy managers. Be wary.

In summary, EMCSs can do many things, but they are only machines. The energy manager must be aware of their limitations as well as their strengths and design their applications accordingly. The major part of the design should occur before selection of the equipment—not afterwards.

9.2.5 Web Based Building Automation Systems

Of all recent developments affecting computerized energy management systems, the most powerful new technology to come into use in the last several years has been information technology—or IT. The combination of cheap, high-performance microcomputers together with the emergence of high-capacity communication lines, networks and the Internet has produced explosive growth in IT and its application throughout our economy. Energy information and control systems have been no exception. IT and Internet based systems are the wave of the future. Almost every piece of equipment and almost every activity will be connected and integrated into the overall facility operation in the next several years[10].

In particular, the future of DDC in Building Automation Systems (BAS) can be found on the Web. Most all BAS manufacturers see the need to move their products to the Internet. Tremendous economies of scale and synergies can be found there. Manufacturers no longer have to create the transport mechanisms for data to flow within a building or campus. They just need to make sure their equipment can utilize the network data paths already installed or designed for a facility. Likewise, with the software to display data to users, manufacturers that take advantage of presentation layer standards such as HTML and JAVA can provide the end user with a rich, graphical and intuitive interface to their BAS using a standard Web browser.

Owners will reap the benefits of Internet standards through a richer user interface, more competition among BAS providers, and the ability to use their IT infrastructure to leverage the cost of transporting data within a facility. Another area where costs will continue to fall in using Internet standards is the hardware required to transport data within a building or a campus. Off-the-shelf products such as routers, switches, hubs, and server computers make the BAS just another node of the IT infrastructure. Standard IT tools can be used to diagnose the BAS network, generate reports of BAS bandwidth on the intranet, and backup the BAS database.

The EMCS of old relied heavily on a collection of separate systems that operated independently, and often with proprietary communication protocols that made expansion, modification, updating and integration with other building or plant information and control systems very cumbersome, if not impossible. Today the BAS is not only expected to handle all of the energy and equipment related tasks, but also to provide operating information and control interfaces to other facility systems, including the total facility or enterprise management system [10].

Measuring, monitoring and maximizing energy savings is a fundamental task of all BAS, and is the primary justification for many BAS installations. Improving facility operations in all areas, through enterprise information and control functions is fast becoming an equally important function of the overall BAS or facility management system. The Web provides the means to share information easier, quicker, and cheaper than ever before. There is no doubt that the Web is having a huge impact on the BAS industry. The BAS of tomorrow will rely heavily on the Web, TCP/IP, high-speed data networks, and enterprise level connectivity. If you have not done so already, it is a good time for energy managers to get to know their IT counterparts at their facility along with those in the accounting and maintenance departments. The future BAS will be here sooner than you think. See Chapter 15 for a complete discussion of this topic.

9.3   COMPUTER UTILIZATION

Computers have become so inexpensive and so powerful that they are used everywhere today, including use in a wide variety of tasks in energy management and energy analysis. The energy manager must be aware of computing capabilities and applications, and must carefully integrate computers into his or her environment. In addition to the direct EMCS applications, there are many other energy management uses for computers. Personal computers can be programmed to perform cash flow analyses, waste heat recovery studies, excess air control studies, and a myriad of other aids. Some desktop computers are capable of running large building simulations or equipment design programs requiring significant data storage and lengthy computing times.

Energy Engineering, the journal of the Association of Energy Engineers, has started publishing an annual Directory of Software for Energy Managers and Engineers. In the 1991 directory there were one hundred and thirty-three different software products for energy applications that were produced by forty-four companies [6]. The editor of this journal commented that there was more interest in the Directory Issue than any other issue produced [7]. The Directory is compiled by companies voluntarily sending in information, so it is not a complete listing of all energy-related computer programs. However, the listings in the 1991 Directory are more than twice as long as the listings in the 1990 Directory, which indicates a growing interest among both the users and the suppliers of these programs.

It is impossible to summarize or list all of the possible energy-related uses for computers, but one does stand out from the rest. That is the use of computers to perform building energy use analysis and simulation studies. In the rest of this chapter we will examine some of the programs that are most commonly available for this task. Building energy use analysis and simulation studies require the input of weather data, operating times, and other energy-consuming parameters such as number and type of lights and equipment, efficiency of various devices, etc. as well as the parameters of the building shell such as wall construction, insulation levels, amount of window area, etc. The computer will then simulate a year of building operation (or whatever cycle is chosen), and develop energy consumption and energy bills. Thus various scenarios involving energy efficiency improvements to the building shell or to building equipment can be fed into the computer and the likely savings identified and estimated. Most of the programs available also contain a financial analysis subroutine that provides the economic decision measures needed to help select the most cost-effective EMOs. Thus, the complete energy management study can be done by the computer.

Some of the better known programs include BLAST 3.0, DOE-2.1D, AXCESS, ASEAM, TRACE, and ECUBE, although there are many others. A call to your local utility, university, and/or energy management consultant can identify which are available in your area. We will discuss the capabilities of BLAST 3.0 and DOE-2.1D. Brief discussions of AXCESS, ASEAM, TRACE, and ECUBE, as well as where to get additional information on those programs, can be found in the Software Directory from reference [6].

9.3.1 BLAST 3.0

BLAST can be used to investigate the energy performance of new and retrofit building design options of almost any type and size [8]. BLAST is the acronym for the Building Loads Analysis and Systems Thermodynamics family of programs. Not only can BLAST calculate the building peak loads using design day criteria, which is necessary for mechanical equipment design, it can also estimate the annual energy performance of a facility. This is essential for the design of solar and co-generation systems and for determining the building's compliance with a design energy budget.

Apart from its comprehensiveness, the BLAST system differs in three key aspects from other similar programs. First, BLAST uses extremely rigorous and detailed algorithms to compute loads, to simulate fan systems, and to simulate boiler and chiller plants. Second, the program has its own user-oriented input language and is accompanied by a library which contains the properties of all materials, wall, roof, and floor sections. Third, BLAST's execution time is short enough to allow many alternatives to be studied economically. In this way, efficient designs can be separated from the inefficient, and proper equipment type, size and control can be determined.

The BLAST Energy Analysis Program contains three major subprograms. First, the Space Load Predicting subprogram computes hourly space loads in a building based on weather data and user inputs detailing the building construction and operation. Next, the Air Distribution System Simulation subprogram uses the computer space loads, weather data, and user inputs describing the building air-handling system to calculate hot water, steam, gas, chilled water, and electric demands of the building and air-handling system. Finally, the Central Plant Simulation subprogram uses the weather data, the results of the air distribution system simulation, and user input describing the central plant to simulate boilers, chillers, on-site power-generating equipment, and solar energy systems; and then computes monthly and annual fuel and electrical power consumption.

Early versions of BLAST were considered difficult to use, but several new methods have been developed to communicate with BLAST. Available methods include BTEXT and Drawing Navigator for Autocad. BTEXT is a text-based scrolling menu program to solicit information about the building model. It builds a special file of building information and can generate BLAST input files. Drawing Navigator for Autocad uses the graphic information accessible in drawings to generate the necessary building geometric data; it passes information into BTEXT for eventual BLAST input file creation. Both preprocessors simplify input file creation [8].

MICRO BLAST 3.0 is available to run on any PC compatible 386 or 486 computer with 20-100 MB of hard drive memory and 4 MB of RAM. Users may obtain access to the BLAST program, and additional information about it from the BLAST Support Office, University of Illinois, 1206 West Green Street, Urbana, IL 61801.

9.3.2 DOE-2.1D

The DOE-2.1D Computer Simulation Program was developed for the Department of Energy (DOE) to perform energy analysis and simulation of plants and buildings [9]. It calculates the hour-by-hour energy use of a building and its life-cycle cost of operation using information on the building's location, construction, operation, and heating, ventilating, and air conditioning system. This program is used to design energy efficient new buildings, analyze energy conservation measures in existing buildings and calculate building design energy budgets. The program is divided into five major subprograms: (1) Building Description Language (BDL), (2) LOADS, (3) SYSTEMS, (4) PLANT, and (5) ECONOMICS.

The Building Description Language (BDL) subprogram allows the user to enter key building design information. The program uses a library of properties of all materials, walls, roof, and floor sections. The user also inputs a description of the HVAC systems, occupancy, equipment, lighting schedules, and other parameters.

The LOADS subprogram computes hourly space loads resulting from transmission gains and losses through walls, roofs, floors, doors, and windows; internal gains from occupants, lighting, and equipment; and in-filtration gains and losses caused by pressure differences across openings. The LOADS calculations are based on ASHRAE algorithms, including the response factor technique for calculating transient heat flow through walls and roofs and the weighting factor techniques for calculating heating and cooling loads.

After the building loads are calculated, the program begins the SYS-TEMS analysis. The SYSTEMS subprogram takes the hourly space loads, along with characteristics of secondary HVAC equipment, the component and control features, and the thermal characteristics of the zone, and determines the actual room temperature and heat extraction or addition rates using ASHRAE algorithms.

The PLANT subprogram uses the building thermal energy load data determined by SYSTEMS and various other user-input operating parameters of the plant equipment to allocate available equipment and simulate their operation. The PLANT program simulates conventional central plants, solar heating and cooling systems, and plants with on-site generation and waste heat recovery. It also permits load management of plant equipment and energy storage. It calculates the monthly and annual cost and consumption of each type of fuel used, the daily electrical load profile, and the energy consumption at the site and at the source.

The ECONOMICS subprogram uses the life-cycle costing methodol-ogy derived from DOE guidelines. Life-cycle costing method investment statistics such as cost savings, savings-to-investment ratio, energy sav-ings, energy savings-to-investment ratio, and discounted payback period are calculated to provide a measure for comparing the cost effectiveness of each case against a reference case.

The MICRO DOE-2.1D program is available for a PC compatible 386 or 486 computer with 2 MB of RAM and at least 20 MB of hard disk space. A newer version, DOE-2.1E, has recently been made available, and a greatly expanded version named DOE-3 is currently under development. The DOE-2 program is used extensively by electric utilities, government agencies, national laboratories, architect/engineering firms, universities, and many private organizations. A version of DOE-2 is used by the State of California to determine building code compliance.

Users may obtain access to the DOE-2.1D program or the MICRO DOE-2.1D program from various commercial vendors. Some of these vendors also offer user support and training for the system. Additional in-formation on DOE-2.1D is available through the Building Energy Analysis Group, Energy and Environment Division, Lawrence Berkeley Laboratory, Berkeley, CA 94720. Information is also available through the Department of Energy, Office of Building Systems Technology, Architectural and Energy Systems Branch, 1000 Independence Avenue, SW, Washington, DC, 20008.

9.4   SUMMARY

In this chapter we have examined control systems and computer applications for energy management. We began with a discussion of the types of controls, including manual, timer, programmable controllers, and computers. Then we discussed each level of control, giving advantages and limitations. Basically, the simpler controls are the least expensive and least robust. The more expensive controls (such as EMCSs) are more robust in that more control activities can be utilized. Computers can be used in other areas of energy management also. Data manipulation, data summary, and large-scale modeling or simulation are among some of the examples of other areas where computers can be utilized. Large-scale computer simulation models of energy systems are available and are quite useful in simulation of system operation, various scenarios of new equipment selection, or use of revised control schemes. BLAST 3.0 and DOE-2.1D are two programs discussed in some depth in the chapter.

References
1.   "Programmable Controllers: A Viable Alternative," *Plant Energy Management*, March 1982, pp. 44-45.
2.   Bynum, Harris D., "What's New in Building Automation: Network Applications Add New Levels of Synergy," *Strategic Planning and Energy Management*, Vol. 9, No. 2, 1989.
3.   Payne, F. William, *Energy Management and Control Systems Handbook*, The Fairmont Press, Atlanta, GA, 1984.
4.   "Guidelines for the Design and Purchase of Energy Management and Control Systems for New and Retrofit Applications," American Consulting Engineers Council, Washington, DC, 1984.
5.   Fisher, David M., "Open Protocol: Today's Options," *Strategic Planning and Energy Management*, Vol. 9, No. 2, 1989, pp. 51-56.
6.   "Directory of Software for Energy Managers and Engineers," *Energy Engineering*, Vol. 88, No. 2, 1991, pp. 6-79.
7.   Williams, Anna Fay, "Editorial," *Energy Engineering*, Vol. 88, No. 2, 1991, p. 5.
8.   Lawrie, Linda K., "Day-to-Day Use of Energy Analysis Software," *Energy Engineering*, Vol. 89, No. 5, 1992, pp. 41-51.
9.   Curtis, R.B., et. al., "The DOE-2 Building Energy Analysis Program," Technical Report LBL-18046, Applied Science Division, Lawrence Berkeley Laboratory, University of California, Berkeley, CA, April 1984.
10. Capehart, B.L., *Editor, Information Technology for Energy Managers — Understand-*

ing Web Based Energy Information and Control Systems, The Fairmont Press, Atlanta, GA, 2003.

Bibliography

Blake, Frederick H., "New Energy Management Controls for Supermarkets," *Energy Engineering*, Aug.-Sept. 1981 PP. 49-54.

Capehart, Barney, Editor, *Encyclopedia of Energy Engineering and Technology*," Taylor and Francis/CRC Publishing Company, Boca Raton, FL, July, 2007.

Capehart, B.L. and Lynne C. Capehart, Editors, *Web Based Information and Control Systems—Case Studies and Applications*, The Fairmont Press, Atlanta, GA, 2005.

Lawrie, Linda K. and T. Bahnfeth, "Guidelines for Evaluating Energy Analysis Software," USACERL Technical Report E89-15, September 1989.

McDonald, J. Michael and Michael B. Gettings, "Military EMCS: Implications for Utilities, Cities, and Energy Services," *Energy Engineering*, Vol. 88, No. 4, 1991, pp 20-35.

Peterson, Kent W., and John R. Soska, "Control Strategies Utilizing Direct Digital Control," *Energy Engineering*, Vol. 87, No. 4, 1990, pp 30-35.

Stewart, Alan F., "Modeling and analysis of HVAC Systems with an EMCS," *Energy Engineering*, Vol. 87, No. 2, 1990, pp 6-17.

Thumann, Albert, Editor, *The Energy Management Systems Sourcebook*, The Fairmont Press, Atlanta, GA, 1985.

Turner, Wayne C., Senior ed., *Energy Management Handbook, Second Edition*, The Fairmont Press, Atlanta, GA, 1993.

Chapter 10

Energy Systems Maintenance

10.0 INTRODUCTION

Maintenance is a critical part of a facility's operation. Properly maintained equipment and processes are necessary to keep the facility functioning at its optimum capability. Unfortunately, the maintenance program is often one of the first victims of any cost-cutting effort. Generally, preventive or scheduled maintenance is cut back or eliminated. Then the maintenance effort is directed more toward repair and replacement than toward keeping the equipment running most efficiently. We have already discussed a number of maintenance measures in prior chapters in this book. In this chapter, we discuss the role of maintenance primarily as it relates to energy costs, and we focus on the maintenance aspects of each major energy-consuming sector of a facility.

Maintenance should be an integral part of any energy management program. Maintenance keeps equipment from failing, helps keep energy costs within reason, helps prevent excess capital expenditures, contributes to the quality of a product, and is frequently necessary for safety. In this chapter, we propose taking a continuous improvement approach to an energy management maintenance program: planning, analysis, action, and monitoring (See Figure 10-1.)

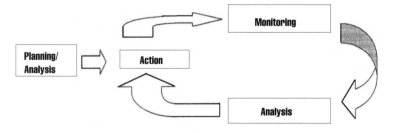

Figure 10-1. Overview of the maintenance energy management function.

10.1 OVERVIEW OF A CONTINUOUS
IMPROVEMENT MAINTENANCE PROGRAM

The first phase, planning, sets the stage for the entire process. In the planning step, the energy management team is formed and the overall goals are established as discussed in Chapter 1. The maintenance staff must be represented on the energy management team so that the overall energy planning will include reasonable estimates of present maintenance costs related to energy and a realistic determination of which of these maintenance tasks need to be done first. The choice and start of these tasks is the first action. Then, these tasks must be monitored and the results must be analyzed and used as the basis for some new action—adding new projects, modifying existing projects, changing deadlines, etc. The new actions are monitored and the results are analyzed, then more new actions are chosen. This monitoring, analysis, action sequence continues and results in continuous improvement of the maintenance function. The details of this process are given below.

10.2 PLANNING

Planning includes these steps, some of which require analysis:

A.   Identify maintenance functions that are energy related.
B.   Estimate the energy-related maintenance costs of the present facility.
C.   Determine the present maintenance state of each of the major energy systems and each major piece of energy consuming equipment within the facility.
D.   Decide which energy-related maintenance tasks must be done immediately and which can be delayed.
E.   Develop a list of ongoing preventive maintenance tasks as well as preventive maintenance that ought to be performed.
F.   Determine initial monitoring procedures.
G.   Choose a set of goals for the maintenance function.

10.2.1 Identify Maintenance Functions That Are Energy-related.
Most process maintenance functions will be energy-related, and many building maintenance functions will also have energy-related components. Indeed, even some maintenance operations such as a nighttime cleaning crew may have energy effects if they turn on lights, heating or air-conditioning that would not otherwise be used. The energy man-

agement team must identify those maintenance activities which can or should contribute most to energy efficiency improvement or which result in significant energy use.

10.2.2 Estimate Present Energy-related Maintenance Costs

Estimating energy-related maintenance costs are an important part of the planning process, in part because such estimates can be used to gain management commitment, in part because they can help to judge how much has been accomplished at the end of the first year of the energy management effort. Energy-related maintenance costs will include both labor and supplies. Labor includes the wages/salaries of maintenance personnel or contractors performing maintenance tasks. It is not necessary to try to isolate the energy-related maintenance personnel from other maintenance personnel; instead, the team should estimate the percentage of time each maintenance worker spends on energy-related maintenance.

The labor component of energy-related maintenance costs includes the costs presently used to keep the facility's energy systems functioning. In brief, the systems include:

Boilers
Steam distribution system
Lighting
HVAC and industrial refrigeration
Building envelope
Hot water distribution
Air compressors and the compressed air distribution system
Electrical distribution
Manufacturing processes

Figure 10-2. The Energy-related Systems.

These systems, including manufacturing processes, are directly related to the use of energy.

- The *labor cost* of maintaining them should be considered energy-related labor costs. The people who maintain the manufacturing processes could also properly be included in this category, but the cost of equipment operators should not.

- The *material costs* associated with energy systems are the costs of the systems themselves, together with any replacement parts. The cost of the manufacturing process equipment itself, and its parts, would probably not be included.

•   When maintenance in any of these systems is performed under contract with a vendor, the *cost of the maintenance contract* should also be considered an energy-related maintenance cost.

Indirect costs that are also affected by energy-maintenance actions are those relating to production and productivity. When a piece of production equipment breaks down because of poor preventive maintenance practices, significant costs arise. Productivity costs may result from poorly maintained equipment such as lighting or ventilation systems. Quality-related costs may arise when poor maintenance of lighting systems makes it difficult to see what needs to be done or when poor maintenance of cooling systems results in defective products. Identifying and estimating these costs may be difficult, but the purpose of collecting them is to have an initial basis for estimating the overall potential savings from the maintenance part of the energy management plan.

The result of this analysis is an estimate of the total facility cost that is attributable to energy-related maintenance functions. This provides a baseline for judging the value of the proposed improvements.

10.2.3 Determine the Present Condition of the Major Energy-related Systems

This step includes a maintenance-directed inspection of each of the major energy-related systems shown in Figure 10-2. The operators of each unit of equipment must be included in the inspection process, since they are more familiar with the problems of operating the equipment than are other members of the energy audit team. Some of the more important problems to look for are listed below. The team collecting the data for each system must make a log of maintenance problems affecting that system, together with the date the conditions were noted.

10.2.3.1 Boilers and the Steam Distribution System

For most companies that use boilers, the boilers consume a large fraction of the total energy bill. They are also important as a source of energy for many process operations and for space heating. One rule of thumb is that every year a boiler control system goes without attention costs an additional 10% in boiler energy costs.

Boiler gauges: These monitoring instruments should be operating and readable. All boilers have gauges for operating steam pressure and the level of water in the boiler, and they often have steam and stack temperature gauges as well as those showing set points. If any gauges are not working, the boiler control system should have professional attention quickly. If there is rust in any of the gauges, the boiler tubes and water piping need to be cleaned to remove scale and to restore acceptable heat transfer rates.

Boiler controls: Boiler controls serve several functions. The amount of combustion air must be very carefully regulated to obtain efficient and pollution-free combustion. If too much air is used, much of the energy in the fuel goes up the stack and heats the outside air. If too little air is used, incomplete combustion takes place, the boiler smokes, EPA gets mad, and an explosion hazard may be created. Therefore, the energy-maintenance program should include a flue gas analysis to see whether the boiler is operating with the right amount of outside air. The results of the analysis may show a need to maintain the control systems better. This test is important and should be done, even if a contractor must be called upon to do it.

The level of water in the boiler must also be carefully regulated to prevent overheating of tubes and to provide sufficient room for steam. The amount of fuel coming into the boiler must be carefully controlled. The flue gas temperature must be monitored. And there are other parameters to be controlled, depending upon the type and maker of the boiler. All of these controls must work well for efficient operation. The control system is also the major defense against boiler explosions.

A boiler analyzer can be purchased for approximately $1500 and is a good investment for most facilities. With an in-house boiler analyzer, the maintenance staff can keep the boiler tuned up to best efficiency with monthly boiler analyses. If the boiler tune-ups are done by outside vendors, the boiler analysis and tune-up should be performed at least twice a year.

Steam traps: Steam traps, described in Chapter 8, separate steam from condensate. This function is important because condensate absorbs heat from steam, thus reducing heat transfer, and because returning hot condensate to the boiler can reduce the amount of fuel needed by the boiler. Steam traps can fail open or shut. If a steam trap fails open, steam passes directly into condensate return lines, and its heat is wasted. If it fails shut, then the steam distribution system has an excess burden of condensate, slowing heat transfer and increasing the weight that the distribution system must carry. The energy cost of a steam trap that is either failed open or failed shut is important, as are possible safety consequences.

Steam traps can usually be checked for proper operation visually with a stethoscope or by the trained use of infrared or ultrasonic equipment. Most vendors of boilers and steam traps will train their customers in techniques for checking steam trap operation and use of relevant monitoring equipment.

Steam lines and condensate lines: Two other important parts of the steam distribution system are the steam lines and condensate return lines. Leaks in steam lines can be very expensive and can reduce both the pressure and the amount of delivered steam, so inspecting and repairing such leaks must be a part of any boiler maintenance program. The calculation method for estimating the Btu loss from steam leaks is given in Chapter 8. Note that a leak equivalent to a hole that is 1/2" in diameter on a 600 psig line can cost as much as $25,000 per month. As with steam traps, infrared (IR) scanning devices are useful for detecting steam leaks.

Returned condensate has three or four advantages over outside water—it is usually hot, thus avoiding some of the fuel used to heat it in the steam cycle, it avoids the cost of purchasing or producing replacement water, and it has been treated, so that the cost of treating it to remove impurities that would otherwise cause scaling is avoided. A fourth cost could be the waste disposal or sewerage charge to drain the condensate into the facility waste water system. Checking for condensate leaks is another part of the IR inspection of the facility that should be part of the energy-maintenance program.

Water treatment: The water entering the system must be treated correctly. Otherwise, scale will build up, heat transfer will be decreased, and tubes will fail. Inspection and monitoring of the water treatment system in accordance with boiler manufacturers' specifications is thus an important part of preventive maintenance of a boiler system.

10.2.3.2 Motors

Motors generally account for at least 50 percent of the electricity use of any facility, and 60-70 percent of the electricity used in a manufacturing plant. Motors are found everywhere—from HVAC systems to process drives to conveyor systems. Since they are such an important part of the maintenance program as well as the energy using system, they are given a special section here.

Motors can experience a variety of problems—worn bearings, misalignment (which can cause worn bearings), voltage imbalance in a three-phase circuit, electrical problems within the motor, and inadequate lubrication. An ultrasonic inspection of a motor can uncover bearing problems as well as other problems associated with the motor. A skilled technician can examine the frequency and amplitude of the ultrasonic spectrum of a motor and tell whether the bearings are operating correctly, whether the motor shaft is out of round, or whether the motor shaft is vibrating back and forth.

Worn bearings: Bearings can wear if a ball or roller develops a flat spot, if one of the bearing races has an imperfection (such as having had the imprint of a burr stamped in by accident at the bearing factory), if particles of dirt or metal get into the lubricant and start to grind, or if lubricant has drained out of the bearing. Worn bearings can make noise and in an extreme case can smoke and cause shafts to seize up.

Lubrication: Each motor should have a log showing when it was last lubricated and how often it should be lubricated. If lubrication is not done often enough to keep all rotating parts covered with lubricant, bearing parts can scrape together and cause major problems. If, however, lubrication is done too often or by inexperienced people, the lubricant pressure can force out the bearing plates, or dirt can be carried into the bearing by the lubricant.

Unbalanced voltage: It is critical to have equal voltages in each phase of a three-phase motor. Otherwise, heat builds up, insulation is degraded, and the life of the motor is reduced. Loss of one of the phases is an extreme example of voltage imbalance and can make a motor fail quickly. Phase voltages can easily be checked with a voltmeter.

Power quality: Irregular voltage spikes and other problems with the power supply can also damage motors. If a motor in one particular location has a history of failure, both the location and the motor should be checked by an expert, preferably either a reputable vendor of motor repair services or someone familiar with motor circuit analysis. Power quality can be checked by facility electricians with true RMS electrical measuring instruments.

Alignment: If the shaft of a motor is connected directly to the shaft of a load, the shafts should be in a straight line. Any angular difference can be a source of bearing wear. If the motor is driving a load through a belt, then the sheaves of the load and the motor must be in a straight line. They cannot have a parallel displacement or a slight angle between them. Sheave alignment problems can be checked by using a yardstick, but direct coupled motors and loads usually require a laser alignment device.

Belts and belt tension: The condition and tension of motor belts should also be checked. A motor belt that is frayed or partially broken can break under a severe load. A belt that is too loose will slip and create heat and will not deliver the full power from the motor to the load. A belt that is too tight will cause increased bearing wear and may also cause misalignment.

Frame and anchor: The motor frame and anchoring should also be checked. The frame should not be cracked, and the anchors should be tight. If a large motor has been giving problems, it may be necessary to use a strobe light to check for frame cracks, since these cracks may only appear during one part of a shaft rotation.

Brush maintenance: Brushes on DC motors should also be checked. Brushes are automatically pushed against motor commutators to compensate for wear. They will, however, ultimately wear out. Maintenance Solutions magazine recommends that they be replaced whenever they are less than 1/4" long.

10.2.3.3 Lighting

Lighting levels can be checked with a light meter to see if they meet IESNA standards (See Chapter 5). The maintenance of the lighting system is important for at least four reasons. First, lighting uses electricity. To have a consistent level of lighting, it is necessary to start with an excess amount and to replace lamps or to clean luminaires when the level of lighting falls below acceptable limits. The excess amount of lighting is the amount necessary to account for lamps failing and for luminaires getting dirty between the time the system is installed and the time it is refurbished. If the system never degraded, the initial installation could provide the level of lighting required; otherwise, the amount required is the average lighting level of the system as it is normally maintained. If lighting maintenance is done continually, there is little or no need for excess lighting capacity, and the electric energy directly used by the lighting system will be lower than with periodic maintenance. The cost of maintenance will, however, be higher than if periodic replacement is used. It takes essentially the same amount of energy to operate a lighting system whether the lighting levels have degraded or not.

A second reason for maintaining the lighting system is that lights give off heat, and unnecessary lighting creates heat that must be removed by the HVAC system.

Third, there is some relationship between lighting and productivity. For example, after a new, more efficient lighting system was installed at the Reno NV post office, worker productivity increased substantially. With the new lighting design, postal workers' at the mail sorting machines increased their output by 6%, and decreased the sorting errors to 1 per 1,000. Working in a quieter and more comfortably lit space, employees did their jobs faster and better.* To ensure productivity gains, the level and

---

*Romm, Joseph and Browning, Bill "Greening the Building and the Bottom Line: Increasing Productivity Through Energy-Efficient Design," Rocky Mountain Institute, Fall 1994.

type of lighting must be chosen appropriately to the tasks involved.

Fourth, lighting maintenance issues involve the time between replacement of lamps, the amount of deterioration of luminaires that is acceptable, and whether lamps should be replaced when they fail (spot relamping) or at a time determined by a group relamping schedule. These and other issues involving lighting are carefully treated in Chapter 5. Information about group relamping and other lighting maintenance can also be found at the General Electric web site, www.ge.com. The US Environmental Protection Agency (EPA) Green Lights program also has a wealth of information on lighting and lighting maintenance at its website, www.epa.gov/GCDOAR/maintain.html#planning.

The importance of appropriate lighting levels is stressed here because it sends a clear message about the importance to management of the entire energy cost reduction program. The dedication of a manager to an energy management program is questionable if an employee can walk into an overlighted manager's office.

10.2.3.4 HVAC and Industrial Refrigeration

In performing an energy audit of a facility, the operating times and temperatures of the HVAC system should be examined to find cost reduction opportunities. A systematic way to look for these opportunities is given in Chapter 6. Any reduction in electricity used for lighting, and any improvements to the building envelope will also reduce the cost of HVAC system operation.

In addition to operating improvements, an energy maintenance audit should look for possible maintenance improvements in the following areas: ductwork, heat transfer surfaces, controls, chillers, and motor and fan operation.

Ductwork: Ductwork should be examined to see whether it is damaged or blocked. If it is torn, conditioned air is escaping. If a duct has been constricted, then either more energy than necessary is being used to move the air, or less air is getting to its intended destination. Another possibility is that insulation within the duct has come loose and is blocking the duct. This can be checked by comparing the velocity of air exiting a duct with the velocity measured at a time when the duct was known to be working. Insulation of the duct should also be checked.

Heat transfer surfaces: These include heating coils, cooling coils, and any surfaces such as fins designed to carry off heat. If heat transfer surfaces are fouled by dirt or debris or are blocked by equipment or pallets, heat transfer performance is degraded, and condensing temperatures and pressures will have to increase to accomplish the same amount of heat transfer as originally designed into the system.

The control system: HVAC and refrigeration controls can be optimized for a given operation by choosing appropriate settings for temperatures, defrost cycle times, and other parameters. The method for finding these values is given in Chapter 6. If a control system itself has problems, such as blowing fuses erratically, then maintenance would include calling in a contractor or someone familiar with the details of the control system.

Cooling towers: In many large chillers, condenser hot water is sent to a cooling tower, where it is sprayed into a system of screens where it cools. It is then returned to the HVAC system. Cooling towers need to be inspected to determine whether the spray nozzles are plugged by scale or debris, whether the water return drain is plugged or restricted, and whether the screens are clean. Significant blockage of nozzles, screens, or the drain can cause a decrease in evaporation efficiency and can lead to more pumping than is necessary.

Dampers: Dampers can jam (or be jammed!) open or shut and interfere with the intended operation of the HVAC system. They can also get dirty and lose their ability to close fully. Inspecting dampers should be a routine part of the in-house inspection of the HVAC system.

Filters: Air filters should be checked regularly. A clogged filter requires more fan energy to filter the same amount of air than a clean filter. People checking air filters should look for other problems at the same time.

Motors and fans: Motors in the HVAC system and in industrial refrigeration systems need to be inspected carefully as described above. Fans should also be inspected to locate excess dirt on the fan blades and to assure that the fan is turning in the correct direction.

HVAC and industrial refrigeration systems are complicated. Consequently, companies frequently find that it is cheaper to have a maintenance contract with a reputable vendor than to train in-house personnel to maintain a system.

10.2.3.5 The Building Envelope

Regular inspections of the building envelope should be part of any maintenance program. In inspecting the building envelope, look for places where heated or conditioned air can get in or out, such as broken windows, and cracks in the wall or roof. One of the authors encountered a school building that had a crack or gap measuring 2-1/2 inches wide, from the bottom to the top of two walls and across the ceiling. This crack opened the building to the outside air and was costing the school a large amount of money. This kind of problem can be caused by incompatible seal materials or by shifting of the foundations, but in any case it should be repaired as part of the maintenance plan.

Doors and windows: Maintenance personnel should also be particularly attentive to cracks and gaps around doors and windows. Cracks can be caulked as part of the maintenance plan, and felt seals can be applied around doors and windows to close gaps.

Roof insulation: If the roof is insulated, the insulation should be checked and repaired routinely. Insulation that is damaged or missing allows heat to escape via conduction in a cold climate, and to come into conditioned spaces in a hot climate. The location of the insulation should be analyzed carefully. Insulating the underside of a roof can cause the roof to retain heat and to damage roof covering such as shingles.

10.2.3.6 Hot Water Distribution

When hot water is used for manufacturing processes or for washing, a maintenance audit should look at temperature gauges, leaks, valves, and insulation.

Temperature controls. Gauges can become inoperable from fouling or bumping. Displays can be checked visually to see if temperatures and flow rates are within the proper range; if not, a more detailed examination is called for. (It may also be that the hot water temperature is higher than needed and is causing unnecessary heat loss. This and other problems associated with the hot water distribution system are discussed in Chapter 8.)

Leaks. Maintenance staff should always look for air, steam, and water leaks. Air and steam leaks are very costly, and leaks in the hot water system may create electrical or slipping hazards as well.

Valves. Most hot and cold water systems have valves which control distribution flow rates. These valves can suffer from scale, worn washers, or worn valve seats. It may be possible to repair these problems on site, but in any case, repair will entail isolating the faucet or valve and any processes that depend upon it. This may be inconvenient for a single repair, and this may be the kind of repair that is done with many other repairs at a time when the entire plant is shut down.

Insulation. Insulation around hot water pipes need to be inspected. Exposed insulation is often susceptible to lift truck damage and to damage by routine bumping, as well as suffering water damage from leaks. When damage is observed, it is important to determine whether any damage has occurred to the piping as well as the insulation. The routine examination and repair of such system components is an important part of energy systems maintenance.

10. 2.3.7 Air Compressors and the Compressed Air Distribution System

Compressed air is used as a source of power and/or as a medium for controls in most facilities. It is often used as a medium for cleaning, but this is usually a very costly and inefficient way to clean things. If the energy maintenance audit finds such inappropriate uses of compressed air, these should be noted and alternate methods should be examined.

Several aspects of the compressed air system are unique—the air compressor, moisture in the air, and the likely presence of leaks.

Air compressor. To check an air compressor, the maintenance staff should examine the following components:

(1)   the heat transfer fins should be clean of dirt and free from debris,
(2)   the compressor gaskets should not make noise or show leaks when the compressor is operating,
(3)   the connections should not leak, and
(4)   the dryer that takes the moisture out of the air should be working correctly. The air dryer is very important because moisture in controls can cause the controls and piping to deteriorate.

Leaks. Compressed air leaks can be expensive, so inspecting for leaks should be a routine part of any maintenance program. Leaks should be suspected if an air compressor runs continuously. It is frequently difficult to hear the noise of an air leak over the noise of surrounding machinery, so ultra-sonic leak detectors are often used. Air leaks can sometimes result in pneumatic controls not working properly.

10.2.3.8 Electrical Distribution

The transformers, junction boxes, wiring, and outlets of the electrical distribution system should be checked carefully. Because of the inherent danger associated with electricity, special care should be taken to avoid electric shock whenever this or any other electrical system is being inspected.

Transformers. Transformers contain dielectric fluid, and this must be sampled annually and checked to see that it has not deteriorated. The fins must be inspected and cleaned, because they remove the heat generated within the transformer. (Squirrels like the heat of transformers, and their nests and nuts can become a problem.) The fence or isolation guard must also be maintained for safety reasons. More information on how to inspect transformers and what safety precautions to take can be obtained from a local electric utility.

Junction boxes. These should be examined for shorts, frayed wire, arcing, and burned insulation. One simple inspection is to feel the wall (if this is possible) where wires enter a junction box to see if wire is hot from overloading. The energy management team may decide that additional circuits are needed if the electrical demand has grown substantially since the plant was first put into operation. IR scanning can also be used.

Wiring. Wiring is unsafe if the insulation is worn or cracked, if the wrong kind of wire is used, or if the diameter is too small for the amperage it is carrying. If the insulation is worn or cracked, a short can develop, so the wire should be replaced. Wire prescribed by the National Electrical Code should be used, in the gauges recommended for the amperage needed and with the kind of insulation recommended for the specific usage.

Outlets. All electrical outlets should be inspected to see if they have been damaged. For safety reasons, only a three wire receptacle should be used for a three-wire plug. This can be a major concern but is becoming less so with the advent of battery-powered hand tools.

Grounding. This is a particularly important item, since no one wants to be the connection between a live wire and a ground. Improper grounding can also damage equipment. The grounding must be done correctly, and only a knowledgeable vendor or a licensed electrician should carry out this inspection.

10.2.3.9 Manufacturing Process Equipment

Each piece of manufacturing process equipment, such as a printing press, an injection molding machine, or a corrugator, has its own maintenance requirements. These are usually described in detail in manuals. One good source for information on maintenance needs of manufacturing equipment is the vendor; another source may be a facilities engineer who has been using and maintaining the equipment.

Motors are a starting place for the inspection of manufacturing process equipment (see the section on motor maintenance above). Next the energy management team needs to check the source of motive power and to trace it to see whether any maintenance actions could reduce the energy use.

10.2.3.10 Waste Reduction

Another place to look is at the generation of scrap and waste. Measures that reduce production waste may also reduce energy consumption. For example, if cooling water has to be treated, it should be reused in the facility instead of being discharged to the sewer system. Paper and wood products companies often use their process scrap for fuel in their boilers instead of disposing of it. Because the maintenance staff has responsibility for waste disposal, they should investigate cost-effective waste reduction measures whether or not they save energy.

10.2.4 Determine Maintenance Tasks to Do Immediately

As the maintenance condition of each major system is determined, specific projects should be identified and listed. Tasks that are safety-related must be done immediately, probably before the energy maintenance survey is completed. The other tasks can be separated into those tasks to be done immediately and those to be done later. One way to classify these tasks is with a Pareto analysis.

Pareto analysis is used to separate out the most significant problems. For example, one of the authors used Pareto analysis for a particularly troublesome oven. Figures 10-3 and 10-4 show the analysis results for the oven problem. To develop these analyses, maintenance logs were first examined and put on a spreadsheet with columns for the maintenance task, the clock time, and the amount of production down time before the repair started. These records were then sorted by task and wording changes made so that similar problem descriptions were grouped together. The records were then sorted by repair type so that the most important problems could be identified. (It would also have been possible to calculate and graph the percentage that each repair contributed to the total number of repair or downtime hours.)

In this situation, the Pareto graphs showed that the heating element was causing much of the repair time and much of the downtime. Further investigation showed that the heating elements were frequently damaged by the operators when the furnace was being charged with parts to be annealed. When this was pointed out to the area supervisor, the problem was corrected.

Pareto analysis helps determine what tasks to do first. When the energy cost of each problem has been estimated, it makes sense to give a high priority to those tasks which save the most energy. In addition to Pareto analysis, other considerations may be important in choosing which of the many items on the task list to do first. Such considerations include the impact on production, impacts on quality, and project visibility. These and other intangible factors must be weighed by the energy management team and used to prioritize the maintenance tasks.

The tasks not chosen for immediate implementation should also be examined. Some of them can be combined, and some of them will ultimately be rejected as not significant. This filtering should be done at this point and the list set aside for future reference.

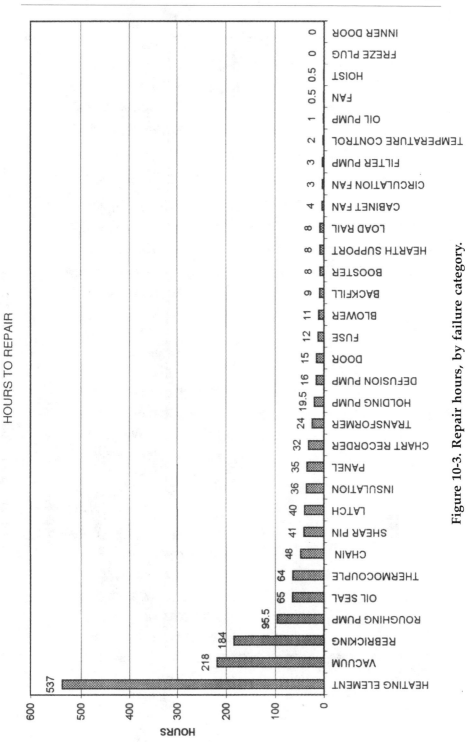

Figure 10-3. Repair hours, by failure category.

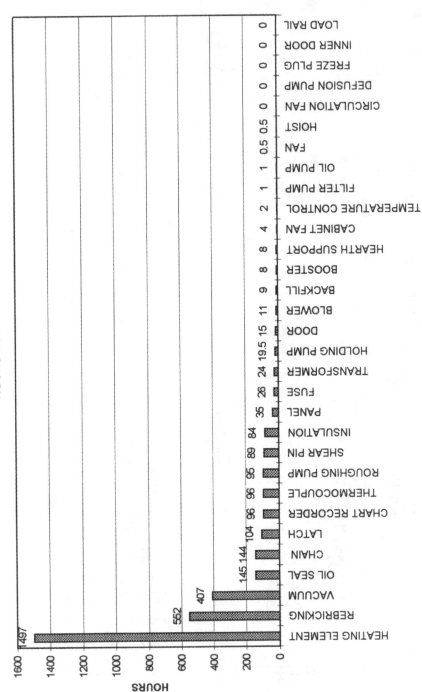

Figure 10-4. Hours down, by failure category.

10.2.5 Develop a List of Ongoing Preventive Maintenance Tasks

These tasks will have been uncovered in the maintenance part of the energy audit. They include the jobs (including monitoring) that are done at regular intervals and the maintenance that is done when the regular monitoring program shows it is needed. A list of maintenance tasks for each piece of equipment should include a description of the skills needed to perform the task, the diagnostic equipment and tools needed, the spare parts required, and the desired intervals between regular preventive maintenance actions. An example list is shown in Figure 10-5. The columns in this list should be structured in the following manner:

Equipment description: This column should clearly identify each machine or energy system component, including its location and function.

Maintenance needed: This column should describe the maintenance task and the kind of equipment needed to perform the task.

Frequency of maintenance action: This will depend upon the amount of use and the environment of the equipment at a given facility. For example, the Pacific Energy Center, part of Pacific Gas and Electric Company, separates motor environments into clean and dry, moderate, and dirty and wet, for example, with a 6-12 month interval between general inspections for a clean and dry environment. For a moderate environment, this is 3-6 months, and for a wet and dirty environment, 1-3 months. (Other guidelines are available from the web site of the Pacific Energy Center and from Pacific Gas and Electric, www.pge.com/pec/index.html www.pge.com/pec/index.html.)

Vendors can provide information about proper maintenance intervals for their equipment and they can often help with questions specific to a facility. They may however tend to be conservative on the theory that too much maintenance will cause fewer problems than too little maintenance. A second source of information is the operating manuals for each piece of equipment. The intervals they recommend may be reliable but they may attempt to meet the needs of everyone using their equipment rather than people in your specific situation, so every facility should tailor the recommended intervals to fit their own needs. A third source of maintenance interval information is the record of the repair information a facility keeps on each piece of equipment. Equipment should be maintained often enough to keep it from breaking down but not so often that the amount of time spent on maintenance is unreasonable.

Skills: A variety of skills are needed to perform maintenance tasks. These range from simple jobs that require a minimal amount of training to jobs that require work done by a certified professional. Some facilities will have maintenance people with the necessary experience. Others may

Equipment description	Maintenance and equipment needed	Frequency	Skills	Mainten-ance Time[A]	Spare Parts Needed	Priority[B]
100 hp motor in NW corner of weaveroom	Check front and rear bearings on motor for hot spots using IR gun	6 months	Trained in IR		Replacement bearings	
	Use yardstick to check sheave alignment.					
	Check motor mountings for tightness	6 months	Basic training			
	Check for wear particles in lubricating oil samples	6 months	Basic training		Lubricating oil	
		Monthly	Basic training			
	Inspect and tighten electrical connections	6 months	Trained or certified electrician			
Lights in main assembly area	Group relamp fluorescent lights in north end of main assembly area	16,000 hrs	Basic training		Lamps and ballasts	
Roof of main warehouse	Visually inspect warehouse roof for leaks	2 months	Basic training		Roofing tar for small leaks	
Steam traps on main boiler return lines	Check steam traps using IR thermography	Weekly	Trained in IR and steam trap operation			
Seals on annealing oven in area A2	Inspect front and rear seals visually	Weekly	Basic training		Sealant	

[A] Maintenance times depend on the specific facility.
[B] Each facility should assign its own priority to the tasks.

Figure 10-5. Maintenance data.

need to hire contractors for non-standard tasks. People can be trained to solve or prevent the most frequently occurring problems, but it is often more cost effective to obtain outside help for the others.

The most sophisticated maintenance skills needed by personnel are those associated with diagnosing the source of problems in malfunctioning equipment and systems. When choosing a contractor for this type of job, chose from those who have had experience with similar repair problems, and make sure they have the right kind of diagnostic equipment. If you have many of the same kinds of problems and find that you will be spending an inordinate amount of money on contractors, it may be a better idea to hire the best person you can find for your staff as a permanent member. Another alternative is to have your own maintenance personnel trained in short courses put on by vendors or by your local technical college. This alternative may come with the added cost of new diagnostic equipment, but the availability of such people on your staff can save a great deal of time and expense on outside contractors.

Repair time: The amount of time needed to repair a problem depends on diagnostic time, the availability of spare parts, the expertise presently on hand to deal with the problem, and many more factors. Diagnostic time depends on the frequency of occurrence of the problem, the amount of experience the assigned technician has in dealing with the particular problem, and the type of diagnostic tools available to help.

Spare parts: Determining what and how many spare parts to keep in inventory is difficult. Some facilities try to implement a Just-In-Time inventory where the only spare parts are those kept on hand for parts that fail, and they are ordered to arrive just before they are needed. This method can be used most successfully when the facility has a planned maintenance program. Planned, or preventive, maintenance avoids having spare parts that tie up large amounts of capital and space, that become obsolete, or that degrade due to rust. If a company is close to a spare parts source, it may be able to rely on the source for inventoried items. The biggest problem is what to do when expensive parts break down unexpectedly. The best approach is to analyze the situation in advance, keep a supply of the most critical items that are known to fail and develop a contingency plan for dealing with the rest.

Priority: Once all the other columns are filled in, the energy management maintenance team should set priorities for each maintenance task. This should reflect information from the Pareto analysis of frequency of problems. It should also relate to the critical nature of a piece of equipment to the overall production effort.

10.2.6. Develop Initial Monitoring Procedures

As the final step in the planning process, it is essential to develop means of monitoring maintenance and its results. This monitoring requires maintenance logs, the purchase and installation of semi-permanent monitoring equipment, and the purchase of portable equipment together with training in its use.

10.2.6.1 Maintenance Logs

At a minimum, these must show the information given in Figure 10-6. Other information can be added to customize the form for a particular company. The form could be filled in with text or with number codes depending on a company's preference.

Equipment maintained	Date	Technician	Equipment condition	Repair time

Figure 10-6. Sample Maintenance Log.

By putting this information on a spreadsheet, it is possible to use the data later in a Pareto analysis to collect together all the maintenance information on one computer in order to analyze maintenance intervals and spare parts needs.

Unfortunately, many maintenance people do not like spread sheets. Although the information is important, they will not collect it if collecting it is more trouble than it appears to be worth. There is a solution—a hand held computer. A computerized system should have the following characteristics:

• It is easy to use by the technician in the field. The technician can record the conditions he/she has found and the actions taken to remedy the problem, without writing anything down on paper.

- A permanent record is kept for each machine or area maintained, and this record can be accessed in the field with a minimum of effort.

- The equipment with the records is lightweight and pocket-portable.

- Data from monitoring equipment can be incorporated into the equipment database easily.

- The system should have a bar-code reader so that the facility can mark all of its equipment with bar-codes. The technician can scan the bar-code on a piece of equipment and call up all past information on the equipment immediately. If bar-codes are used with spare parts, they will also help with inventory control.

- All equipment and labor data are available for more sophisticated analysis at a central site, and the analysis results can be immediately available for use in the field.

The technology for this system* includes

1. A computer capable of handling a large database and many requests for service quickly. Such computers are readily available and not expensive.

2. A portable hand-held transmitter/receiver capable of displaying information sent from the main computer and of sending information back to it in a form that the main computer could understand and analyze. Such equipment is also available now.

3. The software and hardware to tie the computer and the field units together.

The advantages of such a system are many. First, the information is likely to be more accurate and more complete than information from paper-based systems. Second, it is possible to tailor the analysis of a particular machine problem to the particular machine, knowing the history of repairs on that machine. Third, good data can be kept for use in

---

*A system like this has been developed by Datastream (Greenville, SC) (web address: www. datastream.com) and is probably available from other vendors.

spare parts inventory calculations so that this element of repair delay is eliminated. Finally, it is easy for the technicians in the field to use, so its use is more likely than for a more cumbersome system or one based on paper.

10.2.6.2 Semi-permanent Monitoring Equipment

Until recently, this has included electrical submetering and the use of portable recording ammeters. In the last two years, inexpensive battery-powered instruments that record measurements (temperature, relative humidity, light intensity, on/off, open/closed, voltage and events) over time. These data loggers are small, battery-powered devices that are equipped with a microprocessor, data storage and sensor.* Most data loggers utilize turn-key software on a personal computer to initiate the logger and view the collected data.

Using recording equipment gives a baseline for comparison and can be used to show the true amount of energy savings due to the maintenance of a particular unit or area.

10.2.6.3 Portable Equipment

One part of the planning process is the identification of what portable equipment is needed at the facility. This can include equipment for determining temperatures, motor vibrations, voltage and current, light intensity, and stack gas emissions. Monitoring equipment is listed in Figure 10-7. Only enough equipment needs to be purchased to monitor the energy usage associated with the most important production tasks at the facility.

10.2.7 Choose Energy Maintenance Goals

Energy maintenance goals should be established as part of the energy management plan. If the energy management plan is designed with an overall energy savings goal in mind, the maintenance goal might be to provide some percentage of that goal through maintenance activities. Example goals are shown below.

The choice of tasks, the purchase of equipment to monitor energy performance, and the choice of goals c=ompletes the planning process. The first action step (see Figure 10-1) is the implementation of the chosen maintenance tasks.

---

*Such equipment is available from the Onset Computer Corporation. Their web address is www.onsetcomp.com.

Equipment	Function
Digital thermometer with surface and probe attachments, 50-2000°F.	Check stack, process, and room temperatures
IR thermography equipment	1. Check roofs and walls for heat loss areas (missing insulation, unusual sources of heat) 2. Check motor bearings and windings 3. Inspect steam traps 4. Inspect steam lines for leaks, including buried lines 5. Examine oven refractories and seals
Vane anemometer	1. Determine velocity of air from ducts 2. Estimate energy losses and through gaps in building envelope
Pitot tube anemometer	1. Measure flue gas velocities 2. Measure velocity of fluids in ducts
Industrial light meter	1. Measure incident light at workplaces 2. Measure general light conditions
Combustion gas analyzer	1. Measure composition of flue gas to determine boiler efficiency
Vibration analysis equipment	1. Develop baseline spectrum of vibration frequencies for motors 2. Analyze motors for imbalance, bearing faults, improper assembly of parts, misalignment, or improper mounting
Volt-ohm-ammeter	1. Check voltages on all legs of three-phase equipment to analyze for efficiency loss and overheating 2. Check equipment for shorts
Recording ammeter	1. Determine peak kW usage times and amounts 2. Determine routine usage amounts for correlation with production times and amounts
Oil analysis equipment	1. Monitor lubricant condition for water or contaminants 2. Monitor machine condition for component wear prediction

Figure 10-7. Instruments for Maintenance Inspection and Continuous Monitoring.

Goal	Amount
Reduce energy consumption due to poor maintenance	50%
Perform Pareto analysis of plant motor horsepower	90% of total motor hp in plant
Reduce lighting energy costs through group relamping programs	50%
Reduce number and time of unplanned equipment breakdowns	50%
Ensure that boiler controls show correct readings	100%
Ensure that steam traps are functioning correctly	100%
Perform flue gas analyses on each boiler each year	2 or more
Reduce or eliminate the five worst maintenance problems from last year's Pareto analysis	

Figure 10-8. Sample Yearly Goals for an Energy Maintenance Management Program.

10.3 MONITORING PROGRESS

During a preset time period, usually six months, the energy usage is recorded, along with any production problems caused or avoided by the maintenance. Energy costs are noted and are compared with the corresponding costs for a comparable time period. Any complaints are recorded. All the maintenance actions are logged. Enough data are collected that the progress toward the chosen goals can be estimated.

10.4 ANALYSIS

At the end of the 6-month period, the data collected during the monitoring phase are analyzed. The analysis asks four questions:

• **What did we do right, and how can we do it again?**
• **What did we do wrong, and how can we avoid doing it again?**

- **What results have we achieved?**
- **What new actions should we undertake?**

To answer these questions, it is necessary to determine the savings and other accomplishments associated with the maintenance measures that were implemented. Progress toward the goals chosen in the planning process should be measured as closely as possible, and the goals themselves should be refined. Hopefully, there will be some conspicuous successes and few failures. The analysis phase is an attempt to learn from both.

It is also important to compare the projected savings with the actual savings, and to refine the costing methodology to take the differences into account.

In determining the new actions to take, the initial list of maintenance tasks should be reexamined. Any suggestions of new tasks should be included. Then, estimates of savings should be made for each task. It is important at this point to take another look at the maintenance being performed in each area and to determine whether there are new, more cost effective methods of achieving the desired result. Is new equipment available that would make the maintenance task faster or better? Have new maintenance techniques or methods been developed in any area?

Monitoring should also be analyzed during this phase. Is there too much, or too little monitoring? What else should be monitored? Did the results of the data analysis justify the pains taken in its collection? How did the projected savings compare with the actual savings, and can the costing methodology be refined to take the differences into account?

Once these questions have been answered, it is possible to go to the next, or action, phase.

10.5 ACTION

The monitoring phase yields a new list of maintenance tasks and a refined method of estimating their benefits. In the action phase, this list is used as the basis for new tasks to be carried out. As before, deadlines are established, people and budgets are assigned to each task, and some way of monitoring the effectiveness of each task is chosen. The next phases proceed in the same order: action, monitoring, analysis. The ultimate goal of the continuous improvement approach to the maintenance program is to develop and sustain a satisfactory level of energy systems maintenance as a standard part of plant operating policy.

10.6 SUMMARY

In this chapter we have briefly discussed the maintenance component of an energy management program. An effective maintenance program should help reduce energy costs and/or increase the amount of production available from a given amount of energy. Serious thought should be given to the people involved in maintenance: how many, who, and how they should be trained. Sufficient support equipment must be provided for diagnosis and for continued monitoring. Finally, some attention should be given to finding the latest technology that can be used to get information to the technicians in the field.

Bibliography

Capehart, Barney, Editor, *Encyclopedia of Energy Engineering and Technology*," Taylor and Francis/CRC Publishing Company, Boca Raton, FL, July, 2007.

Reliability-Based Maintenance, by CSI, 835 Innovation Drive, Knoxville, TN 37932-2470.

The Datastream web site, www.datastream.com, August, 1999.

The General Electric Corporation web site, www.ge.com, August, 1999.

The MP2 Enterprise System, by Datastream Systems, Inc., 50 Datastream Plaza, Greenville, SC 29605.

The Onset Computer Corporation web site, www.onsetcomp.com, August, 1999.

The Pacific Energy Center web site, www.pge.com/pec/index.html, August, 1999.

Chapter 11

Insulation

11.0 INTRODUCTION

Unwanted heat loss or gain through the walls or roofs of buildings, and heat loss from the pipes, tanks and other equipment in buildings or plants can significantly increase energy use and energy costs. Thermal insulation plays an important role in reducing these energy costs in many situations. Good engineering design of insulation systems will reduce undesirable heat loss or gain by at least 90% in most applications and will often improve environmental conditions at the same time. Consequently, it is highly beneficial to understand insulation theory and applications, and to recognize when cost-effective insulation EMOs can be implemented.

In the first part of this chapter we discuss the theory behind the calculations involving heat transfer and its relation to the amounts or types of insulation used. The two basic areas of insulation applications are buildings and process equipment. In the rest of the chapter we discuss the various insulation types available for buildings and for process applications, and show examples of energy cost savings for several typical insulation EMOs.

Solar energy causes substantial heat gain in buildings, and even some heat gain in process equipment. Unwanted solar heat gains can often be controlled by suitable insulation, but sometimes the solar heat gain is a benefit in reducing use of fuel or electricity for heating. Usually a combination of insulation and architectural features such as overhangs or other blocking systems, films, and sometimes landscaping, is used to control this solar heat gain. The insulation aspects are covered in this chapter, but the direct use of solar heat in buildings and processes will be covered in Chapter 13 on alternative energy sources.

11.1 INSULATION THEORY

In our discussion of the theory of insulation, we cover the topics of heat transfer, thermal conductivity, and thermal resistance. The discussion of heat transfer explains the three ways that heat is moved into or out of a material. Thermal conductivity relates to the material itself and the ease with which heat moves through the material. Thermal resistance is the inverse of thermal conductivity, and relates to how well a given material will block or retard the movement of heat through it. We also develop the heat transfer calculations for flat surfaces and for cylindrical surfaces such as pipes.

11.1.1 Heat Transfer

There are three basic modes of heat transfer. They are conduction, convection, and radiation. These modes are defined as follows:

* *Conduction* is the transfer of heat from a hot side to a cooler side through a dividing medium. The hot side heats the molecules in the dividing medium and causes them to move rapidly, heating the adjacent molecules until the cool side is heated. The transfer stops when the temperature of the hot side equals that of the cool side.

* *Convection* is the transfer of heat between a moving liquid or gas and some conducting surface. Usually the heated fluid rises, causing cooler fluid to come in contact with the conducting surface, which is then heated and rises, etc.

* *Radiation* is the transfer of heat based on the properties of electromagnetic waves so no transfer medium is necessary. For example, the sun heats by radiation.

11.1.2 Thermal Conductivity

The *thermal conductivity* (K) of a material is a physical property that describes the ability of the material to conduct heat. K is measured by the amount of energy (Btus) per hour that can pass through one square foot of surface one inch thick* for a one-degree F temperature difference between the two environments being separated. The units in which K is measured are $(Btu \bullet in)/(ft^2 \bullet h \bullet °F)$. Tables 11-1 and 11-2 present K values for various materials at room temperatures. Table 11-1 contains conductivity values

---

*Be careful; sometimes tables show K values for one-foot thicknesses.

for materials that are good heat conductors, and Table 11-2 has values for good insulating materials. Thermal conductivity will vary with temperature, which can be important for process applications, as will be seen. For K values of common insulating materials at varying temperatures, see Figure 11-1.

The rate of heat transfer is directly proportional to the temperature difference and the thermal conductivity as shown in Equation 11-1:

$$Q \propto K \, \Delta t \qquad\qquad (11\text{-}1)$$

where Q = rate of heat transfer per ft² of surface
          Δt = temperature difference
          K = thermal conductivity

Table 11-1. Thermal conductivity values for various materials at room temperature.

Description	$K \dfrac{\text{Btu} \cdot \text{in.}}{\text{ft}^2 \cdot \text{h} \cdot {}^\circ\text{F}}$
Aluminum (alloy 1100)	1536
Aluminum bronze (76% Cu, 22% Zn, 2% Al)	696
Brass:	
Red (85% Cu, 15% Zn)	1044
Yellow (65% Cu, 35% Zn)	828
Bronze	204
Copper (electrolytic)	2724
Gold	2064
Iron:	
Cast	331
Wrought	418.8
Nickel	412.8
Platinum	478.8
Silver	2940
Steel (mild)	314.4
Zinc:	
Cast	780
Hot-rolled	744

Source: Albert Thumann, *The Plant Engineer's and Manager's Guide to Energy Conservation,* ©1977. Reprinted with permission of Van Nostrand Reinhold Co., New York.

11.1.3 Thermal resistance

The insulating property of a material is generally specified in terms of the *thermal resistance* (R) of the material, also called the R-value. Thermal resistance is related to the K value as follows:

$$R = \frac{d}{K} \tag{11-2}$$

where d = thickness of material
        K = thermal conductivity

Table 11-2.

Material	Description	Conductivity K^a	Conductance $C^{b,c}$
Building boards	Asbestos-cement board Gypsum or plaster board...1/2 in. Plywood Plywood...3/4 in. Sheathing (impregnated or coated) Sheathing (impregnated or coated) 25/32 in. Wood fiber—hardboard type	4.0 0.80 0.38 1.40	 2.25 1.07 0.49

Insulating materials	Blanket and Batt:
	   Mineral wool fibers (rock, slag, or glass)
	   Wood fiber
	
	Boards and slabs:
	   Cellular glass
	   Corkboard
	   Glass fiber
	   Insulating roof deck...2 in.
		
		0.27
		0.25
		
		0.39
		0.27
		0.25
		
			0.18

Masonry materials	Loose fill:
	   Mineral wool (glass, slag, or rock)
	   Vermiculite (expanded)
	
	Concrete:
	   Cement mortar
	   Lightweight aggregates, expanded shale, clay,
	      slate, slags; cinder; pumice; perlite; vermiculite
	   Sand and gravel or stone aggregate
	   Stucco
	
	Brick, tile, block, and stone:
	   Brick, common
	   Brick, face
	   Tile, hollow clay, 1 cell deep, 4 in.
	   Tile, hollow clay, 2 cells, 6 in.
		
		0.27
		0.46
		
		5.0
		
		1.7
		12.0
		5.0
		
		5.0
		9.0
		
			0.90
			0.54

Table 11-2. (*Continued*)

Material	Description	Conductivity K^a	Conductance $C^{b,c}$

Masonry materials (cont'd)	Block, concrete, 3 oval core:
	    Sand & gravel aggregate..............4 in.
	    Sand & gravel aggregate..............6 in.
	    Cinder aggregate...........................4 in.
	    Cinder aggregate...........................6 in.
	    Stone, lime or sand
		
		12.50
			
			1.40
			0.90
			0.90
			0.58

Plastering materials	Cement plaster, sand aggregate
	Gypsum plaster:
	    Lightweight aggregate...1/2 in.
	    Lt. wt. agg. on metal lath...3/4 in.
	    Perlite aggregate
	    Sand aggregate
	    Sand aggregate on metal lath 3/4 in.
	    Vermiculite aggregate
		5.0
		
		1.5
		5.6
		
		1.7
			
			3.12
			2.13
			
			7.70

Roofing	Asphalt roll roofing Built-up roofing...3/8 in.		6.50 3.00
Siding materials	Asbestos-cement, 1/4 in. lapped Asphalt insulating (1/2 in. board) Wood, bevel, 1/2 × 6, lapped		4.76 0.69 1.23
Woods	Maple, oak, and similar hardwoods Fir, pine, and similar softwoods Fir, pine & sim. softwoods 25/32 in.	1.10 0.80	 1.02

[c]Same as U value.
[a]Conductivity given in Btu•in./h•ft^2•°F
[b]Conductance given in Btu/h•ft^2•°F
Source: Extracted with permission from *ASHRAE Guide and Data Book*, 1965. Reprinted with permission from the Trane Co., La Crosse, WI.

To determine the thermal resistance of something composed of several materials, the total thermal resistance (R_{total}) is simply calculated as the sum of the individual components:

$$R_{total} = R_1 + R_2 + ... + R_N \qquad (11\text{-}3)$$

where R_i = the thermal resistance of the ith component,
        i = 1, 2, 3,..., N.

11.1.4 Conductance

The insulating property of a material is often measured in terms of conductance rather than resistance. *Conductance* (U) is the reciprocal of resistance.

Figure 11-1. Thermal conductivities at varying temperatures.

$$U = \frac{1}{R} \tag{11-4}$$

The overall conductance of a total structure is

$$U_{\text{TOTAL}} = \frac{d}{R_{\text{TOTAL}}} \tag{11-5}$$

It is important to note that while the resistances are additive, the conductances are not.

11.1.5 Heat Transfer Calculations

Using the information developed so far, the final heat transfer equations can be constructed as follows:

$$Q = \frac{1}{R_1 + R_2 \dots R_N} \Delta t \qquad\qquad (11\text{-}6)$$

$$Q = \frac{1}{R_{total}} \Delta t \qquad\qquad (11\text{-}7)$$

$$Q = U\Delta t \qquad\qquad (11\text{-}8)$$

$$Q_{total} = UA\Delta t \qquad\qquad (11\text{-}9)$$

where Q_{total} = rate of heat transfer for total surface area involved
        A = area of heat transfer surface

11.1.5.1 Heat Transfer Calculations for Flat Surfaces

The *surface film coefficient* is the amount of heat transferred from a surface to air or from air to a surface per square foot of surface for each degree of temperature difference. Surface film coefficients are usually specified in terms of the surface resistance, as shown in Table 11-3.

Example 11-1:

Assume we have a 1/2-inch-thick uninsulated mild steel tank storing a hot fluid as shown in Figure 11-2. (The K value for mild steel is found in Table 11-1.) The fluid is heated to 200°F, while ambient air is 70°F.

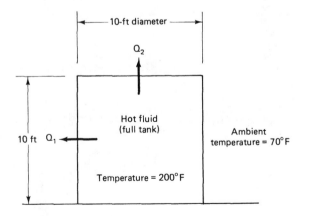

Figure 11-2. Tank storage of hot fluid.

What is the heat loss for the uninsulated tank? Ignore the heat loss to the ground. (Actually, transfer to the ground is a rather significant heat loss, but we will ignore it here.) The tank is 10 feet in diameter and 10 feet tall. Although this cylindrical tank is not truly a flat surface, it is large enough that it can be reasonably approximated as a flat surface. Smaller diameter cylinders such as pipes must be treated in a different way which is explained in the next section.

There are two primary sources of heat loss shown in Figure 11-2. Q_1 is the loss through the walls of the tank, and Q_2 is the loss through the roof. Although the surface film coefficient will vary somewhat for vertical heat flow versus horizontal heat flow, we will ignore that here, and assume that $R_{surface\ coefficient}$ is equal for the roof and walls:

$$R_{total} = R_{tank} + R_{surface\ coefficient}$$

The thermal resistance of the tank itself is quite small and is usually ignored, but we include it below for demonstration purposes; note how small it is compared to the surface resistance:

$$R_{tank} = \frac{d}{K} = \frac{.5}{314.4} = .0016 \; \frac{h \bullet ft^2 \bullet F}{Btu}$$

From Table 11-3, assuming that the surface is dull metal and the surface temperature is 120°F*

$$R_{surface\ coefficient} = .50$$

$$U = \frac{1}{R_{surface} + R_{tank}} = \frac{1}{.5016} = 1.99 \; \frac{Btu}{ft^2 \bullet °F \bullet h}$$

$$A = area = \pi DH + \pi r^2 = \pi(10\ ft)(10ft) + \pi(5\ ft)^2 = 392.7\ ft^2$$

$$Q_{total} = UA\Delta t = \frac{1.99\ Btu}{ft^2 \bullet °F \bullet h} \; (392.7\ ft^2)(200°F - 70°F)$$

$$= 101,591 \; \frac{Btu}{h}$$

---

*We will show you how to check this assumption shortly.

To see the impact that insulation can have on these calculations, assume that aluminum-jacketed fiberglass insulation 1 inch thick is added so that it covers the tank. The new heat loss would be:

$$R_{insulation} = \frac{d}{K} = \frac{1}{.25} = 4$$

Using Figure 11-1, K is found for fiberglass at a mean temperature of around 40°F.

$$U = \frac{1}{R_{tank} + R_{insulation} + R_{surface}} = \frac{1}{.0016 + 4 + .88}$$

$R_{surface}$ is found for aluminum, assuming a surface temperature of around 95°F.

$$U = \frac{1}{4.8816} = .205$$

$$Q_{total} = UA\Delta t = \frac{.205 \text{ Btu}}{\text{ft}^2 \bullet \text{°F}} (392.7 \text{ ft}^2)(130\text{°F}) = 10,465 \frac{\text{Btu}}{\text{h}}$$

$$\text{savings} = 101,796 - 10,465 = 91,331$$

$$\% \text{ savings} = \frac{91,331}{101,796} = 89.7\%$$

Note that this is almost a 90% saving for only 1 inch of insulation. For 2 inches of insulation, the heat loss woud be:

$$Q_{total} = UA\Delta t = \frac{1}{8.8816} (392.7)(130) = 5748 \frac{\text{Btu}}{\text{h}}$$

$$\text{savings} = 101,796 - 5748 = 96,048 \frac{\text{Btu}}{\text{h}}$$

$$\% \text{ savings} = \frac{96,048}{101,796} = 94.4\%$$

Table 11-3. Surface film coefficients, R_s Values[a] (h•ft^2•°F/Btu)

$t_s - t_a$ (°F)[b]	Still air		
	Plain fabric, dull metal, $\varepsilon = .95$	Aluminum, $\varepsilon = .2$	Stainless steel, $\varepsilon = .4$
10	.53	.90	.81
25	.52	.88	.79
50	.50	.86	.76
75	.48	.84	.75
100	.46	.80	.72
	With wind velocities		
Wind velocity (mph)			
5	.35	.41	.40
10	.30	.35	.34
20	.24	.28	.27

[a]For heat loss calculations, the effect of R_s is small compared to R_1, so the accuracy of R_s is not critical. For surface temperature calculations, R_s is the controlling factor and is therefore quite critical. The values presented are commonly used values for piping and flat surfaces.
[b]Note that t_s = surface temperature. Knowing the surface temperature requires measurement or calculation through the concept of thermal equilibrium, which will be discussed.
Source: Courtesy of Manville Corp.

Example 11-1 demonstrates a characteristic of the cost-effectiveness of insulation. The first increment of insulation thickness yields by far the largest energy savings. Each additional increment of thickness increases the energy savings, but at a rapidly decreasing rate. At some point, then, it becomes uneconomical to add any additional insulation. This will be demonstrated later in this chapter.

1.1.5.2 Heat Transfer Calculations for Pipes

For determining the effects of insulating pipes, the calculations become a bit more difficult because the heat flow is in a radial direction away from the pipe through the insulation to a larger surface area. The effect of dispersing the heat over a larger surface is to increase the insulation thickness. This effect is manifested in the calculations through a concept known as *equivalent thickness* d′:

$$d' = r_2 \ln \frac{r_2}{r_1} \qquad (11\text{-}10)$$

where   d′ = equivalent thickness
          r_1 = outside radius of pipe
          r_2 = outside radius of pipe plus insulation

Table 11-4 shows the outside radius of a pipe if the nominal size—or inside diameter—is given.

Table 11-4. Nominal pipe size vs. outside radius.

Nominal pipe size (inches)	Outside radius (inches)	Nominal pipe size (inches)	Outside radius (inches)
1/2	.420	7	3.813
3/4	.525	8	4.313
1	.658	9	4.813
1-1/4	.830	10	5.375
1-1/2	.950	11	5.875
2	1.188	12	6.375
2-1/2	1.438	14	7.000
3	1.750	16	8.000
3-1/2	2.000	18	9.000
4	2.250	20	10.000
4-1/2	2.500	24	12.000
5	2.781	30	15.000
6	3.313		

Example 11-2:

A 6-inch nominal size pipe (meaning a pipe with a six-inch inside diameter) with 2 inches of insulation would have an equivalent thickness of inslation of

$$d' = r_2 \ln \frac{r_2}{r_1}$$

where r_1 = 3.313 from Table 11-4
         r_2 = 3.313 + 2.00 = 5.313

$$d' = 5.313 \ln \frac{5.313}{3.313} = 2.51$$

The equivalent thickness can now be used in the normal equations to determine the effective R value, etc.

Example 11-3:
The same 6-inch pipe carrying fluid at 200°F and insulated by 2 inches of aluminum-jacketed fiberglass in a 70°F ambient area would have a heat loss of:

$$R_{insulation} = \frac{d'}{K} = \frac{2.51}{.25} = 10.04$$

$$Q_{total} = UA\Delta t$$

$$= \left(\frac{1}{10.04 + .88}\right)A\,(200 - 70)$$

$$= \frac{11.9\ \text{Btu}}{\text{ft}^2 \cdot \text{h}}$$

Each linear foot of the 6-inch uninsulated pipe has a surface area of:

$$\frac{[3.313(2)\ \text{in.}]\,\pi(1\ \text{ft})}{12\ \text{in/ft}} = 1.735\ \text{ft}^2$$

To determine the savings from insulating the pipe we must calculate the heat loss for the uninsulated pipe. The heat loss per one-foot length of insulated pipe is:

$$\frac{(11.9\ \text{Btu})}{\text{ft}^2 \cdot \text{h}}\ \frac{1.735\ \text{ft}^2}{\text{ft}} = 20.646\ \frac{\text{Btu}}{\text{ft} \cdot \text{h}}$$

11.1.5.3 Thermal equilibrium and its applications

The concept of thermal equilibrium is important in many types of calculations including the checking of surface temperature assumptions demonstrated earlier. Thermal equilibrium simply says that the total heat flow through a system is equal to the heat flow through any part of the system. For example, a system as shown in Figure 11-3 consists of a wall, two layers of insulation, and an outside surface film.

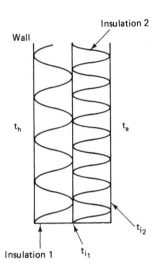

Wall

Insulation 2

t_h

t_a

t_{i2}

Insulation 1 t_{i1}

Figure 11-3. Insulation system.

Thermal equilibrium states that*

$$Q_{total} = \frac{t_h - t_a}{R_{i1} + R_{i2} + R_s} = \frac{t_h - t_{i1}}{R_{i1}} = \frac{t_{i1} - t_{i2}}{R_{i2}} = \frac{t_{i2} - t_a}{R_s}$$

$$= \frac{t_{i1} - t_a}{R_{i2} + R_s} \tag{11-11}$$

By using these equations, any surface temperature or other unknown quantity can be calculated if the heat flow and all other quantities are known.

---

*For purposes here, the inside surface film resistance is assumed to be zero. If not, there would be another set of expressions involving R_s.

Example 11-4:
    The tank example demonstrated in Figure 11-2 assumed a surface temperature of 95°F. If we know the total heat loss, we can check this assumption by setting the total heat loss equal to one of the expressions shown in Equation 11-11 That is:

$$\frac{Q_{total}}{ft^2} = \left(\frac{10{,}500\ Btu}{h}\right)\left(\frac{1}{392.7\ ft^2}\right) = \frac{t_h - t_s}{R_{tank} + R_{insul.}}$$

$$= \frac{200 - t_s}{.0016 + 4}$$

$$t_s = 93.0°F$$

As a furher check:

$$\frac{Q_{total}}{ft^2} = \left(\frac{t_s - t_a}{R_s}\right)\left(392.7\ ft^2\right)$$

$$= \left(\frac{93.0°F - 70°F}{.88}\right) = \left(392.7\ ft^2\right)$$

$$= 10{,}300\ \frac{Btu}{h}$$

We assumed that the surface temperature was 95°F. This temperature results in a total heat loss of 10,300 Btu/hour. This value is close enough to the original value of 10,500 Btu/hour to reasonably validate our assumption.

These calculations can be used for many different objectives. For example, suppose the objective of insulation were to protect personnel and the surface temperature could not be higher than, say, 110°F. The energy manager could take the appropriate expressions as in Equation 11-11 and set the surface temperature equal to 110°F. By back-calculating, the required amount of insulation thickness could be determined.
    Suppose the purpose were to prevent condensation from forming on cold pipes. Then the energy manager could determine the dew point

of ambient air, set the surface temperature equal to that, and back-solve again for the insulation thickness.

In both cases, the energy manager would probably round off to the next highest insulation thickness or even increment up one thickness as a safety measure.

For energy management purposes, it is usually necessary to calculate the heat loss or gain over a year instead of for only an hour as already shown. To do this, it is necessary to sum the hourly heat losses over the number of hours in a year.

Example 11-5:

Suppose the pipe in Example 11-3 carried hot fluid 24 hours/day for 365 days/year. Further, assume that the unit generating the hot fluid is 80 percent efficient using natural gas costing $5.00/10^6 Btu. Find the energy cost savings per foot of pipe from insulating it with 2 inches of aluminum jacketed fiberglass. Assume the pipe is made from mild steel.

The total heat loss for a year for the uninsulated system is:

$$Q_{total} = UA\Delta t \left(\text{assume } t_s \approx 120°F \right)$$

$$= \left(\frac{1}{.50} \right)(1)(130)$$

$$= 260 \left(\frac{Btu}{ft^2 \Sigma \, h} \right)$$

$$Q_{total(\$)} = \left(\frac{260 \, Btu}{ft^2 \Sigma \, h} \right) \left(\frac{1}{.8} \right) \left(\frac{8760 \, h}{year} \right)$$

$$= \$14.24/ft^2$$

Savings for insulating the pipe would be

$$\left(\frac{260 \, Btu}{ft^2 \cdot h} - \frac{11.9 \, Btu}{ft^2 \cdot h} \right) \left(\frac{1}{.8} \right) \left(\frac{\$5.00}{10^6 Btu} \right) \left(\frac{8760 \, h}{year} \right) = \$13.58/ft^2$$

Savings per linear foot of pipe would be

$$\left(\frac{\$13.58}{ft^2}\right)\left(\frac{1.735ft^2}{ft}\right) = \$23.57/ft$$

Consequently, insulation can be a real money saver.

In dealing with systems exposed to outside conditions, it is often helpful to use the degree day or degree hour concept developed in Chapter Two. Degree days are used in the example below.

Example 11-6:
Assume a building wall has an R value of 3. The temperature inside is kept at 65°F during the winter and 75°F during the summer. The plant operates 365 days/year, 24 hours/day. Assuming a heating plant efficiency of 0.80 and a cooling coefficient of performance (COP) of 2.5, what is the cost of energy loss through the wall? Electricity costs $.08/kWh and gas $6.00/10^6 Btu. The plant experiences 4000°F heating days and 2000°F cooling days. The total wall area is 100 ft².

$$\frac{\text{heat loss (\$)}}{yr} = \left(\frac{4000°F \text{ days}}{year}\right)\left(\frac{1 \text{ Btu}}{3 \text{ ft}^2 \cdot h \cdot °F}\right)\left(\frac{24 \text{ h}}{day}\right)(100 \text{ ft}^2)$$

$$\times \left(\frac{1}{.8}\right)\left(\frac{\$6.00}{10^6 Btu}\right)$$

$$= \$24.00/yr$$

$$\frac{\text{cooling gain (\$)}}{yr} = \left(\frac{2000°F \text{ days}}{year}\right)\left(\frac{1 \text{ Btu}}{3 \text{ ft}^2 \cdot h \cdot °F}\right)\left(\frac{24 \text{ h}}{day}\right)(100 \text{ ft}^2)$$

$$\times \left(\frac{1}{2.5}\right)\left(\frac{kWh}{3412 \text{ Btu}}\right)\left(\frac{\$.08}{kWh}\right)$$

$$= \$15.00/year$$

total cost of energy loss per year = ($15.00 + $24.00)/yr = $39.00/year

If the plant described above had a night setback program, then you would have to use degree hours calculated through manipulation of the base corresponding to the hours of setback. For example, if the temperature were setback to 55°F, the heat loss would be calculated using heating degree hours determined from a 55°F base for the respective hours of setback rather than the standard 65°F base for all heating hours.

11.2 INSULATION TYPE

Before an energy manager can select the proper type of insulation for a particular application, he or she must know the properties of various kinds of insulating materials. In this section, the properties of insulation are discussed first, followed by a discussion of the different types of insulation.

11.2.1 Properties of materials used for insulation

Some of the more important properties of materials that would be used to provide insulation properties include the following:

- *Cell structure.* Cell structures are either open or closed. A closed cell is relatively impervious to moisture, especially in a moderate environment, so insulation with a closed cell structure may not need any additional moisture barrier. Open cells pass moisture freely and therefore probably require vapor barriers. For extremely cold applications where a lot of condensation occurs, a vapor barrier is probably required regardless of cell structure.

- *Temperature use.* Different insulating materials react to extreme temperatures in different ways. In some cases, high temperatures can destroy the binders and render the insulation useless. All insulation materials have temperature ranges for which they are recommended. Usually the restriction occurs at the warmest end rather than the coldest end.

- *Thermal conductivity (K).* As mentioned earlier, K values vary with the temperature—sometimes significantly. The energy manager must be familiar with the different types of insulation, their K values, and how the temperature affects the K values. In all cases, the K value chosen to reflect the appropriate conductivity should be that for the mean temperature (t_m) experienced by the insulation:

$$t_m = \frac{t_h + t_s}{2} \qquad\qquad (11\text{-}12)$$

- *Fire hazard.* Fire hazard ratings measure a product's contribution to flame spread and smoke development in a fire. The rating is measured on a flame spread-smoke spread scale where 100/100 is the rating for red oak.

- *Forms.* Insulation is available in a number of different forms. Flexible blankets, batts, rigid board, blocks, and pipe half sections are some of the more popular ones. Insulation is also available in a number of sizes and thicknesses. For example, fiberglass batt insulation with a Kraft paper vapor barrier is available in 15-inch and 23-inch widths for thicknesses of 3.5 and 6 inches. Thicknesses of 12 inches are readily available in many parts of the country.

11.2.2 Common Insulating Materials

Some of the more popular types of materials with a discussion of some of their specific properties are given here. Details are summarized in Table 11-5.

- *Mineral fiber-rock wool.* Mineral fiber insulation is made from molten rock. It is fairly impervious to heat and so can be used in relatively high temperatures (see Table 11-5).

- *Fiberglass.* Probably the most popular type of insulation, fiberglass can be obtained in blankets, batts, boards, and pipe covering. Although organic binders are frequently used which limit temperature ranges somewhat, cell structure is such that the limitations can sometimes be exceeded and still have acceptable results.

- *Foams.* Several types of foam insulation are available; some types have problems meeting fire hazard classifications but have very good K values. Others meet the fire hazard requirements but do not offer very good K values. Foams are particularly applicable to cold applications.

- *Calcium silicate.* A very popular type of insulation for high-temperature use, calcium silicate is spun from lime and silica. It is extremely durable and offers a high thermal resistance.

Table 11-5. Industrial insulation.

Insulation type and form[a]	Temp. range (°F)	Thermal conductivity [Btu•in./h•ft²•°F at T_m (°F)]			Compressive strength (psi) at % deformation	Fire hazard classification or flame spread-smoke developed	Cell structure (permeability and moisture absorption)
		75	200	500			
Calcium silicate blocks, shapes, and P/C	To 1500	.37	.41	.53	100-250 at 5%	Noncombustible	Open cell
Glass fiber blankets	To 1200	.24-.31	.32-.49	.43-.73	.02-3.5 at 10%	Noncombustible to 25/50	Open cell
Glass fiber boards	To 1000	.22	.28	.51-.61			
Glass fiber pipe covering	To 850	.23	.30	.62			
Mineral fiber blocks and P/C	To 1900	.23-.34	.28-.39	.45-.82	1-18 at 10%	Noncombustible to 25/50	Open cell
Cellular glass blocks and P/C	-450 to 900	.38	.45	.72	100 at 5%	Noncombustible	Closed cell
Expanded perlite blocks, shapes, P/C	To 1500	—	.46	.63	90 at 5%	Noncombustible	Open cell
Urethane foam blocks and P/C	(-100 to -450) to 225	.16-.18	—	—	16-75 at 10%	25-75 to 140-400	95% closed cell
Isocyanurate foam blocks and P/C	To 350	.15	—	—	17-25 at 10%	25-55 to 100	93% closed cell
Phenolic foam P/C	-40 to 250	.23	—	—	13-22 at 10%	25/50	Open cell
Elastomeric closed cell sheets and P/C	-40 to 220	.25-.27	—	—	40 at 10%	25-75 to 115-490	Closed cell
MIN-K blocks and blankets	To 1800	.19-.21	.20-.23	.21-.24	100-190 at 8%	Noncombustible	Open cell
Ceramic fiber blankets	To 2600	—	—	.38-.54	.5-1 at 10%	Noncombustible	Open cell

[a]P/C means pipe covering.
Source: Wayne C. Turner, Senior Ed., *Energy Management Handbook*, ©1982. Reproduced by permission of Wiley-Interscience, New York.

- *Refractories-ceramic fiber.* An alumina-silica product, ceramic fibers are available in blankets or felts that can be used alone or added to existing fire brick.

- *Refractories-fire brick.* Fire bricks are made for high-temperature applications. Made of a refractory clay with organic binders which are burned out during manufacture, they offer good thermal resistance and low storage of heat.

- *Others.* Other types of insulation include cellular glass, perlite, and diatomaceous earth. Each has advantages and disadvantages with which the energy manager must become familiar.

11.3 ECONOMIC THICKNESSES

As mentioned in Example 11-1, insulation has an optimum thickness that can be calculated using the principles of engineering economy discussed in Chapter 4. Consider Figure 11-4.

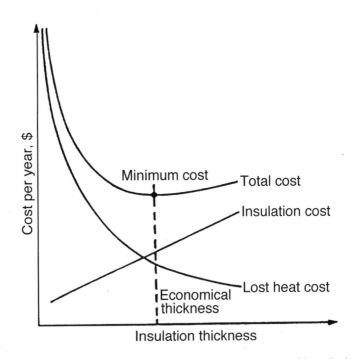

Figure 11-4. Illustration of the economic thickness of insulation.

As thickness of insulation is increased, the cost of material and installation goes up. The cost of lost energy, on the other hand, goes down, but at a decreasing rate. Said in other terms, the energy cost savings also goes up, but at a slower rate of increase than the cost of materials and installation. At some point, then, the total cost, which is the sum of the lost energy cost and the material cost, reaches a minimum point. That amount of insulation is called the *economic thickness.*

As Figure 11-4 shows, the total cost curve is relatively flat in the immediate neighborhood of the economic thickness. This means that the energy manager does not have to use the exact optimum amount of insulation. A small deviation either way will not affect the resulting annual cost very much.

To determine this economic thickness, the energy manager needs to construct cash flow diagrams for the different alternative thicknesses and calculate the annual equivalent cost for each increment. Since the cash flows include future fuel costs, careful handling of inflation is required.

Many simplified charts, graphs, tables, and computer programs are available to help determine optimum thicknesses. The Bibliography includes some of these, but it is fairly simple to develop a program that will quickly calculate equivalent annual cost for each increment.

Figure 11-5 is a printout resulting from a program written at Oklahoma State University. From this printout you can see that the economic thickness for this particular application is 3 inches. However, the total cost for 2 inches and 4 inches of insulation is about equal and is only slightly higher than the total cost for the economic thickness.

11.4 SUMMARY

Reducing heat loss or heat gain by adding insulation is often a cost-effective Energy Management Opportunity. An energy manager must know the different types of insulation available, and their respective advantages and disadvantages in specific applications. Different kinds of insulation are needed for walls and roofs of buildings, as compared to insulation needed for high temperature steam lines. Knowing how to perform heat loss-gain calculations is necessary to properly evaluate the effects of adding insulation, and to calculate the benefits of different levels of insulation. Finally, understanding insulation economics is important for good energy management since the first increments of added insulation produce very cost-effective benefits, while there is a point of diminishing returns of savings for increasing the levels of insulation. The calculation of the optimum economic thickness of insulation allows

Figure 11-5.
Computer printout of insulation economic thickness problem.

ECONOMIC THICKNESS DETERMINATION.

DEPARTMENT OF INDUSTRIAL ENGINEERING AND MANAGEMENT
OKLAHOMA STATE UNIVERSITY

INPUT PARAMETERS

FIRM: PLASTICS, INC.		CONTACT: S.A. PARKER	
SYSTEM: PLANT WALLS		DATE: OCTOBER 10, 198-	
INSULATION: FIBERGLASS BATTS		K-VALUE: 0.3600 BTU·IN/H·FT^2·°F	
AFTER-TAX MARR:	18.0%	GENERAL INFLATION RATE: 10.0%	
FUEL ESCALATION RATE:	20.0%	PRESENT R-VALUE: 1.427 HR·FT^2·°F/BTU	
HEATING DEGREE-HOURS:	112392.0	COOLING DEGREE-HOURS: 153600.0	
HEAT PLANT EFFICIENCY:	75.0%	COOLING PLANT EFFICIENCY: 210.0%	
COST PER MMBTU OF HEAT:	$3700	COST PER MMBTU OF COOLING: $4900	
INCREMENTAL TAX RATE:	48.0%	AVAILABLE TAX CREDIT: 0.0%	
PROJECT LIFE: 10 YEARS		DEPRECIATION LIFE: 5 YEARS	

THICKNESS CALCULATIONS

INSULATION THICKNESS (IN.)	INSTALLED COST ($/FT^2)	ANNUAL ENERGY LOSS OR GAIN (BTU/FT^2)	NPV OF SAVINGS ($/FT^2)	ANNUALIZE COST ($/FT^2)
.00	.00	156271.1	.00	.493
1.00	1.31	53034.6	.52	.378
2.00	1.53	31936.6	.64	.351
3.00	1.65	22847.5	.68	.342
4.00	1.84	17785.7	.62	.355
5.00	2.01	14560.0	.54	.372
6.00	2.15	12324.7	.48	.387
7.00	3.27	10684.4	−.31	.563
8.00	3.43	9429.4	−.41	.584
9.00	3.57	8438.3	−.52	.608

the determination of the greatest amount of insulation to add, while still producing the largest savings in energy cost.

Bibliography

ASHRAE Handbook of Fundamentals, Chapter 20, Thermal Insulation, American Society of Heating, Refrigerating and Air Conditioning Engineers, Atlanta, GA, 2005.

Brand, Ronald E., *Architectural Details for Insulated Buildings*, Van Nostrand Reinhold, New York, NY, 1990.

Capehart, Barney, Editor, *Encyclopedia of Energy Engineering and Technology*," Taylor and Francis/CRC Publishing Company, Boca Raton, FL, July, 2007.

Diament, R.M.E., *Thermal and Acoustic Insulation*, Butterworth, Guilford, Surrey, England, 1986.

Donnely, R.G. et al, *Industrial Thermal Insulation*, Oak Ridge National Laboratory, ORNL/TM-5515, 1976.

Graves, R.S. and D.C. Wysocki, *Insulation Materials, Testing and Applications*, Volume 2, American Society for Testing Materials, Philadelphia, PA, 1991.

McElroy, D.L. and J.F. Kimpflen, *Insulation Materials, Testing and Applications*, Volume 1, American Society for Testing Materials, Philadelphia, PA, 1990.

Parker, Steven A., Insulate, working paper, Oklahoma State University, Stillwater, Okla., 1982.

Silvers, J.P. et al, A Survey of Building Envelope Thermal Anomalies and Assessment of Thermal Break Materials for Anomaly Correction, prepared for Oak Ridge National Laboratory, 1985.

Stamper, Eugene, *National Program Plan for the Thermal Performance of Building Envelope Systems and Materials*, Third Edition, Building Thermal Envelope Council, Washington, DC, 1988.

Thumann, Albert, *The Plant Engineer's and Manager's Guide to Energy Conservation*, Seventh Edition, Fairmont Press, Atlanta, GA, 2006.

Turner, Wayne C., Senior ed., *Energy Management Handbook*, Sixth Edition, Fairmont Press, Atlanta, 2006.

Chapter 12

Motors, Compressed Air Systems*, and Process Energy Management

12.0 INTRODUCTION

In many facilities, energy management is simply a matter of managing the energy required for lighting and space conditioning. In many others, however, energy management is much more complex and involves large motors and controls, compressed air systems, industrial insulation, complex combustion monitoring, unique steam distribution problems, significant amounts of waste heat, etc. Typical facilities offering large energy management opportunities include industrial facilities, manufacturing plants, large office and commercial operations, and government institutions such as schools, hospitals and prisons. Such facilities generally have specialized industrial, commercial or institutional processes that incorporate many of the concepts covered in other chapters. These processes require thorough analytical evaluations to determine the appropriate energy-saving measures. This chapter provides some examples.

The energy manager must be careful in process energy management. Processes can be quite complex, so a full understanding of the entire process is necessary. Examining only one area of a process and making energy use changes to that part may have adverse effects in another area. For example, small changes in heat treatment temperatures or atmosphere can sometimes dramatically decrease the product quality or subsequent workability.

In this chapter we first present a discussion of motors and controls, since they form an integral part of most processes. Next, compressed air systems and air compressors are discussed since they are major suppliers of process energy. Then, some sample case studies of process energy management opportunities are provided.

---

*This section was written by Dr. Diane Schaub while she was the director of the IAC at the University of Florida, Gainesville.

12.1 MOTORS AND ADJUSTABLE SPEED DRIVES

Motor energy use represents well over half of all the electric energy consumed by industrial, commercial and institutional facilities. Motors are found on almost every piece of equipment used to perform a process in manufacturing, mining and agriculture. Even pieces of equipment that perform special functions often have motors as their principal part—for example, chillers and air compressors.

Because of the widespread application of electric motors in almost every facility, they are excellent candidates for improvements to their efficiencies and improvements in their utilization in machines and processes. Just a small improvement in electric motor efficiency can produce significant savings in the energy cost of operating a piece of equipment. The annual cost of operating a motor can often be five to ten times the original purchase price of the motor.

12.1.1 High Efficiency Motors

Electric motors account for about three-quarters of all the electric energy used by industry (Figure 1.6, Chapter One), and almost half of all electric use by commercial facilities. Energy efficient motors are now readily available that are two to eight percent more efficient than the standard motors they would replace. Table 12-1 provides data on the efficiencies and cost premiums for motors in size ranges from 0.75 horsepower to 250 horsepower [2]. Since typical motors last over twenty years, using high efficiency motors offers business and industry substantial energy and dollar savings.

Motor efficiency is a measure of the effectiveness with which electrical energy is converted to mechanical energy. Motor losses occur in five major areas: core losses, stator losses, rotor losses, stray load losses, and windage and friction losses. High efficiency motors are designed and manufactured to reduce these losses. In addition to having lower losses, high efficiency motors also have higher power factors during operation. Cost premiums for high efficiency motors range from 10% to 30%, but since a motor may use 75 times its initial cost in electric energy over its lifetime, the savings potential is great [3]. Many motors in commercial facilities, industries and institutions run 6000 to 8000 hours per year, so very cost-effective paybacks can be achieved.

Table 12-1. Motor data.

hp Rating	Standard Efficiency	High Efficiency	Premium ($)
.75	0.740	0.817	35
1	0.768	0.840	39
1.5	0.780	0.852	48
2	0.791	0.864	56
3	0.814	0.888	73
5	0.839	0.890	69
7.5	0.846	0.902	97
10	0.864	0.910	111
15	0.875	0.916	149
20	0.886	0.923	186
25	0.897	0.929	224
30	0.901	0.931	273
40	0.908	0.934	371
50	0.915	0.938	469
60	0.916	0.940	553
75	0.917	0.944	678
100	0.919	0.950	887
125	0.924	0.952	1,172
150	0.930	0.953	1,457
200	0.940	0.956	2,027
250	0.943	0.956	2,159

Example 12-1: Ace Industries has a 50-hp air compressor that operates at full-load, all day for 365 days per year. If the motor for the air compressor cost $1400, the motor efficiency is 90%, and electricity costs $7.00/kW/month and $0.05/kWh, how much does it cost to operate the air compressor for one year? How much money will be spent to operate the air compressor over a ten-year period?

Solution: If a motor were 100% efficient, there would be an electric load of 0.746 kW/hp, or 37.3 kW. Since the motor is only 90% efficient, its electric load is:

Electric load = (50 hp) × (0.746 kW/hp)/(0.9)
           = 41.44 kW

The annual demand charge is then:

Demand charge = (41.44 kW) × ($7.00/kW/month) ×
             (12 months/year)
          = $3481/yr

The annual energy charge is:

Energy charge = (41.44 kW) × (8760 h/yr) × ($.05/kWh)
          = $18,151/yr

The total electric cost of operating the motor is then:

Total cost = $3841 + $18,151 = $21,992/yr

Since the purchase cost for the motor was $1400, the annual electric cost for operating the motor/air compressor is over fifteen times the initial purchase price.

If the air compressor operates for ten years, the total operating cost will be 10 × $21,992 or $219,920. This is over 150 times the initial purchase price of the motor.

As this example shows, the original purchase cost of a motor can be a small part of the life-cycle cost. Thus, it is important to consider other factors besides the initial cost when buying a new motor for a piece of equipment.

Example 12-2: ACE Industries has been experiencing a period of rapid growth in the success of their products, and they plan to expand their production capacity by building a second plant nearby. They determine that they need another 50-hp air compressor which will also run continuously at full load. They can purchase either the Standard or the Deluxe Model air compressor with the difference being that the Deluxe Model has a high efficiency motor. The motor efficiency for the Standard Model is 91.5%, and for the Deluxe Model it is 93.8%. The additional cost for the Deluxe Model is $470. Is this a good investment for ACE Industries?

ACE Industries also wants to know what kind of "cushion" they have on this decision, since their forecast for new business could be too optimistic. If the new air compressor is only run for two shifts a day, for a total of 5000 hours per year, is the additional investment still worthwhile?

Solution: The electric load for each model of the motor is found, and the difference will be used to compute the savings. The reduction in electric demand is:

Standard Model demand $= (50 \text{ hp}) \times (0.746 \text{ kW}/\text{hp})/(.915)$
$= 40.77 \text{ kW}$

Deluxe Model demand $= (50 \text{ hp}) \times (0.746 \text{ hp})/(.938)$
$= 39.77 \text{ kW}$

Demand savings $= 1 \text{ kW}$

Annual demand cost savings $= (1 \text{ kW}) \times (\$7.00/\text{kW month})$
$\times (12 \text{ months}/\text{yr})$
$= \$84/\text{yr}$

Annual energy cost savings $= (1 \text{ kW}) \times (8760 \text{ h}/\text{yr}) \times$
$(\$.05/\text{kWh})$
$= \$438/\text{yr}$

Total electric cost savings $= \$522/\text{yr}$

We can now compute several economic decision criteria to evaluate how good an investment the additional $470 would be. Using equation 4-1, and the method of section 4.8.3 from Chapter Four gives:

Simple Payback Period $= \$470/\$522/\text{yr} = \underline{0.9 \text{ years}}$

If the motor has a lifetime of ten years running full-time, we can find the ROR as the solution to:

$\$522 \ (P/A, i, 10) = \$470$
ROR $= \underline{111\%}$

If ACE Industries has an investment rate of 12%, we can find the Benefit-Cost Ratio as:

$$B/C = \$522 \times (P/A, 12\%, 10)/\$470$$
$$= 1.11 \times 5.6502 = \underline{6.27}$$

By any one of these three economic evaluation measures, the decision to buy the Deluxe Model with the high efficiency motor is an excellent investment if the motor runs 8760 hours each year. To see how sensitive this result is to the use of the motor, we need to recompute the savings if the motor is only used 5000 hours each year.

The demand savings is the same. The new energy cost savings is:

$$\text{Energy cost savings} = (1 \text{ kW}) \times (5000 \text{ hours/yr}) \times (\$0.05/\text{kWh})$$
$$= \$250/\text{year}$$
$$\text{Total cost savings} \quad = \$84 + \$250 = \$334$$

The new values of the economic measures are:

SPP = <u>1.41 years</u>

ROR = <u>70.7%</u>

B/C = <u>4.02</u>

These are still excellent values for the three economic decision measures. Most companies would still find the investment in the more expensive motor highly attractive, even with the reduced hours of operation of the air compressor.

A number of computer programs have been written to help perform the economic analysis involved in motor selection. One of the best is the MotorMaster program written and made available by the Washington State Energy Office [4]. This program has an extensive data base on motor prices and efficiencies, and is updated twice a year.

12.1.2 Motor Load Factors

The full load horsepower output rating of a motor is stamped on the motor's nameplate. However, just because we find a motor that is stamped 20-hp does not mean that the motor is running at full load - which is 20-hp. A motor is a load driven machine, and will supply only that amount of power needed by the load. For example, a 20-hp motor may be driving a fan that needs only 15-hp. The load on the motor can be expressed as a percent of full load, and this is called the load factor for the motor. In this case, the load factor would be 15/20 or 75%.

From the authors' energy audit experience, few motors run at anywhere near full load. A common assumption made by many energy auditors and analysts is that motor load factors are around 80%. This value is rarely seen in motors other than those specifically sized for known loads in heating, ventilating and air conditioning systems for buildings. In most other applications, motors experience variable loads that average well below 80%. One energy analyst presented data to show that 75% of all motors in his experience have load factors less than 60% [9].

One of the authors of this book has performed over 100 audits of medium-sized manufacturing companies, and the average motor load factors have ranged from about 30-40%. Individual motors, such as found on a wall ventilating fan, may well have load factors of 80%. However, many other pieces of equipment such as some air compressors, conveyors, pumps, dust collector fans, saws, drills and punches have extremely variable load factors which are generally much less than 50%. Pumps and fans with variable loads are usually ideal candidates for use of adjustable speed drives to reduce the energy input when the motor load is low.

To account for the fact that motors typically operate with load factors less than one, the basic motor equation for computing the electrical load must be modified to include a load factor term. NPHP is the nameplate horsepower of the motor.

$$kW = \frac{NPHP \ (0.746 \ kW) \ (LF)}{Efficiency \ x \ hp}$$

For example, if we have a 100-hp motor that is 95% efficient and is running at 60% load, its electrical power consumption is:

$$kW = \frac{100 \ hp \ x \ 0.746 \ kW \ x \ 0.6}{0.95 \ x \ hp}$$

$$= 47.1 \ kW$$

If the motor had been operating at full load, its power consumption would have been 78.5 kW.

12.1.3 Rewinding electric motors

There is at least one other important factor in motor replacement selection, and that is the potential for rewinding a motor that has failed.

There are three options available for a facility that has just experienced a motor failure. One, they can buy a standard efficiency motor to replace the failed one. Two, they can buy a high efficiency replacement. Three, they can send the failed motor out to be repaired, and potentially rewound. Example 12-2 illustrated how cost-effective it can be to buy a more expensive, high-efficiency motor for a piece of equipment that is used heavily.

The cost of rewinding a motor is often substantially less than the cost of purchasing a new motor—whether it is a standard-efficiency model or a high-efficiency model. However, it is fairly common for motors to be damaged during the rewinding process, and to suffer losses in efficiency of 1-2% [5]. Example 12-2 showed the economic impact of a 2.3 percentage point difference in motor efficiency. Often, a 1 percentage point loss in efficiency will result in the cost of the additional electricity being greater than the total cost of rewinding. Thus it is important to consider this factor when replacing a motor. Not all rewinding operations damage motors, but the loss of efficiency is quite common.

12.1.4 Motor drives and controls

Many motor applications require variable speeds, or should use variable speeds to match the actual loads, and thus the area of motor controls is also very important. Adjustable speed drives—or variable speed drives—are motor control systems that reduce the energy input to a motor when it is not fully loaded. These ASDs or VSDs can produce substantial savings in the operational costs of motors, and can often improve the operation of the system that previously used a motor without a speed control.

In addition to improving the efficiency of the motor itself, there are many other opportunities for energy savings in the complete motor system. One opportunity is in the use of solid-state, electronic controls that can provide soft-starts, speed control and power factor correction.

For large motors that have variable loads, the addition of electronic speed controllers—or adjustable speed drives (ASDs)—can be very cost effective. ASDs are electronic devices that vary the speed of a motor to match that of the load being put on the motor. The size of a motor is usually based on maximum load, even though normal design conditions seldom require this full load size. An ASD will reduce the speed of the motor by adjusting the frequency, voltage, or current of the motor input so that the motor performance exactly matches the present load. ASDs are also called variable speed drives (VSDs) or variable frequency drives (VFDs).

Fans and pumps are typical applications where ASDs can improve motor performance. The cube law for fans discussed in Chapter Six (Equation 6-14) showed the potential for substantial energy savings when the speed of a fan is reduced. Since the energy used in many fan and pump applications is proportional to the cube of the flow rate, then small reductions in the required flow rates translate to large savings in energy needed. In addition, many motors are purposefully over-sized to have a safety factor in handling the required load. This over-sizing is not beneficial to energy efficiency, and results in many motors running at conditions that unnecessarily waste energy. Properly sizing motors is probably one of the most cost-effective EMOs that a facility could implement.

ASD costs vary with the size of the motor. Equipment costs range from $150 to $450 per horsepower for ASDs smaller than 50 horsepower, and $100 to $150 per horsepower for larger units [6]. General estimates are that ASDs can save an average of 20-30 percent of the energy used for a typical motor.

Example 12-3: In one industrial application, three cyclone fans used for ventilating stack gases were replaced with one larger fan motor having an ASD. The cost of the new fan, motor, and installation of new ductwork was $18,250. The cost of the ASD was $20,000. The electrical consumption from the new system dropped 500,000 kWh/yr from that of the old fan system. If the facility paid $0.061/kWh, determine the cost-effectiveness of this EMO.

Solution: The electrical cost savings from the new system is found as:

$$\text{Annual savings} = (500,000 \text{ kWh/yr}) \times (\$0.061/\text{kWh})$$
$$= \$30,500$$

The total cost of implementation is the sum of the costs for the new fan, motor, ductwork, installation and the ASD:

$$\text{Total implementation cost} = \$18,250 + \$20,000$$
$$= \$38,250$$

The SPP for this EMO is found using Equation 4-1 from Chapter Four:

$$\text{SPP} = \$38,250/\$30,500/\text{yr} = \underline{1.25 \text{ years}}$$

If the new system had an expected lifetime of 15 years, the ROR can be found as the solution to:

30,500 (P/A, i, 15) = \$38,250
ROR = <u>79.7%</u>

These two measures demonstrate that this was an excellent EMO investment.

12.1.5 Other factors in motor system efficiency

In addition to the efficiency of the motor itself and the use of adjustable speed drives, there are still several other factors that affect the overall motor system efficiency. One of these factors is the mechanism for the transmission of power from the motor to the load. In particular, care should be taken to insure that efficient pulleys, drives, belts and gears are used to couple the motor to the load.

Belts. Many motors are coupled to their loads with belt drives. There are several types of belts, including V-belts, cogged V-belts, and synchronous belts. Standard V-belts are the most common type of drive belt, and have transmission losses that occur because of flexing and slippage of the belt. The cogged V-belts and the synchronous belts are more efficient because they do not allow the same amount of flexing and slippage as the standard V-belts. Motor system performance can be improved by 2-4% with the use of these more efficient belts.

Lubrication. Motor lubrication is also a factor in motor efficiency. There are synthetic lubricants available that reduce the friction losses in motor-driven equipment. Savings of 1-2% are common, and much larger savings are possible for some equipment. Often, however, manufacturers recommend that these synthetic lubricants be used only in new pieces of equipment, so that there is no contamination of the synthetic lubricant.

Maintenance. Motor maintenance is also a factor. This area was discussed in some detail in Chapter Ten on Maintenance. The operating temperature of a motor should be checked periodically, as well as its mechanical and electrical condition. Each motor in a facility should be inspected periodically to determine the condition of its bearings and its pulley and belt alignment if it uses a belt drive. The bearing condition of a motor can be checked with an industrial stethoscope. In this procedure, the stethoscope is used to measure the noise at both ends of a motor that uses chains or belts to drive equipment. In a normal motor, the active end will read 2-3 db louder than the fixed end. If there is no difference or if the fixed end is louder than the active end, the bearings at the fixed end are worn; if the active end is 4-6 db louder than the fixed end, the bearings are badly worn; if the active end is 7-8 db louder, the bearing housing is turned inside the motor; and if the active end is 9 db or more louder than the fixed end, the motor should be replaced immediately. This procedure, developed at Safeway Stores by Mr. Harold Tornberg, has proved practical in industry.

12.1.6 Utility rebates for motors and drives

A general discussion of electrical utility company rebates and incentives was given in Chapter Three. However, since most utilities offer rebates and incentives for electric motor system improvements, it is worth mentioning again. There are generally two forms that the rebates and incentives take for motor systems: an incentive based on kW savings, or an incentive based on the horsepower of the motor involved. The level of the rebate or incentive for a peak load reduction due to a motor system improvement depends greatly on the individual utility. If the utility is working hard to limit its peak demand so that it will not have to add new power generation capability, the incentives may be quite large. Incentives of a few hundred dollars per kW of motor load reduced are quite common.

The incentive may also be related to the horsepower of the motor replaced or the horsepower of the motor that an adjustable speed drive was added onto. For a high efficiency motor replacement, the utility usually has a list of minimum efficiencies that qualify for rebates. If a customer replaces a motor with one that meets the minimum efficiency level, then the customer is automatically eligible for the rebate. This may be in the range of $6 – $15 per hp. Rebates for ASDs generally would qualify for levels of $100 to $300 per hp.

Example 12-4: In Example 12-2, ACE Industries would receive a $6/hp rebate from its electric utility if the company purchases the high efficiency 50-hp motor for the air compressor. How does this rebate influence the cost effectiveness of the purchase decision for ACE Industries?

Solution: The rebate of $6/hp reduces the cost of the motor by:

Cost reduction = ($6/hp) × (50 hp) = $300

Thus, the cost to ACE Industries is now only:

Motor cost = price differential—utility rebate
            = $470 – $300
            = $170

The new SPP for operating the motor 8760 hours is:

SPP = $170/$522/yr = <u>0.33 years</u>

For 5000 hours of operation, the SPP is:

SPP = $170/$334/yr = <u>0.51 years</u>

The utility rebate significantly improves the attractiveness of this EMO. A simple payback period of four to six months for this motor purchase would be considered a highly cost-effective investment by most companies.

12.2 COMPRESSED AIR SYSTEMS

Although most often powered by electrical motors, Compressed Air can essentially be thought of as another source of energy. This similarity stems from the fact that compressed air lines can be designed to allow modular tools to plug into the air lines, just like electrical devices can be powered by tapping into electrical outlets. This use of compressed air to drive pneumatic tools, ranging from nail guns to jackhammers to large drill presses, is ergonomically beneficial. Pneumatic tools are smaller, lighter and more maneuverable compared to electric motor driven models, and they have other operating benefits as well. An additional common use for compressed air is to power pneumatic control systems.

12.2.1 Compressed Air Systems
The typical commercial and industrial compressed air system is composed of:
• One or more in-series or parallel compressors
• An air dryer and air filters
• A receiving tank (for storage)
• Controls

- Piping
- End uses

- Compressed air systems should be thought of as having both a *supply side* and a *demand side*. Figure 12-1 shows a schematic view of a compressed air system with both supply and demand side noted.

- *Improving and maintaining* **peak** compressed air system **performance** requires addressing both the supply and demand sides and how the two interact in order to have dependable, clean, dry, stable air delivered at the proper pressure. A well- planned and balanced system will yield the cheapest and most energy efficient results.

Figure 12-1. Schematic of a Compressed Air System

12.2.2 Components of a Compressed Air System—
Supply Side Perspective

A thorough understanding of the end-use compressed air needs, from both a volume and usage profile perspective, is necessary in order to select the appropriate number and size of air compressors for the supply side. It is rare to find a manufacturing plant that has a constant, uniform use of compressed air throughout the day. Most manufacturing plants have cyclical flow and volume demands due to production schedules, and also need back-up supply, so typically engineers plan for more than one air compressor to meet a facility's needs. A good strategy is to size a compressor for a base load, and have one or more compressors staged to come on-line to meet additional compressed air demand. In designing a compressed air system, altitude, inlet air temperature and relative humidity should be considered as they impact compressor capacity. It may also be helpful to have differing size compressors so that they can be tailored to fit the operating conditions. Additionally, a small compressor or separate booster may be appropriate for off-shift operations or a special high pressure, periodic application.

12.2.2.1 Comparison of Compressors

The vast majority of industrial compressors are of the rotary screw variety, but double acting reciprocating or centrifugal compressors are also available for specific applications. The diverse types of air compressors have various pros and cons associated with their use. The major design differentiation is the method of how the air is compressed, by either positive displacement or dynamically. Positive displacement compressors work by trapping a set amount of air in a compression chamber and mechanically reducing the volume, thereby causing a corresponding increase in pressure. Dynamic compressors operate by a rapidly spinning impeller which increases the air velocity, which in turn is changed into higher air pressure by the design of the impeller discharge and discharge diffusers.

The most common types of positive displacement air compressors are:

- Reciprocating compressors of either a single-acting (power use: 22-24 kW per 100 ft^3/min air) or double-acting variety. A solid cylinder fits snugly into a larger cylinder and moves under fluid pressure to compress the air. The single-acting compressor resembles an automotive piston with compression taking place on the top side of the piston on each revolution of the crankshaft. The double-acting variety uses both sides of the piston for air compression, thereby doubling the capacity and reducing the power usage for the same relative amount of output compressed air. The larger models of this compressor are available both in lubricant-injected or lubricant-free versions; the latter is required for use in the food and pharmaceutical industries, but comes at a cost of increased power usage. The reciprocating type of air compressor is even available in sizes as small as 1 hp, for home workshop applications.

- Rotary screw compressors come in two configurations: with lubricant-injected or lubricant-free cooling, and they can have a single stage or two stages. They operate by having inlet air pass through two intermeshing rotors in a stator housing, discharging out the opposite end. As air comes in the inlet end it is trapped between the lobes and the stator wall and as the rotors turn the trapped volume is continually reduced, thereby increasing the pressure.

— Lubricant-injected rotary screw compressors have the ability to trim to partial loads to meet usage needs which can further save on their already lower power costs (single stage: 18-19 kW/100 ft^3/min, two stage: 16-17 kW/ 100 ft^3/min). The lubricant is injected in the compression chamber during compression and not only lubricates the rotary screws, but also removes the heat of compression, plus the lubricant seals the openings between the meshing rotors and the stator wall (which improves the efficiency over a comparable lubricant-free model). These compressors can be the least expensive and the most compact and lightest for their relative horsepower size. As they are free of vibration, they do not require a special foundation. Single-stage units range in size from 3 hp through 900 hp with discharge pressures from 50 to 250 psig. Two-stage units or twin units have a higher efficiency than would be achievable from one larger diameter single-stage element. The two-stage variety runs on a large first-stage low pressure element and a smaller second-stage high pressure element in series. The twin unit operates with two identical rotary screws mounted in parallel.

— Lubricant-free rotary screw compressors require higher electrical demand (20-22 kW/100 ft^3/min) than their lubricant-cooled counterpart, but assure no lubricant carryover. This may be crucial when ultra-clean air, (such as used in the food industry) is required. There are two types of lubricant-free rotary screw air compressors, the dry type and the water-injected type. With the dry type, the intermeshing rotors are not allowed to touch and maintain their relative position from a lubricated timing gear; they are available in sizes ranging from 25 to 4000 hp; single stages can operate up to 50 psig and two-stage units can operate up to 150 psig. With the water-injected type, a comparable timing gear is used, but water is injected into the compression chamber to remove the heat of compression and to act as a seal in internal clearances. In a two-stage design, there is typically an intercooler located between the stages to remove the heat of compression and an aftercooler after the second stage. Typically there is also lubricant to lubricate the gears and bearings isolated from the compression chamber.

The most common dynamic air compressor is the centrifugal variety:

- Centrifugal air compressors (16-20 kW/ 100 ft³/min) have a continuously flowing air stream accelerated by an impeller operating at speeds up to or exceeding 50,000 rpm. They range from 500 to over 100,000 ft³/min in size, (up to 1000 hp) but are most common in the 1000 to 5000 ft³/min range, and the first cost of these compressors improves as size increases. These compressors deliver lubricant-free air as there is no lubricant in the compression chamber, and do not require a special foundation. They typically are sold in a completely packaged configuration.
- The costs of operating the air compressor alone can be quite significant, not including the auxiliary equipment (discussed below). The electrical costs for each 100 hp compressor can run approximately $25,000 per year, not including maintenance costs.

Example: A 100 hp air compressor operating 8000 hours/year, at 70% load with utility costs of $80/kW-yr, and $0.05/kWh.
Total $ = Energy $ + Demand $
Energy $ = 100hp * 0 .70 (load)* (0.746 kW/hp) * ($.05/kWh) * 8000hr/yr
Demand $ = 100hp * 0.70 * (0.746 kW/hp) * $80/kW-yr
Total $/yr = 20,888 + 4178 = $25,066
Considering that the first cost alone totals about $30,000 to $50,000, it is important to consider life-cycle costs when designing a new air compressor system.

12.2.2.2 Air Dryers

Another component of a compressed air system is the dryer(s). The compressing of air will condense out the moisture from the natural water vapor found in atmospheric air. This liquid water can cause rust problems in the lines or, should compressed air supply lines connect between buildings, freeze in the winter. Compressed air should be dried to a dew point at least 18 degrees F below the lowest ambient temperature of the demand side.

The various types of dryers are:

- Refrigerated: This is the most common type, with both low initial and operating costs. It can be subject to freezing if operating at low capacities. It cannot reduce the moisture level low enough for some applications (minimum dew point around 40 degrees F).
- Regenerative desiccant: Typically operated in tandem between two twin dryers, with one operating and the other regenerating the des-

iccant. The required volume of purge air needed to regenerate can increase the load or even cause an idle compressor to be started. Heaters can be used in place of purge air, but present their own energy penalty.

- Heat of compression: Similar to the regenerative desiccant dryer, this type of dryer is available for lubricant-free rotary screw compressors and utilizes the hot discharge compressed air to regenerate the desiccant. Their efficiency is affected by changing air temperatures and additional heat may be required for low load situations. They are packaged and installed with the compressor.

- Deliquescent desiccant: A dissolvable desiccant is used. Regular replacement of this resource is necessary, requiring labor as well as material costs.

- Membrane-type: A porous membrane separates water vapor from the air and suppresses the dew point. Although there is a low initial cost, these dryers are appropriate only for low-volume applications.

12.2.2.3 Air Filters

In order to deliver clean compressed air, air filters may be installed upstream (inlet air filter) and downstream from the air compressors. The filters remove particulates, some condensate and lubricant. Regular replacement of filter elements, minimally on an annual basis, is necessary to prevent pressure drop which results in a throttling effect. In addition to maintaining inlet air filters, a sufficient supply of inlet air should be available; again, lack of consideration of this supply will result in an unnecessary power loss from throttling.

Downstream air filters are available in particulate, coalescing and adsorption varieties. According to the Compressed Air ChallengeTM, a refrigerant-type dryer may not require a filter before or after it, but a desiccant-or deliquescent-type air dryer requires a pre-filter to protect the drying medium or desiccant from contamination, thereby rendering it ineffective. An after-filter may also be recommended in order to remove desiccant fines in systems with sensitive end-use equipment. To illustrate the importance of filter maintenance, see the following example:

Example (Replacement of a Compressed Air Filter Element)
Assume a 100 hp compressor that operates continuously with an energy cost of 7 cents/kWh, resulting in an annual energy cost of $55,328. As the in-line air filter becomes clogged assume the pressure drop increases to 6 psi across the filter (as compared to a 2 psi pressure drop for a new filter). Consider this 4 psi increase can cost 2% of the

annual required energy, or $1100, as compared to $375 for a new filter element.

12.2.2.4 Air Receivers

Air receivers can be found on either the supply side (immediately after the compressor or the dryer) or on the demand side close to the application end use. Air receivers store compressed air and help cover peak events of short duration. If sized properly, they can greatly reduce the frequent loading and unloading of the compressor, saving both energy and maintenance costs. They also stabilize system pressure which improves performance of the end use.

12.2.2.5 Controls

Another issue that can greatly impact the energy efficiency of air compressors is their control strategy. Start/stop, load/unload and modulating (or throttling) control strategies can be used depending on the facility's compressed air usage profile.

To minimize energy use and to provide precise control, variable speed drives should be considered. The speed of the compressor motor can be operated at a speed that changes dynamically to suit the conditions and load. Variable displacement control is also energy-efficient alternative; it matches displacement to demand without reducing pressure. Additionally controls are available to start and stop individual compressors to match demand needs, and can be programmed to alternate between compressors.

Other components associated with the supply side may include after-coolers or intercoolers (for lubricant-free systems), moisture separators and condensate drains. Depending on the manufacturer, these latter items may be packaged in a single housing with the compressor itself.

12.2.3 Components of a Compressed Air System—
Demand Side Perspective

Besides a downstream air receiver, the demand side consists of the distribution system or piping, and the end-use applications. Correct sizing of the distribution piping is a critical feature in compressed air system design in order to minimize energy costs. Undersized piping can result in a pressure drop that increases energy consumption. Additionally any type of obstruction, restriction or roughness in the delivery system will cause resistance to air flow, and can be an undiagnosed culprit in higher than necessary compressed air set points.

The piping typically consists of rigid metal or plastic piping from the air compressor room to the general area of the end-use equipment. From this point, flexible rubber or plastic tubing is used which may be plumbed directly to the end use, or have a shut-off valve with quick-connect attachment points. This flexible tubing may be subject to being run over by foot or equipment traffic and can wear out over time. As a result, air leaks can grow to epidemic proportions, and greatly increase the demand on the compressor. In fact, it isn't unusual to find a poorly maintained system running a compressor that is only feeding leaks. Some facilities will bury large portions of their distribution piping which make finding and repairing of leaks an expensive proposition. A 3/16" inch hole in a system operating at 100 psig can cost over $5000 a year!

Another operating consideration associated with the demand side is the cost of "normal production". Decisions to add additional applications should undergo a realistic cost evaluation. Consider the following example of an end-use application:

Example (Addition of an End-Use Application)
A quarter inch orifice to operate a pneumatic hand tool with a 100 psig flow will use 63.3 scfm (standard cubic feet per minute). Over a year of constant use, this equates to 33.3 MMcf (million cubic feet) of compressed air. If compressed air generation costs $300/MMcf, then the power cost for this application will be approx. $10,000/year. If we add additional operating costs of $170/MMcf to account for the operator maintaining the compressed air equipment and the maintenance, lubricant and repair costs for the system, we find that the cost of this new application use is over $15,000/year. Compare this with less than $2000/year to operate a comparable electrical tool. In general, it takes about 7 hp of input to the air compressor to deliver 1 hp to the end-use devices.

High costs can also be incurred through artificial demand associated with setting the compressor pressure level higher than is needed. Supplying 20% extra psig will force the system to consume 20% more air flow, resulting in 20% waste. Poor applications, such as stuck condensate drains, personnel use of compressed air for cooling or drying, or sparging (aerating of liquids) also use up precious compressed air.

12.2.4 Estimating Necessary Pressure Setpoint

The pressure set point for the air compressors needs to be computed, and is best done by considering all end-use equipment and considering the recommended pressure settings for each. Sporadically used equipment (for example, intermittent high-pressure bursts of air to shake off baghouse filters) or those that operate at considerably higher pressures than the majority of the equipment may warrant a separate, appropriately sized air compressor (system) located near the application. In this case, a lower overall pressure set point may be used. The energy savings potential from this strategy can be fairly dramatic; for each 1 psi reduction of system pressure above 100 psi, the corresponding energy savings is 0.5%. However, because of natural pressure drops associated with the components of a compressed air system, as well as unrepaired air leaks, the final point is more difficult to find than just dialing in the pressure recommended by the end-use equipment manufacturer. In fact, it is not unusual for plant personnel to reach the desired pressure by trial and error; increasing the set point until equipment operators stop complaining about low pressure.

When possible, pressure measurements should be made after each component of the compressed air system to monitor system performance. Flow or electrical readings can also provide useful performance data. Figure 12-2 shows an example of the pressure drops that can occur along the line.

12.2.5 Costs of Compressed Air

To operate a one hp air motor requires 7 to 8 hp of electrical energy. This large energy penalty, along with a common employee perception that compressed air is essentially a free resource, makes it a challenge to control the costs of compressed air. Inadequate compressor control schemes can cause multiple compressors to run at partial loads, rather than turning them off. Problems with poor maintenance can increase consumption or cause pressure variability. In fact, it isn't unusual to find that compressed air can be the largest end user of electricity.

12.2.6 Opportunities for Energy Savings

The separate components specified for a compressed air system have individual virtues that can improve the overall efficiency of a compressed air system. When applied together and viewed as a system matched with demand usage, economic, operating and energy-savings benefits can all be achieved in tandem. Proper maintenance of the various pieces of equipment to minimize pressure differentials will minimize energy consumption. Resistance to air flow caused by for example, clogged filters, aftercoolers or lubricant separators, results in the need

Estimating Pressure Drop

Measurements can be taken at various points in a compressed air system to monitor the associated pressure drop from each component. The pressure profile shows the lowest pressure seen by the end-uses.

Compressor operating range:	115-105 psig	Air/Lubricant Separator	5 psid
FRL (Filter, regulator, lubricator)	7 psid	Hose and Disconnects	4 psid
Aftercooler	3 psid	Dryer	4 psid
Filter	3 psid	Distribution System	3 psid
	psid = psi drop		

Figure 12-2. Estimating Compressed Air System Pressure Drop (psid = pressure drop in lb/in^2)

to increase the compressor set point. According to the Compressed Air Challenge™, this can involve increasing the drive energy by 1% of connected power for every 2 psi of differential.

Due to the fluctuations in end-use consumption, most compressors run at partial-load or are unloaded during some periods. This should be avoided as it results in a high consumption without proportional creation of compressed air, and with the strategic use of multiple compressors, a compressor(s) can be set to run at full-load and one compressor with good partial-load efficiency can be set to run at partial load to "trim" the system. For multiple compressor installations, it is highly recommended to consider purchasing a sequencer. With some optimizing sequencers, a variable speed compressor is used to trim the air system supply. It is also the first compressor on and last one to turn off.

Another energy savings idea is the use of engineered air nozzles for applications requiring compressed air blow-off following machining operations. These aftermarket nozzle tips are able to use 33% less compressed air for the same blowing capacity, as compared to round orifices. Heat Recovery is yet another energy savings opportunity as 85-90% of compressor input energy is lost as heat, even for an ideal compressor. A fully loaded 100 hp compressor generates 250,000 Btu/hr; equivalent to thirty nine 1,500 watt electric space heaters. The output temperature is not high, but could be used with a heat exchanger for certain applications.

12.2.2.7 Additional Sources of Information

The US Department of Energy's Industrial Technologies Program sponsors compressed air training and AirMaster+ tools through their Best Practices programs. See http://www.oit.doe.gov/bestpractices for more information. The organization charged with actually delivering the compressed air training can be found at: http://www.compressedairchallenge.org

References
Fundamentals of Compressed Air Systems, training manual for the Compressed Air Challenge, 1999.

Advanced Management of Compressed Air Systems, training manual for the Compressed Air Challenge, 1999.

Scales, W., and D. McCulloch, Best Practices for Compressed Air Systems, The Compressed Air Challenge, 2003.

Taranto, T., Compressed Air Management, ConservAIR Technologies Industrial Seminar Series, 1998.

12.3 STEPS FOR PROCESS IMPROVEMENT

Readers who have studied work simplification and improvement may remember the suggested order of changes as (1) eliminate; (2) combine; (3) change equipment, person, place, or sequence; and (4) improve [1].

The same order of change is appropriate for process energy management, but as mentioned earlier, the analyst must understand the entire system and the cascading impacts that changes might effect. In terms of energy management, examples of the preceding changes include the following:

- *Eliminate.* Does that cooling water really need to be there? Sometimes process cooling water is not really necessary; eliminating it saves pumping and chilling costs. Is the paint oven really necessary? Some newer paints will air dry quite well, and paint oven costs can be substantial.

- *Combine.* Machining operations can often be combined with jig and fixture modifications or changes in equipment. This saves the energy used by the additional machines; it also reduces material handling and may save process storage energy. Sometimes combining processes also saves the energy necessary to bring the material back to a required workability.

- *Change equipment, person, place, or sequence.* Equipment changes can offer substantial energy savings as the newer equipment may be more energy efficient. For example, new electric welders are considerably more energy efficient than older ones. Changing persons, place, or sequences can offer energy savings as the person may be more skillful, the place more appropriate, and the sequence better in terms of energy consumption. For example, bringing rework back to the person with the skill and to the place with the correct equipment can save energy.

- *Improve.* Most energy management work today involves improvement in how energy is used in the process because the capital expenditure required is often minimized. Examples include reducing excess air for combustion to a minimum, reducing temperatures to the minimum required (don't forget chilling—maybe the freezer temperature can be increased a few degrees), and removing excess lighting. Improving does sometimes require large amounts of capital. For example, insulation improvements can be expensive, but energy savings can be large, and there can be improved product quality.

12.4 EXAMPLES OF PROCESS ENERGY IMPROVEMENTS

In this section, some examples of abbreviated studies of process energy improvements are presented. They are intended as illustrative examples only and should not be used as general calculation guidelines.

Individual circumstances will vary from these examples, and calculations should be tailored to fit the specific conditions of the facility being studied.

12.4.1 Recuperator for a Large Brick Kiln

Summary:

The two drying kilns at a large brick manufacturing company presently use ambient air for the combustion air. The air intake for the kiln burner could be modified to draw air from the cooling section of the kiln, thus serving as a recuperator. (A recuperator is a device that preheats the combustion air for a boiler, furnace or oven. Preheating the combustion air increases the system efficiency.) This air has the full oxygen content, yet has been heated as it has cooled the hot bricks.

A simple insulated duct could connect the cooling section to the combustion air intake duct. It is recommended that the air drawn from the cooling section be no more than 800°F and the duct be insulated. Usually the higher the temperature of the combustion air, the more efficient the combustion process. Unfortunately, when dealing with temperatures above 800°F, there is a risk of the burners becoming too hot, resulting in a shorter burner life. By using air at 800°F and taking into account the heat loss in the duct, the risk of harming the burners is reduced. Additional controls may also be needed on the air intake motor due to higher air temperatures. Also, a filtering system may be needed if the new air source has unwanted dust particles.

The cost-effectiveness analysis of this EMO follows.

Data:
      Present air intake temperature: 90°F
      Proposed air intake temperature: 800°F
      Airflow*: 3548 cfm/kiln
      Specific heat of air (Cp) at 800°F: 0.259 Btu/lb °F
      Specific heat of air (Cp) at 90°F: 0.240 Btu/lb °F
      Density of air (p) at 800°F: 0.03 lb/ft^3
      Density of air (p) at 90°F: 0.075 lb/ft^3
      Heat loss through ductwork: 30% (using 3-inch hot
          pipe insulation)
      Operating hours: 8760 h/year
      Natural gas cost: $3.30/Mcf

---

*The airflow is calculated from the proposed burner consumption after excess air control is optimized.

Calculations:

In raising the combustion air temperature, the air mass flow rate must remain constant in order to maintain a correct air-fuel mix:

- **Energy Savings:**

Air mass flow rate	$= m$
	$= (3548 \text{ ft}^3/\text{min}) \times (.075 \text{ lb}/\text{ft}^3)$
	$= 266.1 \text{ lb}/\text{min}$

Savings in Btus = (air mass flow) × $(T_2 - T_1)$ × (average specific heat) × (1 − total heat loss)

$$= (266.1 \text{ lb}/\text{min}) \times (800°F - 90°F) \times$$
$$(.250 \text{ Btu}/\text{lb°F}) \times (60 \text{ min}/\text{h}) \times$$
$$(8760 \text{ h}/\text{year}) \times (2 \text{ kilns}) \times (1 -.30)$$

$$= 34{,}756 \times 10^6 \text{ Btu}/\text{year}$$

Gas savings in Mcf = $(34{,}756 \times 10^6 \text{ Btu}/\text{yr}) \times (1 \text{ Mcf}/10^6 \text{ Btu})$
= 34,756 Mcf/year

(Note: This is a conservative estimate for the savings in gas because it does not include the efficiency of the gas heater.)

- **Cost Savings:**

savings in $ = (34,756 Mcf/year) × ($3.30/Mcf)
= $114,695/year

Implementation Cost Data:

Length of duct: 75 ft/kiln
Ductwork: 304 stainless steel at $5.45/lb
  (24-in.-diameter duct of 14-gauge steel: 20 lb/ft)
Insulation for ductwork: $10.00/ft
Engineering design cost: $5000
Labor and contingency: $5000

- **Implementation Cost:**

Implementation cost = [(75 ft) × ($5.45/lb) × (20 lb/ft) +
  (75 ft) × ($10.00/ft)] × (2 kilns) +
  $5000 + $5000
= $27,850

- **Smple Payback Period:**

$$SPP = \frac{\text{implementation cost}}{\text{annual savings}}$$

$$= \frac{\$27,850}{\$114.695/\text{year}}$$

$$= 0.24 \text{ year}$$

Thus, this EMO is highly cost-effective.

12.4.2 Heat Recovery from Compressors to Space Heat a Warehouse
Summary:

An industrial warehouse is heated with natural gas. The shop next to the warehouse has one 75-hp and one 100-hp air compressor. By installing ductwork from the air compressors to the warehouse, the hot air from the compressors can be used to heat the warehouse. Two dampers are required so that air may be exhausted in the summertime. For this system, automatic dampers might be required so that the temperature of the air for space heating will not be too high. The system can then mix cool outside air with the hot compressor exhaust air if needed.

The cost-effectiveness analysis of this EMO follows.

Data:

Compressor size: 100 and 75-hp
Average air temperature before compressor (T_1): 90°F
Average air temperature after compressor (T_2): 110°F
Hot air flow rate from 100-hp compressor: 10,000 cfm
Operation hours: 992 h/year (24 h/day, 5 days/week; 4 h/day,
    1 day/week; 8 weeks/year)
Natural gas cost: $3.30/Mcf
Efficiency of gas heater: 0.80
Percent load on compressor: 75%
Ductwork length: 35 ft each (70 ft for both)

Calculations:

The Btu savings as well as dollar savings from using the hot air from these compressors can be calculated.

• Energy Savings:

$$
\begin{aligned}
\text{Btu savings} \ = \ & \text{(air flow rate)} \times \text{(density of air)} \times \\
& \text{(specific heat of air)} \times \text{(temperature difference)} \times \\
& \text{(load factor)}
\end{aligned}
$$

For the 100-hp compressor,

$$
\begin{aligned}
\text{Btu savings} = \ & (10{,}000 \ \text{ft}^3/\text{min}) \times (.075 \ \text{lb}/\text{ft}^3) \times (.24 \ \text{Btu}/\text{lb}°\text{F}) \\
& \times (110\text{-}90°\text{F}) \times (60 \ \text{min}/\text{h}) \times (992 \ \text{h}/\text{year}) \times (.75
\end{aligned}
$$

load)

$$
= 160.7 \times 10^6 \ \text{Btu}/\text{year}
$$

For the 75-hp compressor, we'll assume that the savings will be about 75% of savings for the 100-hp compressor, so we have

$$
\begin{aligned}
\text{Btu savings} \quad &= (160.7 \times 10^6 \ \text{Btu}/\text{year})(.75) \\
&= 120.5 \times 10^6 \ \text{Btu}/\text{year} \\[6pt]
\text{Total Btus saved} \quad &= (160.7 \times 10^6 + 120.5 \times 10^6 \ \text{Btu}/\text{year}) \\
&= 281.2 \times 10^6 \ \text{Btu}/\text{year} \\[6pt]
\text{Total Mcf saved} \quad &= (281.2 \times 10^6 \ \text{Btu}/\text{year})(1 \ \text{Mcf}/10^6 \ \text{Btu}) \times (1/.80) \\
&= 351.5 \ \text{Mcf}/\text{year} \\
\text{Total \$ saved} \quad &= (351.5 \ \text{Mcf}/\text{year}) \times (\$3.30/\text{Mcf}) \\
&= \$1{,}160/\text{year}
\end{aligned}
$$

Implementation Cost:

Material cost (for the air intake ductwork):
Insulated flexible duct with vinyl-coated spring steel
(or aluminum): $2.90/linear foot
Two dampers: $30 each

Labor cost (for installation of the duct):
Two persons: $20/h each
Time: 8 hours

$$
\begin{aligned}
\text{Total cost} \ &= \ \text{material cost} + \text{labor cost} \\
&= \ \text{(duct cost}/\text{linear ft)} \times \text{(total linear ft)} \\
&\quad + (2 \ \text{dampers}) \times (\$30/\text{damper}) + \text{(number of} \\
&\quad \text{laborers)} \times \text{(number of h worked)} \times \text{(wage}/\text{h)}
\end{aligned}
$$

$$= (\$2.90/\text{linear ft}) \times (70 \text{ ft}) + (\$30/\text{damper}) \times$$
$$(2 \text{ dampers}) + (2 \text{ laborers}) \times (8 \text{ h}) \times (\$20/\text{h})$$

$$= \$583$$

Simple Payback Period:

$$SPP = \frac{\text{implementation cost}}{\text{annual savings}} = \frac{\$583}{\$1160/\text{year}} = 0.5 \text{ year}$$

This EMO is also highly cost-effective.

12.4.3 Installation of an Economizer for a Plastic Plant

Summary:

This EMO considers the installation of an economizer and an associated control system in a plastics plant. To remove the heat being generated by a large injection molder, the plant currently air-conditions 9 months a year (March through November), three shifts a day. The economizer control will read the dry and wet bulb temperatures to determine if the outside air conditions (temperature and humidity) are more desirable than the present return air. When outside air is more desirable (less enthalpy), the economizer cycle will allow the use of cooler outside air to replace the need for additional conditioned air so that the least amount of energy will be expended to get the air to the desired temperature and humidity.

It was assumed that the return air was at 78°F and 50% relative humidity (a conservative estimate), which corresponds to an enthalpy reading of 30 Btu/lb (air). Btu savings occur when the enthalpy of the outside air is less than the return air. The savings can be easily calculated in this manner by using bin data, which give the number of hours of weather experience in a month at a given temperature range for the geographic area where the facility is located. (See Table 12-2 at the end of this chapter for bin data.)

The cost-effectiveness analysis of this EMO follows.

Data:

Electrical energy cost: $.034/kWh
Return air: 78°F, 50% relative humidity
Air flow rate of unit: 7120 ft^3/min
Operation time: 3 shifts (24 h/day Monday-Friday;

4 h/day Saturday) COP of air conditioner = 3.0

Calculations:
• **Energy Savings**

Savings in Btus =   (total savings of Btu•h/lb
air•year)
× (cfm of unit) × (60 min/h)
× (density of air) × (1/COP)

(See Table 12-3 at the end of this chapter for the calculation of the total savings of Btu•h/lb air•year)

$$= \text{(25,858.5 Btu•h/lb air•year)} \times$$
$$\text{(7120 ft}^3/\text{min)} \times \text{(60 min/h)} \times$$
$$\text{(.075 lb/ft}^3 \text{ air)(1/3)}$$

$$= 276.2 \times 10^6 \text{ Btu/year}$$

Savings in dollars $$= \text{(Btu/year)} \times \text{(kWh/3412 Btu)} \times$$
$$\text{(cost/kWh)}$$

$$= \text{(276.2} \times 10^6 \text{ Btu/year)} \times$$
$$\text{(1 kWh/3412 Btu)} \times \text{(\$.034/kWh)}$$

$$= \$2752.29/\text{yr}$$

Implementation Cost Data:
Two 7.5-ton economizer units: $1216 each
Installation cost: $600

Implementation cost   =   2($1216) + $600
= $3032

Simple Payback Period:

$$\text{SPP} = \frac{\text{implementation cost}}{\text{annual savings}} = \frac{\$3032}{\$2752/\text{year}} = 1.1 \text{ years}$$

This EMO is also quite cost-effective.

12.4.4 HPS Relamp of Refrigerated Storage
Summary:
This EMO recommends a major relamping of a refrigerated storage area at a meat packing company. The storage area presently uses 150-

Watt incandescent lamps. Relamping would replace the existing lamps with 70-Watt high pressure sodium (HPS) lamps and maintain the same level of illumination.

High-pressure sodium lighting is one of the most efficient high-intensity discharge (HID) sources and has excellent lumen maintenance over the lifetime of the lamps. The yellow color of HPS lamps has been a limiting factor for many interior applications, but improvements in color rendition have made its application for interior lighting acceptable even for color-critical areas.

The cost-effectiveness analysis of this EMO follows.

Data:
Present lighting:

Type:	incandescent	Initial lumens: 2350
Size:	150 W	Life: 2500 h
	(150-W input to fixture)	
Quantity: 68		Lamp cost: $1.10

Proposed lighting data:

Type:	high-pressure sodium	Initial lumens: 5400
Size:	70 W	Life: 20,000 h
	(88-W input to fixture)	
Spacing-to-mounting-		Fixture cost: $91.55
height ratio: 2.0		Lamp cost: $39.69

Electrical energy cost: $0.034/kWh

Demand cost: $4.97 (June-October)
                $3.29 (November-May)
                $3.99 (average year-round)

Total area: 9800 ft^2
Mounting height: 8.5 ft
Hours of operation: 4576 h/year (16 h/day, 5.5 days/week, 52 weeks/year)
Cooling unit coefficient of performance (COP): 2.5

Calculations:
Present illumination   = (68 lamps)(2350 lumens/lamp)
                       = 159,800 lumens

$$\text{number of HPS lamps required} = \frac{159{,}800 \text{ lumens}}{5400 \text{ lumens/lamp}}$$

$$= 30 \text{ lamps}$$

$$\text{area of lighting/HPS lamp} = \frac{9800 \text{ ft}^2}{30 \text{ lamps}}$$

$$= 326.67 \text{ ft}^2/\text{lamp}$$

$$\text{spacing requirements} = \left(326.67 \text{ ft}^2\right)^{1/2}$$

$$= 18.07 \text{ ft/lamp}$$

$$\text{space–to–mounting–height ratio} = \frac{18.07 \text{ ft}}{8.5 \text{ ft}} = 2.1$$

(This is extremely close to the recommended spacing-to-mounting-height ratio of 2.0 obtained from the manufacturer's catalog.)

- **Energy savings:**

$$
\begin{aligned}
\text{Present kWh} \;=\; & (68 \text{ lamps}) \times (150 \text{ W/lamp}) \times (4576 \text{ h/year}) \\
& (1 \text{ kW}/1000 \text{ W}) \\
=\; & 46{,}675.20 \text{ kWh/year}
\end{aligned}
$$

$$
\begin{aligned}
\text{Proposed kWh} \;=\; & (30 \text{ lamps}) \times (88 \text{ W/lamp}) \times (4576 \text{ h/year}) \\
& \times (1 \text{ kW}/1000 \text{ W}) \\
=\; & 12{,}080.64 \text{ kWh/year}
\end{aligned}
$$

$$
\begin{aligned}
\text{Savings in kWh} \;=\; & (46{,}675.20 \text{ kWh/year} - 12{,}080.64 \text{ kWh/year}) \\
=\; & 34{,}594.56 \text{ kWh/year} \\
=\; & (34{,}594.56 \text{ kWh/year}) \times (3412 \text{ Btu/kWh}) \\
=\; & 118.04 \times 10^6 \text{ Btu/year}
\end{aligned}
$$

- **Demand savings:**

$$
\begin{aligned}
\text{Present kW (demand)} \;=\; & (68 \text{ lamps}) \times (150 \text{ W/lamp-month}) \\
& \times (1 \text{ kW}/1000 \text{ W}) \\
=\; & 10.20 \text{ kW/month}
\end{aligned}
$$

Table 12-2. Mean frequency of occurrence of dry bulb temperature (°F) with mean coincident wet bulb temperature (°F) for each dry bulb temperature range.

Cooling Season

Temperature range (°F)	May — Observ./h, C_p 02 to 09	May 10 to 17	May 18 to 01	May Total observ.	May Mean coincident wet bulb (°F)	June — Observ./h, G_p 02 to 09	June 10 to 17	June 18 to 01	June Total observ.	June Mean coincident wet bulb (°F)	July — Observ./h, G_p 02 to 09	July 10 to 17	July 18 to 01	July Total observ.	July Mean coincident wet bulb (°F)
105-109							0		0	73				0	72
100-104							8	1	9	74		7	0	7	72
95-99		1		1	68		40	7	47	74		51	8	59	74
90-94		8	1	9	71	2	64	25	91	72	1	70	26	97	74
85-89		33	5	38	70	14	61	53	128	71	11	53	43	107	73
80-84	1	58	21	80	68	50	36	70	156	69	34	35	73	142	71
75-79	13	51	46	110	66	86	17	51	154	67	90	20	55	165	70
70-74	53	40	62	155	64	62	10	24	96	63	83	10	34	127	68
65-69	72	29	55	156	61	20	3	8	31	60	26	2	9	37	65
60-64	55	17	32	104	58	5	1	1	7	55	3	0	0	3	60
55-59	33	9	20	62	53	1	0		1	49	0			0	57
50-54	15	2	5	22	48										
45-49	5	0	1	6	43										
40-44	1		0	1	40										
35-39	0			0	35										
30-34															
25-29															

Table 12-2. (Continued)

Cooling Season

Temperature range (°F)	August Observ./h, Cp 02 to 09	10 to 17	18 to 01	Total observ.	Mean coincident wet bulb (°F)	September Observ./h, Gp 02 to 09	10 to 17	18 to 01	Total observ.	Mean coincident wet bulb (°F)	October Observ./h, Gp 02 to 09	10 to 17	18 to 01	Total observ.	Mean coincident wet bulb (°F)
		2	0	2	73		0		0	72		0		0	61
		15	2	17	73		11	0	11	71		6		6	64
	0	37	7	44	73		30	4	34	71		14	1	15	66
	1	63	20	84	73	1	48	13	62	70		30	5	35	65
	7	64	39	110	72	7	43	35	85	69	2	39	14	55	63
	29	38	70	137	71	30	42	56	128	67	13	44	37	94	62
	82	20	64	166	69	67	33	48	148	65	35	46	50	131	59
	84	7	37	128	67	53	18	46	117	62	54	34	53	141	55
	37	2	8	47	63	45	11	26	82	57	49	21	43	113	51
	7	0	1	8	58	27	3	8	38	53	48	8	27	83	47
	1			1	55	7	1	4	12	49	29	4	13	46	42
						3	0		3	46	12	1	3	16	38
											4	1	1	6	34
											1		1	2	29
											1		0	1	26

Table 12-2. (Continued)
Cooling Season

Temperature range (°F)	Nov Observ./h, Cp 02 to 09	Nov 10 to 17	Nov 18 to 01	Nov Total observ.	Nov Mean coincident wet bulb (°F)	Dec Observ./h, Gp 02 to 09	Dec 10 to 17	Dec 18 to 01	Dec Total observ.	Dec Mean coincident wet bulb (°F)	Jan Observ./h, Gp 02 to 09	Jan 10 to 17	Jan 18 to 01	Jan Total observ.	Jan Mean coincident wet bulb (°F)
105-109															
100-104															
95-99															
90-94															
85-89							0		0	60					
80-84							0		0	59					
75-79		7		7	60		0		0	55					
70-74	2	25	4	31	60		2	0	2	53		1		1	56
65-69	6	26	14	46	57		7	1	8	51		7	1	8	53
60-64	11	30	22	63	53	1	17	3	21	51	2	13	4	19	52
55-59	19	35	29	83	49	7	27	14	48	49	4	17	8	29	49
50-54	29	38	42	109	45	11	37	24	72	45	7	29	18	54	45
45-49	43	34	40	117	42	23	40	39	102	41	11	34	24	69	41
40-44	39	20	32	91	37	38	38	45	121	38	29	38	43	110	37
35-39	43	14	31	88	33	46	30	41	117	33	36	31	42	109	33
30-34	26	7	14	47	29	49	20	34	103	29	52	28	35	115	29
25-29	10	3	9	22	24	32	14	22	68	25	39	19	26	84	25
20-24	10	1	3	14	20	23	10	14	47	20	24	14	23	61	20
15-19	1	0	0	1	15	8	4	7	19	16	19	7	10	36	16
10-14	1	0		1	10	7	2	4	13	11	13	7	7	27	11
5-9						3	0	0	3	7	5	3	5	13	6
0-4						0		0	0	3	6	0	2	8	2
-5--1											1	0		1	-2

Table 12-2. (Continued)
Cooling Season

	February					March					April					Annual (total all months)				
	Observ./h, Cp			Total observ.	Mean coincident wet bulb (°F)	Observ./h, Gp			Total observ.	Mean coincident wet bulb (°F)	Observ./h, Gp			Total observ.	Mean coincident wet bulb (°F)	Observ./h			Total observ.	Mean coincident wet bulb (°F)
	02 to 09	10 to 17	18 to 01			02 to 09	10 to 17	18 to 01			02 to 09	10 to 17	18 to 01			02 to 09	10 to 17	18 to 01		
																	2	0	2	73
																	23	3	26	73
																0	108	16	124	73
																3	219	57	279	70
																20	286	128	434	70
							1		1	58						85	292	262	639	66
		0			62		4		5	61						269	259	322	850	64
		1		0	60	0	10	0	12	59						398	245	314	957	61
		3	0	1	57	0	15	1	20	56						326	219	265	810	58
		7	0	3	57	4	24	2	41	54		2		2	65	250	204	219	673	55
		12	1	8	54	10	30	5	61	51		10		11	66	207	188	203	598	51
		15	4	16	52	15	30	13	73	48	0	20	1	25	64	192	195	208	595	46
		22	7	24	49	24	33	21	93	45	1	32	5	48	63	210	180	209	599	42
		29	15	44	45	30	29	28	103	41	10	41	15	84	61	223	159	213	595	38
		26	23	61	42	33	28	36	115	39	31	38	33	108	58	221	132	178	531	34
		30	33	85	38	29	21	38	91	34	39	33	39	113	54	207	85	133	425	29
		32	39	101	33	28	13	42	69	29	41	26	41	104	49	132	57	89	278	25
		37	32	101	29	21	7	28	37	25	41	19	37	90	45	90	38	56	184	20
		41	29	88	25	13	2	20	15	20	35	13	30	70	42	42	17	25	84	16
		29	22	64	20	7	1	9	4	16	26	4	22	42	38	27	9	13	49	11
		23	13	47	16	2	0	3	3	11	12	1	12	17	35	10	3	5	18	6
		12	5	21	12	1	0	1	1	8	3	1	4	3	30	7	0	2	9	2
		5	1	7	7	0		1	0	4	1	1	1	3	26	1	0		1	-2

Proposed kW (demand) =   (30 lamps) × (88 W / lamp-month)
                                              × (1 kW / 1000 W) = 2.64 kW
Savings in kW (demand) =   10.20 kW / month – 2.64 kW / month
                                    =   7.56 kW / month

• **Replacement cost savings:**
Present replacement cost =   (68 lamps) × ($1.10 / lamp) ×
                                              (1 / 2500 h) × (94567 h / year)
                                    = $136.91 / year

Proposed replacement cost =   (30 lamps) × ($39.69 / lamp)
                                                × (1 / 20,000 h) × (4576 h / year)
                                      = $272.43 / year

Savings in replacement cost = $136.91 / year – $272.43 / year
                                        =   – $135.52 / year

• **Savings in reduced refrigeration needs:**
(The reduction in wattage from HPS relamping will also reduce
the amount of heat generated by the lamps that must be removed by the
refrigeration system. This translates to a savings in energy (kWh) but not
a savings in demand (kW) since the air conditioner still operates at its
rated kW.)

Present cost of heat removal   =   (46,675.20 kWh / year) ×
                                                (1 / 2.5) × ($.034 / kWh)
                                      = $634.78 / year
Cost of heat removal with proposed lighting
            =   (12,080.64 kWh / year) × (1 / 2.5) × ($.034 / kWh)
            = $164.30 / year

Total savings in refrigeration cost
            = $634.78 / year – $164.30 / year
            = $470.48 / year

Total savings in $ =   savings in kWh cost + savings in kW
                                    cost + savings in replacement cost +
                                    savings in heat removal cost

                        =   (34,594.56 kWh / year) × ($.034 / kWh) +
                              (7.56 kW / month) × (12 months / year) ×

$$($3.99/kW) - $135.52/year$$
$$+ $470.48/year$$

$$= \quad $1873.15/year$$

Implementation Cost:

Fixtures: (30 fixt.) × ($91.55/fixt.):	$2746.50
Lamps: (30 lamps) × ($39.69/lamp):	$1190.70
Labor (for adding HPS lamps):	
(30 fixt.) × (1 h/fixt.) × ($15/h):	$ 450.00
Labor (incandescent lamp removal):	
(68 fixt.) × (.5 h/fixt.) × ($15/h):	$ 510.00
Total:	$4897.20

Simple Payback Period:

$$\text{SPP} = \frac{\text{implementation cost}}{\text{annual savings}} = \frac{$4897.20}{$1873.15/year} = 2.61 \text{ years}$$

This EMO has a SPP that is over two years, but many companies would still find it an attractive investment.

12.4.5 Sawdust Collection Control System
Summary:

A wood shop at a furniture factory has a sawdust collection system which collects sawdust from five machines. The system uses overhead hoods with vacuum motors, and exhausts the dust and air to the outside through ductwork. The dust collection system operates at all times the plant is in operation. By installing a damper and a control system to shut off the air collection system from any one of the machines not in use, significant savings will be realized.

Each of the five vacuum ducts must have a damper, and a speed control is required for the vacuum motor. Used in conjunction with a programmable controller, the damper corresponding to a specific machine will open that duct when the machine is turned on while the speed of the vacuum motor is increased proportionately. When no machines are in use, the motor is completely shut off. This EMO provides savings both in electricity to run the vacuum motor and in the gas used to heat the air that is being evacuated from the plant.

The cost-effectiveness analysis of this EMO follows.

Data:

Vacuum electric motor: 20-hp
  Operating time of motor: 8.5 hours a day
  Load on motor: 100%
  Efficiency of motor: 88%
Duct system: Exhausts 6500 ft^3 of air per minute
Efficiency of gas heating plant: 70%
Each machine is used approximately 2 hours/day.

Electric cost:
  Demand charge: $4.50/kW/month
  Energy charge: $0.02915/kWh
Natural gas cost: $3.50/10^6 Btu

Calculations:

• Energy Savings:

Electricity cost = demand charge + consumption charge
                  = (20 hp) × (0.746 kW/hp) × (1/.88)
                     × (12 months) × ($4.50/kW/month) +
                     (20 hp) × (0.746 kW/hp) × (1/.88)
                     × (8.5 h/day) × (250 days/year)
                     × ($0.02915/kWh)
                  = $915.55 + $1050.23
                  = $1965.78

Electricity savings = demand savings + energy savings
demand savings   = (motor capacity) × (0.746 kW/hp)
                        × (1/efficiency) × (12 months/year)
                        × (cost/kW/month) × (% demand
                          reduction)

                     = (20 hp) × (0.746 kW/hp) × (1/.88)
                        × (12 months/year) × ($4.50/kW/month)
                        × (3/5)

                     = $549.33/year

(The percent demand reduction is taken to be 3/5 = 60% based on
the assumption that no more than two machines will run at the
same time for the 30 minute demand averaging interval.)

energy savings = (motor capacity) × (0.746 kW/hp)
                   × (1/efficiency) × (6.5 h) ×

$$(250 \text{ days}/\text{year}) \times (\text{cost}/\text{kWh})$$

$$= \quad (20 \text{ hp}) \times (0.746 \text{ kW}/\text{hp}) \times (1/.88)$$
$$\times (6.5 \text{ h}) \times (250 \text{ days}/\text{year}) \times$$
$$(\$0.02915/\text{kWh})$$

$$= \quad \$803.12/\text{year}$$

Total electricity savings $\quad = \quad (\$549.33 + \$803.12)/\text{year}$
$$= \quad \$1352.45/\text{year}$$

Gas cost $\quad = \quad (6500 \text{ ft}^3/\text{min}) \times (8.5 \text{ h}/\text{day}) \times (60 \text{ min}/\text{h}) \times$
$$(100 \text{ heating days}/\text{year}) \times (0.075 \text{ lb}/\text{ft}^3)$$
$$\times (.24 \text{ Btu}/\text{lb}) \times (65\text{-}45°\text{F}) \times$$
$$(1/.7) \times (\$3.50/10^6 \text{ Btu})$$

$$= \quad \$596.70/\text{yr}$$

Gas savings from reduced heat loss
$$= \quad (\text{air volume flow rate}) \times (\text{operating hour}$$
$$\text{reduction}) \times (60 \text{ min}/\text{h}) \times (\text{heating days}/\text{year})$$
$$\times (\text{air density}) \times (\text{specific heat of air})$$
$$\times (\text{temperature difference}) \times (1/\text{eff}) \times$$
$$(\text{gas cost})$$
$$= \quad (6500 \text{ ft}^3/\text{min}) \times (8.5 - 2) \text{ h}/\text{day} \times (60 \text{ min}/\text{h})$$
$$\times (100 \text{ days}/\text{year}) \times (0.075 \text{ lb}/\text{ft}^3)$$
$$\times (.24 \text{ Btu}/\text{lb}) \times (65\text{-}45°) \times (1/.7) \times (\$3.50/10^6 \text{ Btu})$$

$$= \quad \$456.30/\text{year}$$

• **Total Cost Savings:**

Total annual cost savings $\quad = \quad$ electricity cost savings + gas
cost savings
$$= \quad \$456 + \$1352$$
$$= \quad \$1808$$

Implementation Cost Data:

Speed control:	$1500
Programmable controller:	550
Electric dampers (5 at $65)	325
Wire, switches	50
Installation (estimate)	300

Implementation Cost: $2725

Simple Payback Period:

$$SPP = \frac{\text{implementation cost}}{\text{annual savings}} = \frac{\$2725}{\$1808/\text{year}} = 1.5 \text{ years}$$

This EMO is also quite cost-effective.

12.5 TWENTY-FIVE
COMMON ENERGY MANAGEMENT OPPORTUNITIES

Through our combined energy management experiences with over 200 manufacturing plants and a review of the literature, we have found that a number of energy management opportunities (EMOs) have been used time after time. The astute energy manager must become familiar with these opportunities and be ready to apply them (*as well as others*) in energy management work.

Twenty-five of these common changes are summarized below. Most are process modifications, but a few are lighting and space conditioning oriented. The order of listing is not significant; the order would change depending on whether one were listing by frequency of occurrence, amount of savings, financial return on investment, etc.

1. *Switch to energy-efficient lamps.* Switch existing lamps to the energy-efficient ones, such as 34-watt energy-efficient fluorescent lamps for conventional 40-W ones, or replace T-12 lamps with T-8 or T-10 lamps.

2. *Switch to energy-efficient light sources.* Change to more efficient sources usually requiring fixture changes. Change from incandescent lights to fluorescent or from mercury vapor or fluorescent to high-pressure sodium for in-plant lighting, a very frequent conversion.

3. *Use night setback-setup.* Turn temperatures up or down at night when needs are reduced. Examples include large ovens that cannot be turned off, large refrigeration units where night operations involve less infiltration (fewer people going in and out), and space conditioning.

4. *Turn off equipment.* Turn off exhaust fans, ovens, motors, or any other equipment when not needed.

5. *Move air compressor intake to cooler locations.* Move air intakes from hot equipment rooms to cooler (often outside) locations. Efficiency improvements are large and paybacks attractive.

6. *Eliminate leaks in steam and compressed air systems.* Steam and compressed air leaks are very expensive and should be fixed. Technology exists for repairing leaks without shutting the equipment down. Night audits (when noise is minimized) often turn up large numbers of these leaks.

7. *Control excess air.* As shown earlier, careful control of combustion air can lead to significant energy savings.

8. *Optimize plant power factor.* Depending on the utility billing schedule and the company's power factor, large savings may be available through power factor improvement.

9. *Insulate bare tanks, vessels, lines, and process equipment.* Good savings are often available through insulation of process lines and tanks. Condensate return lines and tanks are often not insulated.

10. *Install storm windows, doors, and weather stripping.* Although these are often difficult to justify, sizable savings are sometimes available. This is especially true for large glass exposures in cold climates.

11. *Use energy-efficient electric motors.* When replacement is necessary or for new applications, energy-efficient motors can usually be justified. Electric utilities often provide rebates to customers who replace standard motors with energy-efficient models.

12. *Preheat combustion air.* Recuperators can save large amounts of energy and money. Sometimes they are highly cost effective.

13. *Reduce the pressure of compressed air and steam.* If the pressures have been overdesigned, a reduction will not harm the process. In such cases, large savings are possible.

14. *Insulate walls, ceilings, roofs, and doors.* Industrial plants are frequently poorly insulated. Insulation in dropped ceilings, on roofs or walls, and doors may be cost-justified.

15. *Recover heat from air compressor.* Larger air compressors reject large amounts of heat through air or water cooling. Proper design can allow this waste heat to be used for space conditioning in the winter and to be exhausted in warm weather. Sometimes the payback is very attractive.

16. *Insulate dock doors.* Plastic strips, dock bumpers, vestibules, or air screens all help block infiltration through large dock doors. If the space is heated and/or air-conditioned, the savings can be very large.

17. *Install economizers on air conditioners.* In some areas of the country, economizers can be very attractive. They allow the optimum use of outside air in air conditioning. Sometimes outside air can be used and the air conditioner turned off.

18. *Use radiant heat.* Sometimes infrared heaters can be used to spot-heat rather than heat entire areas. Infrared heat (like the sun) warms objects and people but not space. The payback can be very attractive.

19. *Return steam condensate to the boiler.* Returning hot condensate can yield dramatic savings in energy, water, and water conditioning costs. Return lines should probably be insulated.

20. *Change product design to reduce energy requirements.* Product redesign can often reduce the energy necessary in heat treating, cleaning, coating, painting, etc.

21. *Explore waste heat recovery for space exhaust systems.* Large amounts of exhaust in buildings that are heated and/or air-conditioned offer the potential for waste heat recovery.

22. *Install devices to improve heat transfer in boilers.* Turbulators and other devices designed to reap more energy out of the combustion process are often very cost effective.

23. *Reschedule operations to reduce peak demand.* Sometimes simple changes in equipment scheduling can dramatically reduce demand charges.

24. *Cover open heated tanks.* Covering open heated tanks can often lead to big energy savings. Floating balls, cantilevered tops, and rubber flaps have all been used as covers.

25. *Spot-ventilate or use air filters.* In welding areas or other areas where large amounts of ventilation are required, spot ventilation can often reduce the amount needed. Also, electrostatic or other types of air filters can sometimes allow reuse of the air. Savings are especially large if the space is heated and/or air-conditioned.

12.6 SUMMARY

This chapter has provided a suggested procedure for process improvement that is based on the industrial engineering concept of work simplification. It has also provide a detailed presentation of electric motor and drive system efficiency improvements. These are particularly important because of the large quantity of energy used in business and industry by electric motors. Some case studies of process energy management were also presented. The reader should not consider these examples as typical results but should use them as a starting point for potential analysis. Because process energy management can be intricate and complex, the energy manager must understand the entire process system and consider all of the impacts of any proposed changes.

Table 12-3. Enthalpy data [Btu•h/lbm (air)]

March					
Temperature (°F)					Btu•hr
Dry	Wet	h	$\Delta h_{(30-h)}$	Hours	1 lbm (air)
77	59	25.8	4.2	12	50.4
72	56	24	6	20	120
67	54	22.6	7.4	41	303.4
62	51	21	9	61	549
57	48	19.2	10.8	73	788.4
52	45	17.6	12.4	93	1153.2
47	41	15.8	14.2	103	1462.6
42	38	14.5	15.5	115	1782.5
					6209.5

April

| Temperature (°F) | | | | | Btu•hr |
Dry	Wet	h	$\Delta h_{(30-h)}$	Hours	1 lbm (air)
77	63	28.6	1.4	48	67.2
72	61	27.2	2.8	84	235.2
67	58	25	5	108	540
62	54	22.6	7.4	113	836.2
57	49	19.8	10.2	104	1060.8
52	45	17.6	12.4	90	1116
47	42	16.2	13.8	70	966
42	38	14.5	15.5	42	651
					5472.4

May

| Temperature (°F) | | | | | Btu•hr |
Dry	Wet	h	$\Delta h_{(30-h)}$	Hours	1 lbm (air)
77	66	no savings			
72	64	29.4	.6	155	93
67	61	27.2	2.8	156	436.8
62	58	25	5	104	520
57	53	22	8	62	496
52	48	19.2	10.8	22	237.6
47	43	16.8	13.2	6	79.2
42	40	15.2	14.8	1	14.8
					1877.4

June

| Temperature (° F) | | | | | Btu•hr |
Dry	Wet	h	$\Delta h_{(30-h)}$	Hours	1 lbm (air)
77	69	no savings			0
72	67	no savings			0
67	63	28.5	1.4	96	134.4
62	60	26.5	3.5	31	108.5
57	55	23.2	6.8	7	47.6
52	49	19.8	10.2	1	10.2
47				None	0
42				None	
					300.7

July

Temperature (° F)					Btu•hr
Dry	Wet	h	$\Delta h_{(30-h)}$	Hours	1 lbm (air)
77	70	no savings			0
72	68	no savings			0
67	65	no savings			0
62	60	26.5	3.5	3	10.5
57				None	0
52				None	0
47				None	0
42				None	0
					10.5

August

Temperature (° F)					Btu•hr
Dry	Wet	h	$\Delta h_{(30-h)}$	Hours	1 lbm (air)
77	69	no savings			0
72	67	no savings			0
67	63	28.6	1.4	47	65.8
62	58	25	5	8	40
57	55	23.2	6.8	1	6.8
52				None	0
47				None	0
42				None	0
					112.6

September

Temperature (° F)					Btu•hr
Dry	Wet	h	$\Delta h_{(30-h)}$	Hours	1 lbm (air)
77	67	no savings			0
72	65	no savings			0
67	62	28	2	117	234
62	57	24.6	5.4	82	442.8
57	53	22.2	7.8	38	296.4
52	49	19.8	10.2	12	122.4
47	46	18.4	11.6	3	34.8
42				None	0
					1130.4

October

| Temperature (° F) | | | | | Btu•hr |
Dry	Wet	h	$\Delta h_{(30-h)}$	Hours	1 lbm (air)
77	63	28.6	1.4	55	77
72	62	28	2	94	188
67	59	25.8	4.2	131	550.2
62	55	23.2	6.8	141	958.8
57	51	21	9	113	1017
52	47	18.8	11.2	83	929.6
47	42	16.2	13.8	46	634.8
42	38	14.5	15.5	16	248
					4603.4

November

| Temperature (° F) | | | | | Btu•hr |
Dry	Wet	h	$\Delta h_{(30-h)}$	Hours	1 lbm (air)
77	60	26.5	3.5	7	24.5
72	60	26.5	3.5	31	108.5
67	57	24.6	5.4	46	248.4
62	53	22.2	7.8	63	491.4
57	49	19.8	10.2	83	846.6
52	45	17.6	12.4	109	1351.6
47	42	16.2	13.8	117	1614.6
42	37	14	16	91	1456
					6141.6

$$\text{Total} = \frac{\text{Btu} \cdot \text{h}}{1 \text{ lbm (air)}}$$

March	6,209.5
April	5,472.4
May	1,877.4
June	300.7
July	10.5
August	112.6
September	1,130.4
October	4,603.4
November	6,141.6
	25,858.5

$$\text{Total} = \frac{\text{Btu} \cdot \text{h}}{1 \text{ lbm (air)} \cdot \text{year}}$$

References

1. Benjamin W. Niebel, *Motion and Time Study*, Ninth Edition, James D. Irwin, Homewood, IL, 1993.
2. Steve Nadel et. al., *Energy Efficient Motor Systems*, American Council for an Energy Efficient Economy (ACEEE), Washington, DC, 1991.
3. *Energy-Efficient Motor Selection Handbook*, Bonneville Power Administration, February, 1992. Prepared by the Washington State Energy Office.
4. *MotorMaster*, Washington State Energy Office, Olympia, WA, 1993.
5. Steve Nadel et. al., *Energy Efficient Motor Systems*, American Council for an Energy Efficient Economy (ACEEE), Washington, DC, 1991.
6. *Ibid.*
7. *Compressed Air Buyer's Guide*, Gardner-Denver Compressor Products, Quincy, IL, 1991.
8. *Your Guide to Reciprocating Compressors—A Brief Review of Theory and Practice*, Joy Manufacturing Company, Michigan City, IN, 1987.
9. Robert K. Hoshide, "Electric Motor Do's and Dont's, *Energy Engineering*, Vol. 91, No. 1, 1994.

Bibliography

Brown, Harry L. et al, *Analysis of 108 Industrial Processes*, Fairmont Press, Atlanta, GA, 1985.

Capehart, Barney, Editor, *Encyclopedia of Energy Engineering and Technology*," Taylor and Francis/CRC Publishing Company, Boca Raton, FL, July, 2007.

Dean, Norman L., *Energy Efficiency in Industry*, Ballinger Publishing Company, Cambridge, MA, 1980.

Gandhi, Nikhil, "Industrial Energy Management Success Using IE Techniques," *Proceedings of the Spring 1989 Industrial Engineering Institute Conference*, Toronto, Canada, May, 1989.

Hu, S. David, *Handbook of Industrial Energy Conservation*, Van Nostrand Reinhold Company, New York, NY, 1983.

Metals Fabrication Reference Guide, EPRI Center for Materials Fabrication, Battelle Laboratory, Columbus, OH, 1993.

Optimizing Energy Use in the Process Industries, Volumes 1-4, Electric Power Research Institute, Palo Alto, CA, 1990.

Payne, F. William, *Advanced Technologies: Improving Industrial Energy Efficiency*, Fairmont Press, Atlanta, GA, 1985.

Talbott, E.M., *Compressed Air Systems: A Guidebook on Energy and Cost Savings*, Second Edition, Fairmont Press, Atlanta, GA, 1993.

Turner, W.C., Senior ed., *Energy Management Handbook*, Second Edition, Fairmont Press, Atlanta, GA, 1993.

Turner, W.C., et al., Numerous Energy Analysis and Diagnostic Center Reports, Oklahoma Industrial Energy Management Program, Stillwater, OK, 1980-1983.

Witte, L.C., Schmidt, P.S. and D.R. Brown, *Industrial Energy Management and Utilization*, Hemisphere Publishing Corporation, Washington, DC, 1988.

Chapter 13

Renewable Energy Sources and Water Management

13.0 INTRODUCTION

Renewable energy sources are those sources that replenish themselves and so are essentially inexhaustible, such as solar, wind and biomass energy. While renewable energy sources are not a major percentage of energy sources currently being utilized, their usage is expected to grow substantially, since they are typically less environmentally damaging than traditional energy sources. The future is likely to see more and more utilization of these renewable sources.

In this chapter we examine a selected subset of these potential sources in the following order: solar-active, solar-passive, solar-photovoltaic, wind, and refuse. Emphasis is on applications in the industrial and commercial environment. In the last part of the chapter we discuss management of another vital and renewable resource: water. Water use will likely cause a major crisis someday soon. The energy manager skilled in water management will be prepared to meet this challenge.

13.1 RENEWABLE ENERGY TECHNOLOGY

The largest portion of the contribution by renewable energy today comes from mature technologies that make use of biomass and hydro-power resources. The newer technologies developed over the past two decades are beginning to enter the market and will provide an increasing share of renewable energy supplies in coming decades [1]. The percentage utilization of renewable resources has held relatively constant over the last ten years, and is around 8% of the U.S. national energy supply. Figure 13-1 shows the contributions of the various types of renewable energy sources.

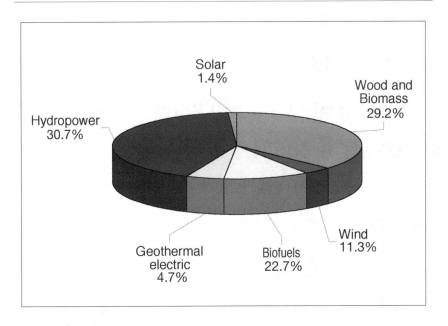

Figure 13-1. U.S. Renewable Energy Supply (2010): 8.167 quads. Source: Monthly Energy Review for March, 2011, Energy Information Agency, Washington, DC, April, 2011.

The contribution of renewable energy to the total energy requirements of the U.S. is expected to grow substantially over the next forty years. A broad-based study of the potential for renewable energy was conducted by five National Laboratories as part of the background for the National Energy Strategy [2]. This study projected that renewable energy sources would provide from 15% to 28% of our total energy supply in the year 2030. As concerns for the environment grow, the benefits of renewable sources in terms of reduced pollution emissions and reduced impacts from energy production will serve as positive factors in the growing use of these renewable energy technologies.

Since the focus of this chapter is on the use of renewable energy sources in commercial buildings and in industry, most of the discussion will be directed toward the use of active and passive solar systems for space heating, water heating, process heating and electricity generation. Wind energy will also be discussed as a source of electricity generation. Biomass and refuse will also be discussed, since they are sources of inexpensive fuel for many industries and some commercial buildings. Few businesses or industries directly operate hydroelectric or geothermal powered electric generation, so those sources will be covered only briefly.

13.2 SOLAR ENERGY

13.2.1 Solar Insulation

Approximately 430 Btu/h/ft^2 of solar energy hits the earth's atmosphere. Because of diffusion in the atmosphere and clouds, this is greatly reduced to somewhere around a maximum of 300 Btu/h/ft^2 on the earth's surface at 40°N latitude. This maximum, of course, only occurs at certain times of the day and year, so the average is significantly less. However, at this rate a set of collectors designed to develop 1×10^6 Btu/h of energy would have to be 3333 ft^2 in size without allowing for cloudiness, variances throughout the day, or collector efficiency. If the collector were tracking the sun throughout the day, it might be able to gather 8×10^6 Btu/day. Assuming it operates 365 days/year, the collector would be able to harvest an absolute maximum of

$$(8 \times 10^6 \text{ Btu/day})(365 \text{ days/year}) = 2920 \times 10^6 \text{ Btu/year}$$

At $\$4.00/10^6$ Btu, this energy would be worth \$11,680/year. The necessary collector space is 3333 ft^2, and the installed cost including controls might be around \$20/ft^2. Therefore, the cost of the proposed collectors would be around \$66,660, making the payback *under ideal conditions* somewhere around 5.7 years. Actual conditions would likely require a significantly larger collector, as will be shown below.

Detailed calculations of the amount of solar energy striking a surface located at a given latitude and tilted at a certain angle require knowledge of several angles, including the solar altitude angle, the solar azimuth angle, and the tilt angle. The values all vary with time of day, month, location, tilt of collector, etc. Tables have been developed to help a person determine the amount of solar energy available [3,4]. Table 13-1 is an example of such a table.

Example 13-1: A solar collector is to be located at Lincoln, Nebraska (approximately 40°N latitude). Find the actual energy available, the value of that energy, and the payback time for the cost of the solar collector.

Solution: Using Table 13-1, look up the data for Lincoln, NE. We find the solar radiation on a surface tilted at 40° to average about 0.6×10^6 Btu/ft^2/year. Assuming the collector is 70% efficient, the energy available is about 0.42×10^6 Btu/ft^2/year. Thus, the 3333 ft^2 collector discussed above would supply about:

Table 13-1. Average solar radiation for selected cities.

| | | Average daily radiation (Btu/day•ft²) | | | | | | | | | | | |
City	Slope	Jan.	Feb.	March	Apr.	May	June	July	Aug.	Sept.	Oct.	Nov.	Dec.
Albuquerque, NM	Hor.	1134	1436	1885	2319	2533	2721	2540	2342	2084	1646	1244	1034
	30	1872	2041	2295	2411	2346	2390	2289	2318	2387	2251	1994	1780
	40	2027	2144	2319	2325	2181	2182	2109	2194	2369	2341	2146	1942
	50	2127	2190	2283	2183	1972	1932	1889	2028	2291	2369	2240	2052
	Vert.	1950	1815	1599	1182	868	754	795	1011	1455	1878	2011	1927
Atlanta, GA	Hor.	839	1045	1388	1782	1970	2040	1981	1848	1517	1288	975	740
	30	1232	1359	1594	1805	1814	1801	1782	1795	1656	1638	1415	1113
	40	1308	1403	1591	1732	1689	1653	1647	1701	1627	1679	1496	1188
	50	1351	1413	1551	1622	1532	1478	1482	1571	1562	1679	1540	1233
	Vert.	1189	1130	1068	899	725	659	680	811	990	1292	1332	1107
Boston, MA	Hor.	511	729	1078	1340	1738	1837	1826	1565	1255	876	533	438
	30	830	1021	1313	1414	1677	1701	1722	1593	1449	1184	818	736
	40	900	1074	1333	1379	1592	1595	1623	1536	1450	1234	878	803
	50	947	1101	1322	1316	1477	1461	1494	1448	1417	1254	916	850
	Vert.	895	950	996	831	810	759	791	857	993	1044	842	820
Chicago, IL	Hor.	353	541	836	1220	1563	1688	1743	1485	1153	763	442	280
	30	492	693	970	1273	1502	1561	1639	1503	1311	990	626	384
	40	519	716	975	1239	1425	1563	1544	1447	1307	1024	662	403
	50	535	723	959	1180	1322	1341	1421	1363	1274	1034	682	415
	Vert.	479	602	712	746	734	707	754	806	887	846	610	373

Location		1	2	3	4	5	6	7	8	9	10	11	12
Ft. Worth, TX	Hor.	927	1182	1565	1078	2065	2364	2253	2165	1841	1450	1097	898
	30	1368	1550	1807	1065	1891	2060	2007	2097	2029	1859	1604	1388
	40	1452	1601	1803	1020	1755	1878	1845	1979	1995	1907	1698	1488
	50	1500	1614	1758	957	1586	1663	1648	1820	1914	1908	1749	1549
	Vert.	1315	1286	1196	569	728	679	705	890	1185	1459	1509	1396
Lincoln, NE	Hor.	629	950	1340	1752	2121	2286	2268	2054	1808	1329	865	629
	30	958	1304	1605	1829	2004	2063	2088	2060	2092	1818	1351	1027
	40	1026	1363	1620	1774	1882	1909	1944	1971	2087	1894	1450	1113
	50	1068	1389	1597	1679	1724	1720	1763	1838	2030	1922	1512	1170
	Vert.	972	1162	1156	989	856	788	828	992	1350	1561	1371	1100
Los Angeles, CA	Hor.	946	1266	1690	1907	2121	2272	2389	2168	1855	1355	1078	905
	30	1434	1709	1990	1940	1952	1997	2138	2115	2066	1741	1605	1439
	40	1530	1776	1996	1862	1816	1828	1966	2002	2037	1788	1706	1550
	50	1587	1799	1953	1744	1644	1628	1758	1845	1959	1791	1762	1620
	Vert.	1411	1455	1344	958	760	692	744	918	1230	1383	1537	1479
New Orleans, LA	Hor.	788	954	1235	1518	1655	1633	1537	1533	1411	1316	1024	729
	30	1061	1162	1356	1495	1499	1428	1369	1456	1490	1604	1402	1009
	40	1106	1182	1339	1424	1389	1309	1263	1371	1451	1626	1464	1058
	50	1125	1174	1292	1324	1256	1170	1137	1259	1381	1610	1490	1082
	Vert.	944	899	847	719	599	546	548	647	843	1189	1240	929
Portland, OR	Hor.	578	872	1321	1495	1889	1992	2065	1774	1410	1005	578	508
	30	1015	1308	1684	1602	1836	1853	1959	1830	1670	1427	941	941
	40	1114	1393	1727	1569	1746	1739	1848	1771	1680	1502	1020	1042
	50	1184	1442	1727	1502	1622	1594	1702	1673	1651	1539	1073	1116
	Vert.	1149	1279	1326	953	889	824	890	989	1172	1309	1010	1109

Source: Reproduced from Reference 2.

$$\text{Solar energy} = (3333 \text{ ft}^2) \times (0.42 \times 10^6 \text{ Btu}/\text{ft}^2/\text{yr})$$

$$= \underline{1400 \times 10^6 \text{ Btu}/\text{year}}$$

At \$4.00/10^6 Btu, the energy value would be:

$$\text{Energy value} = (\$4.00/10^6 \text{ Btu}) \times (1400 \text{ Btu} \times 10^6/\text{yr})$$

$$= \underline{\$5600/\text{year}}$$

The time to pay back the cost of the collector—\$66,660—would be:

$$\text{Simple payback period} = (\$66,660)/(\$5600/\text{yr})$$

$$= \underline{11.9 \text{ years}}$$

Considering the practical factors in this application lengthens the payback time substantially from the original 5.7 years determined earlier.

13.2.2 Solar Collectors

A solar collector is a device used to thermally collect, store, and move solar thermal energy. Essentially, solar collectors are heat exchangers that transfer the energy of incident solar radiation to sensible heat in a working fluid—liquid or air [5]. There are many different types of solar collectors, as shown in Figure 13-2.

13.2.2.1 Flat-Plate Collectors

A flat-plate solar collector generally consists of a shallow metal or wooden box which has a glass or plastic transparent cover, and which contains a black absorption plate that transfers heat to some fluid. The sun's shortwave radiation passes through the transparent cover, enters the collector and heats a fluid (usually water with or without antifreeze, or air). The hot fluid is then moved from the collector to the point of use or to storage for later use. A flat-plate collector almost always faces to the south (in the northern hemisphere) and is tilted at some angle. A typical flat-plate solar collector is illustrated in Figure 13-3.

A typical flat-plate solar collector application is given in Figure 13-4. Here, solar energy heats an ethylene glycol mixture that is pumped to a storage tank. The tank then heats water through a heat exchanger for alternative use as shown. In some applications, such as preheating boiler makeup water, the water itself can be pumped through the collector to a

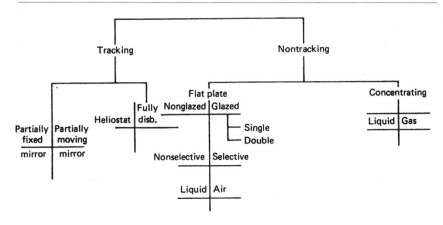

Figure 13-2. Types of solar collectors.

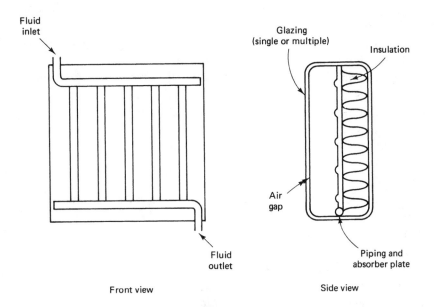

Front view   Side view

Figure 13-3. Flat-plate solar collector.

storage tank or directly to the boiler room. In such applications, care must be taken to prevent freezing; drain down provisions for the solar collector are usually employed.

Glazing:

The flat plate solar collector may be glazed or nonglazed. The most common *glazing* is tempered glass which allows the shortwave radiation

Figure 13-4. Typical flat-plate solar collector application.

of the sun to enter the collector but prevents the longer-waved reradiation from leaving. This produces a greenhouse effect and increases the efficiency of the collector but also increases the cost. Dual glazing would further cut down the heat loss while not appreciably restricting incoming solar energy, and is used for most high temperature flat plate collectors. Unglazed collectors are quite efficient at lower temperature applications such as swimming pool heaters, while glazed collectors are more efficient at higher temperatures. Typical flat-plate collector efficiencies with different glazings are given in Figure 13-5. T_{in} is the temperature of the water coming into the collector, and $T_{ambient}$ is the temperature outside the collector. The graph shows that the heating ability of the collector is greater with glazings that have higher insulating capability.

Selectivity:

The *selectivity* of the collector absorbing surface is an important property that affects a collector's efficiency. The collector must absorb shortwave radiation readily and emit long-wave radiation stingily. Surfaces with high shortwave absorption and low long-wave emittance are *selective surfaces*. Selective surfaces perform better at higher temperatures than nonselective surfaces. A single glazed selective surface collector has efficiencies very similar to a double-glazed nonselective surface collector.

Transport medium:

The *medium* chosen to move the thermal energy from the collector to the point of use or to storage can be either liquid or air. Each has advantages and disadvantages. It is much more difficult and expensive to move thermal energy with air than with liquid. In fact, the horsepower required to move the same amount of thermal energy may be 10 times higher for an air system than for a liquid system. Air systems also have lower heat transfer rates, so the system must be carefully designed to provide a suf-

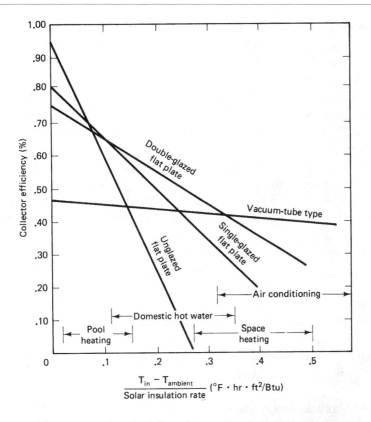

Figure 13-5. Typical flat-plate collector efficiencies

ficiently large heat transfer surface. However, air does not freeze. In liquid systems, ethylene glycol or some other antifreeze must be used, or the system must have well-designed drain down controls. In addition, air systems do not have corrosion problems, and leaks do not present as much of a problem as with liquid systems.

The examples presented thus far have assumed the fluid is a liquid, but many applications are suitable for air. For example, an air solar collector could be used as an air preheater for an industrial furnace or boiler. Although this application is not widespread, it can be useful if the time of solar energy availability and the time of industrial heat use coincide (i.e., the energy is needed when the sun is shining).

13.2.2.2 Concentrating collectors

A need for temperatures of 250°F or higher usually requires a concentrating collector. The surface of a concentrating collector must be highly reflective, enabling concentration of the sun's rays on the heat absorption device. The heat transfer fluid can be a liquid or gas. A concentrating collector is usually also a tracking collector in order to keep the sun's rays focussed on a small surface. A typical design for a parabolic trough-type, tracking collector is shown in Figure 13-6. The collector can track in an east to west direction to follow the daily sun, in a north to south direction to follow the seasons, or both. Concentrating collectors that accurately track the sun's position are more efficient than those that do not track the sun's position as well.

Other types of concentrating, tracking collectors use movable mirrors that can concentrate the solar energy on a small surface that remains fixed. The *power tower* is such an application where the absorption surface or central receiver surface is located in a tower. Tracking mirrors are located on the ground around the base of the tower. These fully tracking mirrors are usually computer-controlled to concentrate the maximum amount of solar energy on the tower. Applications are mainly for steam generation used to produce electric power. This type of application takes a large amount of land area and requires careful maintenance. The most notable power tower is located in California, and is called Solar One [6]. It was operated by Southern California Edison Company up until 1989. This facility had a capacity of 10 MW, and successfully generated electric energy for almost ten years.

13.2.3 Solar Thermal Storage

One of the biggest obstacles to widespread solar utilization is that often the solar energy is not needed when it is available and it is not available when it is needed. For example, maximum heat is usually needed when the sun is not shining, especially at night. Also, solar energy flows cannot readily be regulated. When the sun shines, the collector usually delivers energy at its full capacity. Tracking collectors can be programmed

Figure 13-6. Parabolic trough solar collector

to turn away from the sun, and can be regulated, but this is not generally a cost-effective mode of operation.

For these reasons, many solar applications require some type of storage system. The system must both store energy for later use and regulate energy flows. Figure 13-4 depicts one possible liquid storage system. There are three basic types of storage systems:

- *Liquid*. Liquid storage systems normally utilize water or a water-antifreeze mixture. The storage capability is determined by the sensible heat capacity of the liquid. For water it is 1 Btu/lb/°F.

- *Rocks*. Used for air systems, rock storage uses the sensible heat content of rocks for storage. Typically, airflow is top to bottom for storage and bottom to top for use as needed.

- *Phase change materials*. The preceding systems utilize sensible heat. This system utilizes the larger latent heat in phase changes such as the melting of ice. The required storage volume is smaller, but the cost is higher. Eutectic salts are often used.

13.2.4 Applications of Solar Thermal Systems

Commercial facilities and industry have not yet incorporated the use of solar thermal energy systems on a large-scale basis. There are many solar applications throughout the country, but most replace only a small quantity of the traditional energy supply. The following are some applications of solar thermal energy in business and industry.

- *Solar-heated hot water*. One of the bright spots in the application of solar thermal systems today is solar-heated or augmented hot water. The hot water tank itself is the storage system (or at least a part of it), and the hot water is usually needed the entire year. In some parts of the country, solar water heaters are very cost effective. Hotels, motels and small businesses such as laundries are using solar-heated hot water. In the industrial sector, solar-heated makeup water for boilers and cleaning tanks and solar-augmented process feeds are additional process uses for solar heating.

- *Solar space heat*. Although solar space heating is feasible, active solar collectors for space heating are very seldom cost-effective. Because they are not used all year, they do not often save enough energy to justify their cost. Passive solar applications are often cost effective for minimizing the need for traditional fuel sources in providing space heating. This is discussed in the next section.

- *Solar recuperators.* Industrial furnaces require heat year-round, so combustion air preheating is a likely candidate for solar thermal energy. In installations where outside air might already be used, the application is very simple and probably requires no storage. When the sun shines, the air is preheated; otherwise it is not. However, such systems must be designed carefully or they may cause problems with burners and excess air control.

- *Solar detoxification.* One recent application of solar thermal energy that is rapidly growing in use in commercial and industrial facilities is the detoxification of hazardous wastes. These applications make use of the thermal energy and the high energy photons from solar energy that can more thoroughly decompose and destroy toxic chemicals [7].

- *Solar-heated asphalt storage tanks.* A company which had used portable propane burners to keep asphalt in their storage tanks hot switched to solar collectors and added insulation to the tanks. The energy savings and the convenience made this application attractive.

- *Solar air conditioning.* Solar air conditioning systems use the heat from a solar collector to drive an absorption chiller, and produce cool water or air. The cost effectiveness of these active systems is generally poor at this time, and the use of passive solar features in buildings and structures is a far more successful, and cost effective technology.

13.2.5 Passive solar systems

Passive solar systems result from design strategies and related technologies that use elements of the building structure—primarily glass and thermal mass—and building orientation to heat, cool, shade, and light buildings. Passive solar technologies include direct gain heating, radiative cooling, natural ventilation and economizer cycles, natural lighting, light shelves and shading systems.

13.2.5.1 Use of passive heating

A south-facing glass window serves as a passive solar collector of heat for a building, and the interior of the building serves as the heat storage device. Windows also provide significant amounts of natural lighting. Careful attention to passive solar energy utilization in building design can reduce energy costs significantly.

Example 13.2:

Consider a building located in Fort Worth, Texas that faces south. The building has a total wall area (minus glass) of 2000 ft^2 and a roof area of 3000 ft^2. The R values of the roof and walls are 18 and 12, respectively. There is 300 ft^2 (about 40% of the south wall) of south-facing, double-pane, overhung glass (R-1.85) which permits full sunlight for the 5-month heating season. There are twenty-three 63°F heating days in Fort Worth. The glass transmits 80% of the solar energy hitting it. How much of the building's heating needs does this passive solar feature provide?

Solution: From Table 13-1, for Fort Worth, TX, we find the total solar load on a vertical surface for the months of November, December, January, February, and March to be

$$(80\%)\left(300 \text{ ft}^2\right)[1509(30) + 1396(31) + 1315(31) + 1286(28) + 1196(31)]\frac{\text{Btu}}{\text{ft}^2 \cdot \text{year}}$$

$$= 48.58 \times 10^6 \text{ Btu/year}$$

The heat loss (HL) (assuming no setback) is

$$\text{HL walls} = \left(\frac{1 \text{ Btu}}{12 \text{ h} \cdot °\text{F} \cdot \text{ft}^2}\right)\left(2000 \text{ ft}^2\right)\left(\frac{2363 \text{ °F days}}{\text{year}}\right)\left(\frac{24 \text{ h}}{\text{day}}\right)$$

$$= 9.45 \times 10^6 \text{ Btu/year}$$

$$\text{HL roof} = \left(\frac{1 \text{ Btu}}{18 \text{ h} \cdot °\text{F} \cdot \text{ft}^2}\right)\left(3000 \text{ ft}^2\right)\left(\frac{2363 \text{ °F days}}{\text{year}}\right)\left(\frac{24 \text{ h}}{\text{day}}\right)$$

$$= 9.45 \times 10^6 \text{ Btu/year}$$

$$\text{HL glass} = \left(\frac{1 \text{ Btu}}{1.85 \text{ h} \cdot °\text{F} \cdot \text{ft}^2}\right)\left(300 \text{ ft}^2\right)\left(\frac{2363 \text{ ft}^2\text{days}}{\text{year}}\right)\left(\frac{24 \text{ h}}{\text{day}}\right)$$

$$= 9.20 \times 10^6 \text{ Btu/year}$$

$$\text{total HL} = 28.1 \times 10^6 \text{ Btu/year}$$

Percent heat supplied = 48.6/28.1 = <u>173%</u>

According to these calculations in Example 13-2, placing glass on 40% of the south-facing wall will allow solar energy to supply 170% of the total heat needed for the year. However, the following practical considerations are important:

- The heat loss by infiltration may be as large as the total heat loss through the roof and walls—especially if there is much exhaust air. Thus, the total heating load may be far greater than that initially stated.

- The sun does not shine on some days, so the heating plant will have to be designed as large as it would be without solar aid.

- On bright sunny days, the building might get too hot, so the glass area might have to be reduced or adjustable shading used.

- This building is fairly well insulated. Many manufacturing buildings have little insulation, and thus would need much larger amounts of heat to adequately warm the building.

- All glass is on the south wall. Some glass may be needed on the other walls.

- Fort Worth may not be typical of the rest of the country.

Nevertheless, passive solar energy can contribute a large percentage of the heating required in a facility. Passive solar should especially be considered in the design of manufacturing buildings whose hours of operation normally coincide well with sun hours. At night, the thermostat can be substantially reduced, whereas in homes, night setback cannot be used as readily.

13.2.6 Energy Efficiency in the Design of New Facilities
Building a new facility provides the industrial energy manager with numerous opportunities to incorporate energy efficiency into the facility design. The following sections include ways to use passive solar energy, to avoid unwanted solar loads, and to take advantage of renewable energy opportunities. For completeness, we have also included some other energy efficiency measures that should be considered in the design of a new facility.

13.2.6.1 The land.

Purchase property with good energy-efficiency characteristics. Do not choose property located in energy-intensive spots. Avoid areas that are too windy, areas on the tops of hills, and areas on north slopes. However, these spots may become attractive in the future as locations for utilizing other renewable energy sources.

Choose a location that is near energy supplies to minimize transmission losses as well as costs. Property that is near good transportation facilities will also save energy costs.

13.2.6.2 The building site.

Choose an energy-efficient building site for the facility. You should take advantage of unique spots that can use existing deciduous trees or other natural properties to provide shading from the summer sun and/or windbreaks in the winter. Hills can be utilized as berms to improve insulation instead of spending money to level them.

The building site should be close to major transportation facilities.

13.2.6.3 Facility orientation.

Orient the facility for energy conservation. The building should face south. The shorter dimension should run north to south and the longer, east to west. The manufacturing plant in Figure 13-7 below demonstrates this. This allows minimum sun exposure in the summer on the east and west walls. However, during the winter, because of the lower sun angle, the sun helps to heat the facility.

13.2.6.4 Underground construction.

Consider the use of underground structures. Use a large amount of backfill on northern and western walls. This will protect the facility from cold on the north side and heat on the west.

Partially submerge the entire structure. This will use the thermal mass of the ground to maintain a constant temperature in the facility with less energy input. Some facilities are completely submerged, such as the warehouses in the caverns under Kansas City.

13.2.6.5 Energy-conserving landscaping.

Much of the undesirable heat loss/gain can be prevented through proper landscaping [8,9]. Avoid asphalt or concrete areas around the building as much as possible; grasses, shrubs, and vines are much cooler in the summer.

Place deciduous trees strategically so they offer shade during the

Figure 13-7. Landscaping for energy conservation. (a) Deciduous trees: In summer, allow the early morning sun to penetrate and then protect; in winter, allow the sun to penetrate. (b) Hedge: Catch the late afternoon rays. (c) Deciduous trees: Provide shade in the afternoon. (d) Shrubs or tall hedge: In winter, block northwest winds; in summer, help divert southwest winds around the building.

summer yet allow sunlight to penetrate in the winter. Use vines and shrubs to offer additional shading. In fact, thick shrubs placed close to a building effectively increase the R value of the walls.

Use trees and shrubs as windbreaks and wind diverters. For example, evergreen trees or shrubs at the northwest corner of the building can break the cold winter winds and divert the summer breezes for better utilization. (See Figure 13-7 above)

13.2.6.6 Energy-efficient building envelope.

The energy-efficiency of the building envelope will significantly affect the energy use in a facility. Energy efficiency is easy to incorporate in the initial construction of a building, but very costly to retrofit. The following suggestions should be considered in the initial design:

- **Minimize the wall perimeter area.** Use regular-shaped buildings—square or rectangular. This minimizes the wall area and thus minimizes the heat loss.

- **Insulate the building well.** Install insulation in the ceiling and the walls as well as on the slab (if appropriate). Check for local recommended levels. Figure 13-8 illustrates recommended insulation placement for an underground manufacturing facility.

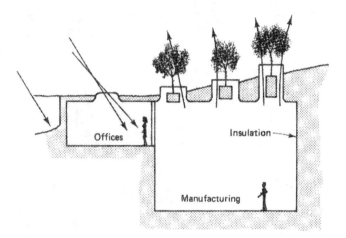

Figure 13-8. Recommended insulation placement

- **Use energy-efficiency considerations in window selection.** Use a minimum amount of glass—well placed. Avoid glass on northern and western sides. As discussed earlier, planning the southern exposure with proper solar influx is best.

Consider installing insulating glass or storm windows. Double or triple glazed windows are cost effective in most parts of the country. Office areas are especially good candidates since they are typically heated and cooled to a greater extent than the production areas.

With proper placement of windows, natural ventilation is also feasible. Therefore, consider the use of windows that can be opened to utilize natural ventilation.

Consider the use of tinted window glass on all walls except the south one. The east and west walls are strong candidates for tinted glass.

Utilize drapes or outside partitions to further insulate windows and to reduce solar load when desired. Several examples of outside partitions are shown in Figure 13-9.

Figure 13-9. Outside partitions.

• **Utilize overhangs or awnings.** On southern exposures, overhangs block the summer sun but allow the winter sun to enter as shown in Figure 13-10. An architect can tell you the proper amount of overhang to allow for sunlight in winter months. It varies with location.

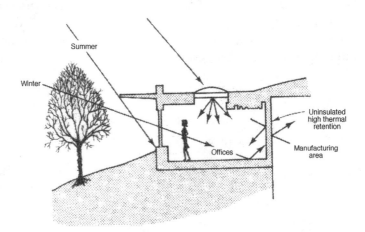

Figure 13-10. Energy conservation through the use of overhangs.

Passive solar systems (also shown in Figure 13-10) help to further reduce energy costs, and they provide attractive warm areas for personnel.

- **Design the roof carefully.** Use light colors in warm climates and dark
  in cold. Design the roof so it can be sprayed in the summer. Be careful
  not to flood the roof—it will leak eventually.

- **Engineer all wall openings for energy efficiency.** Minimize the number of openings. Caulk and weather-strip the doors and windows
  well.

Design the overhead doors with two position openings to match the
truck size and/or use adjustable dock seals. Utilize insulating pads on
the dock doors. Consider using adjustable dock pads.

Consider interlocking the doors and the heating units so that when
the doors are open, the heat is off. Utilize automatic doors or various
types of "see through" materials (plastic strips, plexiglass, etc.) for
doors that must be used frequently. Air curtains are another option for
minimizing heat loss.

Use good pedestrian doors to avoid using large dock doors for pedestrian traffic. Utilize revolving doors or entrance vestibules to minimize air infiltration.

13.2.6.7 Energy-efficient facility layout.
   Locate the facilities within the plant to minimize the energy required
to maintain personnel comfort. Departments with high personnel density
should probably be located in southern exposure areas of the plant and
not in northern or western areas. Figure 13-11 demonstrates one possible
layout considering energy requirements only.
   Avoid or minimize northern or western exposure for dock areas
(shipping and packaging). In Figure 13-11 the shipping and receiving area
is oriented so that the amount of northern exposure is minimized. The
entry area for shipping and receiving should be on the east side.
   Consider departmental or cost center metering of utilities. By doing
this, each cost center can be held accountable for its energy consumption,
and energy can become a part of the budgeting process. This requires extensive preconstruction planning to lay lines and install meters as needed.
Then, arrange the facilities so that energy control is easy (e.g., lights and

motors can be switched off in one location).

Plan the layout so the exhaust air from one area can be used in another; e.g., the hot air at ceiling height in one area may be used as the combustion air for a large furnace.

Figure 13-11. Hypothetical plant layout to minimize energy requirements for space conditioning.

13.2.6.8 Location of process equipment.

Locate all boilers, hot water tanks, and other heated tanks to minimize distribution distances and consequently energy loss in distribution. Design the boiler location and steam distribution system to facilitate return of condensate and/or reflashing or low-pressure steam.

Engineer waste heat recovery systems into the facility design. Waste heat is the one renewable energy source today whose utilization is frequently cost effective. It is much easier to incorporate feedstock preheating apparatus in the initial design rather than retrofitting. Boilers, furnaces, large motors being cooled, lighting fixtures, any cooling fluids, and compressors are just a few of the potential sources for waste heat recovery. Locate the waste heat producing equipment where it can be utilized and where the heat recovery equipment can be installed.

Locate air compressors so they can be maintained easily, they can use fresh cold air, and transmission losses are minimized. Utilize step-up air compressors to be able to reduce plant-wide pressures.

13.3 SOLAR-PHOTOVOLTAICS

Photovoltaics is the direct conversion of sunlight to direct current (dc) electricity through a photocell. Historically, photovoltaics has not been cost effective in competition with fossil fuel or electricity from a grid system, but the needs for photovoltaics in space and the subsequent research coupled with rising costs of traditional energy have pushed photovoltaics ahead. Although still not cost effective for replacing central station electric power plants, photovoltaics is cost effective for many applications where electricity is needed in remote areas. Each year, progress in research and development continually reduces the price of electricity from this source.

Photovoltaic cells are semiconductor devices which can convert the energy in photons of light into dc electrical energy. Most cells are made from single-crystal high-purity silicon and small amounts of trace elements such as boron and phosphorus. These elements are combined into a material with excess electrons (n-type semiconductor) connected to a material with insufficient electrons (p-type region). As long as the cell area is illuminated by light, electrical energy will be produced (see Figure 13-12).

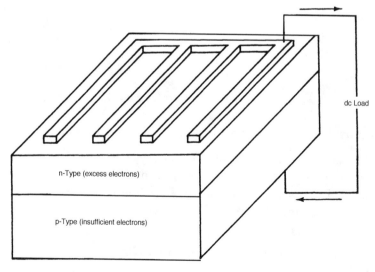

Figure 13-12. Typical photovoltaic cell

The area where the semiconductor materials are joined together can be connected to a battery, or to other cells. Cells are then put together in modules, which can be in the form of flat plates or concentrators. Most cells generate electrical energy of about 0.5 volt and a current that varies with the area of the cell and amount of light. For flat plate modules, typical wattage is about 12.5 W/ft^2 with a conversion efficiency of 5 to 15%. Higher efficiencies can be obtained from concentrator modules.

To achieve higher voltages and currents, solar cells are combined in series and parallel, like batteries. Sets of cells are usually placed in series to generate the necessary voltage, and then multiple strings are placed in parallel to develop the desired current (see Figure 13-13). Because the time of need and the solar intensity do not always match, some type of storage device and voltage regulation is necessary. Normally, a chemical battery storage system fills this need. Finally, a backup generating system is often needed to allow for consecutive cloudy days.

Figure 13-13. Photovoltaic hookups

Generally there are three types (sizes) of photovoltaic systems:

1. *Small* (1-10 kW). These photovoltaic systems are suitable for remote locations or other locations where conventional electricity may be costly. Examples are street lights, irrigation pumps, security lighting at construction sites, communications equipment, and field battery charging.

2. *Medium* (10-1000 kW). These photovoltaic systems could be used by larger industrial facilities and/or remote communities. They could also be used as supplemental or peaking power.

3. *Large* (more than 1000 kW). These photovoltaic systems are used in utility-owned large-scale power-generating stations often located in desert areas.

At their current costs, photovoltaic cells can provide electric energy at 15 to 30 cents per kWh. One of the largest uses of photovoltaic cells today is in supplying remote electric power. Electric utilities are expanding their capabilities with photovoltaic generation. As of October 2007, there were 18 PV solar electric power generating stations in the world with a capacity of 5 MW or larger. Nine of these plants were in Germany; seven were in Spain; one was in Japan; and one was in Portugal. The largest PV plant in the world was the 20 MW Beneixama station in Spain, constructed by City Solar, and located at the Solarpark Beneixama. The largest solar PV plant in the US was the 4.6 MW Springerville Generating Station operated by the Tucson Electric Power Company.

Industry offers a number of applications for photovoltaics, but the actual usage has been limited in size and scope. Ideal applications would have the following characteristics:

1. The equipment operates on dc power.

2. Substantial sunlight is available.

3. The user can tolerate random losses of power.

4. Power is needed when the sun is shining.

Photovoltaics is one of the cleanest, most environmentally benign energy technologies. Many proposals have been made to combine the photovoltaic generation of electric energy with the production of hydrogen through electrolysis. The hydrogen would then be used as the energy supply to be marketed primarily for use in automobiles. Alternatively, the hydrogen could be stored and then burned at a later time to produce electric power when the sun was not shining.

13.4 INCORPORATING SOLAR FEATURES
INTO BUILDING DESIGN—AN EXAMPLE

The new building constructed for the Florida Solar Energy Center (FSEC) provides a good example of some of the areas for saving energy in a facility by utilizing solar energy features. The design objective for this building was to "incorporate the latest solar and conservation building techniques" [10]. Extensive simulations using the DOE-2 program (discussed in Chapter Nine) were conducted to compare alternative design features, and to estimate the energy savings from new design features.

One of the first design decisions was to locate the building on the site with the long axis facing north-south to minimize solar gains and reduce cooling loads, an important consideration in Florida. Both the roof and the walls of the building were coated with a low-absorbance material to further reduce cooling loads. Three-foot wide overhangs are used above the windows to reduce solar heat gains through the glass. Super-windows with spectrally selective coatings and double glazing were also chosen.

Daylighting was used extensively in order to minimize the energy devoted to artificial lighting. To enhance the use of daylighting, all offices are located on the perimeter wall of the main building. The north and south faces of the building utilize an extensive amount of glass. The windows are long, but only 4.5 feet high; the roof overhangs help limit the morning and afternoon solar heat gain, while still providing interior illumination. Above each window is a two-foot high "light shelf" which reflects additional illumination into the building. One of the buildings has a triangular-shaped section, and the windows on its southwest side have fixed shading devices to block direct sun while allowing reflected and diffuse light to enter. The top floors of the main building and the library contain "light wells" to introduce natural daylight to the interior.

Almost 400 photosensors were installed in the facility to measure light levels, and to provide inputs to the dimmable electronic ballasts to control the energy used by the T8 fluorescent lights. The electronic ballasts for the lights allow the power and illumination levels to be varied between 5% and 100% of their maximum output. The DOE-2 simulation showed that daylighting was able to reduce the energy needed for lighting by 46% over the entire year; mid-day reductions averaged 60% for all areas of the building. The payback for the 400 photosensors was less than six months.

The bottom-line energy savings resulting from this new building design using energy efficient and solar features is impressive: the overall energy savings compared to the base-case design was a 70% reduction in utility bills and a peak cooling demand reduction of 88%. Additional solar features such as solar water heating and photovoltaic cells for powering the lighting system are also under investigation. These features may even further reduce the energy use and energy costs for this facility.

13.5 WIND ENERGY

Much has been said about wind energy and its potential. Although the potential in the United States for wind energy is many times greater than the present consumption of electrical energy, this potential will never be obtained due to aesthetics, construction cost and radio-TV interference problems [11]. However, wind energy has proved to be the most cost-effective of the solar technologies, and 1500 MW of electric power generation is located in the state of California.

The power density of wind is given by

$$\frac{P}{A} = \frac{1}{2} \rho V^3$$

(13-1)

where A = area normal to the wind (ft^2)
   ρ = density of air (about .075 lb/ft^3)
   V = velocity of air (mph)
   P = power contained in the wind (Watts)

This can be rewritten as follows (K is a constant for correcting units):

$$\frac{P}{A} = KV^3$$

(13-2)

where K = 5.08×10^{-3}

$$\frac{P}{A} = W/ft^2$$

Unfortunately, only a small percentage of this power can be obtained. It can be shown that theoretically .5926 of the power can be extracted, but practically only 70% or so of that can be obtained. Consequently, only .70 × (.5926) or about 40% of the power is harvestable.

Example 13-3: Find the power in watts per square foot that can be produced from a 10-mph wind. How would this power change if the wind velocity were 20 mph?

Solution: Using Equation 13-2, the power per suare foot is found as:

$$\frac{P}{A} = (5.08 \times 10^{-3})\ (10^3)(.40) = 2\ \frac{w}{ft^2}$$

If the wind speed increased to 20 mph, the power would become:

$$\frac{P}{A} = (5.08 \times 10^{-3}) \times (20^3) \times (.40) = \frac{16.3\ w}{ft^2}$$

These figures demonstrate why it is necessary to find areas with consistent high wind velocities. The difference between a site with 10 mph winds and with 20 mph winds is a factor of 800%.

Just a difference in average wind speed of 3 mph is enough to change the wind characterization of a site. Good, excellent and outstanding wind sites have been described as having average wind velocities of 13, 16 and 19 mph, respectively [12].

Wind speed offers other problems. Too little wind will not initiate power output for most windmills. The wind speed must be above a "cut-in" speed unique to the windmill. Too much wind creates other problems that could conceivably destroy the windmill, so most windmills feather out at some high wind speed.

Industry has done almost nothing to utilize wind. Usually industry is located in congested areas where windmills would not be popular. Furthermore, industry is not likely to develop wind energy as an energy source until it is more cost effective.

13.6 BIOMASS AND REFUSE-DERIVED FUEL

13.6.1 Energy from Biomass

The term *biomass* includes all energy-producing materials that come from biological sources, such as wood or wood wastes, residues of wood-processing industries, food industry waste products, sewage or municipal solid waste, waste from food crops cultivated as energy sources, and other biological materials [13]. Both economic and environmental benefits can be achieved when biomass is used as a source of energy for space heating, process heat, or electricity production. If the biomass comes from a waste product, the cost is usually low; this source may have the added benefit of avoiding the cost of disposing of the waste. Utilizing waste biomass sources not only provides an inexpensive source of fuel, but also solves an environmental disposal problem as well.

Four main technologies are available for converting biomass to usable energy forms. The first is simply burning the biomass to provide direct heat for space and process heating, or for cooking. The second is burning the biomass to produce steam which is then used to generate electrical energy. The remaining two technologies involve converting the biomass to liquid or gas energy forms so that it can be transported to other locations for use.

13.6.2 Energy from Refuse-derived Fuel

Another area for renewable fuel utilization in industry is *refuse-derived fuel*. About 70% of the typical household refuse is combustible, but a higher percentage of industrial wastes is usually combustible. In fact, some industrial waste is a fairly high-quality fuel, as pulp and paper mills have proven for years. Although there are some successful applications of fuel derived from municipal waste, this book is concentrating on industrial—institutional energy management. In the rest of this section we will only discuss industrial-institutional applications.

Combusting waste reduces the volume of the waste by 80% or more which makes it easier to handle. Consequently, one of the big savings for refuse-derived fuel is in reduced disposal costs. In some industries, this disposal savings may be much larger than the economic value of the actual Btu content of the fuel.

Some typical heating values for various types of industrial wastes are shown in Figures 13-14 and 13-15. As shown in these tables, many sources of industrial waste have significant Btu content. Refuse-derived fuel is certainly a fuel source worth considering.

Following is a suggested procedure for analyzing waste fuel sources:

Step 1. Determine the heating value and quantity of the waste.

Step 2. Determine the technical feasibility of utilizing the waste (burning, pyrolysis, anaerobic digestion, etc.). Include necessary pollution control and ash disposal costs.

Step 3. Develop a system design including waste fuel handling, preparation, firing, and disposal. Estimate the cost involved.

Step 4. Perform an economic analysis including all incremental costs identified in step 3, and all savings over conventional fuels. Don't forget the savings in disposal costs and any tax incentives (tax credits or preferred depreciation schemes) from federal and state governments.

Figure 13-14. Wood waste characteristics.
(From *Instructions For Energy Auditors.* U.S. Dept. of Energy, 1978)

	Wood, avg, seasoned (%)	Wood waste, Douglas fir (%)	Hogged fuel, Douglas fir (%)	Sawdust (%)
Proximate analysis:				
Moisture	24.0	35.9	47.2	44.9
Volatile matter	65.5	52.5	42.9	44.9
Fixed carbon	9.5	11.1	8.9	9.5
Ash	1.0	.5	1.0	.7
Ultimate analysis:				
Hydrogen	7.2	8.0		
Carbon	37.9	33.5		
Nitrogen	.1	.1		
Oxygen	53.8	57.9		
Sulfur	–	0		
Ash	1.0	.5		
High heating value (Btu/lb)	6300	5800	4670	4910

Figure 13-15. Heating values of industrial waste fuels.
(From *Instructions For Energy Auditors*. U.S. Dept. of Energy, 1978)

Fuel	Heating value Btu/lb	Btu/ft^3
Solid Fuels:		
Bagasse	3,600–6,500	
Bark	4,500–5,200	
Wood Waste	4,500–6,500	
Sawdust	4,500–7,500	
Coffee grounds	4,900–6,500	
Rice hulls	5,200–6,500	
Corn cobs	8,000–8,300	
Municipal refuse	4,500–6,500	
Industrial refuse	6,600–7,300	
Coal	8,000–24,000	
Liquid fuels:		
Black liquor (pulp mills)	4,000	
Dirty solvents	10,000–16,000	
Gasoline	20,700	
Industrial sludge	3,700–4,200	
Naphtha	20,250	
Naphthalene	18,500	
Oil waste	18,000	
Paints and resins	6,000–10,000	
Spent lubricant	10,000–14,000	
Sulfite liquor	4,200	
Oil	18,500	
Gas fuels:		
Coke, oven gas		500–1,000
Refinery gas		1,200–1,800
Natural gas		1,000

If you go through this procedure at least once for each alternative, you should make the most cost-effective decision.

At some stage, the analysis should include the incorporation of waste from another industry or a municipality so that economies of scale might be obtained. However, extreme care should be taken to ensure that these other sources are reliable in quantity, flow rate, and quality.

The technology for burning waste fuel is rapidly becoming more developed and the equipment is commercially available. Some of the technologies receiving attention include the following:

- Burning raw refuse in water well incinerators

- Shredding, pelletizing, or otherwise preparing fuel and burning in boilers

- Anaerobic digestion where waste is converted to gas in an oxygen-free atmosphere

- Pyrolysis or thermal decomposition of waste in the absence of oxygen

The following paragraphs summarize some applications of refuse-derived fuel:

- A large communications company converted a natural gas-fired boiler to burn methane from a nearby landfill. The plant's consumption of the landfill gas is expected to be between 200,000 and 300,000 million Btu per year. The plant expects to save around $60,000 per year from the use of the landfill gas. The cost of the conversion was $1 million, but it was paid by the supplier of the landfill gas [14].

- A county in North Carolina expanded its operation involving burning municipal waste to cogenerate electric power and steam. A 5 MW steam turbine generator was added to sell electric power to a major state utility, and to sell steam to a chemical and fertilizer firm located nearby. The firm will buy about 100,000 pounds per hour of steam from the county-owned and operated facility. The county calculated the payback time for the new addition at just under seven years [15].

- A dairy farm in Georgia installed a 190-hp engine running on biogas from an anaerobic digester to drive a 55 kW electric generator. Waste heat from the engine was used to increase the effectiveness of the digester by keeping it warm, and to dry the sludge from the digester prior to using it for bedding material. The overall efficiency of the co-generation system was about 60%. The biogas from the digester had an energy content of about 600 Btu per cubic foot—or nearly 60% of the heat content of natural gas [16].

13.7 WATER MANAGEMENT

Water management is the efficient and effective use of water. Like energy management, the main goal is to improve profits or reduce costs. Water management is included in this text on energy management because water and energy utilization are intricately intertwined in most organizations. As water consumption goes up, so does energy consumption, and vice versa. Water management is included in this chapter on renewable energy sources because water is renewable and is often an efficient energy source. For example, water in cooling towers is a very inexpensive source of cooling.

Why worry about water? The sheer volume of water used in this country each day is staggering. Each person in the United States consumes directly and indirectly almost 90 gallons of water per day. At the same time, water costs are going up rapidly, and some sources are drying up (e.g., Midwestern aquifers).

Industry can save dramatic amounts of money and water through attention to its water-consuming equipment. In this section we will demonstrate how to save dollars and gallons of water in industrial water use.

There are three primary types of water users in industry. They are boilers, cooling towers, and process equipment.

13.7.1 Savings from Boilers

In boilers the primary ways to save are the following:

- *Blowdown.* Blowdown should be reduced to the minimum possible, which is determined by the feedwater quality and the amount of condensate return. Reducing blowdown also saves large amounts of energy.

- *Condensate return.* Condensate should be returned to the boiler whenever economically feasible. This saves water by reducing the amount of new makeup water needed and by reducing the amount of blowdown required. Returning the condensate also saves substantial amounts of energy and water treatment costs because condensate is essentially distilled water.

• *Steam leaks and steam traps.* As already demonstrated, steam leaks are extremely expensive energy losses, but significant amounts of water are also lost in steam leaks. An estimate of the amount of water lost can be obtained by dividing the annual loss in Btus by the enthalpy of evaporation. Stuck-open steam traps are steam leaks if the condensate is not returned and are still wasteful even if it is returned.

13.7.2 Water Savings from Cooling Towers
In cooling towers, the primary ways to save are the following:

• *Bleed.* Bleed in a cooling tower is almost identical to blowdown in a boiler. The purpose is to prevent impurity buildup. Bleed should be reduced to a minimum and reused if possible. Sometimes bleed can be used to water lawns or as rinse water makeup. However, careful attention must be paid to the chemicals in the water.

• *Sewerage charges.* Often sewerage charges are based on the amount of water consumed. Yet water consumed in a cooling tower does not go into the sewer. Negotiations with municipalities can often reduce the sewerage charge substantially if large cooling towers are present. Usually about 1% of the flow rate of water must evaporate for each 10°F drop in water temperature.

• *Preventive maintenance.* As baffles become broken or clogged with dirt and slime, the cooling capability drops dramatically. Since a 1°F drop in condensing water temperature can mean a 3 percent savings in electrical chiller input, preventive maintenance of cooling towers is important.

• *Tower water.* Often cooling tower water can be used directly for process cooling instead of using chilled water. When this is possible, large amounts of energy can be saved at the cost of higher water consumption. The trade-off is almost always cost effective.

13.7.3 Savings in Industrial Processes
It is much more difficult to generalize on ways to save in processes

since water can be used in so many different ways in industrial processes. For example, water can be used to cool furnace walls, cool air compressors, wash, rinse, surface-treat, coat, test products, and cool molds and for a wide variety of other uses. The following paragraphs discuss some of the ways water and money can be saved in industrial processes:

- *Use water flow restrictors in shower heads and sinks.* As much as 60% savings in water and energy costs can be realized when flow restrictors are installed.

- *Recycle rinse water.* Often rinse water can be recycled by simple filtering or treatment. In one company, $30,000 and 30×10^6 gal of water were saved annually by simply running rinse water through a sand filter and reusing it.

- *Reuse cooling water.* Often air compressors, small chillers, and other equipment requiring cooling are cooled with once-through cooling water (i.e., water from the tap is run through the equipment one time and dumped into the sewer). In the same room, tap water may be used as boiler makeup water. Because the cooling water is hot, usually about 105°F, if the cooling water is used as boiler makeup water, significant amounts of water and energy can be saved. One company found it could save about $1300 and one million gallons of water per year for each air compressor by reusing the cooling water in other places or by recirculating it through a small cooling tower.

- *Reduce flow rates to the minimum necessary.* Usually, water flow rates are liberally set in washing, coating, or rinsing operations. By setting flow rates at minimum levels, significant water can be saved along with the energy required for pumping.

- *Cover open tanks. Often heated tanks are open at the top.* Floating balls, cantilevered tops, or flexible slit covers can be used to cover the tanks and reduce evaporation and heat loss. One plant saved about $12,000 per year in energy and water costs by covering their heated tanks.

There are many other ways to save dollars and water through water audits, but it is difficult to develop a general list. Each plant and each operation are unique and require individual engineering study. The preceding discussion should serve to stimulate some ideas.

In terms of total water studies, water audits at manufacturing locations uncovered the following potential:

- One plant saved $77,000 annually in energy and water cost and 32 × 10^6 gal of water.

- Another plant saved $20,000 annually in energy and water cost and 12 × 10^6 gal of water.

- A rubber hose manufacturer saved $31,000 in water cost alone and 50 × 10^6 gal of water per year.

- A metal cylinder manufacturer saved about $20,000/year and 36 × 10^6 gal of water.

The potential for real savings is large. In the future, water management will be critical as costs continue to climb and water sources dry up.

13.8 SUMMARY

In this chapter we analyzed alternative energy sources and water management opportunities. First, we examined solar energy options. Active and passive solar systems were studied, and photovoltaics was examined.

Active solar systems were found to be effective but expensive since they normally require backup and storage systems. Passive solar energy, which is cost effective today, offers substantial potential for reducing energy costs for heating; however, backup systems are usually needed here too. Finally, although the future is promising for the reduced cost of photovoltaics, present applications in industry are only for remote sites where other energy sources are too expensive.

Although wind energy is not very promising as an industrial energy source in the near future, refuse-derived fuel is cost-effective in many locations today. Savings result from reduced energy cost and reduced handling and disposal costs.

Water management was discussed in the last section. Water management can yield large dollar and water savings and be cost effective. Because water use is often intertwined with energy use, reductions in the amount of water used can often result in a concurrent energy savings.

References
1. "The Potential of Renewable Energy," An Interlaboratory White Paper, SERI/TP-260-3674, U.S. Department of Energy, March, 1990.
2. *Ibid.*
3. *Climatic Atlas of the United States*, GPO, Washington, D.C., 1968.
4. *Introduction to Solar Heating and Cooling Designing and Sizing*, DOE/CS-0011 UC 59A, B, C, Washington, D.C., Aug. 1978.
5. *Engineering Principles and Concepts for Active Solar Systems*, Solar Energy Research Institute, Hemisphere Publishing Corporation, Washington, DC, 1988.
6. R.W. Hallet and R.L. Gervais, *Design, Startup, and Operational Experiences of Solar One, in Progress in Solar Engineering*, Edited by D. Yogi Goswami, Hemisphere Publishing Corporation, Washington, DC, 1987.
7. D. Yogi Goswami, "Solar Energy and the Environment," *Proceedings of the Energy Systems and Ecology Conference*—ENSEE '93, Cracow, Poland, July 5-9, 1993.
8. John H. Parker, "Uses of Landscaping for Energy Conservation," *Star Project Report 78-012*, Florida Governor's Energy Office, Tallahassee, FL, January, 1981.
9. "Landscaping Can Reduce Energy Bills," *Grounds Keeping*, Intertec Publishing Company, February, 1993.
10. Danny S. Parker, "The Florida Solar Energy Center—Preliminary Analysis of Potential Efficiency Improvements," FSEC-RR-26-91, Florida Solar Energy Center, Cape Canaveral, FL, December, 1991.
11. W.C. Turner, Senior ed., *Energy Management Handbook*, Third Edition, Fairmont Press, Atlanta, GA, 1997.
12. "The Potential of Renewable Energy," An Interlaboratory White Paper, SERI/TP-260-3674, U.S. Department of Energy, March, 1990.
13. Ibid.
14. Amy Gahran, "Landfill Gas Cuts Boiler Fuel Bills at AT&T by $60K/yr.," *Energy User News*, January, 1993.
15. Mark Dunbar, "County Expands Steam Plant, Adds Waste-to-Energy Cogen.," *Energy user News*, October, 1990.
16. Doug M. Moore, "Biogas Powered Cogeneration System at Mathis Dairy," *Industrial Energy Conserver*, Industrial Energy Extension Service of Georgia Tech University, September, 1987.

Bibliography
Capehart, Barney, Editor, *Encyclopedia of Energy Engineering and Technology*," Taylor and Francis/CRC Publishing Company, Boca Raton, FL, July, 2007.

Goswami, D. Yogi, *Alternative Energy in Agriculture,* CRC Press, Boca Raton, FL, 1986.

Goswami, D. Yogi, Editor, *Progress in Solar Engineering*, Hemisphere Press, Washington, DC, 1987.

Instructions for Energy Auditors, U.S. Department of Energy, Washington, D.C., 1978.

Kreider, Jan F., *Solar Design: Components, Systems, and Economics*, Hemisphere Publishing Company, Washington, DC, 1989.

Kreith, Frank, *Economic Analysis of Solar Thermal Systems*, MIT Press, Cambridge, MA, 1988.

Meckler, Milton, *Innovative Energy Design for the 90's*, Fairmont Press, Atlanta, GA, 1993.

Myers, John, *Solar*, research report, Oklahoma Industrial Energy Management Program, Oklahoma State University, Stillwater, Oklahoma, 1981.

Rosenberg, Paul, *Alternative Energy Handbook*, Fairmont Press, Atlanta, GA, 1993.

Chapter 14

Distributed Generation

14.0 INTRODUCTION

Distributed generation (DG) is the generation of electricity at some point or points other than a central station power plant. The source of the electricity can be diesel or gas-fired generator sets, gas turbines, microturbines (small gas turbines), photovoltaic cells, fuel cells, wind, or biomass. The advantages to a facility that installs distributed generation can include reduction in electric costs and increased reliability of electric supply. The advantages to that facility's utility company can include reduction in peak demand and thereby in the demand for adding more generating capacity, and reduction in the need for more transmission and distribution infrastructure. Disadvantages to the user can include, among others, high initial costs and the requirement for developing expertise in an area other than the focus of their business. Disadvantages to the utility can be the disruption of a power grid from electricity that varies in voltage, frequency, or phase from that on the grid, as well as possible loss of planned revenue. These advantages and disadvantages can be far more complex when government regulations are considered. In this chapter, we first discuss the economic motivations for distributed generation. Then we briefly discuss the various technologies that have been used for distributed generation, together with a more complete listing of their costs and benefits. The chapter concludes with a case study of DG costs for a facility.

14.1 THE ECONOMICS OF DISTRIBUTED GENERATION

14.1.1 Benefits to the Facility

The most obvious benefit to a facility that installs distributed generation is the electric cost savings. This savings comes from reduc-

497

tions in peak demand and reductions in total kWh purchased from the electric utility. The savings depend on the electric rate schedule and must be calculated carefully (see Chapter 3). For example, if the utility calculates electric demand as the demand concurrent with the electric utility peak or the demand during a peak load period, generating some of your own power and turning off any unneeded demand at that time can have large and immediate benefits. If the utility calculates the maximum demand in the usual way, then reducing demand during the peak simply changes the peak, and the demand savings may not be nearly so great. In addition, other benefits include the possibility of fewer interruptions in electric service (if the electric source for a facility is switched seamlessly from the electric grid to on-site generation), reduced energy costs for thermal loads if cogeneration is practical, decreased exposure to electricity price volatility, and a possible new source of revenue.

The addition of DG to a facility also opens up the possibility of using a combined heat and power system (CHP) to further increase the cost-effectiveness and efficiency of the DG system. CHP is also known as cogeneration, and produces both electric power for a facility (DG) plus thermal energy for the facility. This thermal energy is most usually in the form of hot water, but can also be in the form of low-pressure steam in some applications, or in the form of very hot combustion gas in cases where a micro or mini gas turbine is used. For facilities that use large quantities of hot water—such as hotels, motels, laundries, hospitals, etc.—DG systems with heat recovery can be significantly more cost effective than DG systems without beneficial heat recovery.

14.1.2 Costs to the Facility

A facility will incur a number of costs when it installs a distributed generation system. According to a study by A. D. Little Co., these costs can include installation costs, operation and maintenance costs, fuel costs, interconnection costs, permitting costs, possible standby charges levied by utilities if utilities are to be used as a source of standby power, competitive transition charges, and emission compliance costs [1].

* **Installation costs**
  The one-time costs for a DG system will include the cost of the DG equipment itself as well as the cost of any necessary modifications to the building that will house that equipment. There may also be some initial expenses for training personnel and

for purchasing any needed diagnostic equipment and portable metering devices. Some facilities, such as hospitals, may be able to charge much of the initial cost against their requirement for emergency power; for them, any cost savings are nice but not necessary to justify the project.

- **Operation and maintenance (O&M) costs—except for fuel**
  The cost of operating the equipment and keeping it in efficient operating condition is an ongoing cost. This cost will include compensating the equipment operators and the maintenance personnel. It will also include the cost of any routine maintenance equipment and supplies not included in the installation costs. It may also include training costs for new employees. A company may sign a maintenance agreement with a local vendor and avoid the cost of some of the maintenance equipment purchases.

- **Fuel cost**
  Although fuel cost could be considered a part of the operating cost, we have identified it separately because it is usually the greatest variable cost. This cost depends upon the type of equipment and the number of hours the equipment is used. Fuel prices may also be volatile, and this volatility can change the desirability of one fuel over another. To mitigate that problem, some facilities install a multi-fuel unit.

- **Interconnection and permitting costs**
  A facility that wants to connect a distributed generation unit to its electric system and to the electric utility network that serves the facility will incur certain connection costs. The utility is concerned that the voltage level, frequency, harmonics, and phase relationships it has established remain in control, and any DG system must have sufficient controls and switching to conform to the utility company standards. There may also be state or federal permitting requirements that must be obtained before the DG system can operate.

- **Safety costs**
  The utility company is concerned about the safety of its installation and repair personnel, and the DG system must be designed with controls so that utility line workers are not confronted with energized circuits when they thought the circuits were not energized. In

particular, islanding is a major issue in both the control and protection of energy systems with DG. Islanding occurs when a section of a feeder is de-energized but the DG device is not disconnected or someone at the facility inadvertently reconnects it. This may cause a dangerous situation for the maintenance crew or cause equipment damage. However, the islanded feeder system will not continue to operate normally unless the load on the islanded section matches the output power of the DG. Also during islanding, the output voltage and frequency may vary with the amount of load on the islanded feeder section. The islanding problem may lead to stability problems in the distribution level. The cost of controls and protection systems to guard against such problems is also an important expense.

- **Standby charges**
  A utility may levy standby charges when a customer generates a significant percentage of its own power and uses the utility primarily for backup power.

- **Competitive transition charges**
  When a company uses distributed generation to reduce its electric demand charge, the utility loses part of the revenue it expected to have to amortize the cost of its own generating equipment. The competitive transition charge is used by the utility to recoup this revenue. Whether this charge is allowed depends on the rules and regulations of the state utility regulatory agency.

- **Emission compliance costs**
  In addition to permitting costs, a facility may have to add emission control equipment if the distributed generation system will exceed air emission standards. Diesel emissions are sometimes treated this way and sometimes given an exception to air quality regulations. NO_x (nitrogen oxide) emissions are of particular concern.

14.1.3 Benefits to Utilities

Perhaps the largest benefit to utilities from DG is the reduction in need to build more generating capacity. If a facility is willing to invest in DG, the kW of that DG is kW capacity that the utility does not need to build. This can represent a significant savings to the utility in terms of capital investment. It also means that the utility does not have to deal with the uncertainty of the power plant permitting process for the DG capacity.

Utilities also benefit when they can reduce or defer their investment in transmission and distribution (T&D) systems. To the extent that DG reduces the need for transmission and distribution lines and other T&D equipment, T&D costs for a new facility can be reduced at the time the T&D system for the site is being planned. Since on-site generation requires little or no distribution equipment from the utility, and the amount of electricity to be transmitted to the site over the major T&D lines of the utility is also reduced, the utilities can invest in smaller T&D systems to serve the facility. In some cases, the utility may also be able to eliminate the need for new investment in T&D lines. In many situations, the major transmission lines connecting parts of the country are at or near capacity. The use of DG can help relieve this capacity problem by providing electricity that does not need to travel along these lines.

Utilities overbuild their capacity so that they have a reserve margin of capacity for unexpected demand. Utility reserve margins can be reduced with DG because some of the peak demand load will be provided by DG units rather than by additional utility capacity.

14.1.4 Costs to Utilities

A utility will incur a number of costs when one of its customers installs a DG unit. According to a study by A.D. Little Co., these costs can include reduced revenues and profits, added controls costs, and stranded costs [2].

- **Reduced revenues and profits**
  Utility revenues and profits are generally reduced when a customer installs a DG unit. This depends, however, on the regulatory system under which the utility operates and on the utility's electric rate schedule. If the utility charges a fixed price, then any time the utility's cost of energy goes above that price, the utility loses money, and reducing the amount of electricity produced can save the utility money. If, however, the utility can pass on any incremental costs to the customer, any reduction in peak demand will reduce the revenue to the utility. The relation between demand and capacity is also very important. If the demand for electricity is very close to the maximum capacity of the utility, the marginal cost to the utility of additional electricity can be very high and the marginal profits very low or even negative; if the capacity far exceeds the demand, the marginal cost is low, the profits are higher, and the utility is less enthusiastic about anything that reduces the peak demand for their electricity.

- **Added controls costs**

  Managing the utility transmission and distribution system to handle distributed generation can be expensive. Problems that must be overcome include phase differences between the distributed generation units and the utilities, harmonics generated by the DG units that may damage equipment on the utility distribution system, and power that flows the wrong way on the utility grid. Controls are available to lock in the phase of the DG unit to that of the utility grid, and other equipment is available to control and minimize harmful harmonics. It is also possible to purchase voltage controls so that power will not flow the wrong way or, if the utility is purchasing power from a DG unit, so that the DG power is of acceptable quality for the grid. All of these controls cost money, but their expense is necessary to insure the quality of the electric power being delivered by the utility.

- **Stranded costs**

  Stranded costs can also be a problem. A stranded cost is a capital cost that cannot be recovered when the income from electricity sales is reduced either by customers using less electricity or by customers leaving the system. The capital costs were originally incurred under the assumption that the capital would generate an acceptable rate of return to stockholders, and stockholders are left holding the bag when some of the expected revenue disappears.

14.2 THE TECHNOLOGIES

The current technologies in use for distributed generation are reciprocating engine generators, fuel cells, photovoltaics, wind, microturbines, and gas turbines. Data in the following discussion are taken, with permission, from Distributed Power Generation, Planning and Evaluation by Lee Willis and Walter G. Scott (Marcel Dekker, 2000).

14.2.1 Reciprocating Engines

More than 90% of all distributed generation is powered by reciprocating piston engines, generally diesel fueled. Due to their general satisfactory performance on small DG units (<250 kW), their availability, and their familiarity, these engines are the most frequently used DG power source today. However, reciprocating engines have some undesirable features, especially when they are used as the power source of DG. They produce noise, vibration and emissions, all of which can be reduced, but such measures increase the initial cost and lead to lower fuel efficiency.

Reciprocating engines turn at speeds of 1200-6000 RPM. Compression ignition (CI) engines, such as diesel engines, have slower speeds than spark ignition (SI) engines (most automobile engines), but CI engines are usually more powerful than SI engines providing more horsepower per unit of size. Other contrasts are shown in Table 14-1 below:

Table 14-1. Comparison of Features of Spark Ignition and Compression Ignition Reciprocating Engines

Spark Ignition:	Compression Ignition:
High RPM (up to 15,000)	Relatively low RPM (up to ~6000)
High maintenance requirements	Relatively low maintenance requirements
Light and easy to burn fuels.	Heavier fuels.
20-43% efficiency.	28-40% efficiency.
Up to ~5,000 HP output	Output may be as much as 50,000HP

14.2.2 Gas Turbines

With three different size categories (Utility, Mini and Micro), gas turbine generators cover a wide range of capacity sizes. The basic principle for the three categories is the same, but due to size differences, their operating characteristics, turbine designs and type of generator used may differ. There are also huge speed differences between different classes of turbines due to their size and capacity differences. These speeds become very significant when the turbine is connected to a generator to produce power.

*   **Utility Gas Turbines:**
    This category includes the largest gas turbines which are most commonly used by electric utility companies. They have a capacity range from 10,000 kVA to 150,000 kVA, are often the size of a building, typically use natural gas or diesel as fuel (with a fuel efficiency of about 37%), and usually need to be removed from service for maintenance every year and a half. They generate AC synchronously and operate at about 1800 RPM. Utility turbine generators produce three-phase electric power, are available at either 50 or 60 Hertz and have a voltage range of 4.2-20 kV. A utility gas

turbine DG unit costs approximately $300 per kW.

- **Mini Gas Turbines:**
  This is the intermediate size gas turbine, with a capacity range of about 800 kVA to 10,000 kVA. This class of gas turbines is not usually produced for DG applications so they have to be modified from their original application. Depending on the purpose of the original design these turbines may be very reliable and robust. They typically use natural gas or diesel as fuel with a fuel efficiency of about 30-35%. They usually need to be removed from service for maintenance every eight months. Usually AC synchronous electric generation is used, with the generator running about 3,600 RPM. A mini gas turbine DG unit costs approximately $450 per kW.

- **Micro Gas Turbines:**
  The smallest gas turbines have capacities from about 20 kVA to 750 kVA and can be as small as a standard refrigerator. Typically natural gas or diesel is used as fuel with a fuel efficiency of approximately 32%. The usual interval between major maintenance is two years. Usually DC electric generation is used, with an inverter to produce AC. AC synchronous generation is not appropriate for these turbines since the speed of the turbine is too high, usually about 70,000 RPM. The use of an AC synchronous generator would require either a gearbox (which would decrease the efficiency) or a specially designed high-speed generator (which would be unreasonably expensive). A micro turbine DG unit costs approximately $700 per kW.

14.2.2.1 Advantages of Gas Turbines:
- Gas turbine powered generators are very light when compared to diesel generators. For a 500 kW unit, a diesel generator is approximately 10 times heavier and 10 times larger than a turbine generator.

- Due to the simple working principle and pure rotary motion, gas turbines provide very high reliability when compared to reciprocating engines.

- Gas turbines respond almost instantly to load variations. In addition, steady state frequency regulation is much better than with other generator types.

- Gas turbines can use a wide variety of fuels—natural gas, propane, diesel or light oil.

- Gas turbines have lower maintenance requirements than reciprocating engines.

- The opportunity for cogeneration is significant. In large and very hot running turbines, the exhaust represents a significant amount of energy. Usually the device used to capture this energy and use it for beneficial purposes slightly decreases the output of the turbine but also works as a muffler/silencer for the turbine.

- The products of combustion in gas turbines are cleaner than the exhaust products of reciprocating engines.

- Gas turbines are less expensive to manufacture than fuel cells.

14.2.2.2 Disadvantages of Gas Turbines:

- Gas turbines require expensive, high temperature materials, so their initial cost is high relative to reciprocating engines.

- Gas turbine maintenance requires extremely skillful and knowledgeable workers. However, the time spent on maintenance is less than for reciprocating engines.

- The fuel efficiency of gas turbines is rather low when compared to well-designed reciprocating engines and fuel cells.

- Gas turbines are more sensitive to ambient conditions than other fossil fuel units.

- Although they are quieter than reciprocating engines, gas turbines still cause a lot of noise and vibration.

Table 14-2 compares the operating characteristics of various sizes of natural gas turbine DG units.

14.2.3 Types of Generators for Reciprocating Engines or Gas Turbines

DG units can be designed and built to produce power at any frequency and voltage, and either single phase or three phase alternating current. Four different types of generators can be employed to produce

Table 14-2. Cost and Operating Data for Gas Turbines

Capacity (kW)	Fuel Efficiency (%)	Consumption (Btu/kWhr)	Availability (%)	Weight (Lb)	Initial Cost* ($)	Operation & Maintenance† ($/yr)	Hourly Fuel Costs¶ ($/hr)
50	30	12,400	97	1,480	42,000	425	7.13
250	32	11,700	96	10,250	176,000	2,100	33.63
750	29	13,000	95	15,000	341,250	10,000	112.125
1,200	30	12,500	96	15,000	510,000	25,000	172.5
6,500	30	12,400	97	94,000	2,700,000	52,000	926.9
12,500	33	11,400	95	165,000	4,750,000	62,500	1,639
18,000	38	9,975	95	340,000	5,720,000	72,000	2,065
22,500	39	9,600	96	450,000	7,540,000	90,000	2,485
250,000	42	9,100	95	2200 t	123,000,000	1,100,000	26,160

*This is the cost of DG unit itself and does not include the other initial costs of the unit such as shipping insurance, site preparation, fuel delivery lines, fuel tank, tools. These other costs can be estimated at 1/3 of the cost of the DG unit itself.

†O&M cost is based on a working period of 3,000 hours/year and includes maintenance, property taxes, annual inspection, yearly insurance, etc.)

¶The fuel cost is calculated using a natural gas price of $11.50/MMBtu electricity from shaft power provided by internal combustion engines or gas turbines. These types of generators are:.

- **AC Synchronous Generators:**
  These generators are efficient, durable, reliable and simple to service, but they need to be driven at constant speed. This constant speed requirement means that additional mechanical or electrical components are needed to provide this constant speed control.

- **AC Induction Generators:**
  These generators are not as efficient as synchronous generators but they are more robust. This characteristic may be very useful for DG applications since low maintenance is one of the most desirable features. Also induction generators do not need governors, because they are automatically synchronizing when connected to the grid. The disadvantage of induction generators is that they require reactive power, which often is drawn from the grid they are connected to. This compounds any problems due to low power factor on the distribution system.

- **DC Generator with DC/AC Inverter:**
  In this system, direct current is generated first, then converted to AC. The drawback to this method is that it uses two separate devices, which will introduce two separate energy losses. Fortunately the DC generation side is very efficient, but the inverter introduces a much more significant energy loss. One advantage of this type of application is the ability to produce power at any frequency. This method also permits the usage of energy storage with batteries. In this approach, the generator is connected to the battery storage unit, and both the generator and the batteries are connected to the AC inverter. Therefore, the generator can temporarily serve higher loads than its rating. This method is very useful for variable speed power sources, such as wind generation.

- **Written Pole Generator:**
  This is a recent but proven technology. These generators are essentially variable speed synchronous generators. They are more expensive than normal synchronous generators but provide better frequency control, therefore reducing the cost of system frequency controls.

- **Piston Generator:**
  The aspects to be considered for the selection of piston generation include:
  —The purpose and capacity of the system.
  —The fuel type and combustion method.
  —Environmental considerations and regulations.
  —Type of generator to be used to convert mechanical energy to electric energy.
  —Possibility of using cogeneration or CHP.
  —Size and weight limits.

Some size, cost and operation data for internal combustion piston driven engines is given Table 14-3 below.

Table 14-3. Purchase and Operating Costs of Some Internal Combustion Piston Driven Units [3]:

Capacity (kW)	Fuel	Con-sumption (Btu/kWh)	Availability (%)	Relative Initial Cost ($)	Relative O&M Cost ($/year)
250	Diesel	15,000	97	109,500	3,900
800	Diesel	14,140	97	333,000	11,000
1050	Diesel	14,000	97	352,000	14,000
50	N. Gas	12,200	97	22,000	1,000
275	N. Gas	13,333	97	119,500	5,000
825	N. Gas	12,140	97	370,000	12,000
1100	N. Gas	12,000	97	482,000	19,974
2500	N. Gas	11,900	96	980,000	29,500
5000	N. Gas	11,100	98	2,050,000	56,000

14.2.4 Fuel Cells

In principle, a fuel cell operates like a battery. Unlike a battery, a fuel cell does not run down or require recharging. It will produce energy in the form of electricity and heat as long as fuel is supplied. A fuel cell consists of two electrodes around an electrolyte. Oxygen passes over one electrode and hydrogen over the other, generating electricity, water and heat. Hydrogen fuel is fed into the "anode" of the fuel cell. Oxygen (or air) enters the fuel cell through the cathode. With the help of a catalyst, the hydrogen atom splits into a proton and an electron, which take different paths to the cathode. The proton passes through the electrolyte. The electrons create a separate current that can be utilized before they return to the cathode, to be reunited with the hydrogen and oxygen in a molecule of water. A pure fuel cell operating on 100% hydrogen produces no pollutants. Its end products are electricity and thermal energy—usually in the form of hot water.

Figure 14-1. Functional Operation of a Fuel Cell

A fuel cell system which includes a "fuel reformer" can extract the hydrogen from any hydrocarbon fuel—from natural gas to methanol, and even gasoline. Since the fuel cell does not rely on combustion, emissions from this type of a system are much less than emissions from the cleanest combustion processes. The reformer does produce CO_2.

Table 14-4 below summarizes the advantages and disadvantages of fuel cells as compared to other types of DG.

<u>14.2.4.1 Some Characteristics of Fuel Cells</u>

• **Heat:**
  Fuel cells operate at internal temperatures of anywhere from 800 C (176 F) to over 6000 C (1112 F), depending on their design, pres-

Table 14-4. Fuel Cells As Compared to Other DG Alternatives

Advantages:	Disadvantages:
• Higher efficiency than the other fossil fuel DG options.	• Very high initial costs. 2-10 times more expensive than the other fossil fuel DG options.
• Low pollution	• Very high maintenance skills needed, which are very rare at the moment.
• Very low noise and vibration	• Fuel cells are very sensitive to fuel quality.
• Simple cogeneration	• Relatively new technology has to be proven by long-term use.
• Modular and quick to install	

surization, catalyst and electrolyte. Those that operate at higher temperatures produce a good deal of "waste heat", which may be put to use in cogeneration applications. Unlike turbine or piston engines, the exhaust of a fuel cell is very clean.

- **Efficiency:**
  The efficiency of fuel cells depends on the catalysts and electrolyte being used, but even a relative low efficiency for a fuel cell (~33%) compares well in this category to turbines and all but the best reciprocating piston DG engines. Manufacturers often quote net fuel-to-electric efficiencies on the order of 45% and up, while the researchers predict that production units eventually could reach up to 60%. However, these quoted efficiencies are always for fuel cells running on pure hydrogen and oxygen, under perfect conditions, producing high-current low voltage DC power, which is hardly the case in real life applications. For practical estimation purposes the efficiency of the unit may be accepted as 5% less than the quoted values.

- **Noise and Vibration:**
  The basic fuel cells have no hot combustion exhaust, no moving parts and no noise sources. However, advanced high capacity fuel cells have air and fuel pumps which may create some noise and

vibration, but the noise and vibration from these components can be decreased to an insignificant amount.

- **Emissions:**
  The exhaust of a well-operated fuel cell consists of little more than pure steam and hot air, and it is nearly free of toxic materials and unwanted pollutants. Fuel cells are so clean with respect to emission and so silent that they are often considered as the "green" power generator.

- **DC/AC Conversion:**
  The output of a fuel cell is DC power and it must be converted to AC power by inverters and power electronics before it can be used by normal AC equipment. The conversion efficiency of DC to AC is on the order of 98%. In addition DC power from the fuel cell is clean and stable enough that it can be directly used to power DC equipment.

- **Maintenance and Operation:**
  Regardless of catalyst or electrolyte types, after an extended period a fuel cell needs to be rebuilt, which essentially means replacing all its internal components. Also some types of fuel cells (for example, those with a phosphoric acid electrolyte) deteriorate with use and lose about 2% efficiency per year. When compared to piston engines, fuel cells need major maintenance less often (approximately once every 5 years) but the maintenance service given to fuel cells is more extensive than for a piston engine, and fewer people are available who are trained in the technology.

14.2.4.2 Types of Fuel Cells:

There are six common types of fuel cells: phosphoric acid, proton exchange membrane, alkaline, regenerative, molten carbonate, and solid oxide fuel cells [43].

- **Phosphoric Acid Fuel Cells (PAFC):**
  These fuel cells are commercially available today. More than 200 fuel cell systems have been installed all over the world—in hospitals, nursing homes, hotels, office buildings, schools, utility power plants, an airport terminal, even a municipal waste dump. Phosphoric acid fuel cells generate electricity at more than 40% efficiency compared to about 35% for the utility power grid in the

United States. Nearly 85% of the steam these fuel cells produce is used for cogeneration. Operating temperatures are in the range of 400°F.

• **Proton Exchange Membrane Fuel Cells (PEMFC):**
These cells operate at relatively low temperatures (about 200°F), have high power density, can vary their output quickly to meet shifts in power demand, and are suited for applications—such as in automobiles—where quick start-up is required. According to the US Department of Energy, "they are the primary candidates for light-duty vehicles, for buildings, and potentially for much smaller applications such as replacements for rechargeable batteries." The proton exchange membrane is a thin plastic sheet that allows hydrogen ions to pass through it. The membrane is coated on both sides with highly dispersed metal alloy particles (mostly platinum) that are active catalysts. Hydrogen is fed to the anode side of the fuel cell where the catalyst helps the hydrogen atoms to release electrons and become hydrogen ions (protons). The electrons travel in the form of an electric current that can be utilized before it returns to the cathode side of the fuel cell where oxygen has been fed. At the same time, the protons diffuse through the membrane to the cathode, where the hydrogen atom is recombined and reacted with oxygen to produce water, thus completing the overall process.

• **Alkaline Fuel Cells (AFC):**
Long used by NASA on space missions, these cells can achieve power generating efficiencies of up to 70 percent. They use alkaline potassium hydroxide as the electrolyte. Until recently they were too costly for commercial applications, but several companies are examining ways to reduce costs and improve operating flexibility.

• **Regenerative Fuel Cells**
Still a very young member of the fuel cell family and mostly a technology for the future, regenerative fuel cells would be attractive as a closed-loop form of power generation. Water is separated into hydrogen and oxygen by a solar-powered electrolyzer. The hydrogen and oxygen are fed into the fuel cell, which generates electricity, heat and water. The water is then re-circulated back to the solar-powered electrolyzer and the process begins again. NASA and others are currently researching these types of fuel cells worldwide.

- **Molten Carbonate Fuel Cells (MCFC):**
  Molten carbonate fuel cells offer high fuel-to-electricity efficiencies (40-50%) and operate at high temperatures (600°C-1000°C). The MCFC typically uses a liquid solution of lithium, sodium and/or potassium carbonates soaked in a matrix as the electrolyte. This type of fuel cell can operate on natural gas or on synthetic gas, and can supply 100 MW or more of electric power.

- **Solid Oxide Fuel Cells (SOFC):**
  The solid oxide fuel cell is another large capacity fuel cell capable of providing up to 50 MW or more electrical output. A solid oxide system usually uses a hard ceramic material with yttrium-stabilized zirconia instead of a liquid electrolyte. Electric efficiencies are in the 45%-60% range and operating temperatures are 650°C to 1000°C. One type of SOFC uses an array of three long tubes; another variation uses a compressed disc that resembles the top of a soup can. Major opportunities are expected in the stationary power source area.

Table 14-5 shows the characteristics of each of the types of fuel cells discussed above.

14.2.5 Photovoltaics

A photovoltaic (PV) cell converts sunlight into electricity. Sunlight is composed of particles of solar energy called photons. These photons contain various amounts of energy corresponding to the different wavelengths of the solar spectrum. When photons strike a PV cell, they may be reflected or absorbed, or they may pass right through. Only the absorbed photons generate electricity. When a photon is absorbed, its energy is transferred to an electron in an atom of the PV cell (which is actually a semiconductor). With this new energy, the electron can escape from its normal position and become part of the current in an electric circuit. By leaving this position, the electron causes a "hole" to form. A special electric property of the PV cell—a built-in electric field—provides the voltage needed to drive the current through an external load (such as a light bulb).

14.2.5.1 Benefits of PV Generation:

PV-generated power offers advantages over diesel generators, primary (one-time use) batteries, and even conventional utility power. The following benefits make PV the power of choice in more cases every day:

Table 14-5. Characteristics of Fuel Cells [3]

	Typical Construction:	Oxidant:	Operating Temperature:	Basic Cell Efficiency*	Feasible Capacity Range:	Typical Application:	Installed Cost: ($/kW)
PEMFC	Plastic-Metal	Air	85°C (~200 F)	~33%	~5-8 kW Up to 250 kW	Car, DG, Space-craft	1400
AFC	Plastic-Metal	Pure O_2	120°C (~250 F)	~35%	~5-10 kW	Car-Other	2700
PAFC	Steel	Air	190°C (~400 F)	~40%	~50-200 kW	DG	2100
MCFC	Titanium	Air	650°C (1200F)	~42%	10kW-2MW†	Large DG	2600
SOFC	Ceramic	Air	1000°C(1832F)	~45%	25kW-25MW†	Larger DG	3000

*These are normally net AC efficiencies and include deductions estimated for reforming, pumping and conversion losses.
†Upper limits are projected values.

- **High Reliability:**
  PV cells were originally developed for use in space, where repair is extremely expensive, if not impossible. PV still powers nearly every satellite circling the earth because it operates reliably for long periods of time with virtually no maintenance.

- **Low Operating Costs:**
  PV cells use the energy from sunlight to produce electricity—the fuel is free. With no moving parts, the cells require little upkeep. These low-maintenance, cost-effective PV systems are ideal for supplying power to communications stations on mountain tops, navigational buoys at sea, or homes far from utility power lines.

- **Environmental Benefits:**
  Because they burn no fuel and have no moving parts, PV systems are clean and silent. This is especially important where the main alternatives for obtaining power and light are from diesel generators and kerosene lanterns. As we become more aware of "greenhouse gases" and their detrimental effects on our planet, clean energy alternatives like PV become more important than ever.

- **Modularity:**
  A PV system can be constructed to any size based on energy requirements. Furthermore, the owner of a PV system can enlarge or move it if energy needs change. For instance, homeowners can add modules every few years as their energy usage and financial resources grow. Ranchers use mobile trailer-mounted pumping systems to water cattle as the cattle are rotated to different fields

- **Low Construction Costs:**
  PV systems are usually placed close to where the electricity is used, requiring much shorter power lines than if power is brought in from the utility grid. In addition, using PV eliminates the need for a step-down transformer from the utility line. Less wiring means lower costs, shorter construction time, and reduced permitting paperwork, particularly in urban areas.

14.2.5.2 Disadvantages of PV Generation
There are also disadvantages to the use of photovoltaic systems, which limit their use and application.

- The power generated is DC so its applications are somewhat limited.

- PV cells only work when the sun is shining. PV-generated energy must be stored for times when there is no sun, or some other backup generation must be available.

- Converting DC from photovoltaic cells to AC may generate undesirable harmonics, and these must be filtered out or otherwise controlled.

14.2.5.3 First Costs of PV Systems:

Initial costs for PV systems typically range from $2,700-$4,000 per kW of peak capacity, with operating costs of 14-18 cents per kWh, all based on the total system cost of a unit operating 24 hours per day.

14.3 ANALYZING YOUR OWN FACILITY FOR POSSIBLE DG APPLICATION

In analyzing your own facility for possible DG application, you need to consider the costs, with and without DG. The following questions should be addressed to help make the DG assessment for your own facility.

- Do you have a significant electric peak demand point that you cannot reduce substantially with demand side management, i.e. by reducing your in-plant load? DG may be one answer.

- Do you have a need for emergency power? It is possible to provide emergency power and to use the same equipment for demand reduction. This way, the burden of paying for the DG equipment is partially offset by the cost of providing emergency power. This is frequently the case for hospitals, and other emergency care facilities.

- Does your electric utility levy fees on your company if you permanently reduce your peak demand? This may offset any savings you might otherwise obtain.

- Do you need a lot of heat as well as electricity? If so, it may be worthwhile to look into cogeneration as part of a DG package.

- Do you have a remote location? Possibly the transmission and distribution costs of getting the electricity there would justify the costs of a DG unit. Farmers and ranchers have had diesel generators for many years.

- Does your plant need continuous electricity, without interruption? A good possibility would be to put in an uninterruptible power supply, together with a DG unit that will come on automatically before the uninterruptible power supply exhausts its stored energy.

14.4 A CASE STUDY

The following case study serves as an overview and as a methodological basis of the economics of installing on-site generation equipment at a typical facility. This case study comes from "Reducing Facility Operating Costs through On-Site Generation of Electricity," by Mr. Gestur Petursson and Dr. Wayne C. Turner [5] and is used with permission from the authors and from Fairmont Press.

14.4.1 Facility Data

Table 14-6 shows us an energy profile of a manufacturing facility in the Midwest. These data will serve as the basis for our case study.

Table 14-6. Facility Data

Parameter	Value
Annual operating hours	8,760 hrs/yr
kW charge	$50/kW-yr
kWh charge	$0.06/kWh
Average kW demand	3,900 kW/mo
Current gas consumption	50,000 MCF/yr
Current gas cost	$3.75/MCF

14.4.2 Technology Cost and Performance Data

Table 14-7 shows the performance data for four DG options. It should be noted that these numbers will change somewhat depending on the existing conditions at each facility.

Table 14-7. Performance Data

Technology	Cost per kW ($/kW)[a]	Avail- ability (%)	Maintenance & Material Cost ($/kWh)	Gas Consumption (MCFR/kWh)	Heat Recovery (Btu/kWh)
Fuel Cell	3,000[b]	97	0.007	0.0095	3,500
Gas Turbine	1,010	98	0.004	0.0156	7,900
Microturbines	650	100[c]	0.005	0.0126	2,500
Engine	640	96	0.008	0.0106	3,300

[a]This number includes the cost for heat recovery units where they are optional.
[b]In the past, incentive programs aimed at encouraging increased usage of fuel cells have reduced this cost to $2,000 per kW.
[c]This number is a result of redundancy in the generator package since it consists of multiple 30 kW units.
[d]This number was arrived at after discussions with gas utilities/brokers and their willingness to reduce prices to those with large and smooth seasonal demands. This reduction amounts to a discount of approximately 6.7 percent, which is a relatively conservative number.

14.4.3 Assumptions

The following assumptions were made for the purpose of providing a uniform base for comparison.

1.   The installation and engineering cost for all units is assumed to be $200/kW
2.   1 MCF = 1MMBtu = 10^6 Btu
3.   Usable heat from heat recovery unit is 60 percent.
4.   Boiler efficiency is 80 percent.
5.   The discounted gas cost is assumed to be $3.50/MCF. This reduces the current gas cost by $0.25/MCF.
6.   Project life is 15 years.
7.   Rate of return is 15 percent.
8.   The total electrical generation of the unit(s) is 1,000 kW.

14.4.4 Methodology

The methodology used in the calculations is outlined below. The data used in the sample calculations are the data for the engine generator. This same methodology is used for each of the four technologies.

kWh Cost Savings   = (Generator electrical output, kW)(Operating hours, hrs/yr)(kWh cost, $/kWh)(Generator availability, %)
= (1,000 kW)(8,760 hrs/yr)($0.06/kWh)(0.96)
= **$504,576/yr**

kW Cost Savings   = (Generator output, kW)(Generator availability, %)(kW cost, $/kW-yr)
= (1,000 kW)(0.96)($50/kW-yr)
= **$48,000/yr**

Gas Cost Savings   = [(Current gas consumption, MCF/yr)(Estimated discount due to increased gas usage, $/MCF)] + [(Recoverable heat, Btu/kWh)(Generator output, kW)(Operating hours, hrs/yr)(Generator availability, %)(Discounted gas cost, $/MCF)(Conversion factor, MCF/Btu)(Percent use of recoverable heat, %)/(boiler efficiency)]
= [(50,000 MCF/yr)($0.25/MCF)] + [(3,300 Btu/kWh)(1,000 kW)(8,760 hrs/yr)(.96)($3.5/MCF)(0.000001 MCF/Btu)(0.60)/(0.80)]
= **$85,348/yr**

Generator Implementation Cost   = (Generator output, kW)[(Generator cost, $/kW)+(Installation and engineering cost, $/kW)]
= (1,000 kW)[($640/kW)+($200/kW)]
= **$840,000**

Generator Gas Consumption Cost   = (Gas consumption, MCF/kWh)(Operating hours, hrs/yr)(Generator availability, %)(Discounted gas cost, $/MCF)(Generator output, kW)
= (0.0106 MCF/kWh)(8,760 hrs/yr)(0.96) (1,000 kW)
= **$311,996/yr**

Annual Maintenance And Material Cost   = (Operating hours, hrs/yr)(Generator availability, %) (Maintenance and material cost, $/kWh) (Generator output, kW)
= (8,760 hrs/yr)(0.96)($0.008/kWh)(1,000 kW)
= **$67,277/yr**

Net Annual Savings = (kWh cost savings, $/yr) + (kW cost savings, $/yr) + (Gas cost savings, $/yr) -(Generator gas consumption cost, $/yr) - (Annual maintenance and material cost, $/yr)

= $504,576/yr + $48,000/yr + $85,348/yr - $311,996/yr - $67,277/yr

= **$258,651/yr**

Simple Payback = (Generator implementation cost, $)/(Net annual savings, $/yr)

= ($840,000)/($258,651/yr)

= **3.25 years**

14.4.5 Results

The following are the results from calculations based on the methodology presented previously.

Table 14-8. Case Study Results.

Technology	Net Annual Savings ($/yr)	Simple Payback (years)	Annual Worth $(A/P,15,15)^a$ ($/yr)
Fuel Cell w/out aid	306,888	10.4	-240,312
Fuel Cell w/aid	306,888	7.1	-69,312
Gas Turbine	251,546	4.8	44,636
Micro turbines	215,472	3.9	70,121
Engine	258,651	3.3	115,011

[a]According to the book *Capital Investment Analysis for Engineering and Management* by Canada et al., this parameter amounts to 0.1710.

14.4.6 Sensitivity Analysis

The following sensitivity analysis is based on one-at-a-time analysis, sometimes known as a break-even analysis. Here the annual worth of the generator unit is set equal to zero. That means your investment will not produce any gains or losses in economical terms. Then by holding all other variables equal, we can calculate how much percent change in each individual variable is needed to affect the financial outcome of the project. In Table 14-9, the three most sensitive variables are highlighted to emphasize the importance of those in terms of risk when it comes to installing on-site generation equipment. In all cases the kWh cost and the gas consumption are two of the three most sensitive variables.

This analysis gives us an idea about which parameters should be included in a multiparameter analysis, which would give an idea of the interaction between two or more variables. Also, by looking at the variables it can be seen that some of them are technical in nature. Thus, there is very little one can do to influence those except to monitor them and make sure that the equipment is delivering according to its specifications. On the other hand, markets control most of the other variables. Therefore, the behavior and available market forecasts of these variables must be investigated carefully before any decisions are made about installing an on-site generator.

Table 14-9. One-At-A-Time Sensitivity Analysis

Parameter	Fuel Cell w/o Aid	Fuel Cell w/Aid	Gas Turbine	Micro Turbines	Engine
kWh cost, $/kWh	+47.2	+13.7	–8.7	–13.3	–22.8
Gas cost, $/MCF	–125.3	–36.2	+15.8	+22.1	+50.7
Heat Recovery, Btu/kWh	+407.9	+88.8	–25.1	–122.0	–157.8
Generator implementation cost, $/kWh	–46.8	–20.3	+25.7	+63.1	+205.2
Maintenance & material cost, $/kWh	–404.2	–116.5	+230.0	+260.0	+264.6
Gas consumption, MCF/kWh	–85.1	–24.5	+9.6	+18.2	+36.9

14.5 CASE STUDY SUMMARY

The main purpose of this case study is to offer a methodology for an economic analysis of on-site generation. Table 14-10 provides an overview of the pros and cons of each of the generation technologies included in the case study calculations.

Table 14-10. Overview of DG Technology

Technology	Pros	Cons
Fuel Cell	1. Very low emissions 2. Exempt from air permitting in some areas 3. Comes in a compete "ready to connect" package	1. High initial investment 2. Only one manufacturer producing commercially available units
Gas Turbine	1. Excellent service contracts 2. Steam generation capabilities 3. Mature technology	1. Requires air permit 2. The size and shape of the generator package is relatively large
Micro turbine	1. Low initial investment 2. High redundancy 3. Low maintenance cost 4. Relatively small size and installation flexibility	1. Relatively new technology 2. Requires an air permit 3. Possible synchronization problems of large installations
Engine	1. Low initial investment 2. Mature technology 3. Relatively small size	1. High maintenance cost 2. Low redundancy

14.5 CONCLUSION

There are many potential benefits to be obtained from implementing distributed generation at commercial, institutional or industrial facilities. Many of the technologies for DG are well established and proven while others are still in the testing and demonstration phase. For some facilities such as server farms and data centers, the benefits from high reliability power can be enormous compared to the cost of a short power outage. Other facilities that have large thermal needs in the form of hot water or steam, as well as large electrical loads, can substantially reduce their energy related operating costs. DG is currently one of the fastest growing areas of energy management, and as the newer technologies become available to the average facility, interest will increase even faster.

References

1. Distributed Generation: Understanding the Economics, An Arthur D. Little White Paper: Arthur D. Little, Inc., 1999
2. Distributed Generation: Policy Framework for Regulators, An Arthur D. Little White Paper: Arthur D. Little, Inc., 1999
3. Distributed Power Generation- Planning and Evaluation by H. Lee Willis & Walter G. Scott (Marcel Dekker, 2000)
4. Definitions are from FuelCell 2000, and are used with their permission.
5. Gestur Petursson and Wayne C. Turner, Reducing Facility Operating Costs Through On-Site Generation of Electricity, Energy Engineering, Vol. 98, No. 5, 2001, pp 65-78.

Bibliography

Audin, L., "Positioning Your Facility for Utility Competition," *Energy Engineering*, Vol. 92, No. 2, 1997.

Bernstein, A.P., Karp, J.M., Schwenck, C.E., " Power Shopping: What Electricity Deregulation Will Mean To You," *Journal of Property Management*, Vol. 62, No. 2, Mar. 1997.

Borbely, A.M., "Distributed Generation: A New On-Site Power Frontier," *Energy User News*, Vol. 22, No. 10, Oct. 1997.

Canada, J.R., Sullivan, W.G., White, J.A., *Capital Investment Analysis For Engineering and Management, 2nd Ed.* New Jersey: Prentice Hall 1996.

Capehart, Barney, Editor, *Encyclopedia of Energy Engineering and Technology,"* Taylor and Francis/CRC Publishing Company, Boca Raton, FL, July, 2007.

Energy Efficiency Improvements Through the Use of Combined Heat and Power (CHP) in Buildings, DOE/EE-0239, October 2000, US Department of Energy by the Oak Ridge National Laboratory.

Hennagir, T., "Distributed Generation Hits Market," *Power Engineering*, Vol. 101, No. 10, Oct. 1997.

Integrated Systems, DOE/EE-0234, September 2000, US Department of Energy by the Oak Ridge National Laboratory.

Mont, J., Petursson, G., Industrial Assessment Report #534, Oklahoma Industrial Assessment Center, 1997.

Chapter 15

Web-based Building Automation Controls and Energy Information Systems*

15.1 INTRODUCTION

A successful energy management program (EMP) has three main components. The first system is the building automation system (BAS) that provides controls for air conditioning, lighting and other systems in each building or facility. The second system is the energy information system (EIS), which is a suite of information technologies that work with the EMS to provide data and information to energy managers and other stakeholders. The final key component is a commitment from both management and staff that collectively participate in the EMP. It is the combination of technology and people that makes an EMP successful and sustainable.

This chapter focuses on the technical systems, the BAS and the EIS, and gets "under-the-hood" to show how these systems work together to effectively control and measure energy usage.

15.2 BUILDING AUTOMATION SYSTEMS

The combination of low cost, high performance microcomputers together with the emergence of high-capacity communication lines, networks and the internet has produced explosive growth in the use of web-based technology for direct digital control building automation systems (BAS) [1]. Many of these current BAS systems use a proprietary information structure and communications protocol that greatly limits the plug and play application and addition of interchangeable components in the system. Control solutions such as BACnet and LonWorks have helped this

---

*This chapter was written by Mr. Paul J Allen, chief energy management engineer, Reedy Creek Energy Services, a division of Walt Disney World Co., Lake Buena Vista, Florida.

situation somewhat, but they have also introduced their own levels of difficulties. The BAS of the future will integrate state of the art information technology (IT) standards used widely on the internet today. These new IT based systems are rapidly overtaking the older BAS systems. Most of the established BAS companies are quickly developing ways to interface their systems using IT standards to allow the use of web browsers such as Internet Explorer and Firefox.

This section examines all facets of a BAS, from field controllers to the front-end interface. The emphasis is on understanding the basic BAS components and protocols first, then examining how a BAS has changed based on the influence of IT standards. Finally, this section also discusses upgrade options for legacy BAS systems and BAS design strategies.

Even though we refer exclusively to the term BAS in this chapter, the building automation controls industry also uses the following terms interchangeably with BAS: direct digital control (DDC), energy management system (EMS), energy management and control system (EMCS), building automation and control system (BACS) and building management system (BMS).

15.2.1 The Basics of Today's BAS

At a minimum, a BAS is used to control functions of a heating, ventilating, and air conditioning (HVAC) system, including temperature and ventilation, as well as equipment scheduling. Even basic BAS are generally expected to perform control functions that include demand limiting and duty cycling of equipment. Additional basic features recording utility demand and energy use, building conditions, climatic data, and equipment status. BAS report outputs can show the facility utility load profiles, trends and operation logs of equipment, and generation of maintenance schedules. Another basic feature of a BAS is to notify service personnel of defective equipment and of sensors out of normal range. Using email or text pages, these alarms notify those responsible individuals so that repairs can be implemented quickly.

More elaborate BAS can integrate additional building systems, such as video surveillance, access control, lighting control and interfacing with the fire and security systems. A BAS might provide a status summary from multiple dedicated systems through a single operator interface, and sometimes to share at least a limited amount of data between controllers. A BAS may not control access to a building or generate security alarms, but it may show the status of door and window switches on the floor plan. Similarly, a single occupancy sensor may control both the HVAC and the lighting systems for a zone. However, in large organizations and campuses today, it is still more common to see dedicated systems for these additional building systems due to divisions in management functional responsibility, code issues, and features/performance of dedicated systems.

Today's BAS can receive and process more sophisticated data on equipment operation and status from such sensors as vibration sensors on motors, ultrasonic sensors on steam traps, infrared sensors in equipment rooms, and differential pressure sensors for filters. Top of the line BAS today also have additional capabilities, such as chiller/boiler plant optimization, time schedule/setpoint management, alarm management and tenant billing to name a few. Most BAS manufacturers today have started to offer some form of web-based access to their existing control systems and are actively developing web-based capability for their future products.

15.2.3 Controller-level Hardware

BAS controllers are used to provide the inputs, outputs and global functions required to control mechanical and electrical equipment. Most BAS manufacturers provide a variety of controllers tailored to suit the specific need. Shown below is a list of the most common BAS controllers:

Communications Interface: Provides the communication interface between the operator workstation and lower-tier controller network. On a polling controller network, a communications interface is used to transfer data between the controllers.

Primary Controller: Provides global functions for the BAS control network that can include, real-time clock, trend data storage, alarms, and other higher-level programming support. Some BAS manufacturers combine all these functions into one primary controller while other manufacturers have separate controllers that are dedicated to each global function.

Secondary Controller: Contains the control logic and programs for the control application. Secondary controllers usually include some on-board I/O and may interface to expansion modules for additional I/O. Inputs include temperatures, relative humidity, pressures, and fan & pump status. Outputs include on/off, and valve/damper control. Also included in this group are application specific controllers that have limited capability and are designed for a specific task. Examples include controllers for VAV boxes, fan coil units or multistage cooling and heating direct expansion (DX) air conditioning systems.

For further reference, the Iowa Energy Center has an excellent web site (http://www.ddc-online.org) that shows a complete overview of the designs, installations, operation and maintenance of most BAS on the market today.

15.2.4 Controller-Level Programming

BAS controllers typically contain software that can control output devices to maintain temperature, relative humidity, pressure, and flow to a desired set point. The software programming can also adjust equipment on-off times based on a time-of-day and day-of-week schedule to operate only when needed.

The software used to program the controllers varies by BAS manufacturer and basically falls into three categories:

1.   Fill-in-the-blank standard algorithms
2.   Line-by-line custom programming
3.   Graphical custom programming.

Fill-in-the-blank: Uses pre-coded software algorithms that operate in a consistent, standard way. The user fills in the algorithm configuration parameters by entering the appropriate numbers in a table. Typically, smaller control devices, like those that control a fan coil or VAV box controller use this type of programming. These devices all work the same way and have the same inputs and outputs.

A few manufacturers have used fill-in-the-blank programming for more complex devices such as air handlers where a variety of configurations can exist. Standard algorithms use the same program algorithm for each individual air handler device. As an example, the chilled water valve for an air-handling unit is programmed using a standard algorithm with only the configuration parameters adjusted to customize it for the particular type of valve output and sensor inputs. The rest of the air-handler components (supply fan, heating coil, outside air damper, etc.) are programmed using the appropriate standard algorithm.

The advantage of fill-in-the-blank standard algorithms is that they are easy to program and are standard. The downside is that if the standard algorithm does not function the way you want, or there is not a standard algorithm available, then the system requires development of a custom program.

Line-By-Line Custom Programming: Control programs are developed from scratch and are customized to the specific application using the BAS manufacturer's controls programming language. In most cases, programs can be re-used for similar systems with modifications as needed to fit the particular application.

The advantage of the line-by-line custom programs is that technicians can customize the programs to fit any controls application. The disadvantage is that each program is unique and trouble-shooting control problems can be tedious since each program must be interrogated line-by-line.

Graphical Custom Programming: BAS manufacturers developed this method to show the control unit programs in a flow chart style, thus making the programming tasks easier to follow and troubleshoot.

Below are some additional issues to consider regarding control unit programming:

- Can technicians program the control units remotely (either network or modem dial-in) or must they connect directly to the control unit network at the site?
- Does the BAS manufacturer provide the programming tools needed to program the control units?
- Is training available to learn how to program the control units? How difficult is it to learn?
- How difficult is it to troubleshoot control programs for proper operation?

15.2.5 Controller-Level Communications Network

The BAS controller network varies depending on the manufacturer. Several of the most common BAS controller networks used today include RS-485, Ethernet, ARCNET and LonWorks.

RS-485: Developed in 1983 by the Electronic Industries Association (EIA) and the Telecommunications Industry Association (TIA). The EIA once labeled all of its standards with the prefix "RS" (Recommended Standard). An RS-485 network is a half-duplex, multi-drop network, which means that multiple transmitters and receivers can exist on the network.

Ethernet: The Xerox Palo Alto Research Center (PARC) developed the first experimental Ethernet system in the early 1970s. Today, Ethernet is the most widely used local area network (LAN) technology. The original and most popular version of Ethernet supports a data transmission rate of 10 Mb/s. Newer versions of Ethernet called "Fast Ethernet" and "Gigabit Ethernet" support data rates of 100 Mb/s and 1 Gb/s (1000 Mb/s). 10-Gigabit Ethernet provides up to 10 billion bits per second.

ARCNET: A company called Datapoint originally developed an office automation network in the late 1970s. The industry referred to this system as ARC (attached resource computer) and the network that connected these resources as ARCNET. Datapoint envisioned a network with distributed computing power operating as one larger computer.

LonWorks: Developed by the Echelon Corporation in the 1990's. A typical LonWorks control network contains a number of nodes that perform simple tasks. Devices such as proximity sensors, switches, motion detectors, relays, motor drives, and instruments, may all be nodes on the network. Complex control algorithms are performed through the LonWorks network, such as running a manufacturing line or automating a building.

15.2.6 Controller-Level Communications Protocol

A communications protocol is a set of rules or standards governing the exchange of data between BAS controllers over a digital communications network. This section describes the most common protocols used in a BAS.

BACnet: Building automation control network is a standard communication protocol developed by ASHRAE specifically for the building controls industry. It defines how applications package information for communication between different building automation systems. The American National Standards Institute has adopted it as a standard (ASHRAE/ANSI 135-2008). The International Organization for Standardization adopted it as ISO 16484-5 in 2003.

LonTalk: An interoperable protocol developed by the Echelon Corporation and named as a standard by the Electronics Industries Alliance (ANSI/EIA-709.1-A-1999). Echelon packages LonTalk on their "Neuron chip" which is embedded in control devices used in a LonWorks network.

Proprietary RS-485: The protocol implemented on the RS-485 network is usually proprietary and varies from vendor to vendor. The Carrier Comfort Network (CCN) is an example of a proprietary RS-485 communications protocol.

Modbus: In 1978, Modicon developed the Modbus protocol for industrial control systems. Modbus variations include Modbus ASCII, Modbus RTU, Intel® Modbus RTU, Modbus Plus, and Modbus/IP. Modbus protocol is the single most supported protocol in the industrial controls environment.

The Internet Protocol Suite: This is a set of communications protocols used for the internet and other similar networks. It is commonly also known as TCP/IP, named from two of the most important protocols in it: the transmission control protocol (TCP) and the internet protocol (IP), which were the first two networking protocols defined in this standard. Most BAS today use communications devices that take advantage of these standards to extend the physical distance of their networks. For further reference, http://www.protocols.com/pbook/tcpip1.htm provides an excellent source of information on this subject.

15.2.7 Enterprise-Level Client Hardware/Software

Normally, a PC workstation provides operator interface into the BAS. The PC workstation may or may not connect to a LAN. If a server is part of the BAS, the PC workstation would need LAN access to the server data files and graphics. Some smaller BAS use stand-alone PCs that have all the BAS software and configuration data loaded on each PC. Keeping the configuration data and graphics in-sync on each PC becomes problematic with this design.

A Graphical User Interface (GUI) is one of the client-side software applications that provide a window into the BAS. The GUI usually includes facility floor plans that link to detailed schematic representations and real-time control points of the building systems monitored by the BAS. The GUI allows technicians to change control parameters such as set points, time schedules, or temporarily override equipment operation. Other client-side software applications include:

- Alarm monitoring
- Password administration
- System setup configuration
- Report generation
- Control unit programming and configuration

15.2.8 Enterprise-level Server Hardware/Software

Servers provide scalability, centralized global functions, data warehousing, multi-user access and protocol translations for a mid to large size BAS. Servers have become more prominent in the BAS architecture as the need has grown to integrate multi-vendor systems, publish and analyze data over an intranet or extranet and provide multi-user access to the BAS. While having a central server on a distributed BAS may seem contradictory, in reality a server does not take away from the stand-alone nature of a distributed control system. Servers enhance a distributed control system by providing functions that applications cannot perform at the controller level. In fact, a BAS may have several servers distributing tasks such as web publishing, database storage, and control system communication.

Servers provide the ability to globally control a BAS. Facility-wide time scheduling, load-shedding, or set point resets are examples of global functions a BAS server can perform. Since these types of functions are overrides to the standard BAS controller-level programs, having them reside in the server requires that steps be taken to insure continued control system operation should the server go down for any length of time. The distributed BAS should have the ability to "time out" of a server override if communications with the server is lost. When the server comes back on line, the BAS should have rules that govern whether the override should still be in effect, start over, or cancel.

15.2.9. BAS Design Issues

Aside from the impact that IT will have on future EMS, there are some fundamental characteristics that Owners have always desired and will continue to desire from a new BAS:

- Single-seat user interface
- Compatible with existing BAS
- Easy-to-use
- Easily expandable
- Competitive and low-cost
- Owner maintainable

There have been several changes made by the BAS industry to help satisfy some of these desires. The creation of open protocols such as LonWorks and BACnet has made field panel interoperability plausible. The development of overlay systems that communicate to multiple BAS vendor systems has made a single-seat operation possible.

There are two strategies available for the design and specification of a BAS for new or existing facilities:

1. Specify a multi-vendor interoperable BAS.
2. Standard on one BAS manufacturers system.

Specifying a multi-vendor interoperable BAS is probably the most popular choice of the facility design community. Using this approach, the engineer's controls design is more schematic and the specifications more performance-based. The engineer delegates the responsibility of the detailed BAS design to the temperature controls contractor since the engineer does not actually know which BAS vendor will be selected. Therefore, the resulting BAS design is by nature somewhat vague and entirely performance-based. The key to making this approach successful is in the details of the performance specification, which is not a trivial task. Competition results from multiple BAS vendors bidding on the entire BAS controls installation. The engineer may further specify that the BAS protocol be open (LonWorks or BACnet) to have the systems be interoperable. Unfortunately, there is currently little commonality between different BAS vendors' low-level panel programming, and different service tool software is needed for each BAS Vendors' system regardless of the open protocol. In the end, even though the BAS meets the original specification, it might not be the same as or even compatible with the existing facility BAS. Thus, the owner might operate this new system as just one of many disparate BAS systems in their BAS portfolio.

Another alternative to the multi-vendor BAS would be to have the owner pre-qualify two or three BAS vendors all using the same standard protocol (generally BACnet) and then allow these selected vendors to competitively bid projects. One vendor is chosen to provide the server front-end software, and the others integrate the projects they win into that front-end. This allows competitive bidding, while keeping the number of service tools and systems the maintenance crews must understand to a manageable number.

The second approach is based on standardizing on one BAS manufacturer's system. To create competition and keep installation cost low, the engineer must create the BAS design as part of the design documents and prescriptively specify all components of the BAS. This allows multiple temperature control contractors to bid on the BAS installation (wire/conduit/sensor actuators)—everything outside of the BAS field panel. Everything inside the BAS field panel is owner furnished. Contractors familiar with the owners' BAS, or the owners' own technicians perform the controller wire termination, programming and startup. This approach is successful when all parties work together. The design engineer must produce a good BAS design. The temperature controls contractor must install the field wire, conduit, sensors and actuators properly. Finally, the BAS Contractor must terminate and program the BAS panel correctly. A successful project is a system that integrates seamlessly with the owners' existing BAS.

15.2.10 Upgrading an Existing BAS

Most users already own and operate a legacy BAS that they might desire to upgrade from a stand-alone BAS to a network–based system [2]. The benefits of a network-based BAS include better standard operational practices and procedures, opportunities to share cost-savings programs and strategies, and wider access to building control processes. The keys to justifying the costs associated with networking a BAS are that it can be done at a reasonable cost and it is relatively simple to implement and operate.

There are three main strategies available when upgrading a BAS from a stand-alone system to a network-based system:

1.   Remove existing BAS and replace with new network-based BAS.

2.   Update existing BAS with the same manufacturer's latest network-based system.

3.   Install a BAS interface product that networks an existing BAS.

The first upgrade strategy is to simply replace the existing BAS with a newer network-based BAS that has been established as a standard within your company. The cost for this option is solely dependent on the size of the BAS that will be replaced. However, this approach might be justified if the existing BAS requires high annual maintenance costs or has become functionally obsolete.

The second upgrade strategy available is to request a proposal from the original BAS manufacturer for their upgrade options. Most BAS manufacturers have developed some form of Ethernet network connectivity. Typically, some additional hardware and software is required to make the system work on an Ethernet network. The cost for this might be very reasonable, or it could be very expensive. It all depends on how much change is required and the associated hardware, software and labor cost to make it all work.

The third upgrade strategy involves the installation of a new network-based system that is specifically designed to interface to different BAS systems **[EDITOR: insert chapter reference here]**. These systems typically have dedicated hardware that connects to the BAS network and software drivers that communicate to the existing BAS controllers. The new BAS interface controllers also have an Ethernet connection so they can communicate on the corporate LAN. Users view the BAS real-time data using web browser software on their PC. The advantage of this strategy is that a multitude of different BAS systems can be interfaced together. The disadvantage is that users must still use the existing BAS software to edit or add new control programs in the existing BAS field panels. Furthermore, software license fees must be maintained for all original BAS along with the new integration software.

15.2.11 Enhancing an Existing BAS with Custom Programs

Users might find they needing their BAS to meet certain functions that are not fully integrated into their BAS. The user could request the manufacturer make the changes to the BAS software, but this is a slow and potentially expensive process. A quicker solution is to use the existing BAS software to obtain the data needed which is then exported to your own custom programs that are designed to perform the desired requirement. Shown below are a few BAS enhancements that use this technique:

Alarm Notification System: When mechanical systems become defective or operate out normal design range, the existing BAS software can be programmed to create an alarm. In most cases, this alarm is either printed on the BAS alarm printer or displayed on the BAS terminal. To enhance the alarm notification beyond the existing BAS functionality, a separate program could be created to read all of the incoming BAS alarms and then notify all responsible service personnel automatically via email or text page. For example, suppose the chilled water temperature exceeded 55F on an operating chiller. The BAS would generate an alarm that would be read by the alarm notification program and automatically emailed/text paged to responsible maintenance department personnel. If desired, the alarm notification program could be designed to interface to the users preventative maintenance program to create a work order for repair. A web-based interface program could also be developed to display a history of the alarms by time and date to let users see the frequency of the alarm occurrence.

Interval Data Collection: Most BAS have the ability to collect data from the sensors and utility meters connected to their system. Typically, the data collection report is stored on the BAS front-end PC for the user to display using a spreadsheet program. An enhancement to this data function would be to create a separate program to automatically read all of the data from these reports into a relational database or to read from the BAS database directly. Another web-based program could be developed to graphically display the collected data quickly an easily. This web-based program could be part of a larger Energy Information System (EIS), which is covered in detail later in this chapter.

Time and Setpoint Reset: A BAS uses time schedules to turn the equipment on/off at scheduled intervals. Likewise, setpoints are used to modulate control devices to maintain specific temperature, relative humidity, flow and other requirements. The time and setpoint schedules will eventually get change from their optimal settings resulting from periodic too hot/too cold calls received by the maintenance department. Adjustments to the time and setpoint schedules are done manually through the BAS software and once changed will stay at those values indefinitely. A better solution is to create a separate program to keep the time and setpoint schedules at their optimal state by automatically resetting them on a daily basis. The program is design to provide the user with a method to manage the time and setpoint schedules for a large campus facility in a master schedule database.

15.2.12 Future Trends in BAS

The future BAS can be found on the web. Most all BAS manufacturers see the need to use web-based system using IT standards. Tremendous economies of scale and synergies can be found there. Manufacturers no longer have to create the transport mechanisms for data to flow within a building or campus. They just need to make sure their equipment can utilize the network data paths already installed or designed for a facility. Likewise, with the software to display data to users, manufacturers that take advantage of presentation layer standards such as HTML, Java and Flash can provide the end user with a rich, graphical and intuitive interface to their BAS using a standard web browser.

Standards help contain costs when new products are developed. While there is a risk of stagnation or at least uninspired creativity using standards, internet standards have yet to fall into this category, due to the large consumer demand for rich content on the internet. A BAS, even at its most extensive implementation, will only use a tiny subset of the tools available for creating content on the internet.

When a BAS manufacturer does not have to concentrate on the transport mechanism of data or the presentation of that data, new products can be created at a lower cost and more quickly. When the user interface is a web browser, building owners can foster competition among manufacturers since each BAS system is inherently compatible with any competitors at the presentation level. All that separates one BAS from another on a web browser is a hyperlink.

Another area where costs will continue to fall in using internet standards is the hardware required to transport data within a building or a campus. Off-the-shelf products such as routers, switches, WiFi, WiMAX, general packet radio service (GPRS), Zigbee and server computers make the BAS just another node of the IT infrastructure. Standard IT tools can be used to diagnose the BAS network, generate reports of BAS bandwidth on the intranet, and backup the BAS database.

Owners will reap the benefits of internet standards through a richer user interface, more competition among BAS providers, and the ability to use there IT infrastructure to leverage the cost of transporting data within a facility.

15.2.13 The Enterprise

Extensible Markup Language (XML): XML is an internet standard that organizes data into a predefined format for the main purpose of sharing between or within computer systems. What makes XML unique is that data tags within the XML document can be custom, or created on the fly;

and unlike HTML, are not formatted for presenting the data graphically. This makes XML a great choice for Machine-to-Machine (M2M) communication. XML data exchanges often utilize SOAP, or "Simple Object Access Protocol" to carry the data. The combination of XML/SOAP is commonly referred to as "web services."

Why is M2M so important? Because the next wave of BAS products will include "hooks" into other internet based systems. BAS systems have done a great job of integrating building-related components together. BACnet, LonWorks, and Modbus provide the capability of connecting together disparate building components made from different manufacturers, so that a lighting control panel can receive a photocell input from a rooftop building controller, or a variable frequency drive can communicate an alarm on the BAS when a failure occurs.

The future will require a BAS to connect to enterprise level systems, not just building level systems. This is where M2M and web Services come into play. Web services can be thought of as plug-ins for your BAS to communicate with a web based system or server. An example of this would be time synchronization. The internet has many time servers that can provide the exact local as well as GMT time. A BAS can have a web service that would plug-in to the BAS, synchronizing all of the time clocks within a facility to the atomic clock in Boulder, Colorado. Another example would be obtaining the outside air temperature from the local weather service. Instead of the BAS just measuring the outside air temperature at a local controller, a web service could provide the outside air temperature, humidity, barometric pressure, and any other weather related data. Now the BAS can make more intelligent decisions on using outdoor air for comfort cooling, determining wet bulb setpoints for cooling towers, or even announcing an alert that a storm is imminent.

More enticing than connecting to weather and time servers is the promise of connecting to a facility's enterprise data. The BAS of the future must become an integral part of the decision making for allocating personnel, budgeting maintenance and upgrades, purchasing energy, and billing those that use the energy. Most larger facilities have departments that provide these types of services, yet the BAS has always stood alone, providing input through exported reports, system alarms, or human analysis. Enterprise level integration would create web services to connect directly to these systems, providing the data necessary to make informed decisions about capital investments, energy, or personnel. See Figure 15-1 for what a BAS might look like in the future.

The good news is that XML and web services have gained the market acceptance to become the standard for enterprise level connectivity. The

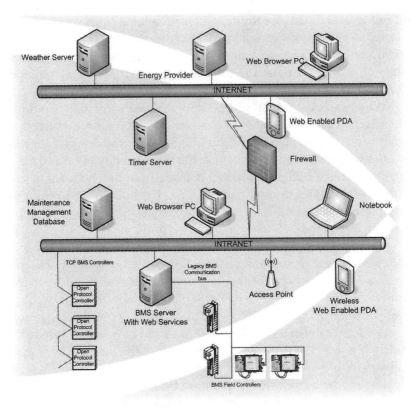

Figure 15-1. Future BAS Network Schematic

bad news is that this is still in its infancy for most BAS vendors. It is a very costly effort to create an enterprise level web service today. Even though web services are supported by Microsoft, Apple, Oracle, and others, they can still be custom solutions, tailored to a specific accounting, maintenance management, or energy procurement system. For web services to become mainstream in the BAS world, common services will need to be created that can be used by all BAS vendors. In addition, for web services to be properly implemented the skill set for BAS programmers and installers will need to include XML and a basic understanding of IP. If facility managers and technicians are to be able to make changes, adjustments and enhancements to their enterprise system, they too will require this skill set.

The future will also need to better define the decision logic and troubleshooting tools when implementing web services. When the BAS sends duplicate alerts to a maintenance management system, where does the logic reside to send only one technician to the trouble call? This is currently undefined. Standard tools for testing scenarios online and offline need to be developed. Even though web services typically rely on XML, which is a self-documenting standard, it can be very verbose. Tools are available to validate XML documents. When a facility decides to change their accounting system to a newer version or a different vendor, will the BAS be able to adapt? Conversion and upgrade tools also need to be considered when defining BAS web services.

Even without all the tools identified, enterprise level connectivity is moving ahead rapidly. The benefits of integrating BAS data within the facility's other systems can outweigh the immediate need for a complete set of tools. web services through XML place the BAS directly into the facility data infrastructure. That is a good place to be for an energy manager wanting to maximize the investment in their BAS.

The BAS of old relied heavily on a collection of separate systems that operated independently, and often with proprietary communication protocols that made expansion, modification, updating and integration with other building or plant information and control systems very cumbersome, if not impossible. Today the BAS is not only expected to handle all of the energy and equipment related tasks, but also to provide operating information and control interfaces to other facility systems, including the total facility or enterprise management system.

15.3 ENERGY INFORMATION SYSTEMS

The philosophy, "If you can measure it, you can manage it," is critical to a sustainable energy management program. Continuous feedback on utility performance is the backbone of an energy information system[3]. A basic definition of an energy information system is:

Energy Information System (EIS): Equipment and computer programs that let users measure, monitor and quantify energy usage of their facilities and help identify energy conservation opportunities.

Everyone has witnessed the continuing growth and development of the internet—the largest computer communications network in the world. Using a web browser, one can access data around the world with a click of a mouse. An EIS should take full advantage of these new tools.

15.3.1 EIS Process

There are two main parts to an EIS: (1) data collection and (2) web publishing. Figure 15-2 shows these two processes in a flow chart format.

The first task in establishing an EIS is to determine the best sources of the energy data. Utility meters monitored by an energy management

Figure 15-2. Energy Information System Schematic

system or other dedicated utility-monitoring systems are a good source. The metering equipment collects the raw utility data for electric, chilled & hot water, domestic water, natural gas and compressed air. The utility meters communicate to local data storage devices by pre-processed pulse outputs, 0-10V or 4-20ma analog connections, or by digital, network-based protocols. Meters are either hardwired directly to the local data storage devices or might use some form of wireless connectivity, such as cell phone technology.

Data gathered from all of the local data storage devices at a pre-defined interval (usually on a daily basis) are stored on a server in a relational database (the "data warehouse"). Examples of relational databases are MS SQL Server and Oracle.*

---

aAny reference to specific products or name brands of equipment, software or systems in this chapter is for illustrative purposes and does not necessarily constitute an endorsement implicitly or explicitly by the authors of this chapter or the others in this book.

15.3.2 Data Collection

Identifying and organizing the energy data sources is the first step in establishing an EIS. Since utility meters can be connected to several disparate data collection systems, there can be several methods used for data collection. Transferring this data in a common format and storing the data in a common EIS relational database is a simple way to pull all this together for further analysis by a web publishing program.

A BAS can provide an excellent utility data collection system. Utility meters that provide a pulse output for each fixed consumption value (i.e. 1 kWh/pulse for electric meters) can be read into a discrete pulsed input point and totalized to record the utility meter consumption. Most BAS have a daily report feature that can automatically generate reports that show the utility meter data at the desired interval (15 minutes, 1 hour, etc.). Once the reports are created and stored on the BAS Server, a separate program can be used to read each report and extract the data into the EIS relational database. On some newer BAS, utility meters are read and the data stored directly into a database that is part of the BAS. The data can be used by the BAS analysis tools or by other enterprise level systems by means of web services or API calls. However, if multiple data collection systems are being used, these data can be copied into a common EIS relational database so that all data from all data collection systems can be stored in one location.

Another approach to collecting utility data is to use a dedicated data acquisition server (DAS). The DAS allows users to collect utility data from existing and new meters and sensors. On a daily basis the DAS uploads the stored data to the EIS Server. Once the data has been transferred to the EIS Server, a program reads the DAS data files and updates the data in the EIS relational database for use by the web publishing program.

The AcquiSuite system from Obvius is typical of the emerging solutions and is a Linux based web server which provides three basic functions:

- Communications with existing meters and sensors to allow for data collection on user-selected intervals
- Non-volatile storage of collected information for several weeks
- Communication with external server(s) via phone or internet to allow conversion of raw data into graphical information

The backbone of the system is a specially designed web server. The DAS provides connectivity to new and existing devices either via the on-board analog and digital inputs or the RS 485 port using a Modbus protocol. The analog inputs permit connection to industry standard sensors for temperature, humidity, pressure, etc and the digital inputs provide the ability to connect utility meters with pulse outputs. The serial port communicates with Modbus RTU devices such as electrical meters from Veris, Square D and Power Measurement Ltd.

15.3.3 Web-publishing

The internet, with the world wide web—or web—has become accessible to all. It has allowed the development of many new opportunities for facility managers to quickly and effectively control and manage their operations. There is no doubt that web-based systems are the wave of the future. The EIS web publishing programs should take full advantage of these web-based technologies.

To publish energy data on the internet or an intranet (a private network that acts like the internet but is only accessible by the organization members or employees), client/server programming is used. The energy data is stored on the EIS server, and waits passively until a user, the client, makes a request for information using a web browser. A web-publishing program retrieves the information from the EIS relational database and sends it to the web server, which then sends it to the client's web-browser that requested the information.

There are many software choices available for the web-publishing process. One method uses a server-side common gateway interface (CGI) program to coordinate the activity between the web-server and the web-publishing program. Using CGI enables conventional programs to run through a web browser.

The web-publishing client/server process for an EIS uses the steps below (See Figure 15-2). This entire process takes only milliseconds depending on the connection speed of the client's computer to the web.

1.   A user requests energy information by using their web browser (client) to send an HTTP (hypertext transfer protocol) request to the web server.

2.   The web server activates the CGI program. The CGI program then starts up the web-publishing program.

3.   The web-publishing program retrieves the information from the relational database, formats the data in HTML (hypertext markup language) and returns it to the CGI program.

4.   The CGI program sends the data as HTML to the web server, which sends the HTML to the web browser requesting the information.

15.3.4 Web-publishing Programming Options

Although there are many web servers available to choose from, two are the most popular by far. Microsoft's Internet Information Services (IIS) comes with Windows server software. Apache web server is a good choice for other operating systems. Any web server needs some configuration to produce web content, especially if it is querying a database. The web-publishing task will likely require custom folders, special access permissions and a default page.

There are many programming alternatives available other than the CGI approach described above. Active Server Pages (ASP and ASP. NET), JavaScript, Java Applets, Java Server Pages, Java Servlets, ActiveX controls, and PHP are a few of the more popular choices available today. Some of these are easier to implement than others. ASP for instance, is a part of IIS so no installation is required. PHP require installation of their respective programs on the web server machine to run. Javascript and VB-Script are somewhat limited in that they are just a subset of the other full fledged programming languages. Most browsers interpret them correctly so no installation is required. Java Server Pages and Java Servlets run on the web server in the same way as ASP but may require some installation depending on the web server used.

After installing the web server, the web-publishing administrator must put a default page in the root directory of the web server. This is the first page users will see in their browser when they type in the web site's internet address. The pages are usually named "default.htm" or "index. htm" but can be anything as long as the web server is configured to treat them as the default page. Next, if CGI is used, the administrator creates a special folder to store the scripts. This is usually called "cgi-bin" or just "scripts." This folder must have permissions specifically allowing the files in the folder to be "executable." In some cases, "write" permissions are required for the folder if the CGI programs write temporary files to it. Other custom folders may be required to organize the web publishing content. Once the web-publishing administrator configures the web server he or she can install and test custom CGI programs and pages. If the CGI program or pages accurately return data from the database then the task of creating custom reports for the energy data can begin.

15.3.5 EIS Implementation Options

Deciding which web server and programming method to use along with configuring and implementing it to create a web publishing system can be quite a task. It really requires an expert in these areas to do a reliable job. Three approaches have evolved to satisfy web-publishing requirements.

1. Use internal resources to accomplish this task. This works well if there are already experienced web programmers available and they have time to work on the project. This makes it easy to customize the web publishing content as needed quickly and cost effectively. Finding time for internal personnel to focus on the project is usually the problem with this option.

2. Hire an outside consultant to do the configuration and programming as needed. This works well if the consultant has a good working relationship with someone internally to facilitate access to the protected systems and help with understanding the data. The consultant must be willing to work for a reasonable rate for this approach to be cost effective. The consultant must also be responsive to requests for support.

3. Purchase and install a somewhat "canned" version of the web publishing software and then customize it to fit the energy data as required. This approach has many possible problems in that the software is usually quite expensive and requires a great deal of customization and support from the outside to make it work well. However, for small simple projects this may be a good fit.

For users, who do not want to invest the time and effort required for this "do-it-yourself" approach, numerous companies provide a complete EIS service for an on-going monthly service fee. The EIS service company provides all of the IT-related functions, including the energy data collection/storage and the web-publishing program. The user accesses the EIS service web site by using a web browser, enters a user ID and password and then uses the available reports/graphs to analyze energy data.

The advantage of this approach is that the user does not get involved with the details and operation of the EIS, but instead is able to work with the EIS service provider to develop the utility data reports most helpful to their operation. The downside to this approach is the on-going monthly service fee that is a function of the amount of data processed—the more meters or bills processed the higher the monthly fee. There may also be additional costs to customize any reporting from the standard reports already created by the EIS service provider. The Building Manager Online service from Obvius* is one of the many choices available to users today.

---

*Jim Lewis, one of the authors of this chapter, is the CEO of Obvius, LLC.

15.3.6 EIS Web-Publishing Example

The Utility Report Cards (URC) is an example of a web-based energy information system that reports and graphs monthly utility data for schools. Each month, a web-based report is automatically generated and e-mailed to school principals and staff as encouragement to examine their school's electricity usage (energy efficiency) and to identify schools with high-energy consumption needing further investigation. The URC also is intended for teachers and students to use as an instructional tool to learn about school energy use as a complement to the energy-education materials available through the U.S. Department of Energy's EnergySmart Schools program (ESS). To see how the URC operates, go to http://www.utilityreportcards.com and click on "URC Live."

The URC was created to help the school staff understand and, therefore, manage their utility consumption and associated costs. The URC allows school principals to become aware of how their school is performing relative to a projected benchmark and to other schools of similar design and capacity. Giving recognition to schools that improve performance from prior-year levels creates a spirit of competition with the opportunity to recognize success. Those schools identified as high-energy users become the focus of attention to determine the reasons for their consumption level and ultimately to decrease the energy used. All of this is done by using the monthly utility data that is provided electronically at minimal or no cost to the schools by the utilities.

15.3.7 Turning Data into Useful Information

The installation of sub-metering is a positive step toward understanding the dynamics of the building systems being measured. However, energy savings are only achieved when the sub-meter data is turned in to actionable information. An EIS that continuously collects the sub-meter data and displays it in easy to understand charts and graphs is the key. By continuously "shining a light" on utility usage at each facility, utility costs are minimized by the actions of those who receive these timely and informative reports. Continuous feedback on utility performance pinpoints problems that result in energy waste which are corrected though BAS programming changes or repair of defective equipment.

Sub-metering utility usage is the most direct method for energy saving measurement. The majority of energy saving retrofit projects are implemented based on engineering calculations of the projected return on investment [4]. As with any projections of ROI, much of what goes into these calculations are assumptions and estimates that ultimately form the basis for implementation. As the folks at IBM used to say, "garbage in—

garbage out," which in the case of energy retrofits means that if any of the assumptions about parameters (run times, set-points, etc.) are wrong, the expected payback can be dramatically in error. The establishment of good baselines (measures of current operations) is the best way to determine the actual payback from investments in energy and sub-metering.

Just as important as building an accurate picture of the current operation is measuring the actual savings realized from an investment. If there is no effective means of isolating the energy used by the modified systems, it may be impossible to determine the value of the investment made. Using monthly utility bills for this analysis is problematic at best since the actual savings achieved can be masked by excessive consumption in non-modified systems.

Consider, for example, a commercial office building whose central chiller plant has an aging mechanical and control structure that provides limited capability for adjusting chilled water temperature. To improve efficiency, the building owner plans to retrofit the system to provide variable speed drives on pumps for the chilled water and condenser water systems along with control upgrades to allow for chilled water set-point changes based on building loads. In the absence of baseline information, all calculations for savings are based on "snap-shots" of the system operation and require a variety of assumptions. Once the retrofit is completed, the same process of gathering snapshot data is repeated and hopefully the savings projected are actually realized. If the building tenants either add loads or increase operational hours, it is difficult if not impossible to use utility bills to evaluate the actual savings.

In contrast, the same project could be evaluated with a high degree of accuracy by installing cost-effective monitoring equipment prior to the retrofit to establish a baseline and measure the actual savings. While each installation is necessarily unique, building a good monitoring system would typically require:

• Data acquisition server (DAS) such as the AcquiSuite from Obvius to collect the data, store it and communicate it to a remote file server.

• Electric submeter(s)—the number of meters would vary depending on the electric wiring configuration, but could be as simple as a single submeter (e.g., Enercept meter from Veris Industries) installed on the primary feeds to the chiller plant. If desired, the individual feeds to the cooling tower, compressors, chilled water pumps, etc. could be monitored to provide an even better picture of system performance and payback.

- Temperature sensors (optional): in most installations, this could be accomplished by the installation of two sensors, one for chilled water supply temperature and the other for chilled water return temperature. These sensors do not provide measurement of energy usage, but instead are primarily designed to provide feedback on system performance and efficiency.

- Flow meter (optional)—a new or existing meter can be used to measure the gallons per minute (gpm). By measuring both the chiller input (kW) and the chiller output (tons) the chiller efficiency can be calculated in kW/ton.

The benefits of a system for actually measuring the savings from a retrofit project (as opposed to calculated or stipulated savings) are many:

- The establishment of a baseline over a period of time (as opposed to "snapshots") provides a far more accurate picture of system operation over time.
- Once the baseline is established, ongoing measurement can provide a highly accurate picture of the savings under a variety of conditions and establish a basis for calculating the return on investment (ROI) regardless of other ancillary operations in the building.
- The presence of monitoring equipment not only provides a better picture of ROI, but also provides ongoing feedback on the system operation and will provide for greater savings as efficiency can be fine-tuned.

15.3.7 Case Study—Retail Store Lighting

A retail store chain in the Northeast was approached by an energy services company about converting some of their lighting circuits to a more efficient design. On paper, the retrofit looked very attractive and the company elected to do a pilot project on one store with a goal to implementing the change throughout the entire chain if it proved successful. The retailer decided to implement a measurement and verification (M&V) program to measure the actual savings generated by comparing the usage before the retrofit (the baseline) and after.

The store had 12 very similar lighting circuits, all of which were operated on a time schedule from a central control panel in the store. Since the circuits were very similar, it was decided that measuring the impact on one circuit would provide a good indication of the savings from the other circuits. The sub-metering equipment consisted of the following:

- An electrical sub-meter was installed on the power lines feeding the lighting circuit.

Figure 15-3. Retrofit Electric Sub-Meter

- A data acquisition server was installed in the store to record, store and upload time-stamped interval data to a remote server for storage and display. The DAS provides plug and play connectivity to the sub-meter and uses an existing phone line or LAN to send data from the store to a remote server on a daily basis.

Figure 15-4. Data Acquisition Server

The remote server was used to monitor consumption before the retrofit and to measure the actual savings.

Figure 15-5 shows the actual kW usage over roughly 24 days. The left side of the chart shows the kW usage for the first 11 days before the retrofit and the average usage is fairly constant at around 1.45 kW. On Feb. 11, the retrofit was performed, as indicated by the drop to zero kW in the

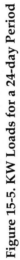

Figure 15-5. KW Loads for a 24-day Period

center of the chart. Immediately after the retrofit (the period from Feb 11 to Feb 15, the kW load dropped to around 0.4 kW, a reduction of over 70% from the baseline load in the left of the graph.

The good news for the retailer was that the retrofit performed exactly as expected and the M&V information obtained from monitoring the energy on this circuit provided clear evidence that the paybacks were excellent. The initial good news, however, was tempered somewhat after looking at the chart. It was immediately evident that this lighting circuit (and the other 11 identical circuits) were operating 24 hours per day, seven days a week. The store, however, operated from 10 AM to 9 PM each day and the lighting panel was supposed to be shutting off the circuits during non-operating hours.

The electrical contractor was called in to look at the system and determined that a contactor in the panel had burned out resulting in continuous operation of the lighting circuits throughout the store. Once the contactor was replaced, the operation of the lighting panel was restored so that the lights were only on during operating hours and shut off during the night, as indicated by the right side of the chart.

This simple chart of energy usage provides an excellent example of two uses of energy information:

Measurement and verification of energy savings—The chart clearly shows the actual energy reduction from the lighting retrofit and the data provided can be used to extrapolate the payback if this same retrofit is applied throughout the chain.

Use of energy information to fine-tune building operations—In addition to the M&V benefits of energy information, this example also shows how a very simple review of energy usage can be used to make sure that building systems are operating properly.

15.4 CONCLUSION

The BAS is used for real-time *control* of building systems. An EIS is used to *measure* the buildings energy usage. The information from the EIS provides feedback to the building operator to make sure the BAS is working properly.

The web provides the means to share information easier, quicker, and cheaper than ever before. There is no doubt that the web is having a huge impact on the BAS industry. The BAS of tomorrow will rely heavily on the web, TCP/IP, high-speed data networks, and enterprise level connectivity. Improving facility operations in all areas, through enterprise information and control functions is fast becoming an equally important function of the overall BAS or facility management system.

Historically, hardware, software and installation of EIS has been prohibitively expensive and has limited implementation to those commercial and industrial facilities that could afford to pay for custom systems integration services. These costs have fallen dramatically as companies leverage the enormous investment in the internet to provide the building owner with tools that make do-it-yourself data acquisition a cost effective reality. Hardware and software designed specifically for data acquisition and using available tools such as TCP/IP, HTTP and Modbus put valuable energy information literally at the fingertips of today's facility owners and provide an excellent method for measurement and verification of energy saving projects.

Web integration of BAS and EIS are inevitable, so if you have not done so already, it is a good time for energy managers to know their IT counterparts. Getting a good handle on the technical-side of things can be a daunting task. A successful energy manager will find a way to master their BAS and EIS.

At the same time, it is important to remember that commitment from people (management and staff) is the most important aspect of a successful energy management program. Once all three components are working together, the energy-saving results are significant and sustainable.

References
1.   Barney Capehart, Paul Allen, Rich Remke, David Green, Klaus Pawlik, *IT Basics for Energy Managers—The Evolution of Building Automation Systems Toward the Web*, Information Technology for Energy Managers, The Fairmont Press, Inc., 2004
2.   Paul Allen, Rich Remke, Steve Tom, *Upgrade Options for Networking Energy Management Systems*, Information Technology for Energy Managers—Vol II, The Fairmont Press, Inc., 2005
3.   Barney Capehart, Paul Allen, Klaus Pawlik, David Green, *How a Web-based Energy Information System Works*, Information Technology for Energy Managers, The Fairmont Press, Inc., 2004
4.   Jim Lewis, *The Case for Energy Information, Information Technology for Energy Managers*, The Fairmont Press, Inc., 2004

Software References
RS-485, http://www.engineerbob.com/articles/rs485.pdf
Ethernet, http://www.techfest.com/networking/lan/ethernet.htm
ARCNET, http://www.arcnet.com
LonWorks & LonTalk, http://www.echelon.com/products/Core/default.htm
BACnet, http://www.bacnet.org/
Modbus, http://www.modbus.org/default.htm
XML, http://www.xml.com
Iowa Energy Office, http://www.ddc-online.org
TCP/IP, http://www.protocols.com/pbook/tcpip1.htm

Chapter 16

Creating Green Buildings*

16.0 INTRODUCTION

Buildings and structures play an important role in meeting human needs. They provide shelter and can provide healthy environments. They are important components of sustainable communities. However, buildings can also provide unhealthy conditions and stress the environments in which they are built. Buildings consume land, require excavation for construction, can modify wildlife habitats, change rainwater runoff patterns, change landscapes, and require a cornucopia of materials. They also absorb and radiate heat, need paving for pedestrians and vehicles and, most importantly, consume resources—including energy. Energy consumption varies with building size, design and climate conditions. Chicago's Sears Tower alone consumes more electricity than Rockford, Illinois, a city with a population over 150,000 (Rifkin 2006).

In the U.S., buildings represent 39% of primary energy use (including production), 70% of electrical consumption, 12% of potable water use, and generate 136 million tons of construction waste annually (USGBC October 2006:1-3). If there were a way to reduce electrical demand in buildings by only 10%, most of the new electrical generating facilities scheduled for construction in the U.S. over the next 10 years would not be required. The carbon associated with the energy used in U.S. buildings "constitutes 8% of the current global emissions—equal to the total emissions of Japan and the United Kingdom combined."[1]

Buildings have an economic life span. Many are often financed for 30 years or more and have been designed to serve for 40-50 years. Over time, buildings are subject to deterioration, exposure to the elements, changes in use and occupancy, changes in building standards, and changes in own-

---

*This chapter was authored by Dr. Stephen A. Roosa, Account Executive, Energy Systems Group in Louisville, KY.

16-1. Moving Buildings—The Ultimate Recycling Project

ership. With this in mind the idea of a 'sustainable building' may seem to be a contradiction.

However, buildings and building materials can often be recycled and reused. Older structures can be updated to serve new purposes, recycling their existing infrastructure. Perhaps the ultimate recycling project is moving a building to another location rather than demolishing it. In such cases, a substantial portion of the energy and material embedded in its original construction can be reused.

"A design that is green is one that is aware of and respects nature and the natural order of things; it is a design that minimizes the negative human impacts on the natural surroundings, materials, resources, and processes that prevail in nature. It is not necessarily a concept that denies the need for any human impact, for human existence is part of nature too. Rather, it endorses the belief that humankind can exist, multiply, build and prosper in accord with nature and the earth's natural processes without inflicting irreversible damage to those processes and the long-term habitability of the planet."

Grumman (2003:3)

The idea of developing sustainable buildings stands in striking contrast to typical standard construction practices. Traditional architectural design processes are quite ossified while integrated design approaches are rarely standard procedure. *Sustainable* construction involves changes in the fundamental design processes, refocusing on-site development, rethinking materials and building components, considering environmental impacts, and determining how healthy interior environments can be configured. To further complicate the process of designing and developing green buildings, Huber (2005:216) concludes that "…building sustainability strategies have to consider basic societal conditions. Building regulations, financing models, rental legislation, environmental legislation, etc., significantly determine the construction design. The solution is approached by an eco-efficient optimization which considers the basic societal conditions."

Knowing that buildings require substantial inputs of resources to construct and maintain, what can be done to reduce their environmental impacts and energy requirements? The solutions include but are not limited to:

1) Changing land development practices;
2) Designing buildings with attention to improved construction standards;
3) Upgrading and reusing existing structures;
4) Providing more efficient buildings and higher quality equipment;
5) Changing the physical arrangement and configuration of buildings;
6) Carefully selecting construction materials; and
7) Harvesting on-site resources.

This chapter considers how buildings can employ green practices and how building technologies are incorporated into the design of buildings. "Green building programs" will also be discussed. These programs include efforts to establish energy and environmental standards for existing buildings and for new construction.

16.1 LAND DEVELOPMENT PRACTICES

Land development practices have contributed to adverse environmental consequences and changes in infrastructure that have increased energy and water consumption. Large-scale developments such as manufacturing centers and residential subdivisions constitute a type of "terraforming," which cause permanent changes to their environs. The growing appetite of these structures for energy and environmental resources has contributed to ecosystem disruption and has forced us to rethink how buildings are sited and constructed.

While development is associated with localized environmental disruptions, scientific assessments of construction impacts have yielded mixed results. We know that construction is energy intensive and resources are consumed in the process of construction. Often, initial costs are minimized, and life-cycle cost analysis is not used in the selection of materials and mechanical systems. When initial costs are minimized, maintenance, energy, and water costs typically increase. Development that favors suburbanization and automobile biased transportation systems also increases resource costs. Green development practices coupled with design alternatives for structures with environmentally friendly and energy efficient attributes have now become a feasible alternative. When these are addressed in the site selection and design of buildings, life-cycle costs are typically reduced.

Sustainable development focuses on the built environment: buildings are seen as the primary building blocks of the urban infrastructure. Green construction practices provide alternatives. It is understood that "being green" implies a commitment to environmental protection and natural resource conservation."[2] If buildings can be constructed in a manner that is less environmentally damaging and more energy efficient, then they can be called "green" buildings. Being green means that construction needs to be "eco-efficient." Life-cycle cost assessments should be made in selecting building systems and construction components. "Eco-efficiency is achieved by the delivery of competitively-priced goods and services that satisfy human needs and quality of life, while progressively reducing ecological impacts and resource intensity throughout the life-cycle to a level at least in line with the earth's carrying capacity" (DeSimone and Popoff 1997:47).

In regard to the built environment, architectural designers have recently renewed their emphasis on designing healthy buildings. Fundamental design issues such as site orientation, daylighting, shading, landscaping, more thermally cohesive building shells, and more energy efficient mechanical and electrical systems are getting renewed attention.

Drivers include experiences with "sick building syndrome" and illnesses linked to systemic malfunctions such as chemical exposures, legionnaire's disease, asthma and asbestosis. Medical conditions have been linked to mold exposures that result from indoor environmental conditions. These served to jolt architects and engineers into re-establishing the importance of indoor environmental conditions in general and indoor air quality in particular when designing their buildings. With concerns mounting as to product safety and liability issues involving the chemical composition of materials, manufacturers have began to mitigate the potential adverse impacts of construction building materials upon occupants. Concurrently, resource availability and waste reduction became issues influencing both building design and the design of construction components.

Decisions as to what types of buildings to construct, what construction standards apply, and what sorts of materials to use in the construction of buildings are typically made locally. Those who influence decisions regarding the physical form of a proposed structure include the owners, builders, developers, contractors, architects, engineers, planners and local zoning agencies among others. With so many parties involved in the decision making process, planning for green buildings can be a difficult, decentralized and often divisive process. In addition, all involved must abide by regulations that apply to the site and structure being planned. Planning ordinances and building codes often vary from one locale to another, further complicating the process. For these reasons, making green buildings requires a team of cooperative individuals. It is a comprehensive process that involves careful design and selection of building components and systems.

What is alarming is that past professional practice within the U.S. building industry has only rarely gauged the environmental or energy impact of a structure prior to its construction. Building types, especially for commercial construction, are often standardized, regardless of differentiators such as local climate conditions, geography and site conditions. This lack of differentiation contributes to increased resource consumption. Green buildings are an attempt to improve the planning and design of new structures and major renovations.

16.2 THE CONCEPT OF GREEN BUILDINGS

There was a time in the U.S. when most construction material was obtained locally. Indigenous materials included accessible timber, fieldstone, locally quarried rock, adobe, thatch, slate, clapboard, and cedar shakes. Since construction materials were costly to manufacture (most were hand-made) and troublesome to transport, most components of demolished structures were reused in some manner.

Early in U.S. history, one- and two-room log houses were the norm. The central heating system was a drafty fireplace with a chimney constructed of local stone. When possible, design features, learned by trial and error, were added in an attempt to optimize thermal comfort. Examples included architectural features that would control lighting and to temper the indoor environment with shading devices, take advantage of breezes, and carefully size and orient fenestration. Rainwater was often collected from roofs. Such were the humble beginnings of green construction practices.

Buildings today are infinitely more complex—and sustainable building practices can be even more so. Accepting the notion that sustainable, environmentally appropriate, and energy-efficient buildings can be labeled "green," the degree of "greenness" is subject to multiple interpretations. The process of determining which attributes of a structure can be considered "green" or "not green" can be viewed as inconclusive and subjective. A green design solution appropriate for one locale may be inappropriate for others, due to variations in climate or geography. Complicating the process, there are no clearly labeled "red" edifices with diametrically opposing attributes. While it is implied that a green building may be an improvement over current construction practice, comparison is often unclear and confusing. It can often be perplexing as to what sort of changes in construction practice, if imposed, would lead to greener, more sustainable buildings. At times, markets adjust and provide materials, components and products so that greener buildings can arise. Since standards are often formative and evolving, gauging the degree of "greenness" risks the need to quantify often subjective concepts. One attribute of green construction practices is the attempt to preserve and restore habitat that is vital for life, or to become "a net producer and exporter of resources, materials, energy and water rather than being a net consumer."[3] In Pennsylvania, the Governor's Green Government Council defines a green building as "one whose construction and lifetime of operation assures the healthiest possible environment while representing the most efficient and least disruptive use of land, water, energy and resources."[4]

There are qualities of structures, such as reduced environmental impact, improved indoor air quality and comparatively lower energy usage, which are widely accepted as evidence of green construction practices. For example, using recycled materials that originate from a previous use in the consumer market, or using post-industrial content that would otherwise be diverted to landfills, are both widely accepted green construction practices.

Grumman (2003:4) describes a green building by saying that "…a green building is one that achieves high performance over the full life cycle." Performance can be defined in various ways. "High performance" can be interpreted widely, and often in "highly" subjective ways. In an attempt to clarify, Grumman further identifies a number of attributes of green buildings:

- Minimal consumption—due to reduction of need and more efficient utilization—of nonrenewable natural resources, depletable energy resources, land, water, and other materials as well.

- Minimal atmospheric emissions having negative environmental impacts, especially those related to greenhouse gases, global warming, particulates, or acid rain.

- Minimal discharge of harmful liquid effluents and solid wastes, including those resulting from the ultimate demolition of the building itself at the end of its useful life.

- Minimal negative impacts on site ecosystems.

- Maximum quality of indoor environment, including air quality, thermal regime, illumination, acoustics, noise and visual aspects.

While goals and attributes of green construction practices are readily identified, developing construction standards to achieve such goals is another matter. High performance sounds like added value, for which a premium is likely to be paid. In fact, adding green building features is likely to increase construction costs by 1% to 5%, while lowering operating expenses and life-cycle costs. Energy-efficient electric motors may be slightly more expensive to install but the savings can be substantial. An estimated "97% of the life-cycle costs of a standard motor goes to energy costs and only 3% to procurement and installation."[5]

Buildings are further differentiated by the desires of the owners, the skills and creativity of their design teams, site locations, local planning and construction standards, and a host of other conditions. What may be a green solution for one building might be inappropriate if applied to another.

There are a multitude of shades of green in green building construction practices. Qualities of green construction practices have been variously identified.[6] Some qualities focus on exterior features and others on the types of green materials that are used. The 2006 International Energy Conservation Code (IECC), for example, requires that energy efficient design be used in construction and provides effective methodologies. Its focus is on the design of energy-efficient building exteriors, mechanical systems, lighting systems and internal power systems. The IECC is being included in building codes across the U.S. and in many other countries. ASHRAE 90.1 (Energy Standard for Buildings Except Low-Rise Residential Buildings) deals with the energy-efficiency of buildings and HVAC components. Other standards concern improved air quality and better ventilation systems. Green construction opportunities are further described in the *ASHRAE Green Guide*. Green buildings may also be designed to focus carbon impact and some are classified as "zero-carbon," meaning that they have a net zero carbon emissions impact.

16.3 COMPARATIVE BUILDING ENERGY PERFORMANCE

Making buildings stingy in their use of energy is one aspect of green construction. Thankfully, there are many opportunities to improve energy performance in residential, industrial, and commercial buildings. Advances in construction practices have yielded striking results and provide the means to create more efficient buildings.

The author conducted a one-month (December 2006) heating and electrical system assessment of two occupied residences in the mid-western U.S. during the same time period. The homes are about 16 km (10 miles) apart. Only energy consumption was considered in this the assessment. Both residences are two story buildings, located in the same metropolitan area, served by the same utility companies, and on the same rate structures for electricity and natural gas. Neither used alternative energy sources. Both residences were heated with natural gas with thermostats set at approximately 22°C (72° F) for the period of comparison and neither used temperature setback controls during the assessment period. Both used high-efficiency (approximately 90-94%) natural gas fired, forced-air furnaces to heat interior spaces.

One building was a frame residence constructed in 1910 that was approximately 212 m^2 (2,280 ft^2). This frame residence had limited insulation, single pane windows, electric water heating and used primarily incandescent lighting. The other residence (designed by the author) was a brick home in a suburban location constructed in 2001. Both residences used the conventional construction practices available at the times of their construction. The newer 427 m^2 (4,600 ft^2) brick residence was designed with an expanded south-facing façade, reduced northern exposure, extra insulation in walls, an exterior infiltration wrap, insulated ceilings and foundation, high efficiency windows, natural gas water heating, extensive use of compact fluorescent lighting, and other features. To take advantage of topography and the thermal moderation available from the earth, it was set into the slope of a hill, minimizing the northwestern exposure and maximizing the solar gain from the southwest. The total energy bills for the older 212 m^2 frame structure for the period of study totaled $432, equating to $2.24/m^2 (19 cents/ft^2). The total energy bills for the newer 427 m^2 brick residence for the period totaled $275, equating to $.58/m^2 (5.2 cents/ft^2).

This comparison provides interesting findings. For the newer brick structure, total energy costs were 36% less than the older frame residence for the period of study despite the fact that it was twice as large. The energy costs for the newer residence (based on a unit area) were roughly 74% less for the period. The older frame residence used 556 kWh of electricity and 1,001.3 m^3 (353.6 ccf) of natural gas while the newer residence used 1,182 MWh of electricity and 519.9 m^3 (183.6 ccf) of natural gas. The newer residence consumed twice as much electricity and half as much natural gas during the study period. The increase electrical use was likely due to greater areas of space for lighting. The lower natural gas usage was likely the result of the improved thermal envelope.

While not a scientific study, the analysis provides empirical evidence that energy costs can be lower when energy efficiency technologies, using conventional construction practices, are incorporated in the design of residences.[7] It also demonstrates how the consumption of source fuels in buildings of similar function can vary significantly.

16.4 ENERGY STAR BUILDINGS

In an effort to provide information to improve the energy efficiency of buildings in the U.S., agencies of the central government co-sponsored the development of the Energy Star™ program. This program provides "technical information and tools that organizations and consumers need to choose energy-efficient solutions and best management practices" (USEPA 2003). The type of technical information available includes information about new building designs, green buildings, energy efficiency, networking opportunities, plus a tools and resources library. Energy Star offers opportunities for organizations and governments to become partners in the program. The program offers guidelines to assist organizations in improving energy and financial performance in an effort to distinguish their partners as environmental leaders. The multi-step process involves making a commitment, assessing performance, setting goals, creating an action plan, implementing the action plan, evaluating progress and recognizing achievements.[8]

Expanding on their success, Energy Star™ developed a building energy performance rating system which has been used for over 10,000 buildings. It does not claim to be a green building ranking system but rather a comparative assessment system that focuses on energy performance, an important component of green building technologies. The rating system is based not on energy costs but on source energy consumption. In the view of the Energy Star program, "the use of source energy is the most equitable way to compare building energy performance, and also correlates best with environmental impact and energy cost."[9] Using the Energy Star rating system, buildings are rated on their energy performance on a scale of 1 to 100 when compared to similar structures constructed in environments that experience similar weather conditions. Buildings that achieve a rating of 75 or greater qualify for the prized Energy Star label.[10]

16.5 GREEN CONSTRUCTION MATERIALS AND METHODS

Material and product recycling has a long history and green construction has its roots in byproduct recycling. During World War II, strategic materials, such as steel and aluminum, were recycled and reused to manufacture military equipment. After the end of the war, recycling programs fell into decline. Beginning anew in the 1970s, metals such as aluminum, copper and steel began to be recycled. By the 1980's, construction site wastes, such as steel frame windows and their glass panels, were being recycled rather than being sent to landfills. By the 1990's landfill space became more costly. In addition, once-flared natural gas from landfills began to be seen as a potential energy resource rather than a waste by-product.

Clayton (2003:486) defines a "sustainable" building philosophy as "to design, build and consume materials in a manner that minimizes the depletion of natural resources and optimizes the efficiency of consumption." Green construction practices have several categorical commonalities:

- Green buildings are designed to reduce energy usage while optimizing the quality of indoor air. These buildings achieve energy reductions by using more insulation and improved fenestration, by optimizing the energy usage of mechanical and electrical building subsystems, and by the use of alternative energy.

- There is an emphasis on reducing the costs of energy used to transport material to the construction sites. One means of achieving reductions in transportation costs is to use materials that have been locally manufactured. This provides the added benefit of supporting local employment and industries.

- There is a focus on using recycled construction materials (such as re-using lumber from demolished structures) or materials made from recycled products (such as decking materials that use recycled plastics). The idea is to reduce the amount of virgin material required in the construction.

- There is a preference for materials that are non-synthetic, meaning that they are produced from natural components such as stone or wood, etc. This often reduces the number of steps required for product manufacture and may also reduce the use of non-renewable resources employed: for example, reducing the use of oil required in the production of plastics. Extracted metals such as aluminum and copper may also be preferred as they can be more easily reused once the building's life cycle is completed.[11]

- There is a mandate of green construction to avoid the use of materials that either in their process, manufacture or application, are known to have environmentally deleterious effects or adversely impact heath. Examples include lead in paint or piping, mercury thermostats, and solvents or coatings that may outgas fumes and carcinogens.

- In an effort to reduce water from municipal sources and sewage treatment requirements, green buildings are often designed to harvest rainwater by using collection systems. Rainwater may be used for irrigation, toilets or other non-potable requirements. In addition, green buildings use technologies, such as flow restriction devices, to reduce the water requirements of the building's occupants.

- The quantities of construction wastes are reduced. This occurs by strategically reducing wastes generated during construction and by reusing scrap materials whenever possible. The goal is to reduce the amount of material required for construction and to reduce the quantities of scrap material that must be trucked to a landfill.

New industries and entire product lines have emerged in an effort to provide construction materials that meet green building standards. There are numerous creative examples. Beaulieu Commercial has brought to market a carpet tile backing made from 85% post-consumer recycled content using recycled plastic bottles and glass, yielding carpet backing that is 50% stronger than conventional carpet backing. The Mohawk Group manufactures carpet cores using a recycled plastic material rather than wood. This change saves the equivalent of 68,000 trees annually. There are products on the market (e.g., ProAsh), made with fly ash—a once wasted by-product—which has been recycled into concrete. Armstrong now offers to pick up old acoustical tile from renovation projects and will deliver them to one of their manufacturing sites to be recycled into new ceiling tiles. PVC products became available to satisfy piping needs in domestic applications for drainage systems rather than copper, lowering the weight of products and their components. The installation times are reduced by eliminating the need to sweat pipes, which lowers labor costs. While PVC has its own associated environmental issues, it is now being used for exterior trim due to its durability. Many wood products are certified and labeled if they have used environmentally appropriate growing and harvesting techniques in their production. Pervious paving systems, that allow vegetation to grow and reduce the heat absorption from paving, are also available. The list of available green technologies and products seems endless. Triple pane, low "E" window glazing, solar voltaic roof shingles, waterless toilets and solar powered exterior lighting systems are among

16-2. Construction Waste

the green building products that are being used in building systems.

Green construction has provided opportunities to introduce new product lines and the movement to green construction practices offers ready markets for the products. Hydrotech manufactures a green roof system that provides a balance between water drainage and soil retention thus allowing roof gardens to flourish.

To learn which products are greener than others, computer software tools are available to assist in determining their environmental impacts. These programs also help ascertain product life-cycles.[12]

16.6 RATING SYSTEMS FOR GREEN BUILDINGS

While green building components are available, incorporating them into green buildings requires forethought, engineering, and creative design. Green building standards have been developed by both private and governmental organizations, all bent on finding ways to assess green construction practices. Comparing the degree of "greenness" from one building to the next is difficult. One solution is the use of categorical rating systems in an effort to reduce subjectivity. The development of green building attributes or standards by private organizations recognizes that deci-

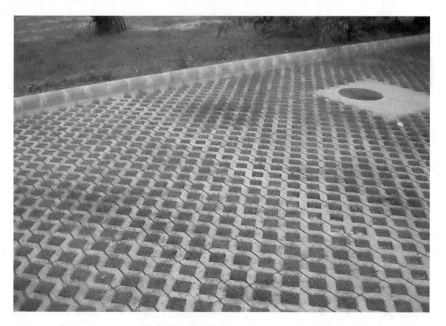

16-3. Pervious Paving System in Central Hungary

sions are to be based on stakeholder consensus. These stakeholders are often from widely-diverse industries and geographic locations.

Developing a rating system for green buildings is both difficult and challenging. According to Boucher (2004), "the value of a sustainable rating system is to condition the marketplace to balance environmental guiding principles and issues, provide a common basis to communicate performance, and to ask the right questions at the start of a project." Rating systems for sustainable buildings began to emerge in the 1990s.

Perhaps the most publicized of these first appeared in the U.K., Canada, and the U.S. In the U.K., the Building Research Establishment Environmental Assessment Method (BREEAM) was initiated in 1990. BREEAM™ certificates are awarded to developers based on an assessment of performance in regard to climate change, use of resources, impacts on human beings, ecological impact and construction management. Credits are assigned based on these and other factors. Overall ratings are assessed according to grades that range from pass to excellent (URS Europe 2005).

The International Initiative for a Sustainable Built Environment (IISBE), based in Ottawa, Canada has a Green Building Challenge program which is now used by more than 15 participating countries. This collaborative venture provides an information exchange for sustainable building initiatives and has developed "environmental performance assessment systems for buildings" (iiSBE 2005). The IISBE has created one of the more widely used international assessment systems for green buildings.

The U.S. Green Building Council (USGBC), an independent non-profit organization, grew from just over 200 members in 1999 to 3,500 members by 2003 (Gonchar 2003), and to 7,200 members by 2006. The core purpose of the USGBC "is to transform the way buildings and communities are designed, built and operated, enabling an environmentally-friendly, socially-responsible, healthy and prosperous environment that improves the quality of life" (USGBC 2006:1).

Prior to the efforts of organizations like the USGBC (established in 1995), the concept of what constituted a "green building" in the U.S. lacked a credible set of standards. The USGBC's Green Building Rating System has a goal of applying standards and definitions which link the idea of high performance buildings to green construction practices. The program developed by the USGBC is called "Leadership in Energy and Environmental Design" (LEED™). Sustainable technologies are firmly established within the LEED project development process. LEED loosely defines green structures as those that are "healthier, more environmentally

Figure 16-4. LEED Certified Building—Lincoln Hall, Berea, KY[13]

responsible and more profitable" (USGBC 2004).

The LEED Green Building Rating System is a consensus developed and reviewed standard, which allows voluntary participation by diverse groups of stakeholders interested in its application. The LEED system is actually a set of rating systems for various types of green construction projects, and ranks projects as part of the labeling process. LEED rating systems are available for new construction, existing buildings, commercial interiors, core and shell development projects, residences, and neighborhood development.[14]

Developed by the U. S. Green Building Council (USGBC), LEED rating systems evolved to help "fulfill the building industry's vision for its own transformation to green building" (USGBC 2004). The first dozen pilot projects using the rating system were certified in 2000. By 2006 there were 27,200 LEED Accredited Professionals, 3,539 LEED registered projects in 12 countries, and 484 projects that have completed the certification process (USGBC, October 2006). Additional LEED rating systems are being developed and separate rating programs for educational facilities, healthcare and retail new construction will soon be available. The LEED rating system is poised to become the new international standard for green buildings.

16.7 THE LEED-NC RATING SYSTEM

LEED for new construction and major renovations (LEED-NC) is perhaps the most widely adopted LEED program to date. LEED-NC provides an example of how rating systems for green buildings are structured. This section describes how LEED works, discusses the influences that shaped the development of LEED-NC for new construction and major renovations, and reviews how new projects are scored.

LEED-NC is the USGBC's standard for new construction and major renovations. It is used primarily for commercial projects such as office buildings, hotels, schools, institutions, etc. The rating system is based on an assessment of attributes and an evaluation of the use of applied standards. Projects earn points as attributes are achieved, and as evidence has been provided that the requirements of the standards have been followed. Depending on the total number of credits (points) a building achieves upon review, the building is rated as Certified (26-32 points), Silver (33-38 points), Gold (39-51 points), or Platinum (52 or more points) (USGBC 2003:6). Theoretically, there are a maximum of 69 achievable credits. However, in real world applications, gaining certain credits sometimes hinders the potential of successfully meeting the criteria of others. While achieving the rating of Certified is more easily accomplished, obtaining a Gold or Platinum rating is rare and requires both creativity and adherence to a broad range of prescriptive and performance-based criteria.

The LEED process involves project registration, provision of documentation, interpretations of credits, application for certification, a technical review, rating designation, award and appeal. Based on variables such as project square footage and USGBC membership status, registration fees for the process can range up to $7,500 per building (USGBC 2004).

To apply for the LEED labeling process, there are prerequisite project requirements which earn no credits (points). There are also categories of initiatives and attributes that do earn credits. The prerequisite requirements are interesting. For example, in the Sustainable Sites category, prerequisite procedures must be followed to reduce erosion and sedimentation. In the category of Energy and Atmosphere, procedures are required for building systems commissioning, minimal energy performance standards (e.g., adherence to ANSI/ASHRAE/IESNA Standard 90.1-2004, Energy Standard for Buildings Except Low-Rise Residential Buildings, or the local energy code if more stringent) and there must be verification that CFC refrigerants will not be used or that their use will be phased out. In addition, there are prerequisite requirements outlining mandates for storage and collection of recyclable material, mandates for minimum

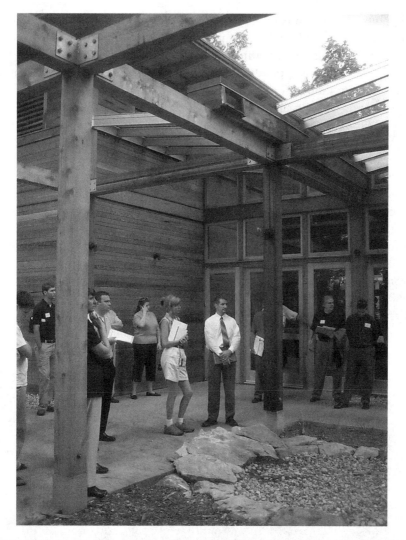

16-5. LEED Certified Interpretive Center at Bernheim Forest, Shepherdsville, KY

indoor air quality performance (the requirements of ASHRAE Standard 62.1-2004, Ventilation for Acceptable Indoor Air Quality must be adhered to) and a requirement that occupants who do not smoke tobacco will not be exposed.

In addition to the prerequisite requirements, the LEED process assigns points upon achieving certain project criteria, or complying with certain standards. The total points are summed to achieve the determined appropriate rating. Projects can achieve points from initiatives within the following sets of categories: Sustainable sites (14 points); water ffficiency

(5 points); energy and atmosphere (17 points); materials and resources (13 points); indoor environmental quality (15 points).

Using a LEED Accredited Professional (1 point) to assist with the project earns a single point (USGBC November 2002). Additional points are available though application for creativity in the innovation and design process (maximum of 4 points).

Within each category, specific standards and criteria are designed to meet identified goals. In the category of sustainable sites, 20.2% of the total possible points are available. This category focuses on various aspects of site selection, site management, transportation, and site planning. The goals of this category involve reducing the environmental impacts of construction, protecting certain types of undeveloped lands and habitats, reducing pollution from development, conserving natural areas and resources, reducing heat island impacts, and minimizing external light pollution. Site selection criteria are designed to direct development away from prime farmland, flood plains, habitat for endangered species, and public parkland. To encourage higher development densities, a point is awarded for projects that are essentially multi-story. If the site has documented environmental contamination or has been designated by a governmental body as a brownfield, another point is available.

16-6. Recharging Station for Electric Vehicles

In regard to transportation, as many as four points are available for locating sites near publicly available transportation (e.g., bus lines or light rail), providing bicycle storage and changing rooms, providing for alternatively fueled vehicles, and for carefully managing on-site parking.

Two points in this category can be obtained by limiting site disturbances and by exceeding "the local open space zoning requirement for the site by 25%" (USGBC 2002). Points are awarded for following certain storm water management procedures, increasing soil permeability and attempting to eliminate storm water contamination. Potential urban heat island effects are addressed by crediting design attributes such as shading, underground parking, reducing impervious surface areas, and using high albedo materials such as reflective roofing materials, or for using vegetated roofing systems. Finally, a point is available for eliminating light trespass from the site.

Water efficiency credits comprise 7.2% of the total possible points. With the goal of maximizing the efficient use of water use and reducing the burden on treated water systems, points are credited for reducing or eliminating potable water use for site irrigation, capturing and using rainwater for irrigation, and using drought-tolerant or indigenous landscaping. The LEED-NC standard also addresses a building's internal water consumption. Points are available for lowering aggregate water consumption, and also for reducing potable water use. Reducing wastewater quantities or providing on-site tertiary wastewater treatment also earns points.

The category energy and atmosphere offers the greatest number of points, 24.6% of the total possible. These include improving equipment calibration, reducing energy costs, supporting alternative energy, reducing the use of substances that cause atmospheric damage, and offering measurement and verification criteria. Optimizing the energy costs of regulated energy systems can achieve a maximum of ten points. To assess the result, project designs are modeled against a base-case solution which lacks certain energy-saving characteristics.

Interestingly, the unit of measure for evaluating energy consumption and performance in order to achieve credits is not a measure of energy (e.g., kilocalories or million Btus) but dollars. Points are awarded in increments as the percentage of calculated dollar savings increases. In addition to the ten points for energy cost optimization, a maximum of three additional points are available for buildings that use energy from a site-generated renewable energy source. Purchased green power is allocated a single point if 50% of the electrical energy (in kWh) comes from a minimum two-year green power purchasing arrangement. This category provides points for additional commissioning, and for eliminating the use of HCFCs and halon gases. Measurement and verification (M&V), a means of validating equipment performance and associated energy use, is allowed a point, but only if M&V options B, C and D, as outlined in the 2001 edition of the International Performance Measurement and Verification Protocol (IPMVP), are used.

The Materials and Resources category represents 18.8% of total possible points. This category provides credit for material management, adaptive reuse of structures, construction waste management, resource reuse, use of material with recycled content, plus the use of regionally manufactured materials (certain renewable materials, and certified wood products). A point is earned for providing space in the building for storage and collection of recyclable materials such as paper, cardboard, glass, plastics, and metals. A maximum of three points is available for the reuse of existing on-site structures and building stock. The tally increases with the extent to which existing walls, floor, roof structure, and external shell components are incorporated into the reconstruction.

LEED-NC 2.1 addresses concerns about construction waste by offering a point if 50% of construction wastes (by weight or volume) are diverted from landfills and another point if the total diversion of wastes is increased to 75%. A project composed of 10% recycled or refurbished building products, materials and furnishings gains an additional two points. Another two points is available in increments (one point for 5%, two points for 10%) if post-consumer or post-industrial recycled content (by dollar value) is used in the new construction. To reduce environmental impacts of transportation systems, a point is available if 20% of the materials are manufactured regionally (defined as being within 500 miles or roughly 800 km of the building site) and an added point is scored if 50% of the materials are extracted regionally. A point is available if rapidly renewable materials (e.g., plants with a 10 year harvest cycle) are incorporated into the project. Another point is earned if 50% of the wood products are certified by the Forest Stewardship Council.

The category of Indoor Environmental Quality allows 21.7% of the possible total points available. Goals include improving indoor air quality, improving occupant comfort, and providing views to the outside. With ASHRAE Standard 62-2001 as a prerequisite, an additional point is available for installing CO_2 monitoring devices in accordance with occupancies referenced in ASHRAE Standard 62.1-2004, Appendix C. A point is also available for implementing technologies that improve upon industry standards for air change effectiveness or that meet certain requirements for natural ventilation. Systems that provide airflow using both under-floor and ceiling plenums are suggested by LEED documentation as a potential ventilation solution. Points are available for developing and implementing indoor air quality (IAQ) management plans during construction and also prior to occupancy. The IAQ requirements include using a Minimum Efficiency Reporting Value (MERV) 13 filter media with 100% outside air flush-out prior to occupancy.

There are points available for use of materials that reduce the quantity of indoor air pollutants in construction caused by hazardous chemicals and by volatile organic compounds in adhesives, sealants, paints, coatings, composite wood products, and carpeting. A point is offered for provision of perimeter windows and another for proving individual controls for conditioned airflow and lighting in half of the non-perimeter spaces. Points are available for complying with ASHRAE Standard 55-1992 (Thermal Environmental Conditions for Human Occupancy), Addenda 1995, and installing permanent temperature and humidity control systems. Finally, points are gained for providing 75% of the spaces in the building with some form of daylighting and for providing direct line of sight vision for 90% of the regularly occupied spaces.

In the category of Innovation and Design Process, 7.2% of total possible points are available. Innovation credits offer the opportunity for projects to score points as a result of unusually creative design innovations, such as substantially exceeding goals of a given criteria or standard.

Despite the complexity of the scoring system and the challenges of constructing a LEED building, major corporations are lining up to use the LEED process for their buildings. Genzyme Corporation's headquarters building in Cambridge, Massachusetts features 18 indoor gardens, thermostatic controls in every office, and mirrors that reflect sunlight into an atrium. The company estimates that sick leave among employees is 5% lower than what is normally anticipated and that 58% of building occupants feel they are more productive in their new $140 million headquarters building (Palmeri 2006:96).

16.7.1 Assessing LEED-NC: Strengths

The LEED-NC process has numerous strengths. Perhaps the greatest is its ability to focus the owner and the design team on addressing site energy and environmental issues early in the design process. The LEED design process brings architects, planners, energy engineers, environmental engineers, and indoor air quality professionals into the program at the early stages of design development. The team often adopts a targeted LEED rating as a goal for the project. A strategy evolves based on selected criteria required and points required to achieve the rating. The team members become focused on fundamental green design practices that are often overlooked when traditional design development processes are pursued.

Furthermore, the LEED program identifies the intents of the environmental initiatives. The program requirements are stated in LEED documentation and acceptable strategies are suggested. Scoring categories directly address the criterion associated with the related energy costs and environmental concerns. When appropriate, the LEED-NC program defers to engineering and environmental standards developed outside of the USGBC. The components of the program accommodate local regulations.

The educational aspects of the program, which succinctly describe select environmental concerns, cannot be understated. Professionals must be accredited. Obtaining LEED accreditation can be challenging for professionals who have not been educated in green buildings requirements. LEED documentation and manuals concisely present information on LEED construction requirements. Case study examples, when available and pertinent, are described in the LEED literature. A web site (www.usgbc.org) provides updated information on the program and clarifies LEED procedures and practice. To expedite the process of documenting requirements, letter templates and calculation procedures are available to accredited professionals. Training workshops sponsored by the USGBC are instrumental in engaging professionals with a wide range of capabilities. These workshops also provide a forum to explain how the rating system works.

These strengths bring credibility to the LEED evaluation process. Advocates of the LEED rating system hope it will be the pre-eminent U.S. standard for rating new green construction. To its credit, LEED is becoming a highly regarded standard and it continues to gain prestige. Nick Stecky, a LEED Accredited Professional, firmly believes that the system offers a "measurable, quantifiable way of determining how green a building is" (Stecky 2004).

Requirements for buildings to meet LEED standards are taking root. The Washington D.C. City Council passed The Green Building Act of 2006 which is unique in that it applies to both public and non-public sector buildings. It calls for all new development in the city to conform to the U.S. Green Building Council's LEED standard beginning in 2008 for publicly financed buildings. Beginning in 2008, public buildings with more than 10,000 ft² will be required to be certified LEED Silver. The law requires the city's mayor to review LEED for Schools, and also circulate rules that require public schools to fulfill or exceed LEED for Schools (or use a similar rating system that mandates full-building commissioning). Private buildings in Washington, DC, will have to meet or exceed the new standards beginning in 2012. In addition, all nonresidential buildings will have to be LEED-certified.[16]

16.7.2 Assessing LEED-NC: Weaknesses

Despite its strengths, the LEED-NC has observable weaknesses and is not without its critics. Auden Schendler of the Aspen Skiing Corporation comments that the LEED process is too costly, too bureaucratic, fails to reward the best environmental options, and requires too much documentation (Palmeri 2006:96). He believes that for final project approval, the fees for architectural consultants and computer modeling can add as much as $50,000 to the cost of a 930 m^2 (10,000 ft^2) building (Palmeri 2006:96).

The LEED-NC registration process can sometimes be burdensome, and has been perceived as slowing down the design process and creating added construction cost.[17] Isolated cases support these concerns.[18] However, there are few comparative studies to validate claims of significant cost impact. Seemingly minor changes in the design of structures or the selection of materials can have major effects on construction costs, while being unrelated to green construction. Alternatively, there are a few case studies that suggest that there is no construction cost impact as a result of the LEED certification process. Most indicate costs ranging from 1% to 5% more than traditional construction. However, savings resulting from the use of certain LEED standards (e.g., reduced energy use) can be validated using life-cycle costing procedures.

LEED-NC fails as a one-size-fits-all rating system for new construction. There are regions where the program has been successful and others where the program has not yet been validated. Many of the LEED Platinum and Gold buildings constructed to date are located in California, Washington state, Oregon, Pennsylvania and Michigan. Other areas of the country have no or few buildings that have obtained higher levels of certification. Given the differences in geography and climate in the U.S., the lack of empirical examples leaves open the question as to whether or not the program will prove to be universally applicable.

There are other valid concerns in regard to the use of LEED-NC. In an era when many standards are under constant review, standards referenced by LEED are at times out of date. The ASHRAE Standard 90.1-1999 (without amendments) is referenced throughout the March 2003 revision of LEED-NC. However, ASHRAE 90.1 was revised, and republished in 2001. Currently, LEED—NC 2.2 has been updated to include ASHRAE Standard 90.1-2004 as the referenced standard.

Since design energy costs are used to score Energy and Atmosphere points, and because energy-use comparisons are base-lined against similar fuels, cost savings from fuel switching is marginalized. In such cases, the environmental impact of energy use remains unmeasured since energy units are not the baseline criteria. Using only the dollar cost of energy as the criteria (nicknamed the "green-back" approach) is not necessarily related to green construction and therefore questionable. At some point in the future, source energy may actually be priced inversely to its carbon intensity. There is no standardized energy modeling software commercially available that is specifically designed for assessing LEED buildings. LEED allows most any energy modeling software to be applied and each has its own set of strengths and weaknesses when used for energy modeling purposes. It is possible for projects to comply with only one energy-usage prerequisite, applying a standard already widely adopted, and still become LEED certified. This allows certification of buildings that are not energy efficient. In fact, it is not required that engineers have specialized training or certification to perform the energy models.

Though LEED aspires to international stature, its documentation currently lacks System International (SI) unit conversions, which limits its applicability and exportability to countries outside the U.S.

A number of the credits or points offered by the rating system seem questionable. For example, using extensive areas of glazing to improve daylighting may substantially increase energy use in some regions, offsetting the sustainability of the design. Increased daylighting may reduce electrical requirements for interior lighting. However, if a daylighting system is not carefully designed, cooling and heating loads can be increased. While indoor environmental quality is touted as a major LEED concern, indoor mold and fungal mitigation practices which are among the most pervasive indoor environmental issues, are not addressed by LEED and are not necessarily resolvable using the methodologies prescribed. It would seem that having a LEED accredited professional on the team would be a prerequisite rather than a valid credit.

LEED projects in locations with abundant rainfall or where site irrigation is unnecessary can earn a point by simply documenting a decision not to install irrigation systems, leaving the implication that scoring systems are difficult to regionalize. The ability of the point system to apply equally to projects across varied climate classifications and zones is also questionable and unproven.

Finally, the LEED process is not warranted and does not necessarily guarantee that in the end, the owner will have a "sustainable" building. While LEED standards are more regionalized in locations where local zoning and building laws apply, local regulations can also pre-empt green construction criteria. Of greater concern, is that it is possible for a LEED certified building to *devolve* into a building that would lack the qualities of a certifiable building. For example, the owners of a building may choose to remove bicycle racks, refrain from the purchase of green energy after a couple of years, disengage control systems, abandon their M&V program after the first year, fail to re-commission or maintain equipment and control systems, or remove recycling centers... yet retain the claim of owning a LEED certified building.

Despite the growing popularity of the LEED rating systems, only a small percentage of projects actually apply for the process in the U.S. The planned construction of a new $450 million dollar University of Kentucky hospital in Lexington, Kentucky presents an opportunity for a green building... yet the university is not seeking LEED certification. Typical of many building operators, Bob Wiseman, the University of Kentucky president for Facilities Management, states that "we are producing much greener buildings but haven't gone through the formal certification process yet... but (someday) we will."[19]

16.8 Measurement and Verification is Vital

Increasingly, measurement and verification (M&V) is being used for green building projects. M&V refers to the process of identifying, measuring and quantifying utility consumption patterns over a period of time. Measurement and verification can be defined as the set of methodologies that are employed to validate and value proposed changes in energy and water consumption patterns that result from an identified intervention (e.g., set of energy conservation measures) over a specified period of time. This process involves the use of monitoring and measurement devices and applies to new construction and existing buildings and facilities.

Measurement and verification methodologies are used for LEED projects, performance based contracts, project commissioning, indoor air quality assessments and for certain project certifications. By establishing the standards and rules for assessment criteria, the concept of measurement and verification is a key component of energy savings performance contracts. In performance contracts, the performance criterion of a project is often linked to guaranteed cost saving that are associated with the facility improvements.

Technologies and methodologies are available to measure, verify and document changes in utility usage. Tools are available in the form of M&V guidelines and protocols that establish standards for primary measurement and verification options, test and measurement approaches, and reporting requirements. Using procedures identified in the guidelines and protocols, a measurement and verification plan is developed to validate savings and to serve as a guide as the process unfolds.

The process of measurement and verification typically involves five primary steps: 1) performing the pre-construction M&V assessment; 2) developing and implementing the M&V Plan; 3) identifying the M&V project baseline; 4) providing a post-implementation report; and 5) providing periodic site inspections and M&V reports.

The theoretical basis for measurement and verification in regard to assessments of resource usage over comparative periods of time can be explained by the following equation:

**Change in Resource Use$_{(adj)}$ = Σ Post-Installation Usages
+/– Σ Adjustments – Σ Baseline Usages**

Baseline usages represent estimates of "normal" utility usages prior to implementation of any cost savings improvements. Adjustments are changes in resource use that are not impacted by an intervention and are considered exceptional. The term intervention refers to the implementation of a project that disrupts "normal" or "projected" utility usage patterns. Examples of these interventions include electrical demand reduction measures and energy and water conservation measures. Post-installation usage refers to resource consumption after the intervention has been performed. Using this formula, negative changes in resource use represent declines in adjusted usage while positive changes represent increases in adjusted usage.

The International Performance Measurement & Verification Protocol (IPMVP, April 2003) is the most widely used M&V protocol.[20] Its standards will be used as an example. The measurement and verification options in the IPMVP provide alternative methodologies to meet the requirements for verifying savings. The four measurement and verification options described in the IPMVP are summarized as follows:

Option A: Partially measured retrofit isolation.

Using Option A, standardized engineering calculations are performed to predict savings using data from manufacturer's factory testing (based on product lab testing by the manufacturer) and a site investigation. Select site measurements are taken to quantify key energy related variables. Variables determined to be uncontrollable can be isolated and stipulated (e.g., stipulating hours of operation for lighting system improvements).

Option B: Retrofit isolation of end use, measured capacity,
measured consumption

Option B differs from Option A, as both consumption (usage) and capacity are measured (output). Engineering calculations are performed and retrofit savings are measured by using data from before and after site comparisons (e.g., infrared imaging for a window installation or sub-metering an existing chiller plant).

Option C: Whole meter or main meter approach

Option C involves the use of measurements that are collected by using the main meters. Using available metered utility data or sub-metering, the project building(s) are assessed and compared to base-lined energy usage.

Option D: Whole meter or main meter with calibrated simulation

Option D is in many ways similar to Option C. However, an assessment using calibrated simulation (a computer analysis of all relevant variables) of the resultant savings from the installation of the energy measures is performed. Option D is often used for new construction, additions and major renovations (e.g., LEED-NC certifications for new construction).

Depending on site conditions and the technologies being used, each approach has discrete advantages and disadvantages. For example, in cases where facilities have main meters in place, Option C may be preferred. In new construction, Option D is the favored alternative.

With recent advancements in monitoring and measurement technologies, it is possible for energy engineering professionals to log and record most every energy consumption aspect of the energy conservation measures they implement. Examples include the use of data loggers, infrared thermography, metering equipment, monitors to measure liquid and gaseous flows, heat transfer sensors, air balancing equipment, CO_2 measurement devices and temperature and humidity sensors. Remote monitoring capabilities using direct digital controls (DDC), fiber optic networks and wireless communication technologies are also available. Chapter 15 discussed the development and use of most of these technologies.

Measurement and verification costs vary as a function of the methodology, the complexity of the monitoring, and the period of time that M&V needs to be performed. As applied monitoring technologies evolve and become accepted by the marketplace, costs for installed monitoring equipment will continue to decline as the capabilities of monitoring technologies continue to improve.

16.9 MEASUREMENT AND VERIFICATION FOR LEED PROJECTS

While an M&V credit is available for LEED projects, there is no prerequisite requirement for M&V or for a credentialed measurement and verification professional be part of the M&V plan development or the review process. In fact, LEED projects are not *required* to undergo the rigor of M&V—*it is optional*. To obtain a LEED credit, there is no requirement that M&V be performed to state-of-the-art standards—*only that it meets IPMVP 2001 requirements*. Regardless, measurement and verification is vital to the success of projects. Without a requirement that LEED projects perform M&V for an extended period of time, it is difficult to determine if predictive pre-construction energy modeling was accurate, or if predicted cost savings reductions were actually achieved. The lack of mandates to determine whether or not LEED buildings actually behave and perform as intended from an energy cost standpoint is a fundamental weakness of the LEED process. Prerequisite commissioning does not resolve this. Without M&V, any projected life-cycle cost savings resulting from the project cannot be accurately validated, risking illusionary energy cost savings.

There are a few problems with how the LEED certification process handles M&V. LEED excludes Option A as an acceptable M&V solution. Option A was updated in the 2003 version of the IPMVP and there was no rational argument to exclude Option A as an acceptable M&V alternative. Another problem with LEED's use of the IPMVP is that no standardized requirements are provided for calibration when Option D is used. The calibration methodologies vary depending on the software, who is using it and how it is being used. Without a mandated M&V requirement based on actual performance data for a minimum period of two to three years, predictions of energy cost savings cannot be validated and verified.

16.10 GREEN CONSTRUCTION IN SCHOOLS

Schools are important. A "Green School" is a "high-performance" school. More and more, the professionals who design and construct new schools or major school additions are aware of their responsibilities. Between the ages of 5 and 18, students will spend roughly 14,000 hours of their lives in them. High performance schools require green construction practices. High performance schools provide comfort and a healthy environment for students and staff. They use energy and other resources efficiently and have lower maintenance costs. High performance schools involve a commissioning process, are environmentally responsive, are safe and secure, and feature stimulating architecture (Eley 2006:61)

To spearhead the "Green Schools" effort, standards are being developed and employed. A number of recently developed manuals are available that provide guidance on how to implement green construction practices. The Sustainable Buildings Industry Council (SBIC) has produced a *High Performance Schools Resource and Strategy Guide*[21] to show school building owners and operators how they can initiate a process that will result in better buildings—ones that provide students with better learning environments.

For new construction, kindergarten-12th grade (K-12) school systems in New Jersey, California and elsewhere have adopted their own sustainable building standards. Kentucky is an example of one state that is developing new standards for "Green and Healthy Schools." The USGBC is developing a LEED rating system for educational structures.

16.10.1 New Jersey

New Jersey has codified its construction practices for schools in a document entitled *21st Century Schools Design Manual*, developed by the New Jersey Schools Construction Corporation. Its performance objectives are structured to create schools that are healthy and productive, cost effective, educationally effective, sustainable, and community centered.[22] The manual establishes a set of 24 comprehensive design criteria for schools and mandates that design teams consider the following categories of issues in new construction:[23]

Engineering Aspects

Energy Analysis	Renewable Energy
Life Cycle Costing	Efficient HVAC
Efficient Building Shell	Efficient Electrical Lighting
Water Efficiency	Thermal Comfort
Commissioning	Indoor Air Quality

Architectural Design

Acoustic Comfort	Flexibility and Adaptability
Stimulating Architecture	Environmentally Responsive
Daylighting	Site Planning
Environmentally Preferable Material & Products	Learning Centered Design

Other Considerations

Information Technology	Visual Comfort
Accessibility	Safety and Security
Community Involvement	Catalyst for Economic Development
Community Use	

Each topic covered in the manual provides a set of recommendations and identifies applicable standards. Interestingly, the manual's recommendations are similar to LEED-NC prerequisite requirements and elective credits. Schools meeting the standard can be considered to be high performance, LEED-like facilities—while avoiding the rigor and costs of LEED certification. The manual is a call for integrated design solutions to establish sustainable design as a cost effective means of achieving high performance schools in New Jersey.

16.10.2 California

The Collaborative for High Performance Schools (CHPS) is a non-profit organization established in 2000 to raise the standards for school facilities in California. Goals include improving the quality of education by facilitating the design of learning environments that are resource efficient, healthy, comfortable and well lit—amenities required for a quality educational experience (Heinen 2006:1). The CHPS program for high performance schools has been adopted by 14 school districts in California. The standard was recently updated and reissued as the *2006 CHPS Criteria* and applies to new construction, major renovation, and additions to existing school facilities (Heinen 2006:1). The CHPS offers a *Best Practices Manual* that details sustainable practices and resources available for the planning,

16-7. Solar Collector Array at Twenhofel Middle School, KY

design, criteria, maintenance and operations, and commissioning of high performance schools.

16.10.3 Kentucky

To implement sustainable technologies to help improve education in schools, Kentucky has developed its "Green and Healthy Schools Program," a voluntary program to encourage green standards for schools.

Twenhofel Middle School in northern Kentucky uses a number of technologies to reduce energy and water consumption. The building shell provides extensive use of insulation and high performance fenestration. The school uses a geothermal heat pump system for heating and cooling. Rainwater collected by a metal roof and drainage system is treated and used for non-potable needs. A central computer control system manages the energy used in the building. In addition, solar panels, installed on the roof, collect energy to generate a portion of the school's electrical requirements.

The building makes extensive use of daylighting in classroom areas. Sensors in the classrooms detect light levels to allow fixtures to adjust light output in response to the daylight available in classrooms. There is

16-8. Display in Lobby of Twenhofel Middle School, KY

a touch screen monitor in the lobby of the school that allows students to monitor information concerning the water collection system, solar output and the geothermal loop. Science programs use the school as a learning laboratory. Students often provide guided tours to explain the sustainable technologies that were incorporated into the design of the facility.

16.11 CONCLUSIONS

Buildings are resource intensive in their construction and operation. Buildings are also complex systems. Today, buildings can be constructed with features that allow them to use less energy and consume fewer resources. The ideal of sustainable buildings is a response to the energy and environmental impacts which accompany every one of them. Developing a green building project is a balancing act and requires a series of trade-offs. It involves considering how buildings are designed and constructed—at each stage of the project delivery process.

Standards are constantly evolving. The 2006 International Energy Conservation Code (IECC), for example, requires that certain energy efficient design methodologies by used in construction. The Code "addresses the design of energy-efficient building envelopes and installation of energy-efficient mechanical, lighting and power systems through requirements emphasizing performance."[26] It is comprehensive and provides regional guidelines with specific requirements for each state in the U.S. New construction materials and products are available that offer new design solutions.

There are many opportunities to include green design features and components in buildings to make them more sustainable. Yet, there are differences in the standards for green construction. While energy assessment systems for buildings (the Energy Star program, for example) typically focus the analysis on source energy, the USGBC's LEED program considers the cost of energy as its primary rating criteria. A number of assessment systems for sustainable buildings are now being used throughout the developed world. LEED-NC is becoming a widely-adopted standard for rating newly constructed "green" buildings and projects in the U.S. and elsewhere. LEED projects as rated as Certified, Silver, Gold or Platinum using a credit-based system. LEED projects are credited for design attributes, costs of energy, environmental criteria and use of green building standards.

In the U.S., the LEED-NC program has proven successful by offering its credit-based rating system for green buildings. Its popularity is gaining momentum. A total of 49 localities and 17 state governments now encourage the use of green building practices, policies and incentives. Their number is growing. The USGBC estimates that 5%—almost $10 billion—of current nonresidential construction in the U.S. is seeking certification (Palmeri 2006:96). According to Richard Fedrizzi of the USGBC, "this movement has created a whole new system of economic development... *We are at a tipping point.*"

Perhaps LEED's greatest strength is its ability to focus the owner and her design team on energy and environmental considerations early in the design process. Its greatest weaknesses are its focus on energy costs rather than energy use and its lack of mandated requirements for measurement and verification of savings. Today, there are over 3,500 projects that have applied for LEED certification. Due to the program's success in highlighting the importance of energy and environmental concerns in the design of new structures, it is likely that the program will be further refined and updated in the future to more fully adopt regional design solutions, provide means of incorporating updated standards, and offer programs for maintaining certification criteria.

Establishing rating systems for sustainable structures, such as BREEM and LEED, is difficult due to the often subjective concepts involved, the evolving nature of standards, and the local variability of construction practices. Future research will hopefully respond to concerns about increased construction costs, and actual energy and environmental impacts. Measurement and verification has an important role to play, as it outlines procedures that can be followed to verify utility cost avoidance from energy and water saving projects.

In the states of New Jersey, California and Kentucky, standards are being used and developed for high performance schools. High performance schools have many qualities that include a fresh look at the architectural, engineering and educational aspects of school design.

Many green building technologies such as high efficiency windows, solar arrays and day-lighting applications are easy to find when walking by or through a building. On the other hand, it is discouraging to owners that many important engineered features of green buildings are hidden from view in the mechanical rooms and spaces not visible to the ordinary visitor. "Achievements in sustainable building design often go unnoticed by people who visit, work or study" in a green building (Ling 2006:1). Examples of these technologies include computer control systems to manage energy and water use, rainwater collection systems, lighting control systems, under-floor air flow systems, etc. Many green technologies used in structures require a trained eye to observe.

This chapter considered sustainable construction, the features of green buildings, and the importance of resource and energy use in green construction. There are a growing number of buildings, especially new ones which incorporate many aspects of green design. It is likely that the use of green construction techniques and technologies will continue to expand. In the future, more examples of green buildings will be available.

References
1. Koomey, J., Weber, C., Atkinson, C., and Nicholls, A. (2003:1209). Addressing energy-related challenges for the U.S. buildings sector: results from the clean energy futures study. *Energy Policy*. 29. Elsevier. This study provides an analysis of building sector energy use in 1997 and projects usage, carbon and dollar savings potential for the following 20 years based on two scenarios.
2. Bowman, A. (2003, August 28-31). From *Green politics in the city*. Presentation at the annual meeting of the American Political Science Association. Philadelphia.
3. Governor's Green Government Council. What is a green building? *Building Green in Pennsylvania*. http://www.gggc.state.pa.us/gggc/lib/gggc/documents/whatis041202.pdf, accessed 18 January 2007.
4. Ibid.
5. Siemens AG. (2007). Combat climate change – less is more. http://wap.siemens-mobile.com/en/journal/story2.html, accessed 3 July 2007.
6. The ASHRAE green guide is a primer for green construction practices.
7. The average outside air temperature during the study period was 46 degrees F. The assessment was based on billings provided by the local utility. It is likely that the substantially larger electrical usage experienced by the newer building was also due, in part, to two desktop computers which operated almost continuously during the study month, and to the larger area of interior lighting. The newer brick building also had 11 outside light fixtures while the older frame building only had 2.
8. USEPA. Superior energy management creates environmental leaders, Guidelines for energy management overview. www.energystar.gov/index.cfm?c=guidelines.guidelines_index, accessed 3 January 2007.
9. USEPA. Be a leader – change our environment for the better. Portfolio manager overview. www.energystar.gov/index.cfm?c=spp_res.pt_neprs_learn, accessed 3 January 2007.
10. USEPA. Portfolio manager overview. www.energystar.gov/index.cfm?c=evaluate_performance.bus_portfolimanager, accessed 3 January 2007.
11. Lincoln Hall, a LEED building at Berea College in Berea, Kentucky, actually uses copper downspouts.
12. International Design Center for the Environment. eLCie - Building a sustainable world. www.elcie.org, accessed 4 July 2007.
13. This building was a LEED renovation project performed by Berea College under LEED – NC for new construction prior to the LEED – EB program for existing buildings.

14. For a listing and explanation of the various rating systems developed for LEED, see the U.S. Green Building Council rating systems website. www.usgbc.org/Display/Page.aspx?CMSPageID=222, accessed 2 January 2007.

15. This picture was taken in the parking area of Jewish Community Center Southwest, located in Brooks, KY.

16. American Institute of Architects. (2007, 4 January). D.C. requires private buildings to go green. *The AIA Angle.*

17. Yudelson, J. (2006, 17 July). *Engineers and green buildings – Get green or be left behind.* http://www.igreenbuild.com/_coreModules/content/contentDisplay.aspx?contentID=2480, accessed 4 July 2007. Yudelson cites a survey by Turner Construction in 2005 involving 665 executives that indicated 68% believed "higher construction costs were the main factor discouraging construction of green buildings." In addition, 64% of the respondents cited a "lack of awareness of the benefits of green buildings" and 50% cited long payback periods associated with green construction practices.

18. Kentucky's first LEED-NC school, seeking a Silver rating, was initially estimated to cost over $200/ft2 ($2,152/m2) compared to the local standard costs of roughly $140/ft2 ($1,505/m2) for non-LEED construction.

19. State hopes to create more "green buildings." *The Courier-Journal*, 26 December 2006.

20. A new 2007 edition of the IPMVP has been developed by the Efficiency Valuation Organization (EVO). See http://www.evo-world.org/index.php?option=com_content&task=view&id=60&Itemid=79.

21. Available for purchase from the SBIC bookstore by accessing http://www.sbicouncil.org/store/index.php.

22. See *High performances schools New Jersey.* www.njhighperformanceschools.org/performance.html, accessed 7 January 2007.

23. See the 21st century schools design manual, New Jersey Schools Construction Corporation, 30 September 2004, available at http://www.hp-snj.org/PDF/what/Design_Manual.pdf, accessed 8 January 2007.

24. Located in northern Kentucky, this LEED certified middle school is in the Kenton County School District.

25. This touch-screen display allows students to view the status of the building's geothermal heating and cooling system, the rainwater harvesting and storage system, the natural lighting and the solar-electric generation systems.

26. International Code Council (2003). *International Energy Conservation Code.* p. iii.

Bibliography

Boucher, M. (2004). Resource efficient buildings—Balancing the bottom line. *Proceedings of the 2004 world energy engineering congress*, Austin, Texas, September 22-24 2004; Atlanta: Association of Energy Engineers.

Capehart, Barney, Editor, *Encyclopedia of Energy Engineering and Technology*," Taylor and Francis/CRC Publishing Company, Boca Raton, FL, July, 2007.

Clayton, M. H. and Radcliffe, N.J. (1996). *Sustainability—A systems approach.* Westview Press. Edinburgh.

DeSimone, L. and Popoff, F. (1997). *Eco-efficiency—The business of sustainable development*. Cambridge MA: The Massachusetts Institute of Technology Press.

Eley, C. (2006, May). High performance school characteristics. *ASHRAE Journal*. 48.

Gonchar, J. (2003, January) Green building industry grows by leaps and bounds. *Engineering News-Record*.

Grumman, D.L. (ed.) (2003). *ASHRAE green guide*. Atlanta: ASHRAE.

Heinen, K. (2006, 1 September). *New standard approved for high performance schools*. Press release, the Collaborative for High Performance Schools.

Huber, H. (2005). Sustainable buildings—The systems perspective. *Climate Change—Energy Awareness—Energy Efficiency*. Fourth international conference, June 8-10, 2005. Visegrad, Hungary.

iiSBE (2005). *International initiative for a sustainable built environment*. http://iisbe.org/iisbe/start/iisbe.htm, accessed February 2005.

IPMVP New Construction Subcommittee (2003). International Performance Measurement & Verification Protocol, Vols. 1, 2 & 3. April 2003.

Ling, J.A. (2006). High performance buildings can teach sustainability 24/7. *Proceedings of the 2006 world energy engineering congress*. Atlanta: Association of Energy Engineers.

Palmeri, C. (2006, 11 September). The green stamp of approval. *Business Week*. p. 96.

Rifkin, J. (2007, 17 December). The risks of too much city. *Washington Post*. p.8.

Stecky, N. (2004, 22-24 September). Introduction to the ASHRAE Greenguide for LEED, *Proceedings of the 2004 world energy engineering congress*, Austin, Texas. Atlanta: Association of Energy Engineers.

URS Europe (2005). http://www.urseurope.com/services/engineering/engineering-breeam.htm, accessed February 2005.

U.S. Environmental Protection Agency (2003). *Understanding the AQI*. Epa.gov/air/data/gosel.html.

U.S. Green Building Council (2002, November). *Green building rating system for new construction and major renovations Version 2.1*.

U.S. Green Building Council (2004). *LEED green building rating system*.

U.S. Green Building Council (2006). *About USGBC*. www.usgbc.org/DisplayPage.aspx?CategoryID=1, accessed 1.3.07.

Acknowledgments

Portions of this chapter that outline measurement and verification methodologies are from Shirley J. Hansen's *Performance Contracting—Expanding Horizons, Second Edition* in the chapter entitled "An ESCO's Guide to Measurement and Verification" by Stephen A. Roosa. Permission granted by The Fairmont Press, Copyright 2006.

Additionally, this chapter is republished in its entirety from Stephen A. Roosa's *The Sustainable Development Handbook* in the chapter entitled "Sustainable Buildings." Permission granted by The Fairmont Press, Copyright 2007.

Chapter 17

Green House Gas Emissions Management*

INTRODUCTION

Greenhouse gases are those gases in the atmosphere that allow heat in the form of direct solar radiation to pass through them as the solar radiation comes to the Earth's surface and is absorbed and then some is re-radiated at a lower frequency. Some of this re-radiated heat is absorbed by the lower atmosphere which then re-radiates part of it back to the earth and the rest toward the outer atmosphere. Thus, some of the Earth's re-radiation is blocked from going toward the outer atmosphere, and stays in the lower atmosphere to increase the Earth's temperature. This produces the "greenhouse effect," and is why the greenhouse gases are given their name.

The major man-made greenhouse gases in the Earth's atmosphere are carbon dioxide, methane, nitrous oxides, and small amounts of chlorinated or fluorinated elements. By far the greatest amount of man-made carbon dioxide comes from our combustion of fossil fuels. Therefore our interest in energy management, energy efficiency, energy conservation, and renewable energies is closely related to the effects on the production of these greenhouse gases. As with controlling our use of energy where the first step is to measure our use of energy, Dr. Woodroof explains that the first step in controlling our production of greenhouse gases is the measurement and accounting of what we are producing in the form of greenhouse gases. And the second step is to show how we control and reduce our production of those greenhouse gases.

This chapter discusses the fundamentals of green house gas (GHG) emissions management. The chapter consists of three parts:

---

*Part I of this chapter is from DOE/EIA x012, Greenhouse Gases, Climate Change, and Energy, 2008 from the US Energy Information Agency, Washington, DC. Parts II and III are authored by Dr. Eric Woodroof, President of Profitable Green Solutions, www.ProfitableGreenSolutions.com

Part I—A basic description of the interrelated issues of *Greenhouse Gases, Climate Change, and Energy* from the US Energy Information Agency and its 2008 report by that same name [1]. This report explains the detailed mechanism of the greenhouse effect, the types of greenhouse gases produced, and the data on production of greenhouse gases particularly related to energy use.

Part II—Measuring and reporting GHGs, emphasizing Why Report? How to Report, and Should you Report? Several protocols are discussed which direct or require reporting formats and standards for GHGs.

Part III—GHG reduction fundamental strategies, emphasizing elimination of GHG production, reduce the production of GHGs or reuse/recycle materials or products that have GHGs as a result of their initial use, mitigate the production of GHGs by changing fuels or using renewable energy sources, or offset the production of GHGs by purchasing offset contracts for planting trees

17.1 PART I—GREENHOUSE GASES, CLIMATE CHANGE, AND ENERGY

What Are Greenhouse Gases?

Many chemical compounds found in the Earth's atmosphere act as "greenhouse gases." These gases allow sunlight to enter the atmosphere freely. When sunlight strikes the Earth's surface, some of it is reflected back towards space as infrared radiation (heat). Greenhouse gases absorb this infrared radiation and trap the heat in the atmosphere. Over time, the amount of energy sent from the sun to the Earth's surface should be about the same as the amount of energy radiated back into space, leaving the temperature of the Earth's surface roughly constant.

Many gases exhibit these "greenhouse" properties. Some of them occur in nature (water vapor, carbon dioxide, methane, and nitrous oxide), while others are exclusively human-made (like gases used for aerosols).

Why Are Atmospheric Levels Increasing?

Levels of several important greenhouse gases have increased by about 25 percent since large-scale industrialization began around 150 years ago (Figure 17-1). During the past 20 years, about three-quarters of human-made carbon dioxide emissions were from burning fossil fuels.

Concentrations of carbon dioxide in the atmosphere are naturally regulated by numerous processes collectively known as the "carbon cycle" (Figure 17-2). The movement ("flux") of carbon between the atmosphere

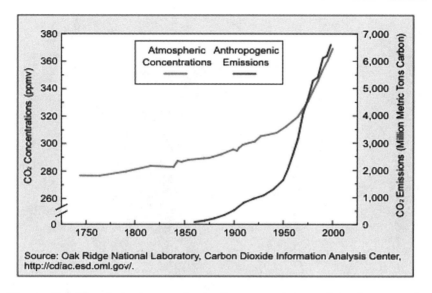

Source: Oak Ridge National Laboratory, Carbon Dioxide Information Analysis Center, http://cdiac.esd.oml.gov/.

Figure 17-1. Trends in Atmospheric Concentrations and Anthropogenic Emissions of Carbon Dioxide

and the land and oceans is dominated by natural processes, such as plant photosynthesis. While these natural processes can absorb some of the net 6.1 billion metric tons of anthropogenic carbon dioxide emissions produced each year (measured in carbon equivalent terms), an estimated 3.2 billion metric tons is added to the atmosphere annually. The Earth's positive imbalance between emissions and absorption results in the continuing growth in greenhouse gases in the atmosphere.

What Effect Do Greenhouse Gases Have on Climate Change?

Given the natural variability of the Earth's climate, it is difficult to determine the extent of change that humans cause. In computer-based models, rising concentrations of greenhouse gases generally produce an increase in the average temperature of the Earth. Rising temperatures may, in turn, produce changes in weather, sea levels, and land use patterns, commonly referred to as "climate change."

Assessments generally suggest that the Earth's climate has warmed over the past century and that human activity affecting the atmosphere is likely an important driving factor. A National Research Council study dated May 2001 stated, *"Greenhouse gases are accumulating in Earth's atmosphere as a result of human activities, causing surface air temperatures and sub-surface ocean temperatures to rise. Temperatures are, in fact, rising. The changes observed over the last several decades are likely mostly due to human activities, but we cannot*

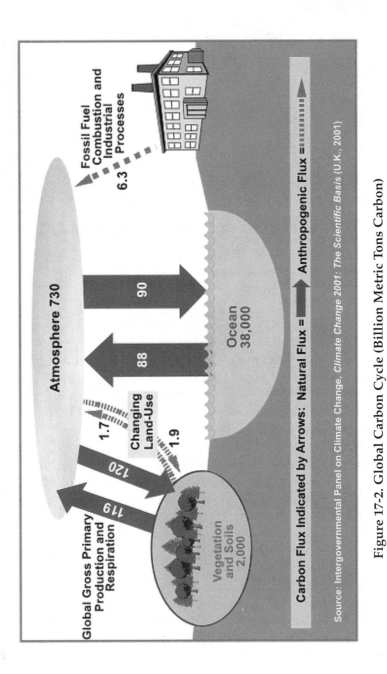

Figure 17-2. Global Carbon Cycle (Billion Metric Tons Carbon)

rule out that some significant part of these changes is also a reflection of natural variability."

However, there is uncertainty in how the climate system varies naturally and reacts to emissions of greenhouse gases. Making progress in reducing uncertainties in projections of future climate will require better awareness and understanding of the buildup of greenhouse gases in the atmosphere and the behavior of the climate system.

What Are the Sources of Greenhouse Gases?

In the U.S., our greenhouse gas emissions come mostly from energy use. These are driven largely by economic growth, fuel used for electricity generation, and weather patterns affecting heating and cooling needs. Energy-related carbon dioxide emissions, resulting from petroleum and natural gas, represent 82 percent of total U.S. human-made greenhouse gas emissions (Figure 17-3). The connection between energy use and carbon dioxide emissions is explored in the box on the reverse side (Figure 17-4).

Another greenhouse gas, methane, comes from landfills, coal mines, oil and gas operations, and agriculture; it represents 9 percent of total emissions. Nitrous oxide (5 percent of total emissions), meanwhile, is emitted from burning fossil fuels and through the use of certain fertilizers and industrial processes. Human-made gases (2 percent of total emissions) are released as byproducts of industrial processes and through leakage.

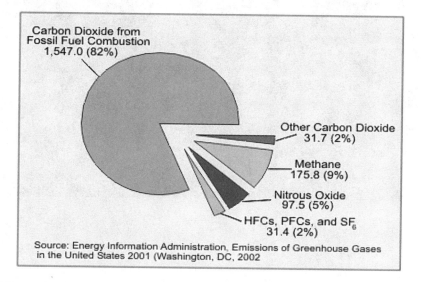

Figure 17-3. U.S. Anthropogenic Greenhouse Gas Emissions by Gas, 2001 (Million Metric Tons of Carbon Equivalent)

Figure 17-4. U.S. Primary Energy Consumption and Carbon Dioxide Emissions, 2001

What Is the Prospect for Future Emissions?

World carbon dioxide emissions are expected to increase by 1.9 percent annually between 2001 and 2025 (Figure 17-5). Much of the increase in these emissions is expected to occur in the developing world where emerging economies, such as China and India, fuel economic development with fossil energy. Developing countries' emissions are expected to grow above the world average at 2.7 percent annually between 2001 and 2025; and surpass emissions of industrialized countries near 2018.

The U.S. produces about 25 percent of global carbon dioxide emissions from burning fossil fuels; primarily because our economy is the largest in the world and we meet 85 percent of our energy needs through burning fossil fuels. The U.S. is projected to lower its carbon intensity by 25 percent from 2001 to 2025, and remain below the world average (Figure 17-6).

National Energy Information Center (NEIC)
Energy Information Administration, EI-30
Forrestal Building, Room 1E-226
Washington, DC 20585
Telephone: (202) 586-8800 FAX: (202) 586-0727
E-Mail: infoctr@eia.doe.gov
For more information, see EIA's Environmental Web Site: http://www.eia.doe.gov/environment.html.

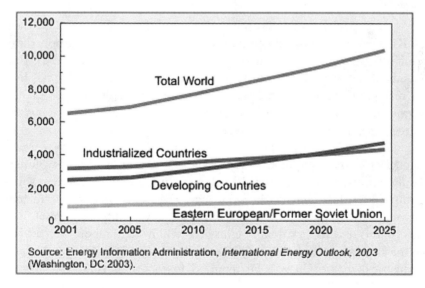

Figure 17-5. World Carbon Dioxide Emissions by Region, 2001-2025 (Million Metric Tons of Carbon Equivalent)

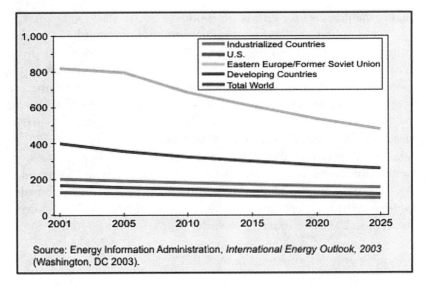

**Figure 17-6. Carbon Intensity by Region, 2001-2025
(Metric Tons of Carbon Equivalent per Million $1997)**

17.2 PART II: MEASURING AND REPORTING GHGS

WHY Measure GHGs?

If companies want to say they are "green," they need to measure their GHG inventory among other environmental performance indicators (how much they recycled, etc.). In addition to "backing up" their marketing claims, if new legislation affects them—they will be in a better position to manage GHG emissions. Many companies have embraced "green" reporting as a marketing advantage because their customers are buying "green" products and services over "non-green" companies.

From a legal standpoint, some industries (and US government agencies) are now required to report their 2010 GHG emissions. Similar to the annual reports required by the Internal Revenue Service, some companies will have to begin reporting GHG emissions for 2010 (actual reports must be filed in 2011). In addition, any facility that emits over 25,000 tons of direct, stationary emissions will have to report.* According to the EPA estimates, less than 15,000 facilities (mostly utilities and heavy industries) in the US will be required to report. Beyond "reporting"

---

*25,000 Metric Tons of Carbon Dioxide emissions is roughly the equal to the emissions from 2,300 homes or 4,600 vehicles.

emissions, many US government agencies, must also "reduce" their emissions according to Executive Order 13514.

HOW to Measure

Several tools from different organizations are available to quantify your GHG emissions. The good news is that because all of these tools comply with the Kyoto Protocol, they are mostly the same despite being from different organizations. Examples include the *General Reporting Protocol* guidelines from TheClimateRegistry.org or the ISO Standards, etc.. *These "guideline" documents are analogous to the IRS guidelines for reporting your taxes. Basically, if you follow the guidelines (or have a consultant do this for you), you should be able to satisfy the reporting requirements.*

If you are just concerned with the EPA regulations, you only need to worry about your emissions from stationary combustion sources (gas-fired turbines, large boilers, etc.). However, if you want to begin to report a complete GHG emissions inventory, you will have to include emissions from your fleet (mobile source emissions), as well as emissions from refrigerant leaks and other "process emissions." All of the emissions previously listed are called *"Scope I"* emissions because they occur from assets that your company owns.

Another category of emissions is called *"Scope II."* These are classified as "indirect" because the emissions don't occur from your assets, but they do occur at the electric generator (usually owned by your utility). For many offices and light commercial businesses, Scope II emissions will comprise more than 75% of the GHG emissions. *In most places that require GHG reporting, Scope I and II emissions reporting are mandatory.*

A final category of emissions is called *"Scope III,"* which represents the emissions from activities that you may not be in control of, such as the emissions that occur from assets that you don't own, but are related to your activities. For example, if your company assembles a product that has parts made by other companies, the emissions from these companies to make your parts would be called "Scope III." Another example would be the emissions from private cars that are driven by students to get to a University. These emissions occur because the university exists, but the university really has no control over the emissions of its students' cars.

Should You Report?

Unless you are a large utility in the US, a federal agency or have enough emissions to exceed the EPA limit, most formal reporting in the US is *voluntary*. So for those companies that are seeking a strategic advantage, many will measure their emissions and not formally report, but analyze the emissions data for internal improvement opportunities.

If your company wants to publically report its emissions, usually you will report your GHG inventory to a "registry" or a non-profit entity or even a trading platform, which keeps all of this data. When you report emissions formally, there is usually a verification step that is required, which is performed by a 3rd party. *The verification process is similar to when a company reports its income and a 3rd party accountant checks the books to make sure that they are accurate.*

As mentioned previously, reporting emissions is a very similar process to reporting taxes. However, beyond reporting your formal emissions, there is much more you can do such as marketing your year-to-year improvement. For example, you can claim Scope III emissions progress from efforts such as recycling or setting up an employee carpool program, etc..

In addition, by incorporating reporting data into the decision-making process, you can have better accounting for future regulations (perhaps in your specific industry). *Many clients have added "emissions impact" as a criterion for evaluating future projects (along with financial impact, ROI, etc.). In other words, projects can win or lose approval based on their impact to a company's annual emissions.*

At this time it is hard to predict what future emissions reporting requirements will exist. However it is likely to become more common as the Securities and Exchange Commission as well as several states (and other industry associations) have asked some companies to begin reporting so that consumers can better identify risks associated with specific companies. For these reasons, it is worth watching to see how the legislation evolves.

17.3 PART III: GHG REDUCTION FUNDAMENTALS

Introduction
After you have measured your emissions, the next task is to manage and reduce them. "GHG Reduction Strategies" can be summarized by the following steps: Eliminate, Reduce, Mitigate and Offset. Knowing why can save you a lot of money.

Step 1: Eliminate Unneeded Processes
As Yoda said, "We must unlearn what we have learned." Many "green" solutions involve "undoing a solution to a problem that doesn't really exist anymore." *For example, a fax machine was a brilliant solution for*

the 1980s, solving a problem of quick document delivery. Today, the internet is a much better solution. Yet although the fax machine is obsolete, many of us still have one (or three) and we pay for their electric energy, toner, paper and replacement costs (not to mention the phone line service costs). If we can eliminate the fax, we will save all those costs and associated "footprints." You can take this same approach to business processes—keep asking, "Do we really need this process?." If the process yields only small benefits, consider eliminating it to make your business simpler while saving emissions.

Step 2: Reduce/Reuse/Recycle

After eliminating unneeded processes, the next step is to implement the three famous "Rs." Beyond the usual recycling context, applying the *reduce* principle to energy consumption is one of the most cost-effective ways to cut emissions. There are literally thousands of highly profitable ways to reduce energy consumption by leveraging new technologies in lighting, HVAC and other building systems. In addition, maintenance savings or behavior modification savings (getting employees/tenants to use less energy) can be equally as valuable—yielding 15-30% savings with almost an immediate payback.

An example from a client: One of the largest cell phone providers in the US redesigned their shipping boxes for small mobile phones. The new boxes used a different type of cardboard that was stronger and thinner. By making the walls of the boxes thinner, they were able to cut the amount of cardboard per box, which reduced their raw material consumption of cardboard (Incidentally, reducing or recycling does reduce upstream emissions and you can account for the reductions.) The new boxes also took less energy to transport and they could fit more of the boxes within a given shipment from the factory to the distribution center, which saved even more. Plus—employees were inspired by the project and then suggested even more resource conservation ideas. They asked management to change the company's policy of providing employees with bottled water (one serving per bottle). They switched to 5 gallon bottles located in central locations and employees felt that they were "walking the talk" of working "green." This approach reduced the amount of plastic waste (which also had an upstream emissions impact).

Mitigate

After you have reduced/optimized the amount of energy/resources you are using, the next step is to mitigate the way that you use energy. If you are using fossil-based fuels, then you can move towards getting energy from "clean" sources. In many regions, you can choose to buy "green" power from your utility. Alternatively, you can install your own renewable power sources such as solar and wind on your property. However, in most traditional buildings it is unlikely that you will be able to generate enough energy to meet your consumption requirements.

Via financing vehicles such as a power purchase agreement, you can get solar on your roof at no up-front cost. With a PPA, the installation costs are financed (just like your mortgage) such that your savings (avoided annual electricity purchases from the utility) are greater than the finance payment. Many of the solar installations have utilized this method to capture the "green" benefits (employee morale, company image, etc.) while avoiding the upfront costs and yielding positive cash flow as well as reduced risk from utility price spikes.

Offset

If it is unfeasible to mitigate your fossil fuel energy consumption on your property or you can't buy "green power" from your utility, then you can buy an "offset," which basically means that you pay the costs of someone else's mitigation efforts and you get the "carbon credit" portion of your investment. Although this is a relatively new market, you can buy credible offsets on the internet and immediately eliminate your carbon footprint each year. Compared to direct mitigation efforts (installing solar on your own roof), offsets are relatively inexpensive, but the downside is that you have to buy offsets each year to counterbalance your footprint and you are dependent on someone else's actions/credibility/costs to be sure that the offset projects actually occur and you are not being scammed.

In the US, there are two primary offset products. One is a "carbon credit," which equals one metric ton of CO_2 avoided/absorbed (as from a forestry or methane re-capture project). The second product sold in the US is a renewable energy credit (REC), which represents one MWh produced from renewable energy (as from a wind or solar farm). Regardless of the offset product you purchase, they are easy to buy on the internet, but buy from a reputable brand!

In many cases, due to increased competitive forces, offsets can be less expensive than buying "green power" from your local utility. *For example, in Texas, I can choose to pay an additional 3 cents per kWh to buy green power, or I can buy a Wind REC for about .7 cents per kWh. Therefore I can get the same result and save 2.3 cents per kWh, which* (Association of Energy Engineers (2010), *The Carbon Reduction Manager Workbook from the CRM Training Course*, Fairmont Press, Atlanta) *is about a 20% savings off my electric bill.*

One more thing on RECs: they have different amounts of carbon savings for each geographic region in the US, so you could save even more.

For example, a REC from California will be worth about 878 lbs of CO_2, while a REC from Michigan is worth about twice as much carbon! If they are selling at the same price, you can get more "bang for your buck" if you know where to look! I call this approach "REC speculation" and we talk about it in some of our classes. It has a huge savings potential for US facility managers who are trying to reduce their carbon footprint.

In Conclusion

Reduction strategies coupled with proper GHG accounting (as discussed in Part II) can be a powerful combination and yield many benefits for your organization/building/business. Just knowing a few principles can save your organization a lot of money! You can also download a free spreadsheet to get a rough estimate of your carbon emissions (basic estimate) and key strategies at www.FreeCarbonAudit.com.

Bibliography

Woodroof, E. (2009), *The Green Facility Handbook*, Fairmont Press, Atlanta.

The Climate Registry (2008), General Reporting Protocol for GHG Emissions, May 2008.

Roosa, S. and Jhaveri, A. (2009), Carbon Reduction: Policies, Strategies and Technologieshttp://www.fairmontpress.com/store/cbrandindex.cfm?ID=198, Fairmont Press, Atlanta.

US EPA (2009), Mandatory Reporting of Greenhouse Gas Emissions Final Rule, United States Environmental Protection Agency, epa.gov.

Woodroof, E. (2008), "Overcoming the Barriers to Energy and Green Projects," *Strategic Planning for Energy and the Environment*, Vol 27(3).

Appendix One

Study Questions and Problems

CHAPTER ONE

Questions

1.1 What is energy management? Is energy conservation the same as energy efficiency in an effective energy management program?

1.2 Why is there an increasing interest in energy management?

1.3 In the concept of energy management, distinguish between an energy management steering committee and an energy management technical committee. Should they be combined into one committee or not?

1.4 In your opinion, what is the single most important ingredient for a successful energy management program?

1.5 You have recently been hired as a consultant to develop an energy cost accounting system for a medium-sized job shop plant involved in metal working. Discuss your approach to this project. State some of your first activities.

1.6 Discuss the relationship between a good energy accounting system and an effective energy management program.

Problems

1.1 For your university or organization, list some energy management projects that might be good "first ones," or early selections.

1.2   Again for your university or organization, assume you are starting
       a program and are defining goals. What are some potential first-
       year goals?

1.3   If you were a member of the upper level management in charge of
       implementing an energy management program at your university
       or organization, what actions would you take to reward participat-
       ing individuals and to reinforce commitment to energy manage-
       ment?

1.4   Perform the following energy conversions and calculations:

       a)   A spherical balloon with a diameter of ten feet is filled with
             natural gas. How much energy is contained in that quantity
             of natural gas?

       b)   How many Btu are in 200 therms of natural gas? How many
             Btu in 500 gallons of #2 fuel oil?

       c)   An oil tanker is carrying 20,000 barrels of #2 fuel oil. If each
             barrel of fuel oil will generate 550 kWh of electric energy in
             a power plant, how many kWh can be generated from the oil
             in the tanker?

       d)   How much coal is required at a power plant with a heat rate
             of 10,000 Btu/kWh to run a 6 kW electric resistance heater
             constantly for 1 week (168 hours)?

       e)   A large city has a population which is served by a single elec-
             tric utility which burns coal to generate electrical energy. If
             there are 500,000 utility customers using an average of 12,000
             kWh per year, how many tons of coal must be burned in the
             power plants if the heat rate is 10,500 Btu/kWh?

       f)   Consider an electric water heater with a 4500 watt heating
             element. Assuming that the water heater is 98% efficient, how
             long will it take to heat 50 gallons of water from 70 degrees F
             to 140 degrees F?

1.5   A person takes a shower for ten minutes. The water flow rate is 3 gallons per minute, and the temperature of the shower water is 110 degrees F. Assuming that cold water is at 65 degrees F, and that hot water from a 70% efficient gas water heater is at 140 degrees F, how many cubic feet of natural gas does it take to provide the hot water for the shower?

1.6   An office building uses 1 million kWh of electric energy and 3000 gallons of Number 2 fuel oil per year. The building has 45,000 square feet of conditioned space. Determine the Energy Use Index (EUI) and compare it to the average EUI of an office building.

1.7   The office building in Problem 1.6 pays $65,000 a year for electric energy and $3300 a year for fuel oil. Determine the Energy Cost Index (ECI) for the building and compare it to the ECI for an average building.

1.8   As a new energy manager, you have been asked to predict the energy consumption for electricity for next month (February). Assuming consumption is dependent on units produced, that 1000 units will be produced in February, and that the following data are representative, determine your estimate for February.

Last year	Units produced	Consumption (kWh)
January	600	600
February	1500	1200
March	1000	800
April	800	1000
May	2000	1100
June (vacation)	100	700
July	1300	1000
August	1700	1100
September	300	800
October	1400	900
November	1100	900
December	200	650
(1-week shutdown)		
January	1900	1200

1.9   For the same data as given in Problem 1.8, what is the fixed energy consumption (at zero production, how much energy is consumed and for what is that energy used)?

1.10 At the Gator Products Company, fuel switching caused an increase in electric consumption as follows:

	Expected energy consumption	Actual energy consumption after switching fuel
Electric/cooling degree days	75×10^6 Btu	80×10^6 Btu
Electric/units of production	100×10^6 Btu	115×10^6 Btu

The base year cost of electricity is $15.00/$10^6$ Btu, while this year's cost is $18.00/$10^6$ Btu. Determine the cost of fuel switching, assuming there were 2000 cooling degree days and 1000 units produced in each year.

CHAPTER TWO

Questions

2.1   Which performance measure should be used in setting up an audit procedure for a series of buildings: Btu/ft^2/year or Btu/year? Discuss the reasons for your decision.

2.2   Sketch a graph similar to Figure 2-3 for electric energy consumption for a building in your geographic location.

2.3   What information does the Bin Weather Data provide that the HDD and CDD data does not? Can you obtain HDD from Bin Weather Data? Explain.

2.4   Discuss some of the advantages and disadvantages of using a portable computer to prompt the auditor for the data needed in a facility audit.

2.5   Describe a representative energy management team for a county school system. For a city government. For a newspaper.

Problems

2.1   Select a building and perform some of the initial audit steps so that you can become familiar with the basic audit process. Collect energy cost data for the building for one year, plot that data, and analyze it. Collect data on the building layout, operating hours and equipment contained in the building. Note preliminary areas for EMO's, and determine which EMO is most likely to produce the greatest savings.

2.2   Compute the number of heating degree days associated with the following weather data.

Time Period	Temperature
Midnight - 4:00 AM	20 F
4:00 AM - 7:00 AM	15 F
7:00 AM - 10:00 AM	18 F
10:00 AM - Noon	22 F
Noon - 5:00 PM	30 F
5:00 PM - 8:00 PM	25 F
8:00 PM - Midnight	21 F

2.3   Select a specific type of manufacturing plant (e.g. metal furniture, plastic injection molding, laser medical devices, electronic circuit boards, etc.) and describe the kinds of equipment that would likely be found in such a plant. List the audit data that would need to be collected for each piece of equipment. What particular safety aspects should be considered when touring that plant? Would any special safety equipment or protection be required?

2.4   Section 2.1.2 provided a list of energy audit equipment that should be used. However, this list only specified the major items that might be needed. In addition, there are a number of smaller items such as hand tools that should also be carried. Make a list of these other items, and give an example of the need for each item. How can these smaller items be conveniently carried to the audit? Will any of these items require periodic maintenance or repair? If so, how would you recommend that an audit team keep track of the need for this attention to the operating condition of the audit equipment?

2.5   Section 2.2 discussed the point of making an inspection visit to a facility at several different times to get information on when certain pieces of equipment need to be turned on and when they are unneeded. Using your school classroom or office building as a specific example, list some of the unnecessary uses of lights, air conditioners, and other pieces of equipment. How would you recommend that some of these uses that are not necessary be avoided? Should a person be given the responsibility of checking for this unneeded use? What kinds of automated equipment could be used to eliminate or reduce this unneeded use?

2.6   An outlying building has a 25 kW company-owned transformer that is connected all the time. A call to a local electrical contractor indicates that the core losses from comparable transformers are approximately 3% of rated capacity. Assuming that the electrical costs are ten cents per kWh and $10.00/kW/month of peak demand, that the average building use is ten hours/month, and that the average month has 720 hours, estimate the annual cost savings from installing a switch that would energize the transformer only when the building was being used.

CHAPTER THREE

Questions

3.1   Recently, there has been a trend across the country for utilities to charge more for demand but keep consumption billing about the same (or even reduce the charges). Discuss why this may be occurring.

3.2   Discuss why demand control during peaking months may be more profitable than during nonpeaking months. How might a ratchet clause affect this?

3.3   Discuss ways a manufacturing company might prepare for natural gas curtailments to minimize their impacts.

3.4   Discuss why some managers have failed to analyze and understand their energy rate schedules.

3.5   Do you think a company should periodically analyze its energy rate schedules to see if a change is in order? Explain.

3.6   Discuss why a utility does not pay as much (buy-back rates) for electricity generated by cogeneration, wind, and solar as it charges its customers for the electricity it generates.

3.7   Discuss the advantages and disadvantages of a time-of-day electric rate to residential customers. Examine the time-of-use rate shown in Figure 3-6. What actions could a residential customer on this time-of-day rate take to reduce his on-peak use of electricity?

Problems

3.1   In working with Ajax Manufacturing Company, you find six large exhaust fans running constantly to exhaust general plant air (not localized heavy pollution). They are each powered by 30-hp electric motors with loads of 27 kW each. You find they can be turned off periodically with no adverse effects. You place them on a central timer so that each one is turned off for 10 minutes each hour. At any time, one of the fans is off, and the other five are running. The fans operate 10 h/day, 250 days/year. Assuming the company is on the rate schedule given in Figure 3-10, what is the total dollar savings per year to the company? The company is on service level 3 (distribution service). Neglect any ratchet clauses. (There will be significant heating savings since conditioned air is being exhausted, but ignore that for now.)

3.2 A large manufacturing company in southern Arizona is on the rate schedule shown in Figure 3-10 (service level 5, secondary service). Their peak demand history for last year is shown below. They have found a way to reduce their demand in the off-peak season by 100 kW, but the peak season demand will be the same (i.e., the demand in each month of November through May would be reduced by 100 kW). Assuming they are on the 65% ratchet clause specified in Figure 3-10, what is their dollar savings? Assume the high month was July of the previous year at 1150 kW. If the demand reduction of 100 kW occurred in the peak season, what would be the dollar savings (i.e., the demand in June through October would be reduced by 100 kW)?

Month	Demand (kW)	Month	Demand
Jan.	495	July	1100
Feb.	550	Aug.	1000
March	580	Sept.	900
April	600	Oct.	600
May	610	Nov.	500
June	900	Dec.	515

3.3 In the data for Problem 3.2, how many months would be ratcheted, and how much would the ratchet cost the company above normal billing?

3.4 In working with a company, you find they have averaged 65% power factor over the past year. They are on the rate schedule shown in Figure 3-10 and have averaged 1000 kW/month. Neglecting any ratchet clause and assuming their demand and power factor is constant each month, calculate the savings for correcting to 80% power factor. How much capacitance (in kVARs) would be necessary to obtain this correction? Assume they are on transmission service, PLY (level 1).

3.5 A company has contacted you regarding their rate schedule. They are on the rate schedule shown in Figure 3-10, service level 5 (secondary service), but are near transmission lines and so can accept service at a higher level (service level 1) if they buy their own transformers. Assuming they consume 300,000 kWh/month and are billed for 1000 kW each month, how much could they save by owning their own transformers. Ignore any charges other than demand and energy.

3.6   In working with a brick manufacturer, you find for gas billing that they were placed on an industrial (priority 3) schedule (see Figure 3-12) some time ago. Business and inventories are such that they could switch to a priority 4 schedule without many problems. What is the savings? They consume 7000 Mcf of gas per month for process needs and essentially none for heating.

3.7   Calculate the electric bill for a customer with a January consumption of 140,000 kWh, a peak 15-minute demand during January of 500 kW, and a power factor of 80%, under the electrical schedule of the example in Section 3.6. Assume that the fuel adjustment is $0.01/kWh.

3.8   Compare the following residential time-of-use electric rate with the rate shown in Figure 3-6.

Customer charge: $8.22/month

Energy charge:
     On-peak energy $0.123/kWh
     Off-peak energy $0.0489/kWh

On-peak hours:
     Summer: Noon to 9:00 pm
         May 15th to October 15th
             (Including weekends)
     Winter: 7 am to 11 am; 6 pm to 10 pm
     January 2nd to February 28th
         (Excluding weekends)

Off-peak hours:
     All other hours

This rate charges less for electricity used during off-peak hours—about 80% of the hours in a year—than it does for electricity used during on-peak hours.

Sample time-of-day electric rate.
     (Courtesy Gainesville Regional Utilities, FL)

3.9   A small facility has 20 kW of incandescent lights and a 25 kW motor that has a power factor of 80%. What is the power factor of the combined load? If they added a second motor that was identical to the one they are presently using, what would their power factor be?

3.10 A utility charges for demand based on a 30 minute synchronous averaging period. For the load curve shown below for Jones Industries, what is their billing demand and how many kWh did they use in that period?

3.11 The A1 Best Company has a steam demand of 6,500 lb/hour and a consumption of 350,000 lbs during the month of January. Based on the hypothetical steam rate in Figure 3-13, determine their steam consumption cost for the month.

3.12 A1 Best also purchases chilled water with the rate schedule of Figure 3-13. During the month of July, their chilled water demand was

485 tons, and their consumption was 250,000 ton-hours. What was their monthly cost? What was their Btuh (Btu/hour) equivalent for the average chilled water demand?

CHAPTER FOUR

Questions

4.1 The early part of the chapter refers to avoided energy costs. Why is this term more correct than reduced energy costs?

4.2 Why would a company require a higher rate of return for energy management projects than for other projects? If you don't understand the answer to this question, then you will have a difficult time defending your projects against these arguments.

4.3 How would you defend the use of an economic performance measure that did not include the time value of money?

4.4 Which are more important in the budget decisions for a state — economic criteria or non-economic criteria? Under what circumstances does one group of criteria predominate?

4.5 Should the equivalent uniform annual cost method be the only method used in comparing projects of unequal service lives?

4.6 What are some good sources for inflation rate projections?

Problems

4.1 The Orange and Blue Plastics Company is considering an energy management investment which will save 2500 kWh of electric energy per year at $0.08/kWh. Maintenance will cost $50.00 per year, and the company's discount rate is 12%. How much can they spend on the purchase price for this project and still have a Simple Payback Period of two years? Using this figure as the cost, what is the return on investment (ROI), and the Benefit-Cost Ratio (BCR)? Assume a life of 5 years for the project.

4.2   A new employee has just started to work for Orange and Blue
      Plastics, and she is debating whether to purchase a manufactured
      home or rent an apartment. After looking at apartments and
      manufactured homes, she decides to buy one of the manufactured
      homes. The Standard Model is the basic model that costs $20,000
      and has insulation and appliances that have an expected utility cost
      of $150/month. The Deluxe Model is the energy efficient model
      that has more insulation and better appliances, and it costs $22,000.
      However, the Deluxe Model has expected utility costs of only $120/
      month. If she can get a 10% loan for 10 years to pay back the entire
      amount for either home, which model should she buy to have the
      lowest total monthly payment including the loan and the utility
      bill?

4.3   The A1 Best Company uses a 10 hp motor 16 hours per day, 5 days
      per week, 50 weeks per year in its flexible work cell. This motor is
      85% efficient, and it is near the end of its useful life. The company
      is considering buying a new high efficiency motor (91% efficient)
      to replace the old one instead of buying a standard efficiency mo-
      tor (86.4% efficient). The high efficiency motor costs $70 more than
      the standard model, and should have a 15 year life. The company
      pays $7.00/kW per month and $0.06/kWh. The company has set a
      discount rate of 10% for their use in comparing projects. Determine
      the SPP, ROI and BCR for this project. The company's discount rate
      is 10%.

4.4   Craft Precision, Incorporated must repair their main air condition-
      ing system, and they are considering two alternatives.

      (1)   purchase a new compressor for $20,000 that will have a future
            salvage value of $2000 at the end of its 15 year life; or

      (2)   purchase two high efficiency heat pumps for $28,000 that will
            have a future salvage value of $3000 at the end of their 15 year
            useful life.

      The new compressor will save the company $6500 per year in
      electricity costs, and the heat pumps will save $8500 per year. The
      company's discount rate is 12%. Using the BCR measure, which
      project should the company select? Is the answer the same if Life
      Cycle Costs are used to compare the projects?

4.5   There are a number of energy-related problems that can be solved
       using the principles of economic analysis. Apply your knowledge
       of these economic principles to answer the following questions.

       a)   Estimates of our use of coal have been made that say we have
             a 500 years' supply at our present consumption rate. How
             long will this supply of coal last if we increase our consump-
             tion at a rate of 7% per year? Why don't we need to know
             what our present consumption is to solve this problem?

       b)   Some energy economists have said that it is not very impor-
             tant to have an extremely accurate value for the supply of a
             particular energy source. What can you say to support this
             view?

       c)   A community has a 100 MW electric power plant, and their
             use of electricity is growing at a rate of 10% per year. When
             will they need a second 100 MW plant? If a new power plant
             costs one million dollars per MW, how much money (in to-
             day's dollars) must the community spend on building new
             power plants over the next 35 years?

4.6   A church has a gymnasium with sixteen 500 Watt incandescent
       ceiling lights. An equivalent amount of light could be produced
       by sixteen 250 Watt PAR (parabolic aluminized reflector) ceiling
       lamps. The difference in price is $10.50 per lamp, with no difference
       in labor. The gymnasium is used 9 months each year. How many
       hours per week must the gymnasium be used in order to justify the
       cost difference of a 1-year payback? Assume that the rate schedule
       used is that of Problem 3.8, that gymnasium lights do contribute
       to the peak demand (which averages 400 kW), and that the church
       consumes enough electricity that much of the bill comes from the
       lowest cost block in the table.

4.7   Find the equivalent present worth of the following 6-year project
       using the depreciation schedule in Table 4-6 (MACRS percent-
       ages for 3-, 5-, and 7-year property); purchase and installation cost,
       $100,000; maintenance per year, $10,000; energy saving per year,
       $45,000; salvage value, $20,000. Assume that the minimum attrac-
       tive rate of return is 12%/year. Assume that the corporate tax rate

is 34%, and that the equipment has a 5-year life for tax purposes. What is the after-tax ROR or IRR for this project?

4.8   Calculate the constant dollar, after-tax ROR or IRR for Problem 4-7, if the inflation rate is 6%.

4.9   Find the equivalent constant dollar after-tax present worth of the following 6-year project using the depreciation schedule in Table 4-6: purchase and installation cost, $100,000; maintenance per year, $10,000, increasing at 5%/year; energy saving per year, $45,000, increasing at 8%/year; salvage value, $20,000, increasing at 6%/year; and the Consumer Price Index (CPI) projected to increase at 6%/year. Assume that the minimum attractive constant dollar rate of return is 12%/year. Assume that the corporate tax rate is 34%, and that the equipment has a 5-year life for tax purposes. What is the constant dollar, after-tax ROR or IRR for this project?

CHAPTER FIVE

Questions

5.1   How does lighting affect worker productivity?

5.2   What factors are important in selecting a lamp for a manufacturing plant in which color rendition and finely detailed tasks are important?

5.3   What factors affect the amount of light reaching the work plane?

5.4   How would you convince the management of a facility to switch to group relamping when they have a large number of relatively new lamps that have been installed through spot relamping?

5.5   Why wouldn't you automatically specify the lamp with the greatest efficacy for every application?

Problems

5.1   When performing an energy survey, you find twelve 2-lamp F40T12 security lighting fixtures turned on during daylight hours (averaging 12 hours/day). The lamps draw 40 Watts each, the ballasts draw 12 Watts each, and the lights are currently left on 24 hr/day.

How much can you save by installing a photocell? What is the payback period of this investment? Costs: energy use = $0.055/kWh; power demand = $7.00/kW per month; lamps = $1.00 each; photocell = $85 installed.

5.2   You count 120 4-lamp F40T12 troffers that contain 34 Watt lamps and two ballasts. How much can you save by installing:
a.   3 - F40T10 lamps (at $15/fixture)?
b.   3 - F32T8 lamps and an electronic ballast (at $40/fixture)?

Assume the same energy costs given in Problem 5-1, and that ballasts draw an additional 15% wattage. What is the simple payback period and what is the return on investment for each alternative? The lights are on 8760 hours per year, and the life of the lighting system is 7 years.

5.3   You see 25 exit signs with two 20-Watt incandescent lamps each. How much can you save by replacing the two 20-Watt bulbs with a 7-Watt CFL? The 20-Watt incandescent lamps have a 2500 hour lifespan and cost $3.00 each. The 7-Watt CFLs have a 12,000 hour lifespan and cost $5.00 each and require the use of a $15 retrofit kit. Assume the same energy costs given in Problem 5-1.

5.4   An old train station is converted to a community college center, and a train still passes by in the middle of the night. There are eighty-two 75-Watt A19 lamps in surface-mounted wall fixtures surrounding the building, and they are turned on about 12 hours per day. The lamps cost $0.40 each and last for about one week before failure. How can this problem be solved, and how much money can you save in the process? Assume electricity costs 8 cents per kWh.

5.5   During a lighting survey you discover thirty-six 250-Watt mercury vapor cobrahead streetlights operating 4300 hours per year on photocells. How much can you save by replacing these fixtures with 70-Watt HPS cutoff luminaires? There is no demand charge, and energy costs $0.055 per kWh.

5.6   You find a factory floor that is illuminated by eighty-four 400-Watt mercury vapor downlights. This facility operates two shifts per day for a total of 18 hours, five days per week. What is the savings from retrofitting the facility with eighty 250-Watt high pressure sodium

(HPS) downlights? Assume that the lights are contributing to the facility's peak demand, and that the rates given in Problem 5-1 apply. What will happen to the lighting levels?

5.7   An office complex has average ambient lighting levels of 27 footcandles with four-lamp F40T12 40-Watt 2'×4' recessed troffers. They receive a bid to convert each fixture to two centered F32T8 lamps with a specular reflector designed for the fixture and an electronic ballast with a ballast factor of 1.1 for $39 per fixture. What will happen to the lighting levels throughout the space and directly under the fixtures? Will this retrofit be cost-effective? This lighting is used on-peak, and electric costs are $6.50 per kW per month, and $0.05 per kWh. What is your recommendation?

5.8   An exterior loading dock in Chicago uses F40T12 40-Watt lamps in enclosed fixtures. They are considering a move to use 34-Watt lamps. What is your advice? Based on what you have learned from previous chapters, what other information, sources, and approaches are needed to fully answer the questions with confidence?

5.9   A turn-of-the-century power generating station uses 1500-Watt incandescent lamps in pendant mounted fixtures to achieve lighting levels of about 18 footcandles in an instrumentation room. They plan on installing a dropped ceiling with a 2'×4' grid. How would you recommend they proceed with lighting changes. What will be the savings if they have a cost of 6 cents per kWh?

5.10 A meat-packing facility uses 100-Watt A19 lamps in jarlights next to the entrance doors. These lamps cost $0.50 each and last for 750 hours. What would be the life-cycle savings of using 13-Watt compact fluorescent lamps in the same fixtures? The CFLs cost $15.00 each, and last 12,000 hours. The lights are used on-peak, 8,760 hours per year, and electricity costs 8 cents per kWh. The MARR of the facility is 15%.

5.11 A retail shop uses a 1000-Watt mercury vapor floodlight on the corner of the building to illuminate the parking lot. Some of this light shines out into the roadway. What problems can you anticipate from the light trespass off the lot? How would you recommend improving the lighting? How much can you save with a better lighting source and design? Use electric costs from Problem 5-7, and assume the light does not contribute to the shop's peak load.

5.12 A commercial pool uses four 300-Watt quartz-halogen floodlights. What are the energy, power, and relamping savings from using two 250-Watt HPS floodlights? What will happen to the lighting levels? The lights do contribute to the facility's peak load, and the electric rates are those of Problem 5-7.

5.13 You notice that the exterior lighting around a manufacturing plant is frequently left on during the day. You are told that this is due to safety-related issues. Timers or failed photocells would not provide lighting during dark overcast days. What is the solution?

5.14 A manufacturing facility uses F96T12HO lamps to illuminate the production area. Lamps are replaced as they burn out. These fixtures are about 15 years old and seem to have a high rate of lamp and ballast failure. How can you solve these problems?

CHAPTER SIX

Questions

6.1 You are performing an energy audit in the winter on a school with a dual-duct HVAC system. In one room, you note that the temperature is 55 degrees F although the thermostat is set at 70 degrees. Explain how this discrepancy could be caused by each of the following: (a) dampers, (b) grilles, (c) fans, (d) filters or ductwork, (e) the boiler, or (f) the control system.

6.2 What factors other than those discussed in the text should be considered in determining the heating and cooling requirements for a building?

6.3 In a split-system air conditioner, the compressor unit is outside and the evaporator unit is inside. The two units are connected with refrigerant lines. Which one of the lines should be insulated? Why?

6.4 A refrigerant-to-water heat exchanger is sometimes added to an air conditioner to provide hot water using the waste heat from the compressor unit. Where is this heat exchanger connected, and what is its effect on the efficiency of the air conditioner?

6.5   A friend of yours says he bought a heat pump that is 200% efficient. Is this possible? Explain.

Problems

6.1   Estimate the total heating load caused by a work force of 22 people including 6 overhead personnel, primarily sitting during the day; 4 maintenance personnel and supervisors; and 12 people doing heavy labor. Assume that everyone works the same 8-hour day.

6.2   If the HVAC system that removes the heat in Problem 6.1 has a COP of 2.0 and runs continuously, how many kW will this load contribute to the electrical peak if the peak usually occurs during the working day? Assume that the motors in the HVAC system are outside the conditioned area and do not contribute to the cooling load.

6.3   Answer Problem 6.2 under the assumption that 8 of the 12 people doing heavy labor and 2 foremen-maintenance personnel come to work when the others are leaving and that 3000 Watts of extra lighting are required for the night shift.

6.4   A heated building has six 8 × 10 inch window panes missing on the windward side. The wind speed has been measured at 900 ft/min, and the location has 6000 heating degree days/year. (a) Calculate the total number of Btu lost through these windows per year. (b) If the heat is supplied by a boiler, and the heat generation and transmission efficiency is 60%, estimate the cost of leaving the windows broken if gas costs $0.50/therm.

6.5   You have measured the ventilation in a large truck bay and have found that you are using 12,000 cfm. An analysis shows that only 8000 cfm are required. Measurements at the fans give the total electrical consumption of the ventilation system as 16.0 kW at current cfm rates. You are currently ventilating this area 16 hours each day, 250 days each year, including the times of peak electrical usage. Your monthly electric rates are $.045/kWh and $12.00/kW per month of demand. Assuming that your power factor is 90% and that your marginal electrical costs are at the least expensive rates, what is the amount of annual savings that you can expect by the proposed reduction in ventilation rates?

6.6   After implementing the improvements suggested in Problem 6.5, you decide to analyze the value of having the second shift come in just as the first shift is leaving, thereby reducing the amount of time that ventilation is needed by 1 hour each day. How much annual savings do you expect this measure to achieve?

6.7   Suppose the HVAC system in Problem 6.2 needs to be replaced. Compare the cost of running the present system with the cost of a new system with a COP of 3.0. The more efficient system costs $100 more than a replacement that has the old efficiency. Using the SPP method of analysis, which system would you recommend? If the life of the HVAC system is ten years, what is the ROI for the additional cost of the more efficient system? If the company's investment rate is 10%, what is the discounted Benefit/Cost Ratio for this investment? Assume electricity costs 8¢ per kWh, and the HVAC system operates the equivalent of 2000 hours per year at full load.

6.8   ACE Industries has a plant in Nebraska (40° N latitude) with a building that has three, 5×10 foot windows facing South. The windows are single pane glass, one-eighth inch thick. The building is air conditioned with a unit that has an EER = 8.0, and the plant pays $0.08 per kWh for electricity. The plant manager is considering installing interior shades for each of these three windows. The shades would give the windows shading coefficients of about 0.2, and would cost about $150 per window. Would you recommend that the plant manager authorize the investment decision?

6.9   A window air conditioner is rated at 5000 Btu/hour, 115 volts, 7.5 amps. Assuming that the power factor has been corrected to 100%, what is its SEER? How many kWh are used if the unit runs 2000 hours each year at full load? What is the annual cost of operation if electric energy costs 7.5 cents per kWh? How many kWh would be saved if the unit had an SEER of 9.1? How much money would be saved? Compute three economic performance measures to show whether this more efficient unit is a cost-effective investment. The low efficiency unit costs $200, the higher efficiency unit $250, and each unit lasts ten years. Use a MARR of 15%.

6.10 On an energy audit visit to the Orange and Blue Plastics Company, the chiller plant was inspected. Readings on the monitoring gauges showed that chilled water was being sent out of the plant at 44°F and being returned at 53°F. The flow rate was 6000 gallons of water per minute. How many tons of chilling capacity was the plant supplying?

CHAPTER SEVEN

Questions

7.1 Why don't hospitals (usually) use their plastic, paper, and cloth wastes as fuel for their own boilers?

7.2 A company has offered to pay your company $10/T if you use its PVC waste as fuel. What problems do you see that would need to be addressed before using this type of fuel?

7.3. Your present system has 10% flue gas oxygen. A new control system is proposed that would cost $250,000 but would reduce this to 3%. How much annual fuel cost would be necessary to justify this expense if your company demands a maximum payback period of 3 years? Assume that the combustion efficiency curve is the same as Figure 7-11.

7.4 What concentration of CO is poisonous to humans? How can this answer change the desirability of excess air?

7.5 In Section 7.4.1.4, a series of questions were asked to illustrate the uncertainties associated with burning any new fuel. Make a list of 10 such questions that should be asked when deciding whether to use the industrial waste most prevalent in your area as a fuel.

7.6 Many of the basic costs given in the example of Section 7.4.2 are subject to change. How do you (a) estimate the range of parameter values, (b) express the resulting economic evaluation in ways that are clear to the public, and (c) incorporate the range of values into your decision-making process?

7.7   What could go wrong with the waste-fired boiler proposal in Section 7.4.2, and what additional data do you need at this time to prevent these problems?

7.8   What is a reasonable value for the minimum rate of return which would make a waste-fired boiler attractive?

Problems

7.1   Suppose that a reputable firm has been advertising a new burner system that would enable the boiler in Section 7.5 to operate at 8% excess air before CO was detected in the flue gas. Why would this system be worth examining? Quantify your answer.

7.2   In Table 7.2, a waste-burning boiler was described. Assume the capacity of this boiler is 28,000 lb/h. Suppose that these figures are 5 years old, that your company is contemplating the purchase of such a boiler, and that it is planned to save twice the energy amounts and have twice the capacity of the given boiler. The energy cost has been inflating at 10% per year, base construction costs have been inflating at 6%/year, the basic inflation rate of the economy has been 5%, and without inflation the cost of constructing a unit is $R^{.73}$ multiplied by the cost of the existing unit, where R is the ratio between the capacity of the proposed unit and the capacity of the present unit. The tax rate of the company is 34%. The unit is subject to the 5-year depreciation schedule shown in Table 4-6. What is the after-tax present worth of the first 5 years of cash flows associated with this investment if the company uses a constant-dollar after-tax rate of return of 8% on this kind of investment?

7.3   The choice of an optimum combination of boiler sizes in the garbage-coal situation is not usually easy. Suppose that health conditions limit the time that garbage, even dried, can be stored to 1 month. Use the initial costs given in the accompanying table, and assume that the municipality and your company have supplies and needs for energy, respectively, as given in the table labeled "Data for Problem 7.3." Suppose that all the other costs for this problem are the same as in Section 7.2. What is the optimum choice now?

Costs for Problem 7.3

Capacity, 750 psi	Initial costs: trash-fired boiler	Initial costs: coal-fired boiler
50,000 lb/h	—	$1,800,000
100,000	—	3,500,000
150,000	$6,250,000	5,100,000
200,000	8,640,000	6,900,000
250,000	10,870,000	8,900,000
300,000	13,000,000	11,000,000

Data for Problem 7.3

Month	Garbage needed (tons)	Garbage available (tons)
January	23,000	13,500
February	23,000	13,500
March	21,600	16,500
April	19,500	18,000
May	14,100	18,900
June	9,500	19,500
July	7,600	22,500
August	9,500	21,000
September	10,800	21,000
October	13,500	18,000
November	18,400	15,000
December	24,300	18,600

CHAPTER EIGHT

Questions

8.1   Conceivably, most of the heat in the flue gas should be recoverable.
      What are the practical considerations that limit the amount of heat
      recovery?

8.2 How can a gas log for a home be fueled by LP gas and yet have no vent to the outside? What efficiency does this indicate? What precautions are necessary with such a gas log?

8.3 What are some of the ways suggested by Figure 7-10 for improving the boiler system?

8.4 In problem 8.4 in the following section, it is unlikely that the pressure of the entire steam system will be 350 psig because of steam drops. Does this make the use of Grashof's formula invalid if 350 psig is used for P?

8.5 If cogeneration produces both electric power and process heat so efficiently, why don't you see more cogeneration facilities in operation?

8.6 A citrus processing plant uses gas heat to dry pulp for cattle feed. Would this be a possible application for cogeneration?

Problems

8.1 An audit of a 600 psi steam distribution system shows 50 wisps (estimated at 25 lb/h each), 10 moderate leaks (estimated at 100 lb/h each), and 2 leaks estimated at 750 lb/h each. The boiler efficiency is 85%, the ambient temperature is 75°F, and the fuel is coal, at $65.00/ton and 14,500 Btu/lb. The steam system operates continuously throughout the year. How much do these leaks cost per year in lost fuel?

8.2 Steam enters a heat exchanger at 567°F and 1200 psia and leaves as water at 300°F and 120 psia. How much heat is exchanged per pound of entering steam?

8.3 What would be the potential savings in the example of Section 7.5 if the amount of boiler blowdown could be decreased to an average rate of 3000 lb/h, assuming that it remained at 400°F? How much additional heat would be available from the 3000 lb/h of blowdown water for use in heating the incoming makeup water? Assume 100% of the heat could be used. Calculate the combined cost savings of these two measures using a fuel cost of $65.00/ton for 14,200 Btu/lb coal.

8.4   Suppose that you are preparing to estimate the cost of steam leaks in a 350-psig system. The source of the steam is 14,200 Btu/lb coal at $70.00/ton, and the efficiency of the boiler plant is 70 percent. Hole diameters are classified as 1/16, 1/8, 1/4, 3/8 and 1/2 in. Develop a table showing the size of the orifice, the number of pounds of steam lost per hour, the cost per month, and the cost for an average heating season of 7 months.

8.5   A 300 foot long steam pipe carries saturated steam at 95 psig. The pipe is not well insulated and has a net heat loss of about 50,000 Btu per hour. The plant Industrial Engineer suggests that the pipe insulation be increased so that the heat loss would be only 5000 Btu per hour. If this change is made, how many pounds per hour of steam does this Energy Management Opportunity save?

8.6   Tastee Orange Juice Company has a large boiler that has a 450 ft^2 exposed surface that is at 225°F. This boiler discharges flue gas at 400°F and has an exposed surface for the stack of 150 ft^2. Calculate the heat loss from the boiler for these two sources.

8.7   In Section 8.2.1.1 two methods were given to estimate the energy lost and cost of steam leaks. What is the relationship of the wisp, moderate leak, and severe leak as defined by Waterland to the hole sizes found from Grashof's formula for 600 psia steam? In other words, find the hole sizes that correspond to the wisp, moderate leak and severe leak.

CHAPTER NINE

Questions

9.1   In the section on demand control, the discussion said that some loads must be recovered (i.e., run later) and some not. Give an example of a load that must be fully recovered, one that does not need any recovery, and one that may need partial recovery.

9.2   What uses of computers in energy management can you think of that are not discussed in this chapter?

9.3   You have just finished auditing a large supermarket that operates 16 hours per day. The supermarket has substantial glass exposure to the outside and substantial lighting for display purposes. Outside lights are used for parking and security. Forgetting any change of light sources, what control schemes would you recommend?

9.4   Someone once said that improperly maintained timers can cost more energy than they save. Section 9.2.2 discuss several examples of this problem. What other possibilities can you come up with?

9.5   Discuss examples of loads whose start-stop times can be optimized as in Figure 9-2(d).

Problems

9.1   Ugly Duckling Manufacturing Company has a series of 12 exhaust fans over its diagnostic laboratories. Presently, the fans run 24 hours per day, exhausting 600 cfm each. The fans are run by 2-hp motors with load factors of 0.8 and efficiencies of 80%. Assuming the plant operates 24 hours per day, 365 days/year in an area of 5000°F heating degree days and 2000°F cooling degree days per year, how much will be saved by duty-cycling the fans such that each is off 10 minutes/hour on a rotating basis? At any time, two fans are off and 10 are running. The plant pays $.05/kWh and $5.00/kW for its electricity and $5.00/10⁶ Btu for its gas. The heating plant efficiency is .80, and the cooling COP is 2.5. Assume that the company only approves EMO projects with a two year or less simple payback period. How much will they be willing to spend for a control system to duty-cycle the fans?

9.2   Profits, Inc. has a present policy of leaving all of its office lights on for the cleaning crew at night. The plant closes at 6:00 pm and the cleaning crew works from 6:00 - 10:00 pm. After a careful analysis, the company finds it can turn off 1000 fluorescent lamps (40 W each) at closing time. The remaining 400 lamps leave enough light for the cleaning crew. Assuming the company works 5 days/week,

52 weeks/year, what is the savings for turning these lamps off an extra 4 hours/day? The company pays $.06/kWh and $6.00/kW for electricity. Peaking hours for demand are 1:00-3:00 pm. Assume there is one ballast for every two lamps and the ballast adds 15% to the load of the lamps. What type of control system would you recommend for turning off the 1000 lamps? (Manual or automatic? Timers? Other sensor?)

9.3   In problem 9.2, assume that the plant manager has checked on the lighting situation and discovered that the cleaning crew does not always remember to turn the remaining lights off when they leave. In the past year, the lights have been left on overnight (8 hours) an average of twice a month. One of the times the lights were left on over a weekend (56 hours). How much did it cost the company in extra charges not to have the lights on some kind of control system? What type of control system would you recommend and why?

9.4   Therms, Inc. has a large electric heat-treating furnace that takes considerable time to warm up. However, a careful analysis shows the furnace could be turned back from a normal temperature of 1800°F to 800°F, 20 hours/week and be heated back up in time for production. If the ambient temperature is 70°F, the composite R value of the walls and roof is 12, and the total surface area is 1000 ft^2, what is the savings in Btu for this setback? (Heat loss equations are given in Chapter 11.). How could this furnace setback be accomplished?

9.5   Obtain bin data for your region, and calculate the savings in Btu for a nighttime setback of 15°F from 65 to 50°F, 8 hours per day (midnight to 8:00 am).

9.6   Petro Treatments has its security lights on timers. The company figures an average operating time of 1 hour per day can be saved by using photocell controls. The company has 100 mercury vapor lamps of 1000 Watts each, and the lamp ballast increases the electric load by 15%. If the company pays $.06/kWh, what is the savings? Assume there is no demand savings. The photocell controls cost $10.00 apiece and each lamp must have its own photocell. It will cost the company an average of $15.00 per lamp to install the photocells. Determine the simple payback period for this EMO. Would you recommend it to the company?

9.7 CKT Manufacturing Company has an office area with a number of windows. The offices are presently lighted with 100 40-Watt fluorescent lamps. The lights are on about 3000 hours each year, and CKT pays $0.08 per kWh for electricity. After measuring the lighting levels throughout the office area for several months, you have determined that 70% of the lighting energy could be saved if the company installed a lighting system with photosensors and dimmable electronic ballasts and utilized daylighting whenever possible. The new lighting system using 32 Watt T-8 lamps and electronic ballasts together with the photosensors would cost about $2500. Would you recommend this change? Explain the basis for your answer.

CHAPTER TEN

Questions

10.1 What routine preventive maintenance tasks should be performed for a residential gas furnace? Do you think they are performed very often? If not, why not?

10.2 What criteria should be used in determining priorities of repair maintenance projects? How would you weight these criteria?

10.3 With two other people, walk through a church or some building with systems in need of repair, and list specific repair jobs. Then make a list of criteria to be used in weighting these jobs, and weigh each job against each criterion. Then multiply the criteria weights by the job weights to get a weight for each job. Does the resulting ranking make sense? If not, find some way to improve this system.

10.4 What are the training needs and costs of maintenance personnel?

10.5 Why is safety training especially important for maintenance personnel?

Problems

10.1 In determining how often to change filters, an inclined tube manometer is installed across a filter. Conditions have been observed as follows:

Week	Manometer reading	Filter condition
1	.4 in water	Clean
2	.6	Clean
3	.7	A bit dirty
4	.8	A bit dirty
5	.8	A bit dirty
6-9	.9	Dirty
10-13	1.0	Dirty
14-18	1.1	Dirty
19-23	1.2	Very Dirty
24	1.3	Plugged up: changed

Based on this table, give a range of times for possible intervals for changing filters.

10.2 You have been keeping careful records on the amount of time taken to clean air filters in a large HVAC system. The time taken to clean 35 filter banks was an average of 18 min/filter bank and was calculated over several days with three different people-one fast, one slow, and one average. Additional time that must be taken into account includes personal time of 20 minutes every 4 hours. Setup time was not included. Calculate the standard time for filter cleaning, assuming that fatigue and miscellaneous delay have been included in the observed times.

10.3 Your company has suffered from high employee turnover and production losses, both attributed to poor maintenance (the work area was uncomfortable, and machines also broke down). Eight people left last year, six of them probably because of employee comfort. You estimate training costs as $10,000/person. In addition, you had one 3-week problem that probably would have been a 1-week problem if it had been caught in time. Each week cost approximately $10,000. All these might have been prevented if you had a good maintenance staff. Assuming that each maintenance person costs $25,000 plus $15,000 in overhead per year, how many people could you have hired for the money you lost?

10.4 A recent analysis of your boiler showed that you have 15% excess combustion air. Discussion with the local gas company has revealed that you could use 5% combustion air if your controls were maintained better. This represents a calculated efficiency improvement of 2.3%. How large an annual gas bill is needed before adding a maintenance person for the boiler alone is justified if this person would cost $40,000/year?

10.5 Your steam distribution system is old and has many leaks. Presently, steam is being generated by a coal-fired boiler, and your coal bill for the boiler is $600,000/year. A careful energy audit estimated that you were losing 15% of the generated steam through leaks and that this could be reduced to 2%. What annual amount would this be worth, considering energy costs only?

10.6 Group relamping is a maintenance procedure recommended in Chapter Five. Using data from Chapter Five, construct a graph which plots maintenance costs per hour and relamping interval expressed as a percentage of the lamps rated life against the total relamping cost. Can you construct such a graph that will provide the answer to the question of whether group relamping is cost-effective for a particular company?

CHAPTER ELEVEN

Questions

11.1 Give examples of heat transfer by radiation, conduction, and convection.

11.2 Infrared heaters heat by radiation. Why are they recommended for large open areas or areas with a lot of air infiltration?

11.3 Discuss whether insulation actually stops heat loss or only slows it down.

11.4 Demonstrate why the R value of a metal tank itself is usually ignored and the surface resistance Rs is used.

11.5 If it is necessary to calculate an effective insulation thickness for pipes, why isn't it necessary to do the same for cylindrical tanks?

11.6 Discuss why the concept of thermal equilibrium is important.

Problems

11.1 A metal tank made out of mild steel is 4 feet in diameter, 6 feet long, and holds water at 180°F. What is the heat loss per year in Btu? The tank holds hot water all the time and is on a stand so all sides are exposed to ambient conditions at 80°F. If the boiler supplying this hot water is 79% efficient and uses natural gas costing $5.00/10^6 Btu, what is the cost of this heat loss? Assume there is no air movement around the tank.

11.2 Ace Manufacturing has an uninsulated condensate return tank holding pressurized condensate at 20 psig saturated. The tank is 2.5 feet in diameter and 4 feet long. Management is considering adding 2 inches of aluminum-jacketed fiberglass at an installed cost of $.60/ft². The steam is generated by a boiler which is 78% efficient and consumes No. 2 fuel oil at $7.00/10^6 Btu. Energy costs will remain constant over the economic life of the insulation of 5 years. Assume that the tank is on a stand, and that the facility MARR = 15%. Ambient temperature is 70°F. Use Rs = .42 for the uninsulated tank. The tank is utilized 8000 hours/year. Calculate the present worth of the proposed investment.

11.3 Your plant has 500 ft of uninsulated hot water lines carrying water at 180°F. The pipes are 4 inches in nominal diameter. You decide to insulate these with 2-inch calcium silicate snap-on insulation at $1.00/ft² installed cost. What is the savings in dollars and Btu if the boiler supplying the hot water consumes natural gas at $6.00/10^6

Btu and is 80 percent efficient? Ambient air is 80°F, and the lines are active 8760 hours/year.

11.4 Given a wall constructed as shown in Figure 11-6, what is the cost of heat loss and heat gain per ft² for a year? Heating degree days are 4000°F days, while cooling degree days are 2000°F days. Heating is by gas with a unit efficiency of .7. Gas costs $6.00/10⁶ Btu. Cooling is by electricity at $.06/kWh (ignore demand costs), and the cooling plant has a 2.5 seasonal COP.

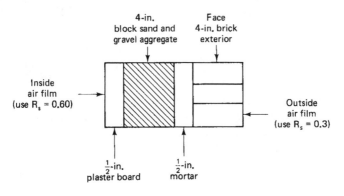

Figure 11-6

11.5 A 6-inch pipe carries chilled water at 40°F in an atmosphere with a temperature of 90°F and a dew point of 85°F. How much fiberglass insulation with a Kraft paper jacket is necessary to prevent condensation on the pipes?

11.6 A building consists of four walls that are each 8 feet high and 20 feet long. The wall is constructed of 4 inches of corkboard, with 1 inch of plaster on the outside and 1/2 inch of gypsum board on the inside. Three of the walls have 6×4 foot, single-pane windows with R = 0.7. The fourth wall has a 6×4 window and a 3×7 foot door made out of one inch thick softwood. The roof is constructed of 3/4 inch plywood with asphalt roll roofing over it. What is the R value of one of the walls with just a window? What is the R value of the wall with the window and the door? What is the R value of the roof? If the inside temperature of the building is regulated to 78 degrees F by an air conditioner operating with a thermostat. The air conditioner has an SEER = 8.0. If the outside temperature is 95 degrees F for one hour, how many Btus must that air conditioner remove in order to keep the inside temperature at 78 degrees F? How many kWh of electric energy will be used in that one hour period by the air conditioner?

11.7 Repeat Problem 11.6 with the single-pane windows replaced with double-paned windows having an R value of 1.1.

11.8 While performing an energy audit at Ace Manufacturing Company you find that their boiler has an end cap of mild steel that is not insulated. The end cap is six feet in diameter and two feet long. You measure the temperature of the end cap as 250°F. If the temperature in the boiler room averages 90°F, the boiler is used 8760 hours per year, and fuel for the boiler is $6.00 per million Btu, how many dollars per year can be saved by insulating the end cap? What kind of insulation would you select? If that insulation cost $300 to install, what is the simple payback period for this EMO? Assume the boiler efficiency is 80%.

11.9 Assume the tank in problem 11.1 is a hot water tank that is heated with an electrical resistance element. If this were a hot water tank for a residence, it would probably come with an insulation level of R-5. A friend says that the way to save money on hot water heating is to put a timer or switch on the tank, and to turn it off when it is not being used. Another friend says that the best thing to do is to put another layer of insulation on the tank and not turn it off and on. What is the most cost effective solution? Assume that there are four of you in the residence, and that you use an average of 20 gallons of hot water each per day. Assume that you set the water temperature in the tank to 140 degrees F, and that the water coming into the tank is 70 degrees F. You have talked to an electrician, and she says that she will install a timer on your hot water heater for $50, or she will install an R-19 water heater jacket around your present water heater for $25. Assume that the timer can result in saving three-fourths of the energy lost from the water heater when it is not being used. If electric energy costs $.08 per kWh, what is the most cost effective choice to make between these two alternatives?

CHAPTER TWELVE

Questions

12.1 Give an example of how the design of the layout of a manufacturing operation could influence the energy consumed by the facility.

12.2 Flagging Industries has a purchasing manager who says it is always cheaper to have a motor rewound than to buy a new one. How would you convince the purchasing manager that this is not always the best decision for the company?

12.3 JumpStart Manufacturing Company has a production line that is mechanized, and the drive motors are manually switched off and on to control the speed of the line. Motors and drives usually last about six months. Can you think of a process improvement for this operation?

12.4 Tiger City Bakeries has a large oven whose excess heat is presently being vented outside. What uses for this waste heat can you think of for the bakery?

12.5 Reducing waste streams often has a benefit of improved process energy efficiency. Give at least one example of a waste stream in a manufacturing plant that could be reduced or eliminated, and would have an energy efficiency benefit.

Problems

12.1 Florida Electric Company offers financial incentives for large customers to replace their old electric motors with new, high efficiency motors. Crown Jewels Corporation, a large customer of FEC, has a 20-year-old 100-hp motor that they think is on its last legs, and they are considering replacing it. The motor has a load factor of 0.6. Their old motor is 91% efficient, and the new motor would be 95% efficient. FEC offers two different choices for incentives: either a $6/hp (for the size motor considered) incentive or a $150/kW (kW saved) incentive. If Crown Jewels buys the new motor, which one of these incentives should they ask for?

12.2 During an energy audit at Orange and Blue Plastics Company you saw a 100-hp electric motor that had the following information on the nameplate: 460 volts; 114 amps; three phase; 95% efficient. What is the power factor of this motor? (Hint-See Eq. 6-11 in Chapter Six)

12.3 Ruff Metal Company has just experienced the failure of a 20-hp motor on a waste-water pump that runs about 3000 hours a year. Using the data in Table 12-1, determine whether Ruff should purchase the high efficiency model or the standard model motor. Find the SPP, ROI and B/C ratio, assuming the new motor will last for 15 years, and the company's investment rate is 15%. The demand charge is $7.00 per kW per month, and the energy charge is $0.05 per kWh.

12.4 A rule of thumb for an air compressor is that only 10% of the energy the air compressor uses is transferred into the compressed air. The remaining 90% becomes waste heat. If you have seen a 50-hp air compressor on an audit of a facility, but you do not have any measurements of air flow rates or temperatures, how would you estimate the amount of waste heat that could be recovered for use in heating wash water for metal parts? Assume the efficiency of the motor is 91.5%.

12.5 Orange and Blue Plastics has a 150-hp fire pump that must be tested each month to insure its availability for emergency use. The motor is 93% efficient, and must be run for 30 minutes to check its operation. The facility pays $7.00/kW for its demand charge and $.05/kWh for energy. During your energy audit visit to Orange and Blue, you were told that they check out the fire pump during the day (which is their peak time), once a month. You suggest that they pay one of the maintenance persons an extra $50 a month to come in one evening a month to start up the fire pump and run it for 30 minutes. How much would this save Orange and Blue Plastics on their annual electric costs?

12.6 Our "rules of thumb" for the load of a motor and air conditioner have implicit assumptions on their efficiencies. What is the implied efficiency of a motor if we say its load is 1 kW per hp? What is the implied COP of an air conditioner that has a load of 1 kW per ton?

12.7 During an audit trip to a wood products company, you note that they have a 50-hp motor driving the dust collection system. You are told that the motor is not a high efficiency model, and that it is only 10 years old. The dust collection system operates about 6000 hours each year. Even though the motor is expected to last another five years, you think that the company might be better off replacing the motor with a new high-efficiency model. Provide an analysis to show whether this is a cost-effective suggestion.

CHAPTER THIRTEEN

Questions

13.1 What is a selective surface? How and why does it affect the efficiency of a solar collector?

13.2 Why would phase change materials be popular for thermal storage in solar applications where space is limited?

13.3 Describe the refuse stream of a typical university. State its probable Btu content.

13.4 What renewable energy source is most popular today? Why?

13.5 Discuss some hindrances facing wider-spread utilization of solar energy in industry.

13.6 Is water in short supply in your area? What measures are being taken to insure the adequacy of the water supply?

Problems

13.1 In designing a solar thermal system for space heating, it is determined that water will be used as a storage medium. If the water temperature can vary from 80°F up to 140°F, how many gallons of water would be required to store 1×10^6 Btu?

13.2 In designing a system for photovoltaics, cells producing 0.5 volts and 1 ampere are to be used. The need is for a small dc water pump drawing 12 volts and 3 amperes. Design the necessary array but neglect any voltage-regulating or storage devices.

13.3 A once-through water cooling system exists for a 100-hp air compressor. The flow rate is 3 gal/min . Water enters the compressor at 65°F and leaves at 105°F. If water and sewage cost $1.50/10^3$ gallons and energy costs $5.00/10^6$ Btu, calculate the annual water savings (gallons and dollars) and annual energy savings (10^6 Btu and dollars) if the water could be used as boiler makeup water. Assume the water cools to 90°F before it can be used and flows 8760 hours/year. Assume a boiler efficiency of 70%.

13.4 A large furniture plant develops 10 tons of sawdust (6000 Btu/ton) per day that is presently hauled to the landfill for disposal at a cost of $10/ton. The sawdust could be burned in a boiler to develop steam for plant use. The steam is presently supplied by a natural gas boiler operating at 78% efficiency. Natural gas costs $5.00/10^6$ Btu . Sawdust handling and in-process storage costs for the proposed system would be $3.00/ton. Maintenance of the equipment will cost an estimated $10,000/year. What is the net annual savings if the sawdust is burned? The plant operates 250 days/year.

13.5 Design an energy-efficient facility (location on site, layout, building envelope, etc.) for an existing factory whose operation is familiar to you. Do not be constrained by the existing facility.

13.6 At 40°N latitude, how many square feet of solar collectors would be required to produce each month the energy content of a) one barrel of crude oil? b) one ton of coal? c) one therm of natural gas? Assume a 70% efficiency of the solar heating system.

13.7 Using Table 13.1, determine whether Portland, OR, New Orleans, LA, or Boston, MA have the greatest amount of solar energy per square foot of collector surface? Assume each collector is mounted at the optimum tilt angle for that location.

13.8   A family car typically consumes about 70 million Btu per year in fuel. How many gallons of gasoline is this? Using the maximum Btu contents shown in Table 13-15, how many pounds of corn cobs would it take to equal the Btus needed to run the car for one year? How many pounds of rice hulls? Of dirty solvents?

13.9   Determine the power outputs in Watts per square foot for a good wind site and an outstanding wind site as defined in Section 13.5.

13.10 How much difference—in percent—is there between these two sites?

Index